The Future of Phylogenetic Systematics: The Legacy of Willi Hennig

Willi Hennig (1913–76), founder of phylogenetic systematics, revolutionised our understanding of the relationships among species and their natural classification. An expert on Diptera and fossil insects, Hennig's ideas were applicable to all organisms. He wrote about the science of taxonomy or systematics, refining and promoting discussion of the precise meaning of the term relationship, the nature of systematic evidence, and how those matters impinge on a precise understanding of monophyly, paraphyly, and polyphyly. Hennig's contributions are relevant today and a platform for the future. This book focuses on the intellectual aspects of Hennig's work and gives dimension to the future of the subject in relation to Hennig's foundational contributions to the field of phylogenetic systematics. Suitable for graduate students and academic researchers, this book will also appeal to philosophers and historians interested in the legacy of Willi Hennig.

David Williams is a research scientist at the Natural History Museum, London specialising in diatom systematics-taxonomy. He has published over 200 journal papers and is the author or editor of nine books. His research interests include the systematics and biogeography of diatoms and theoretical studies related to advances in cladistics.

Michael Schmitt is a retired adjunct professor of zoology at Ernst-Moritz-Arndt-Universität Greifswald, Germany and was recently appointed President of the German Society for History and Philosophy of Biology. He is the author of *From Taxonomy to Phylogenetics – Life and Work of Willi Hennig*, the only biography of Willi Hennig.

Quentin Wheeler is President of the State University of New York College of Environmental Science and Forestry, Syracuse, New York. He is the author or editor of six books and wrote the 'New to Nature' feature in *The Guardian* newspaper. His research interests include the morphology, taxonomy and phylogeny of beetles, systematic biology theory, and the role of taxonomy in biodiversity exploration and conservation.

The Future of Phylogenetic Systematics

The Legacy of Willi Hennig

EDITED BY

DAVID WILLIAMS
Natural History Museum, London

MICHAEL SCHMITT
Ernst-Moritz-Arndt-Universität, Germany

QUENTIN WHEELER
College of Environmental Science and Forestry, USA

University Printing House, Cambridge CB2 8BS, United Kingdom

Cambridge University Press is part of the University of Cambridge.

It furthers the University's mission by disseminating knowledge in the pursuit of education, learning and research at the highest international levels of excellence.

www.cambridge.org
Information on this title: www.cambridge.org/9781107117648

First published 2016

Printed in the United Kingdom by TJ International Ltd. Padstow Cornwall

A catalogue record for this publication is available from the British Library

Library of Congress Cataloguing in Publication data
Names: Williams, David M. (David Mervyn), 1954–, editor. | Schmitt, Michael, 1949– editor. | Wheeler, Quentin D., 1954–, editor.
Title: The future of phylogenetic systematics : the legacy of Willi Hennig / edited by David Williams, Natural History Museum, London, Michael Schmitt, Ernst-Moritz-Arndt-Universitäat, Germany, Quentin Wheeler, State University of New York, USA.
Description: New York, New York : Cambridge University Press, 2016. | Includes bibliographical references and index.
Identifiers: LCCN 2016014632 | ISBN 9781107117648 (hardback : alk. paper)
Subjects: LCSH: Cladistic analysis. | Hennig, Willi, 1913–1976.
Classification: LCC QH83 .F88 2016 | DDC 578.01/2–dc23
LC record available at https://lccn.loc.gov/2016014632

ISBN 978-1-107-11764-8 Hardback

Contents

Contributors

Leandro C.S. Assis, Departamento de Botânica, Instituto de Ciências Biológicas, Universidade Federal de Minas Gerais, Brazil

Andrew V.Z. Brower, Department of Biology, Middle Tennessee State University, USA

Robin Bruce, London, UK

Malte C. Ebach, School of Biological, Earth and Environmental Sciences, University of New South Wales, Australia

Torbjørn Ekrem, Museum of Natural History and Archaeology, Norwegian University of Science and Technology, Norway

Anthony C. Gill, Macleay Museum, University of Sydney, Australia,

Gonzalo Giribet, Museum of Comparative Zoology, Department of Organismic and Evolutionary Biology, Harvard University, USA

Jaakko Hyvönen, University of Helsinki, Plant Biology (Biocenter 3), Finland

Nobuhiro Minaka, Ecosystem Informatics Division, National Institute for Agro-Environmental Sciences, Japan

Randall D. Mooi, The Manitoba Museum, MB Canada and Department of Biological Sciences, University of Manitoba, MB Canada

Gareth Nelson, School of Botany, University of Melbourne, Australia

Sophie Pécaud, Centre Atlantique de Philosophie, Université de Nantes, France

Norman I. Platnick, Division of Invertebrate Zoology, American Museum of Natural History, USA

Stéphane Prin, Muséum national d'histoire naturelle, France

Olivier Rieppel, Science and Education, Integrative Research Center, The Field Museum, USA

Michael Schmitt, Ernst-Moritz-Arndt-Universität, Allgemeine und Systematische Zoologie, Greifswald, Germany

Ole Seberg, Natural History Museum of Denmark, University of Copenhagen, Denmark

Per Sundberg, University of Gothenburg, Sweden

Pascal Tassy, Muséum national d'Histoire naturelle, Centre de Recherches sur la Paléobiodiversité et les Paléoenvironnements (CR2P), France

Charissa S. Varma, Department of History and Philosophy of Science, University of Cambridge, UK

Quentin Wheeler, College of Environmental Science and Forestry, State University of New York, USA

Ward C. Wheeler, Division of Invertebrate Zoology, American Museum of Natural History, USA

Rainer Willmann, Johann-Friedrich-Blumenbach-Institute of Zoology and Anthropology and Zoological Museum, Göttingen University, Germany

David Williams, Department of Life Sciences, The Natural History Museum, UK

Willi E.R. Xylander, Senckenberg Museum für Naturkunde Görlitz, Germany

René Zaragüeta Bagils, Sorbonne Universités, UPMC Univ Paris 06; UMR 7205 ISyEB CNRS-MNHN-UPMC-EPHE Institut de Systématique, Evolution, Biodiversité, Paris, France

Foreword
Willi Hennig and systematics:
a personal view

NORMAN I. PLATNICK

In September of 1973, I joined the staff of the American Museum of Natural History in Manhattan. I had spent the previous five years doing graduate work on spider systematics, first at Michigan State University, then at Harvard University. At both institutions, I had taken courses on the principles of systematics and evolutionary biology; the one at Michigan State had been taught by entomologist Roland Fisher, and the ones at Harvard were given by Ernst Mayr, Steve Gould, and entomologist John Burns. Although access to the spider collections at both institutions allowed me to learn a lot about spiders, what little I learned about "how to do systematics" was actually gleaned mostly from studying the papers of my professors (Dick Sauer and Herb Levi) and colleagues, not from courses covering systematic theory.

At the American Museum, I was assigned an office that had been occupied by my two predecessors as curator of the world's largest spider collection: Willis Gertsch, who had spent his entire career at the Museum and who was largely responsible for building the collection, and Oxford naturalist John Cooke, who spent only a few years in New York. The office next door was occupied by entomologist Pedro Wygodzinsky. Among other groups, Wygo (as he preferred to be called) worked on Diptera, and he was therefore thoroughly familiar with the work of Willi Hennig; he had met Hennig, discussed systematics with him, and had even translated into English a large paper by Hennig on the biogeography of New Zealand Diptera (Hennig 1966a). Wygo was fluent in five languages, and had prepared that translation while on his honeymoon (Schuh and Herman 1988)!

It was Wygo's custom to start each weekday morning with a visit to each of the curators in the department, so my first morning at the museum, Wygo came into my office for his daily chat. Among other things, that first morning, he asked me if I had read Hennig's (1966b) book yet. I replied that I had not; in fact, the only time I had ever heard Hennig's name mentioned at Harvard was in a seminar I'd attended that had been given by coleopterist Phil Darlington. Darlington's talk was about "cladism" and its shortcomings; he had made Hennig's work sound so ill-conceived and misdirected as simply to be a waste of anyone's time.

The next morning, Wygo came in again, with the tiny cup of ungodly strong coffee he favored, and he asked me again: "Have you read Hennig's book yet?" This time, I got the message!

Over the ensuing months, I discovered that although Wygo was personally rather retiring, and not one to suggest, in public, that his fellow systematists might benefit by paying attention to Hennig, the museum had others on its staff that were far less reticent. I had the great privilege and pleasure of meeting and getting to know folks like Gary Nelson, Donn Rosen, Gene Gaffney, and (some months later, when he arrived at the museum), Toby Schuh. Gary had spent time in Stockholm, and had found the work of the Swedish entomologist Lars Brundin (1966), one of Hennig's earliest "bulldogs."

I wouldn't presume to speak for any of those folks, but for me, Hennig's writings were nothing short of a revelation. Once one worked through Hennig's somewhat arcane terminology (which was perhaps even more intrusive in the English translations than in the original German) and grasped his arguments, it became easy to see "how to do systematics." In a nutshell, it became easy to see why some taxonomic groups "work" and others don't. Hennig had realized that if life has diversified over time, then there has to be a hierarchy of taxonomic characters that is an exact mirror of that history, but that not all observed similarities among organisms directly reflect that history. There are similarities that are synapomorphies – features that were acquired by the common ancestor of a given group of organisms and were therefore passed on to all the descendants of that common ancestor. There are other similarities that do not characterize a group in this way, and if a proposed group has no putatively synapomorphic character(s) that support it, then there is no evidence that the group is not simply artificial. Every synapomorphic resemblance among organisms fits at some node on the cladogram connecting them all, and it fits at only one such node. Any resemblances that do not fit in this way are just not synapomorphic.

In some cases, it quickly became obvious why a proposed group doesn't work. Vertebrata, for example, has at least one obvious synapomorphy, the presence of a vertebral column. Invertebrata, in contrast, is merely an assortment of taxa that do not belong to the Vertebrata. They have no vertebral column, but that similarity does not unite them as a group; if it did, we would have to include plants in the Invertebrata, as they also lack a backbone.

Presence/absence characters are obvious examples of what might be called Hennigian winnowing. Everything else being equal, a group united by the presence of a feature unique to them is a likely candidate to be real, natural, and monophyletic. A group united only by the absence of such a feature is likely to be false, artificial, and not monophyletic.

But Hennig's winnow went far beyond the "Greek/barbarian" dichotomy that was pointed out as early as Plato, because it also applies to character transformation. In 1973, for example, it was obvious to all that spiders (the order Araneae) form a real, or monophyletic, group; they are united by multiple unique characters, including the peculiarly modified pedipalps of males, which are used to transfer sperm to the females. However, there was no consensus about the higher classification of spiders. Some workers, for example, recognized two suborders of spiders, one in which the chelicerae, or jaws, move in an up-and-down orientation, and another in which the chelicerae move in a side-to-side orientation.

With Hennig's winnow, it quickly became evident that this classification is false. Chelicerae that move side-to-side are indeed unique to one (huge) group of spiders, but chelicerae that move in an up-and-down orientation are not unique to a subgroup of spiders; in fact, they are found in all the closest relatives of spiders (other chelicerates like whipscorpions, etc.). There turned out to be no characters which support the lumping of those spiders with the plesiomorphic state of the cheliceral character into one group; on the contrary, there were putatively synapomorphic characters indicating that some of these spiders are actually more closely related to higher spiders than they are to the liphistiids, one of the other groups with the plesiomorphic cheliceral orientation. Application of Hennig's winnow therefore supported a different cladogram, in which tarantulas and their allies (Mygalomorphae) are more closely related to typical spiders (Araneomorphae) than they are to the Liphistiidae. The group including just liphistiids plus mygalomorphs (the Orthognatha) doesn't work; it is the analog of Invertebrata within spiders (Platnick and Gertsch 1976). In other words, transformation of the cheliceral orientation – from vertical to horizontal – appears to be a synapomorphy uniting a group, but the absence of that transformation does not.

So we came to see that all characters, even the presence/absence ones, can be viewed as transformations of some other character, and that those transformations combine to produce a character hierarchy that provides the evidence for choosing one taxonomic hierarchy over another. In effect, it is only the synapomorphic presence of one or more character transformations that can delineate a group. And the task of taxonomists thus became clear: to find the synapomorphies that allow us to construct a single hierarchy uniting all taxa.

In my view, Hennig's winnow provided a clear understanding of the past (some proposed groups are supported by evidence – putative synapomorphies – whereas others are not) and an equally clear imperative for the future: our job, as systematists, is to find the synapomorphies that support every node in the cladogram of life.

From this, one might suspect that work done during the ensuing decades would demonstrate this clarity, but that does not appear always to have been the case. Workers like Steve Farris showed that one could construct a matrix of similarities (characters x taxa), of the sort that had been favored by pheneticists, and analyse it parsimoniously to produce the equivalent of Hennig's character schemes (clado-grams), rather than simple phenograms (branching diagrams reflecting only raw, or "overall," similarity). This was an important discovery, in that it showed clearly why phenetics, and phenograms, are wrong (see, for example, Farris 1979).

However, the fascination with matrices soon became bound up with the illusory notion of "total evidence." One can certainly sympathize with the desire to include, for example, multiple morphological and molecular data sources in a single, "total evidence" analysis, but one can never actually have all the evidence, and rigorous hypothesis testing demands that newly available sources of data should have the opportunity to call into question prior results, even when considered independently from previous datasets.

The emphasis on "total evidence" unfortunately led some workers to the operational principle that the way to discover Hennig's hierarchy is simply to code every conceivable observation of specimens into a matrix, and trust that parsimony (in this case, the congruence among these close-to-random observations) will find the right cladogram (i.e. the right hierarchy), even if the signal (i.e. any synapomorphies that might happen to be included) is vastly outweighed by the noise.

Congruence among different characters is, of course, the best evidence that a group is real, and combinability of that group with other similarly supported groups into a single cladogram is the best evidence that the real hierarchy has been found, as the probability of finding congruent characters, in a random sample of characters of all organisms, or combinable groups, in a random sample of groups of all species, by chance alone, is infinitesimally small. However, real congruence (for example, between two actual synapomorphies) could easily be overwhelmed, by chance alone, given a large enough sample of noise, in any particular matrix involving just a limited number of taxa.

It is thus scarcely surprising that as matrices have grown to include more and more "characters" that are unlikely to be synapomorphies of any group, so too has the need to apply methods of character weighting that attempt to minimize the damage they do, or that as matrices have come to be dominated by such pseudo-characters, even the heaviest weighting functions may still produce results that seem patently nonsensical to anyone who knows the taxa involved. To my mind, such pseudocharacters are "evidence" mostly that some systematists desire to dis-play the largest possible matrix, in the hopes that quantity will somehow substitute for quality, and that finding new potential synapomorphies requires vastly more work than does finding mere noise.

Nevertheless, as some of the proponents of the "stuff as much irrelevant trivia as possible into the matrix, so maybe we can find a secondary signal" approach have indicated, they purposefully "no longer 'group by synapomorphy' " (Nixon and Carpenter 2012: 225). That is their loss, but it is also a loss to any biologist unwary enough to use their results, under the misapprehension that they are necessarily the results Hennig would have obtained.

Hennig, although he worked on some of the most diverse groups of organisms on the planet, never used a matrix, nor needed to. He knew what he was looking for: synapomorphies, and the groups they delineate. In spiders, for example, the goal of "total evidence" has sometimes led to the construction of matrices that are dominated by similarities that Hennig would likely have scoffed at. Many of those spider gigomatrices (gigantic, garbage-in, garbage-out matrices), for example, are chock-full of "characters" involving leg spines; to date, no one has found a way to homologize individual leg spines, and without such individual homologies, observations about leg spines are about as likely to contribute real synapomorphies as are many other such superficial "similarities" in other groups of organisms (scale counts in fishes, for example).

The advent of DNA sequence data has also clouded what should have been a clear perspective. Today, those data are most often analysed by likelihood methods that no longer associate actual character state changes with nodes on a cladogram, but rather merely provide an estimate of the most likely cladogram, given a particular set of data and a particular model of evolution. Since there is seldom, if ever, any reason to believe that the chosen model actually applies across the entire gamut of taxa being considered, there is seldom, if ever, any reason to believe that the hierarchy chosen is the one that Hennigian analysis would provide. Genomics is likely to provide some outstandingly useful synapomorphies, but in my view, they are likely to be higher-order similarities, involving changes in gene order and expression, rather than low-level similarities among endless saturated, paralogous sequences of nucleotides, each of homology fully as dubious as the leg spines of spiders or scale counts of fishes.

So what does the future hold? Only time will tell, but my prediction is that the farther systematists depart from Hennig's clear vision, the harder it will be to obtain the actual tree of life.

References

Brundin, L. (1966). Transantarctic relationships and their significance, as evidenced by chironomid midges. *Kungliga Svenska Vetenskapsakademiens Handlingar, Fjärde series*, 11, 1–472.

Farris, J.S. (1979). The information content of the phylogenetic system. *Systematic Zoology*, 28, 483–519.

Hennig, W. (1966a). The Diptera fauna of New Zealand as a problem in systematics and biogeography. *Pacific Insects Monograph*, 9, 1–81.

Hennig, W. (1966b). *Phylogenetic systematics*. Urbana, IL: University of Illinois Press.

Nixon, K.C. and Carpenter, J.M. (2012). More on homology. *Cladistics*, 28, 225–226.

Platnick, N.I. and Gertsch, W.J. (1976). The suborders of spiders: a cladistic analysis (Arachnida: Araneae). *American Museum Novitates*, 2607, 1–15.

Schuh, R.T. and Herman, L.H. (1988). Biography and bibliography: Petr Wolfgang Wygodzinsky. *Journal of the New York Entomological Society*, 96, 227–244.

Introduction

DAVID WILLIAMS, MICHAEL SCHMITT
AND QUENTIN WHEELER

There's a story concerning the 37th American President Richard Nixon and his first meeting in 1972 with the Chinese Prime Minister Chou En-Lai in Beijing. Stuck for discussion topics and recalling that Chou En-Lai was an enthusiastic student of French history, Nixon thought to ask him what he considered to be the impact of the French Revolution on western civilization. After reflecting for a while on the question, Chou En-Lai simply replied: "It's too early to tell." Whatever truth there is in the details of this story (and the details do vary from version to version), the sentiment expressed is of some general significance to most revolutions and their perceived impact. This book concerns itself with Willi Hennig, the founder of phylogenetic systematics, the impetus behind the transformation of "taxonomy to phylogenetics," more generally thought of as the "Cladistic Revolution" in biology. If asked, "What was the impact of the cladistics revolution on systematics?", replies might vary. Some might consider the revolution, if it was one, well and truly over. After 30 years or so of discussion, systematists/taxonomists, having transformed their art, George Gaylord Simpson's view of the taxonomic process (a "useful art"; Simpson 1961: 110), into a science, can get on with their job, happy in the knowledge that fundamentals have been established: data are available, and with DNA, in large enough quantities to satisfy requirements for even the most poorly known taxonomic group, criteria have been provided, computer programs written – all that one needs to do is get some specimens, follow a protocol, find the phylogeny and then get on with more interesting questions. Or maybe the best that can be said of the revolution's

The Future of Phylogenetic Systematics: The Legacy of Willi Hennig, eds. D. Williams, M. Schmitt and Q. Wheeler. Published by Cambridge University Press.

Fig I.1 Willi Hennig, 12th August, 1976, probably the last portrait. Courtesy of Gerd Hennig, Tübingen.

effect is: "It's too early to tell." While this book concerns itself with some biography and some history of that revolution, it is not a history book.

The year 2013 was the centenary of the birth of (Emil Hans) Willi Hennig (April 20, 1913–November 5, 1976; Fig I.1), who exerted perhaps one of the strongest influences on systematic (comparative) biology since the publication of Charles Darwin's *On the Origin of Species* (1859), with the publication of his book *Phylogenetic Systematics* (1966a; it has been reprinted twice, in 1979 and 1999). Hennig was an entomologist, and his primary interest was the classification of Diptera (the true flies). In his extensive *Die Larvenformen der Dipteren* (published in three volumes from 1948 to 1952), much of the groundwork for his subsequent entomological studies was laid – the first volume of that trilogy contained an essay on theoretical issues in zoological taxonomy (*Theorie der Zoologischen Systematik*, Hennig 1948: 2–22). Hennig's dedication to empirical studies forced him to consider a host of issues pertinent to biological classification in general, which resulted in a book-length treatment of the subject, *Grundzüge einer Theorie der phylogenetischen Systematik* (1950, now one of the rarest and most expensive twentieth century taxonomic texts), principles he applied to the phylogeny of insects (Hennig 1953).

Prior to the publication of *Phylogenetic Systematics*, Hennig published two other items of general interest: *Die Dipteren-Fauna von Neuseeland als systematisches und tiergeographisches Problem* (1960, translated into English by Petr Wygodzinsky as *The Diptera fauna of New Zealand as a problem in systematics and zoogeography* but not published until 1966b), a study on the union of biogeography with systematics, and a short summary of phylogenetic systematics in English published in the

Annual Review of Entomology (Hennig 1965), both are still worth reading today. But it was, of course, the book *Phylogenetic Systematics*, translated into English from a German text by two zoologists, D. Dwight Davis and Rainer Zangerl, that had the biggest impact on systematic biology, a book that presented a detailed account in English Hennig's views on systematic methodology. Here one could engage with Hennig on the meaning of taxonomic relationship and the nature of systematic evidence, and how these matters impinged on the understanding of commonly used phylogenetic terms such as monophyly and homology – matters hardly settled today and still topics of interest and discussion for some (Nelson 2011, 2014, Ebach et al. 2013, Vanderlaan et al. 2014, and the contributions in Hamilton 2014).

Reception of *Phylogenetic Systematics* was somewhat guarded and many reviewers "generally failed to perceive the possibilities in applying Hennig's approach..." (Rosen et al. 1979); reactions from those with a vested interest in alternative approaches to classification – primarily those who had created and promoted the "New Systematics" (Huxley 1940) – were outright hostile (Philip Darlington, Ernst Mayr and George Gaylord Simpson), but "That criticism was often based on unwarranted misunderstandings by those who for no good reason considered themselves experts..." (Brundin in Wanntorp 1993: 365).

Fortunately, others did see possibilities, most notably the entomologist Lars Brundin (1907–93; see Wanntorp 1993), who specialised in chironomid midges (the nonbiting midges) and their "transantarctic relationships." Brundin's efforts were realised in a 500-page monograph, which he undertook as "an attempt to deepen the understanding of a major biogeographical problem on the basis of a phylogenetic study of some holometabolous insect groups of the dipteran order" (Brundin 1966: 5), a companion, in spirit if nothing else, to Hennig's *Die Dipteren-Fauna von Neuseeland als systematisches und tiergeographisches Problem* (Hennig 1960). Like Hennig's major opus, Brundin's monograph was published in 1966. It was this monograph that exerted a major influence on ichthyologist Gareth Nelson (Schultze 2005, Nelson 2014) who discussed the ideas presented therein with a group of (palaeo)ichthyologists – among them Colin Patterson, Roger Miles, Philipe Janvier and Niels Bonde – who, in the 1960s, were at one time or another studying fossil fishes at the Naturhistoriska Riksmuseet, Stockholm, where Brundin happened to be employed as an entomologist at the time, and at the Natural History Museum in London. Years later, Colin Patterson, a palaeoichthyologist at the Natural History Museum, noted that Brundin's monograph was "at first sight an unlikely place to find enlightenment" (Patterson 2011: 124) but "The first 50 pages ... are still a wonderfully clear and strong statement of Hennig's ideas."

As noted elsewhere (Williams and Ebach 2008, 2009), the remarkable century between 1866, marked by the publication of Haeckel's *Generelle Morphologie der Organismen* (Haeckel 1866, notable for the inclusion of the very first collection of phylogenetic "oak" trees, all eight reproduced in Pietsch 2010), and 1966, with the publication of Hennig's *Phylogenetic Systematics*, bookends for a century of examining the impact and effects of Darwinian evolution on the principles and practice of systematic biology and its results (classification). In short, that period represents the long slog in which comparative biology attempted to come to terms with evolution and its role in classification, that dialogue having its beginnings with Haeckel and the search for ancestral taxa, "enlightenment" coming primarily from the fossil record (which is not the same thing as the fossils themselves, of course).

Brundin and Hennig were indeed unlikely places "to find enlightenment" as neither needed nor extensively used the fossil record, good or otherwise, to investigate and tease apart problems in systematics and biogeography (Hennig did publish extensively on the fossil flies found in amber). The challenge to Haeckel's "palaeontological phylogeny" began primarily through the recognition of the concept of paraphyly, a term coined by Hennig to broadly meaning those uncharacterisable groups of species that were often regarded to be ancestors of one kind or another; followed by the realization that searching for ancestor – descendant sequences of taxa, fossils or otherwise, was largely a futile endeavour; and that the empirical realities of searching, instead, for taxon sister groups (relationships) rather than ancestor – descendant sequences, the critical value of the homology concept (relationships) and the necessity of recognising classification (relationships) as an independent science – ideas that today seem to have been "temporarily forgotten" in the rush to acquire more and more information as if merely accumulating data will solve problems.

Cladists, cladistics and cladism received a lot of criticism over the years, some of it just (and self-inflicted), some of it not; but much of it, as Brundin noted, misguided or misunderstood, even among cladists themselves. But like all exciting scientific disciplines, they develop piecemeal, haphazardly even, some "advances" becoming momentarily fashionable, then fading into the background destined for justifiable oblivion, while other genuine advances slumber only to be re-discovered in years to come. For example, taxonomy (classification) as a valid pursuit was first recognised and discussed in biology some 200 years ago by Augustin Pyramus de Candolle (1778–1841) in his *Théorie élémentaire de la botanique*, published in 1813 (a 2013 anniversary that went largely unnoticed). *Systematics and Biogeography*, written by Gareth Nelson and Norman Platnick addressed the same issues as de Candolle from a post-cladistics, post-Hennig perspective. A few years back when trying to sum up the achievements of *Systematics and Biogeography*, Malte Ebach and David Williams wrote, in a piece as yet unpublished, "Cladistics has been seen by some as a reaction to phylogeny reconstruction, or at least Haeckel's paleontological version of it. *Systematics and Biogeography* was a detailed critique of Haeckel's legacy as well as an attempt to revive natural classification, as outlined in Candolle's *Théorie élémentaire de la botanique*." Is Hennig a bridge between Haeckel and Nelson and Platnick, with *Systematics and Biogeography* (1981) a *Théorie élémentaire* for the twenty-first century? There are many other examples, many other perspectives, and many other books and articles written as a result of the Cladistic Revolution (for a selection of recent books see Wägele 2005, Williams and Ebach 2008, Schuh and Brower 2009, Wiley and Lieberman 2011, Wheeler 2012, Baum and Smith 2012, Wilkins and Ebach 2013), the variety of interpretations in these contributions testimony to the value of Hennig's phylogenetic systematics.

It was timely, then, that the first biography of Hennig, *From Taxonomy to Phylogenetics – Life and Work of Willi Hennig* was published in 2013 (Schmitt 2013), that the Willi Hennig Society (the Hennig Society was founded in 1980 with the expressed purpose of promoting the field of Phylogenetic Systematics, with James S. Farris, creator of Hennig86, an early parsimony computer program (Farris 1988), as its first president) organised a plenary talk at their 2013 annual meeting (given by Schmitt: "Willi Hennig's way from taxonomy to phylogenetics"), and the Systematics Association, Linnean Society and the Natural History Museum, London, hosted a daylong meeting (27 November) to celebrate Hennig's 100th anniversary – this book is the result of that symposium.

Here we wish to say a few words about the three sponsors of that meeting. The Linnean Society has been especially significant in the early dissemination of Hennig's ideas, sponsoring the first symposium to explore the implications of *Phylogenetic Systematics* in the context of a group of organisms other than insects (Greenwood et al. 1973), assisting and encouraging the first reprinting of the book *Phylogenetic Systematics* (the first reprinting in 1979 had a new preface, Rosen et al. 1979) and, in 1974, awarding Hennig their prestigious gold medal.

The Systematics Association travelled a somewhat different path when dealing with *Phylogenetic Systematics*. The Association, created in 1937, could "count[s] among its members enthusiastic exponents of what may perhaps be called 'The New Systematics' which seeks largely to supplement the traditional methods of the museum and the herbarium," as Arthur Tansley noted in his introduction to the first Systematics Association symposium, *The Reciprocal Relationship of Ecology and Taxonomy* (Tansley 1939: 401). This may be the first mention in print of the New Systematics but it certainly points to its strong link with the Association, who eventually enabled the publication of the edited book entitled *The New Systematics* (Huxley 1940). *The New Systematics* was an attempt to bring what was thought to be known about species evolution into classification, derived from new areas of study, particularly cytology, ecology and genetics; but it was almost wholly focused on species evolution and so of little use to classification in general. Nevertheless, it shifted the focus of systematics almost exclusively to species. In many ways, Hennig's methods were to tease out justification for "the traditional methods of the museum and the herbarium" and to counter the excesses of the New Systematics and its obsession with genetics and species (Wheeler 1995). In 1964, the Association published a small book entitled *Phenetic and Phylogenetic Classification* (Heywood and McNeill

1964). In the introduction Vernon Heywood noted that "Taxonomy is in an extraordinary situation today. It is poised on the edge of a far greater revolution than that promised by the New Systematics of the 1940s" (Heywood 1964: 1). Influential though *Phenetic and Phylogenetic Classification* was, there was no mention of cladistics or phylogenetic systematics, its revolution was fixed on the phenetic phase of taxonomy, which may not have entirely vanished from the numerical perspective. Still, systematists are good at promising revolutions.

A glance at the first 13 Systematics Association publications (1960–79), cladistics is not even mentioned. The Systematics Association's influence on the development of cladistics can be appreciated in some of the edited books derived from symposia held in the early 1980s, such as *Problems of Phylogenetic Reconstruction* (Joysey and Friday 1982; some papers given at that meeting were published separately in a special issue of the *Zoological Journal of the Linnean Society* as *Problems in Phylogenetic Reconstruction* (Patterson 1982), testifying to the continued good relations between these two organisations), *Evolution, Time and Space: The Emergence of the Biosphere* (Sims et al. 1983, five additional papers given at this meeting were published separately in various issues of the Linnean Society journals) and the handbook *Cladistics: A Practical Course in Systematics* (1992, the 2nd edition appearing in 1998 with a modified title and a different roster of authors, *Cladistics: The Theory and Practice of Parsimony Analysis*; a 3rd edition is on its way). The latter book stemmed primarily from a workshop organised by staff of the Natural History Museum, London, the third sponsor, whose role in the dissemination of cladistics became the most controversial during the 1980s, events that have been covered in detail by Hull (1988) and Williams and Ebach (2008), and more recently, with a focus on the controversy in the Museum's exhibitions, in Guillé (2015).

This present book can hardly do justice to every aspect of Hennig's impact on systematic biology, least of all can it do justice to his considerable body of entomological work, for which see Andersen (2001), Meier (2005), Engel and Kristensen (2013) and Schmitt (2013). Herein are 19 chapters, which cover a broad range of topics, most of a general nature: species, evolution, biogeography, networks and trees, the principle of dichotomy, alongside a few historical and biographical chapters to allow context for the development of these ideas.

The primary aim of this book is to act as a beacon to the future, as well as a light on the past.

Acknowledgements

We have endeavored to convey our gratitude above to the Systematics Association, the Linnean Society, and the Natural History Museum, London, for supporting the symposium that generated many of the contributions herein. Some of the text above has appeared in a modified form in the Systematics Association's newsletter *The Systematist* (Williams, D.M. 2013. Willi Hennig and the Cladistic Revolution. Notes for the meeting "Willi Hennig (1913–1976): His Life, Legacy and the Future of Phylogenetic Systematics." *The Systematist*, 35, 6–11), while other parts were first given in a presentation at the 75th Anniversary of the Systematics Association by DMW ("A history of the Systematics Association"); both are used here with permission from the SA.

References

Andersen, N.M. (2001). The impact of W. Hennig's 'phylogenetic systematics' on contemporary entomology. *European Journal of Entomology*, 98, 133–150.

Baum, D.A. and Smith, S.D. (2012). *Tree Thinking: An Introduction to Phylogenetic Biology*. Greenwood Village, CO: Roberts and Co.

Brundin, L. (1966). Transarctic relationships and their significance, as evidenced by chironomid midges. With a monograph of the subfamilies Podonominae and Aphroteniinae and the Austral Heptagyiae. *Kungliga Svenska Vetenskapsakademiens Handlingar, Fjärde series*, 11, 1–472.

Candolle, A.P. de (1813). *Théorie élémentaire de la botanique*. Paris: Déterville.

Darwin, C. (1859). *On the Origin of Species*. London: John Murray.

Ebach, M.C., Williams, D.M. and Vanderlaan, T.A. (2013). Implementation as theory, hierarchy as transformation, homology as synapomorphy. *Zootaxa*, 3641, 587–594.

Engel, M.S. and Kristensen, N.P. (2013). A history of entomological classification. *Annual Review of Entomology*, 58, 585–607.

Farris, J.S. (1988). Hennig86 version 1.5. Computer program and manual, published by the author.

Forey, P.L., Humphries, C.J., Kitching, I.J., et al. (1992). *Cladistics: A Practical Course in Systematics*. Oxford: Oxford University Press.

Greenwood, P.H., Miles, R.S., and Patterson, C. (ed.) (1973). *Interrelationships of Fishes*. London: Academic Press.

Guillé, G.C. (2015). *El neuvo esquema expositivo del Museo de Historia Natural de Londres, 1968–1981. Una perspectiva histórica*. Thesis doctoral. Centro de Estudios de Historia de la Cienca.

Haeckel, E. (1866). *Generelle Morphologie der Organismen: Allgemeine Grundzüge der organischen Formen-Wissenschaft, mechanisch begründet durch die von C. Darwin reformirte Decendenz-Theorie*. Berlin: Reimer. 2 volumes.

Hamilton, A. (ed.) (2014). *The Evolution of Phylogenetic Systematics*. Berkeley, CA: University of California Press.

Hennig, W. (1848–1952). *Die Larvenformen der Dipteren* (Teil 1, 1948; Teil 2, 1950; Teil 3, 1952). Berlin: Akademie-Verlag.

Hennig, W. (1950). *Grundzüge einer Theorie der phylogenetischen Systematik*. Berlin: Deutsche Zentralverlag. [Reprinted 1980 Koenigstein: Otto Koeltz.]

Hennig, W. (1953). Kritische Bermerkungen zum phylogenetischen System der Insekten. *Beiträge zur Entomologie*, 3, 1–85.

Hennig, W. (1960). Die Dipteren-Fauna von Neuseeland als systematisches und tiergeographisches Problem. *Beiträge zur Entomologie*, 10, 221–239.

Hennig, W. (1965). Phylogenetic systematics. *Annual Review of Entomology*, 10, 97–116.

Hennig, W. (1966a). *Phylogenetic Systematics*. Urbana, IL: University of Illinois Press. [Reprinted 1979, 1999.]

Hennig, W. (1966b). The Diptera fauna of New Zealand as a problem in systematics and zoogeography. *Pacific Insects Monograph*, 9, 81 pp. [Translation of Hennig 1960, with additional footnotes.]

Heywood, V.H. (1964). Introduction. In *Phenetic and Phylogenetic Classification. A Symposium*, ed. V.H. Heywood and J. McNeill. London: Systematics Association. Publication no. 6, pp. 1–6.

Heywood, V.H. and J. McNeill (ed.) (1964). *Phenetic and Phylogenetic Classification. A Symposium*, London: Systematics Association. Publication no. 6.

Hull, D.L. (1988). *Science as Process: An Evolutionary Account of the Social and Conceptual Development of Science*. Chicago, IL: University of Chicago Press.

Huxley, J. (ed.) (1940). *The New Systematics*. Oxford: Oxford University Press.

Joysey, K.A. and Friday, A.E. (ed.) (1982). *Problems of Phylogenetic Reconstruction*. London: Academic Press.

Kitching, I., Forey, P.L., Humphries, C.J. and Williams, D.M. (1998). *Cladistics: The Theory and Practice of Parsimony Analysis*, 2nd edition. Oxford: Oxford University Press.

Meier, R. (2005). Role of dipterology in phylogenetic systematics: the insight of Willi Hennig. In *The Evolutionary Biology of Flies*, ed. D.K. Yeates and B.M. Weigmann. New York: Columbia University Press, pp. 45–62.

Nelson, G.J. and Platnick, N.I. (1981). *Systematics and Biogeography: Cladistics and vicariance*. New York: Columbia University Press.

Nelson, G. (2011). Resemblance as evidence of ancestry. *Zootaxa*, 2946, 137–141.

Nelson, G. (2014). Cladistics at an earlier time. In *The Evolution of Phylogenetic Systematics*, ed. A. Hamilton. Berkeley, CA: University of California Press, pp. 139–149.

Patterson, C. (ed.) (1982). Problems in phylogenetic reconstruction. *Zoological Journal of the Linnean Society*, 74, 197–344.

Patterson, C. (2011). Adventures in the fish trade [edited and with an introduction by David M. Williams and Anthony C. Gill]. *Zootaxa*, 2946, 118–136.

Pietsch, T.W. (2010). *Trees of Life: A Visual History*. Baltimore, MD: John Hopkins University Press.

Rosen, D.E., Nelson, G. and C. Patterson. (1979). Foreword. In *Phylogenetic Systematics*, ed. W. Hennig. Urbana, IL: University of Illinois Press, pp. vii–xiii.

Schmitt, M. (2013). *From Taxonomy to Phylogenetics: Life and Work of Willi Hennig*. Leiden: Brill.

Schuh, R.T. and Brower, A.V.Z. (2009). *Biological Systematics: Principles and Applications*, 2nd edition. Ithaca, NY: Cornell University Press.

Schultze, H.-P. (2005). The first ten symposia on early/lower vertebrates. *Revista Brasileira de Paleontologia*, 8, v–xviii.

Simpson, G.G. (1961). *Principles of Animal Taxonomy*. New York: Columbia University Press.

Sims, R.W., Price, J.H. and Whalley, P.E.S. (1983). *Evolution, Time and Space: The Emergence of the Biosphere*. Systematic

Association Special Volume 38. London: Academic Press.

Tansley, A.G. (1939). Introduction. *Journal of Ecology*, 27, 401–402.

Vanderlaan, T.A., Ebach, M.C. and Williams, D.M. (2014). Defining and redefining monophyly: Haeckel, Hennig, Ashlock, Nelson and the proliferation of definitions. *Australian Systematic Botany*, 26, 347–355.

Wägele, J.-W. (2005). *Foundations of Phylogenetic Systematics*. Munich: Verlag Dr. Friedrich Pfeil.

Wanntorp, H.-E. (1993). Lars Brundin 30 May 1907–17 November 1993. *Cladistics*, 9, 357–367.

Wheeler, Q.D. (1995). The 'Old Systematics': classification and phylogeny. In *Biology, Phylogeny and Classification of Coleoptera: Papers Celebrating the 80th Birthday of Roy A. Crowson*, ed. J. Pakaluk and S.A. Slipinski. Warsaw: Muzeum i Instytut Zoologii PAN, pp. 31–62.

Wheeler, W.C. (2012). *Systematics: A Course of Lectures*. Chichester: Wiley-Blackwell.

Wiley, E.O. and Lieberman, B.S. (2011). *Phylogenetics: Theory and Practice of Phylogenetic Systematics*. Hoboken, NJ: John A. Wiley and Sons.

Wilkins, J.S. and Ebach, M.C. (2014). *The Nature of Classification: Relationships and Kinds in the Natural Sciences*. New York: Palgrave Macmillan.

Williams, D.M and Ebach, M.C. (2008). *The Foundations of Systematics and Biogeography*. Berlin: Springer.

Williams, D.M. and Ebach, M.C. (2009). What, exactly, is cladistics? Re-writing the history of systematics and biogeography. *Acta Biotheoretica*, 57, 249–268.

1

Mission impossible: the childhood and youth of Willi Hennig

WILLI E.R. XYLANDER

1.1 Introduction

Willi Hennig, the father of phylogenetic systematics, was born on 20 April 1913 in Dürrhennersdorf, Germany, a small village in the mountain region of the Saxonian part of Upper Lusatia (Oberlausitz). He died on 5 November 1976 at the age of 63 in Ludwigsburg in Baden-Württemberg, Germany. Schmitt has recently written a comprehensive biography of Hennig (Schmitt 2013a).

Hennig was an acknowledged specialist in the taxonomy of flies and their larvae (Diptera), head of the Department for Phylogenetic Research at the Staatliches Museum für Naturkunde Stuttgart, author of several textbooks and, during the last years of his life, honorary professor of zoology at the Eberhard Karls University, Tübingen (Eberhard Karls Universität Tübingen).

This chapter provides an overview of Hennig's childhood and youth up until he started his academic studies at the University of Leipzig (1932). I will focus on the socio-economic background of his family and the fortunate circumstances that enabled his career to overcome numerous disadvantages.

1.2 The family of Willi Hennig

Willi Hennig's mother, Maria Emma Hennig née Groß, was born on 12 June 1885 in the manor Nieder-Gebelzig in Upper Lusatia, located close to Weißenberg about

The Future of Phylogenetic Systematics: The Legacy of Willi Hennig, eds. D. Williams, M. Schmitt and Q. Wheeler. Published by Cambridge University Press.

30 km west of Görlitz. She was the illegitimate daughter of a servant and suffered all of her life from the social stigma of her parents never having married. Emma worked as a servant and later a factory worker but quit immediately before Willi Hennig, her first son, was born. Emma Hennig was described as extremely moody and socially awkward, getting into trouble with her neighbours and others very easily. Rudolf, her second son, in his unpublished, handwritten autobiography[1] noted that she sometimes locked herself in a room for days to avoid any contact, even with her own family and children. On the other hand she was extremely involved with her three sons, careers and projected her ambitions onto them.

Willi Hennig's father, Karl Ernst Emil Hennig, was a railway worker, born on 28 August 1873 in Kaana in Upper Lusatia, the fifth of eight children, the son of a farmer and a local representative. In contrast to his wife, he was described as very balanced, quiet and diligent. His personality and ambition is reflected in his career, where he started as a normal railway worker eventually becoming a foreman in the 1920s.

Emil and Emma Hennig were married in Dürrhennersdorf on 19 November 1911. It was here that all three of their boys were born: Willi, in 1913 (Fig 1.1), Fritz Rudolf, in 1915, and Karl Herbert, in 1918 (Vogel and Xylander 1999). In Dürrhennersdorf, the Hennigs lived in a house located directly beside the rails close to the elementary school (Fig 1.2). The house had a ground floor and attic room, estimated to be not larger than 60 square meters. The financial situation of the family was bad as they lived on a single (small) salary. But the relationship among the three brothers was extraordinary close.[2]

Willi Hennig was baptized on 25 May 1915. He was described as introverted and quiet, avoiding all conflicts, but diligent and very ambitious at school. He seems to have had poor health during his youth as indicated by the rather high number of non-attendance days at school (Vogel and Xylander 1999).

Willi's brother Rudolf was, in contrast, wild and extroverted, later becoming a parson in Oberoderwitz where he died in 1990. The youngest brother, Herbert, was very vivacious but small and thin. He became a commercial clerk but died during World War II.

1.3 Willi Hennig and his time in Upper Lusatia

Hennig started elementary school (Volksschule Dürrhennersdorf, Fig 1.2) at Easter 1919, and from the very beginning was a good pupil. He left that elementary school

---

[1] Handwritten biography by Rudolph Hennig, brother of Willi Hennig. Photocopy and digital copy preserved in the Willi-Hennig-Archiv, Senckenberg Museum für Naturkunde Görlitz; Original with Katharina Linke, Oberoderwitz.

[2] Ibid.

Fig 1.1 Willi Hennig with his mother and father, roughly late summer/early autumn 1913; this is probably the oldest photo of Hennig that exists. Original: K. Linke, Oberoderwitz, Digital copy: Willi-Hennig-Archiv, Senckenberg Museum für Naturkunde Görlitz (SMNG).

in autumn 1919, as his family moved to Taubenheim where they lived together with Willi's grandmother, Marie Groß, the mother of Emma Hennig (Vogel and Xylander 1999). The children called Emma 'Mamma' but called their grandmother Marie Groß 'Mutter', which means mother (Schmitt 2013a). In Taubenheim, the Hennig family lived in a slightly larger railway workers' home, again just few meters from the rails. Here Hennig attended the local Kirchschule (Vogel and Xylander 1999).

In October 1921, the Hennig family moved to Oppach, where they lived on the first floor of the railway station. At that time Oppach was a small industrial town of around 3000 inhabitants. Moving to Oppach meant yet another change of school and the time here had a lasting impact on Hennig's personal development.

He attended primary school in Oppach from 2 November 1921 until 19 March 1927. His supervisor and teacher was Kantor Lehmann, who taught German, mathematics, singing and religion (Fig 1.3). Although Hennig was top of his form, his fellow pupils remember him as quiet, reserved and unpretentious. He had few friends, but was accepted among his classmates. The inventory of the Oppach primary school shows that Hennig had many days of non-attendance due to his impaired health.

Fig 1.2 The Hennig family home in Dürrhennersdorf (arrow) where Willi Hennig was born; it is located close to the rails and in the vicinity of the elementary school (bottom right). Slide and digital copy: Willi-Hennig-Archiv, Senckenberg Museum für Naturkunde Görlitz (SMNG).

The primary school in Oppach still exists and was re-named 'Willi-Hennig-Grundschule Oppach' on 31 August 2013 thus honouring the famous son of this small German town (*Amtsblatt der Gemeinde Oppach*, August 2013). Hennig was confirmed here in April 1927 by parson Mr Böhm (Fig 1.3).

1.4 Social and familiar circumstances of Willi Hennig during his childhood and youth

To become the scientist who would develop a theory that had a revolutionary impact on the thinking of systematists, evolutionary biologists and taxonomists (see e.g. Schmitt 2010, 2013a, Xylander 2013a), Hennig had to overcome several social and family disadvantages during his childhood and youth.

His family's economic circumstances were difficult, at a time when there were no regular scholarships for talented pupils, and attending a secondary school (which would incur fees) was hardly possible for the child of a worker. Furthermore, Oppach had no secondary school (gymnasium), so Hennig would have to travel every day to Löbau, the nearest city that had a gymnasium, stay there in a furnished room, or attend a boarding school. His family could afford

Fig 1.3 Confirmation of the class of Willi Hennig on 10 April 1927. Willi is the boy on the right, directly beside the parson, Mr Böhm; on the left is Kantor Mr Lehmann. Original: K. Linke, Oberoderwitz, Slide and digital copy: Willi-Hennig-Archiv, Senckenberg Museum für Naturkunde Görlitz (SMNG).

neither of these options. Additionally, at the primary school Hennig attended, several subjects, which constituted a prerequisite for attending any secondary school at the age of 14, were not taught: Hennig had no Latin, French or English lessons at his primary school. His socially isolated, labile mother, who (to some extent) destabilized the family, was a further disadvantage; his own introverted personality made it difficult for him to promote his talents; finally, last but not least, the political and economic conditions in Germany in the late 1920s were generally difficult because of the national economic breakdown and the political destabilization of the Weimar Republic. Also, at that time, the inflation of the German mark was dramatic (as described, for example, by Schmitt 2013a). Furthermore, simultaneously nationalistic as well as Nazi and communist organizations tried to destabilize the elected democratic government. In summary, for a boy like Hennig the chances of attending secondary school and then pursuing an academic career were little better than zero.

But Hennig had some positive cards to play: he was extraordinarily intelligent and ambitious; his father, with his balanced nature, counteracted the moody personality of his mother; he had the good fortune that the former military physician Dr Reinhold Seifert could offer Hennig lessons in those subjects not taught in his primary school as Seifert had retired from active service most probably due to a psychological handicap.[3]

It was, then, a sequence of extremely lucky circumstances that enabled Willi Hennig to attend secondary school and further his scientific career:

1. The youngest daughter of the factory owner Fabian refused to visit the Lyceum (Lyzeum, the secondary school for girls at that time) because she did not want to leave her family to continue her education in a neighbouring larger city. Her father, therefore, organized private lessons from Dr Reinhold Seifert.[4]

2. Hennig was able to attend Seifert's lessons for a reduced fee (his family could not have afforded more) and so received the additional education in mathematics and French (and probably Latin) that were prerequisites for the exams to enter secondary school. Dr Seifert also introduced Hennig to collecting beetles (Schmitt 2013a) and may have thus initiated his lifelong interest in entomology.

3. In the spring of 1927, just a few weeks before Hennig finished his primary school, the federal school (Landesschule) re-opened and increased its capacity at its new site in Dresden-Klotzsche. This school provided scholarships for extraordinarily talented pupils living in the state of Saxony so as to open their way to a university career; such scholarships were especially valuable for pupils whose families were not so well off and would otherwise have been excluded from further academic opportunities.

4. Oppach was part of Saxonian Lusatia, not more than 15 km from the border that divided Lusatia into a western Saxonian and an eastern Prussian part. If Hennig's parents had lived a few kilometres further east (or the borderline drawn elsewhere during the congress of Vienna), a scholarship and attendance at the Saxonian Landesschule Klotzsche would not have been possible.

5. Hennig undertook the exams and achieved extraordinarily good results. Now the gate into secondary school was open. On his trip to the examinations in Dresden, he was accompanied by his mother. When returning from the daytrip (euphoric from success) they learnt that during their absence grandma Marie Groß had died from an accident. She was cleaning the windows when she fell from a chair, cut a blood vessel and died soon after the doctor had arrived.[5] If the accident had happened a few days before, the grieving family may possibly never have attended the exams.

But in the end everything turned out well: Hennig started secondary school Easter 1927.

---

[3] Minutes of an interview of the Author and Jürgen Vogel with Mrs Fabian, Oppach. 15 October 1996. Original: Willi-Hennig-Archiv, Senckenberg Museum für Naturkunde Görlitz.

[4] Ibid.

[5] Handwritten biography by Rudolph Hennig.

1.5 Support and supervision: the time at the Landesschule Klotzsche

The Landesschule Klotzsche was a boarding school where pupils lived in a family environment, in groups of six pupils (of different ages) supervised by a teacher acting as tutor. During his time in secondary school Hennig changed tutors several times: when he entered the Landesschule (in the form UIII A, Untertertia) he was first supervised by Dr Költzsch,[6] then later by Maximilian Rost (from 1928, which contradicts the unpublished autobiography of Rudolph Hennig,[7] cited in Schmitt 2013a; Rost is visible in Fig 1.4, upper arrow) who taught Hennig biology; his final tutor was Willy Matthes.[8] Hennig's brother Rudolph later became a pupil at the Landesschule and joined Hennig's group when it was supervised by Maximilian Rost (see also Schmitt 2013a). From the 'Klotzsche-period' few photographs exist (Figs 1.4 and 1.5).

Hennig was a very good and ambitious student, skipping one form (OIII), finishing school a year earlier than the usual time period for the Landesschule.[9] On the other hand, he was reported as 'an absolutely normal pupil' (Schmitt 2013a; see Figs 1.4 and 1.5), rapidly being acknowledged as 'the scientist' in his form and nicknamed 'Orang'.

He appears to have already been interested in entomology when he entered Klotzsche, his passion for insects probably initiated by Dr Seifert was also noted by Hegewald.[10] As an example, Hennig had noted a beetle (cockchafer) pest problem in 1927 caused by *Melolontha hippocastani* rather than the more common *M. melolontha*, the event written up in an article he published in *Mitteilungen aus der Landesschule Dresden* some years later in 1931(Hennig 1931). Hennig presented a list of species which he had documented on the Landesschule property and announced an exhibition of his own collection (Hennig 1931). The six-page paper documents a wide range of insect groups (Saltatoria, Blattoidea, Hemiptera, Lepidoptera, Coleoptera, Hymenoptera and a few Diptera, altogether more than 70 species). The majority of the records mentioned were made with reference to the year of collection, 1930. Furthermore, Hennig had at that time

---

[6] Letter by Rudolf Hegewald, Nossen, to the Author, 1 February 2000. Hegewald entered the Landesschule the same day as Hennig also starting in Dr Költzsch's group. In his letter he describes Hennig as a pupil between 1927 and 1931. He wrote 'ging Willi akribisch seiner Leidenschaft nach: der Beschäftigung mit Schmetterlingen usw. In seinem Pult […] bewahrte er sorgfältig seine Kästchen mit den aufgespießten Schmetterlingen usw' (Willi followed his passion with scientific precision: dealing with butterflies and others. In his desk he accurately kept his boxes with needled butterflies and others). Original: Willi-Hennig-Archiv, Senckenberg Museum für Naturkunde Görlitz.

[7] Handwritten biography by Rudolph Hennig.

[8] Handwritten biography by Rudolph Hennig.

[9] Letter from Dr Paul Dribbisch.

[10] Letter by Rudolf Hegewald.

Fig 1.4 Willi Hennig (larger arrow) with other pupils and teacher Maximilian Rost (small arrow), Easter 1929, in Klotzsche, at the entrance to the school garden. Courtesy of Rudolf Hegemann, Nossen. Original: Willi-Hennig-Archiv, Senckenberg Museum für Naturkunde Görlitz (SMNG).

developed a passion for butterflies, the taxon best represented in his school collection (Schmitt 2013a, 2013b).

It was most likely Maximilian Rost, Hennig's teacher in biology, who introduced his young pupil to two zoologists at the Museum für Tierkunde in Dresden: the young ornithologist Wilhelm Meise (1901–2002) and the older entomologist Fritz van Emden (1898–1958) (Schmitt 2013a). Although it is not clear which of the two first made contact with Hennig, both zoologists supported him with respect to their specific fields of research.

This introduction was another lucky event in Hennig's life: it was his first exposure to the field of taxonomy in the museum, where he found engaged specialists willing to share their expertise with a talented pupil and still open to discussing the frontiers of knowledge and the deficiencies in their field of work. So Meise and van Emden (both in different fields) laid the foundations for Hennig's lasting interest in systematics and entomology.

Hennig worked with Meise on the flying dragons *Draco* and *Dendrophis*, producing two papers from these joint studies (Meise and Hennig 1932, 1935, Xylander 2013a). But, as Meise noted (Schmitt 2013a), Hennig's major interest was already

Fig 1.5 Willi Hennig (arrow), Christmas celebration of Department 10 of the Landesschule, 1927 Courtesy of Rudolf Hegemann, Nossen. Original: Willi-Hennig-Archiv, Senckenberg Museum für Naturkunde Görlitz (SMNG).

entomology and he often disappeared to the upper floor of the Museum collection building (at that time located in the famous Dresden Zwinger) where Fritz van Emden had his office and collection.

Probably inspired by these contacts, Hennig produced an essay on the relevance of systematics for modern zoology, an essay that was published posthumously (Hennig 1978), showing his interest in entomology and especially taxonomy. Regardless of details, contact with Meise and van Emden sealed Hennig's enthusiasm that became his professional passions that he followed throughout his life.

1.6 Mission fulfilled: reflections from Hennig's childhood and youth into his scientist life

Hennig was born into a worker's family with socio-economic circumstances that would normally have never allowed him to pursue an academic career. However, he found his way into science via a sequence of nearly unbelievably lucky events, as well as having a precocious talent, the encouragement and support of his family, teachers and others with whom he made contact, mostly by chance rather than any strategic planning – but if a chance arose, Hennig took it with admirable consequences.

Even after his ideas had become acknowledged more widely in the scientific community, he remained the shy, remote person he had been in (or become, due to) his childhood and youth. He was neither an eloquent speaker nor a charismatic academic teacher giving fascinating lectures for his students. Rather, he was, as reported by the Professor Claas Naumann (former director of the Museum Koenig, Bonn, *pers. comm.* to the author) and Dr Paul Dribbisch.[11] Paul Dribbisch, one of his students in 1953 described him as, 'gehemmt' (inhibited), 'sehr zurückhaltend' (very reserved), 'unsicher im Verhalten' (insecure in his behaviour), 'aber sehr selbstsicher im Stoff' (but very self-conscious with regard to his subjects).[12] Later, during his honorary professorship in Tübingen Hennig avoided all teaching engagements with large numbers of students and stuck to seminars with a small number of participants and preferred an iterative approach to phylogenetic questions. Then he was able to present convincingly (and in a more relaxed way) his wide expertise and the application of his theory in modern systematics and development of phylogenetic trees.

In the wider public arena, such as large congresses, Hennig was never able to garner much attention for his comments or to use his authority, again indicating that he was not a "man of the spoken word" but a scientist who found his own way, shy and remote, requiring his 'followers' to advertise and promote his ideas (and interesting parallel to the personality of Charles Darwin). Thus, in Germany, Peter Ax (Xylander 2013b, 2016), Otto Kraus, Wolfgang Dohle and Walter Sudhaus, developed his ideas, while internationally Lars Brundin, Gareth Nelson, Steve (James) Farris and Ed Wiley – to name a few – developed, published and, most of all, taught their students the scientific ideas behind Hennig's phylogenetic systematic (Xylander 2013a).

This was the definitive success of Willi Hennig's 'mission impossible'.

Acknowledgements

I would like to thank the Linnean Society of London for the invitation to attend the Willi Hennig Symposium held in November 2013. Furthermore, I would like to thank my colleagues Dr Michael Schmitt, who encouraged me to develop my

---

[11] Letter from Dr Paul Dribbisch, Wilhelmshorst, to the Author, 20 July 2002, 4 pp. describing his impressions of Hennig as a student in Potsdam in 1953, with some personal impressions and remarks. Also included are three photocopies of the handwritten draft from lectures given by Henning on animal phylogeny showing cladograms of metazoan evolution, 'fishes' and Tetrapoda. Original: Willi-Hennig-Archiv, Senckenberg Museum für Naturkunde Görlitz.

[12] Ibid.

interests in the early biography of Willi Hennig, and Jürgen Vogel who first drew my attention to the places of birth and childhood of Willi Hennig in Upper Lusatia and accompanied me on many trips to meet people who knew Willi Hennig in the 1920s and 1930s. Last, but not least, I would like to address my sincere thanks to the niece of Willi Hennig, Mrs Katharina Linke, Oberoderwitz, who made available many sources from her family archive. Copies of the cited material are deposited in the Willi-Hennig-Archiv of the Senckenberg Museum for Natural History Görlitz (SMNG).

References

Amtsblatt der Gemeinde Oppach (August 2013). www.oppach.de/amtsblatt/2013/August2013.pdf

Hennig, W. (1931). Einiges über die Insekten des Landesschulgebietes. *Mitteilungen aus der Landesschule Dresden*, 8, 1–6. [Original: Willi-Hennig-Archiv, Senckenberg Museum für Naturkunde Görlitz.]

Hennig, W. (1978). Die Stellung der Systematik in der Zoologie. *Entomologica Germanica*, 4, 193–199.

Meise, W. and Hennig, W. (1932). Die Schlangengattung *Dendrophis*. *Zoologischer Anzeiger*, 99, 273–297.

Meise, W. and Hennig, W. (1935). Zur Kenntnis von *Dendrophis* und *Chrysopelea*. *Zoologischer Anzeiger*, 109, 138–150.

Schmitt, M. (2010). Willi Hennig, the cautious revolutioniser. *Palaeodiversity*, **3** suppl., 3–9.

Schmitt, M. (2013a). *From Taxonomy to Phylogenetics: Life and Work of Willi Hennig*. Leiden: Brill.

Schmitt, M. (2013b). Willi Hennig: Entomologe und Phylogenetiker.

Entomologische Nachrichten und Berichte, 57, 78–80.

Vogel, J. and Xylander, W.E.R. (1999). Willi Hennig. Ein Oberlausitzer Naturforscher mit Weltgeltung. Recherchen zu seiner Familiengeschichte sowie Kinder- und Jugendzeit. *Berichte der Naturforschenden Gesellschaft der Oberlausitz*, 7/8, 145–155.

Xylander, W.E.R. (2013a). Willi Hennig (1913–1976) – Wissenschaftliche Bedeutung, Leben und Werk. Eine Würdigung anlässlich seines 100. Geburtstages am 20. April 2013. *Der Sekretär*, 13 (1), 3–14.

Xylander, W.E.R. (2013b). Obituary: Prof. Dr. Peter Ax. *Organisms, Development and Evolution*, 13, 1–3.

Xylander, W.E.R. (2016). From the interstitial to phylogeny of the animal kingdom. In: Peter Ax, the promoter of phylogenetic systematics.*Peckiana* 11, 7–19.

2

Willi Hennig: a shy man behind a scientific revolution

MICHAEL SCHMITT

2.1 Willi Hennig's life

Willi Hennig, the man who created a 'new paradigm' (Kühne 1978) in biological systematics, was born on 20 April 1913 into humble circumstances (Xylander 2015). His father was a railway worker, his mother was illegitimate. He was the first of three sons. The family lived in the village of Dürrhennersdorf in the southeast of Saxony, moving *c.* 10 km west to Taubenheim in 1919, and from there *c.* 3 km northeast to the small town of Oppach in 1921. These three places are located in an area of about 10 x 5 km; here, Hennig received his primary school education. He was an excellent pupil, and in addition to the lessons taken at school he learned from private tuition his mother had organised (Xylander 2015). An extremely lucky circumstance allowed his acceptance as a pupil at the Landesschule Dresden in 1927, a boarding school that offered places at reduced, or even completely waived, fees for boys from lower social classes. He performed superbly here as well, so much so that he skipped a year and passed the final examination in 1932. The same year he entered the University of Leipzig and enrolled in zoology, botany and geology (Fig 2.1).

Already, during his time in primary school, he developed a certain interest in nature, which was fostered by teachers at the Landeschule and, later, especially by two scientists at the Dresden State Museum of Zoology: Wilhelm Meise (1901–2002) and Fritz van Emden (1898–1958) (Schmitt 2013: 20–25, Xylander,

The Future of Phylogenetic Systematics: The Legacy of Willi Hennig, eds. D. Williams, M. Schmitt and Q. Wheeler. Published by Cambridge University Press.

Fig 2.1 Willi Hennig as a student, 1933. Courtesy of Gerd Hennig, Tübingen.

this volume). As a student, Hennig, guided by these two scientists, studied and published on reptiles and flies. Before he received his doctoral degree in 1936, he had already published eight scientific papers with more than 900 printed pages. For his doctoral thesis, he studied the copulatory apparatus of Diptera Cyclorrhapha (Hennig 1936a). Originally, Hennig was to be supervised by Johannes Meisenheimer (1873–1933), who had authored a two-volume work on sex and gender in animals (Meisenheimer 1921, 1930). However, because of Meisenheimer's untimely death in February 1933, his successor Paul Buchner (1886–1978), the founder of symbiosis research in Germany, supervised Hennig's PhD. Remarkably, Hennig published 17 papers between completing his PhD and starting paid work, contributions that comprise 402 pages, all on the morphology and anatomy of Diptera.

Due to the 1933 National Socialist racist laws, Fritz van Emden was expelled from the Museum, and his successor, Klaus Günther (1907–75), continued friendly interactions with Hennig. The two men soon became lifelong friends (Fig 2.2). Günther, 6 years older than Hennig, played an important part in his scientific, and especially philosophical, development. Hennig found a position at the Deutsches Entomologisches Institut (DEI, German Entomological Institute), first on a grant, then, from October 1938, as an assistant. From the beginning of his employment at the DEI until the end of World War II, Hennig published 41 papers with a total of 805 pages, nearly all of these on the morphology and taxonomy of Diptera.

Fig 2.2 Klaus Günther (left) and Willi Hennig, 1954. Courtesy of Gerd Hennig, Tübingen.

In May 1939, Hennig and Irma Wehnert (1910–2000), whom Hennig had met at the university, married (Fig 2.3). The couple lived in Berlin during their marriage; they had three sons, Wolfgang, Bernd and Gerd.

Hennig was soldier during World War II, serving in Poland, France, Denmark and Russia, where he was severely wounded in January 1942. After recovery, he worked at the Military Medical Academy as an entomologist from July 1942. He was sent to Greece and Italy to study insect pests, especially those transmitting malaria. When the German army surrendered in Upper Italy on 2 May 1945, Hennig was taken prisoner of war by British troops and placed in a camp near Abano Terme. During the following months in British captivity, he jotted down the manuscript of the *Grundzüge einer Theorie der phylogenetischen Systematik*. The book was only published in 1950, and marks a turning point in biological systematics (Hennig 1950).

Hennig was released from British captivity in October 1945 and managed to reach his wife and their three sons in Leipzig, in the Soviet zone of Germany. His wife lived at her parents' place after leaving Berlin because of the bombing during the war. Hennig found a job at the Leipzig University as a stand-in for his doctoral supervisor, Paul Buchner, who did not return to Germany but stayed on the island of Ischia, where he had lived at various times before and during the war. Although the position at Leipzig University was not temporary and would have guaranteed him

Fig 2.3 Irma Wehnert and Willi Hennig, 1934. Courtesy of Gerd Hennig, Tübingen.

sufficient income to support his family, he returned to the DEI as soon as he had the chance, in 1947.

The DEI had moved *c.* 170 km north-west from Berlin-Dahlem to Blücherhof manor in Mecklenburg in 1943. Hennig worked there from time to time but had also an office in Berlin-Dahlem (in the American sector) at the Biologische Zentralanstalt für Land und Forstwirtschaft (Central Institution for Agriculture and Forestry) until 1950, when the DEI returned to Berlin; its return, however, was to Friedrichshagen, in the Soviet sector. As the Hennig family lived in Berlin-Steglitz, in the American sector, Hennig had to commute daily to the DEI, which took him *c.* 90 minutes each way.

As an entomologist at the DEI, Hennig published 41 titles, among them three books (the *Grundzüge* and the two volumes of *Taschenbuch der Zoologie*, Hennig 1957a, 1959) totalling more than 3100 pages. During this period, he did not exclusively publish on Diptera but, in addition, wrote several publications on more general topics. He wrote in greater detail on his method of Phylogenetic Systematics (e.g. Hennig 1947, 1949, 1957b, Hennig et al. 1953), and replied to criticism (e.g. Hennig 1955). He must have worked on an extensive revision of the *Grundzüge*, because he later stated that he submitted the manuscript for translation to the University of Illinois Press in 1961 (Schmitt 2013: 84).

As a reaction to the erection of the Berlin Wall on 13 August 1961, Hennig left the DEI immediately and found temporary employment with the Technische Universität (TU) of Berlin (West), which lasted 3 years. In 1963, the director of the Staatliches Museum für Naturkunde in Stuttgart (SMNS) offered him a position as head of a department of phylogenetic research, which Hennig accepted. He clearly preferred to stay in Berlin, and never became a confirmed citizen of Suebia (the region in the south-west of Germany where Stuttgart is located; Schmitt 2013: 97). However, he saw no chance of finding a position at an institution in West Berlin where he had access to an entomological collection and library. In Stuttgart – or rather Ludwigsburg, where the collections were then housed, *c.* 10 km north of Stuttgart – he had no curatorial duties but could work on the museum's collection as he pleased. Here, he worked with increasing interest on amber fossils. By the time he died from a heart attack on 5 November 1976, he had published another 5000 pages in 54 papers and books, among them the revolutionary work *Phylogenetic Systematics*, published in 1966, and *Stammesgeschichte der Insekten* published in 1969. He left numerous unfinished manuscripts, which were published posthumously by his eldest son Wolfgang, partly in collaboration with Gerhard Mickoleit (Tübingen).

Hennig had been appointed as professor of zoology by the Brandenburgische Landeshochschule at Potsdam in 1951. He became an honorary professor at the University of Tübingen in 1970. He received the Fabricius medal of the German Entomological Society in 1954, the gold medal of the Linnean Society of London in 1974 and the gold medal of the American Museum of Natural History (New York) in 1975. He had become a corresponding member of the Finnish Entomological Society in 1955, a member of the German Academy of Natural Scientists Leopoldina in 1959, a foreign member of the Royal Swedish Academy of Sciences in 1972, a corresponding member of the American Entomological Society in 1963 and its honorary member in 1976, and an honorary member of the Society of Systematic Zoology in 1976. He had been awarded an honorary doctorate (Dr. h.c.) by the Freie Universität Berlin in 1968.

He is buried in the Bergfriedhof in Tübingen (Fig 2.4).

2.2 Willi Hennig's personality

All descriptions of Hennig's personality by contemporaries concur in characterising him as introverted, shy, cautious or self-effacing (Schmitt 2010, 2013: 101–105). As a child, he preferred to read books or study insects than play with his brothers. As Vogel and Xylander (1999) discovered in a 1921 report from one of his teachers, Hennig had 37 days of absence in less than a year, which probably indicates that he was rather frequently ill. In his late fifties, he suffered at least two heart attacks (Schlee 1978) and, before 1976, he must have suffered another (Schmitt 2013: 97). This probably explains why he often referred to his feeble health when refusing invitations to present lectures or participate in symposia. However, there were certainly two additional reasons for his reluctance to attend congresses and to not give talks. On the one hand, he had such enthusiasm for his empirical studies that he was

Fig 2.4 Tombstone of Willi and Irma Hennig on the 'Bergfriedhof', Tübingen. Original, 16 February 2012.

reluctant to interrupt them. He even had problems sparing time to write theoretical or methodological papers. On the other hand, he was almost unable to talk publicly to groups of more than three people. He liked meeting colleagues and having discussions in small groups, especially if his counterparts were genuinely interested in his work (Senglaub 2001).

In the course of his professional career, Hennig was an academic teacher at four universities (Leipzig, Potsdam, TU Berlin, Tübingen). He gave lectures, courses and seminars. During his 13 semesters at the University of Tübingen he gave four seminars of 2 hours per week, one jointly with Gerhard Mickoleit. This is not just an indication of a high motivation to teach. He supervised or co-supervised six doctoral theses during his time at the University of Tübingen. As far as is documented in the archive of the SMNS, Hennig exchanged more than two letters with only one of these doctoral students and met this student only a few times (Schmitt 2002).

Hennig never showed any signs of humour in scientific affairs, be it in correspondence or publications, in contrast to his friend Klaus Günther or, for example, Carl Linnaeus and other renowned role models. However, in private, Hennig could exhibit a cheerful side. He very much liked Klaus Günther's sense of humour (typical for Berliners). He also enjoyed teasing colleagues and friends by letting them guess where he had spent his holidays, for example, when he had returned with a tanned skin. He played a game with Gerd von Wahlert in 'who can cite more classical authors' (preferably in their original languages, only exceptionally in German), and could laugh heartily at jokes with the ladies in the departmental administration (Schlee 1978). He exchanged letters full of wit and jokes with Klaus Günther and other colleagues who were close to him, such as his former teacher Willy Matthes, and his professor at the University of Leipzig, Arno Wetzel (Schmitt 2013: 103).

This all indicates that Hennig was not at all a revolutionary person. Nevertheless, some biographical factors could have fostered his openness towards scientific innovations. As discussed earlier (Schmitt 2010), it might be expected that Hennig would be rather more conservative than innovative, according to Sulloway's (1996) thesis that firstborns tend to be more conforming and traditional than those born later in a series of siblings. However, Sulloway also found that firstborns of lower social classes were nearly as open to innovations as those born later of all classes, provided that they were older than 21 when their parents died. This applies to Hennig: he was 34 when he lost his father, 52 when his mother died. Interestingly, Sulloway reports that 'shyness' is an additional influence enhancing the receptiveness of firstborns for scientific innovations. Since Hennig was certainly shy, this feature possibly helped making him a scientific 'rebel.'

2.3 From order to phylogeny

Hennig's publications prior to 1936 are purely morphological and taxonomic. He discussed 'Verwandtschaft' (relatedness) in order to group taxa and mentioned geographical distribution and morphological resemblance as indicators of relatedness. However, in several publications in that year (e.g. Hennig 1936b), he pondered on the weakness of 'similarity' as an argument for 'relatedness' and emphasised that a classification should always be made according to phylogenetic relatedness ('Phylogenetische Verwandtschaft'), and not according to morphological differentiation (Hennig 1936b: 172). However, although he had developed a certain idea for a method of revealing phylogenetic relatedness by finding 'progressive characters in common' ('gemeinsame fortschrittliche Merkmale'; Hennig 1936b: 170), he made no attempts to analyse the phylogeny of a group of species or to erect a substantiated phylogenetic tree. In his doctoral thesis (Hennig 1936a: 363), he referred to the phylogenetic trees of other authors but he neither questioned their empirical basis nor provided any alternatives. It is evident that whenever he mentioned 'Verwandtschaftsgruppen' (groups of related individuals/species), he did so in the context of sorting taxa into them.

In 1943, Hennig explicitly stated that 'the system of animals is, in my opinion, quasi[1] a theory on the phylogenetic relationships among the forms of animals'

---

[1] Quasi is a direct translation from the German. Hennig could have written 'is nothing but a theory' or simply 'is a form of presenting the phylogenetic relationships.' By 'gleichsam' (= quasi), Hennig avoided sounding strict.

('Das System der Tiere ist nach meiner Auffassung gleichsam eine Theorie über die phylogenetischen Beziehungen der Tierformen', Hennig 1943: 143). The subject of the introduction to his *Grundzüge einer Theorie der phylogenetischen Systematik* (Hennig 1950: 1–96) as well as the short pre-published sections from it (Hennig 1947, 1949) is neither phylogeny nor evolution. Hennig was much more focussed on classifying, ordering, and ranking. In this early period of phylogenetic systematics, he used phylogenetic relatedness as a means to form natural groups. He was not especially interested in the real history of those organisms he studied. Even the remainder of the *Grundzüge* treats taxonomy, Linnean ranks and hierarchies and only marginally addresses the subject of phylogeny. When Hennig reviewed his own book in 1952, he did not mention a new method for tracing the phylogenesis of organisms. Instead, he argues at length in favour of a strictly genealogical concept of 'phylogenetic relationship', emphasising the advantages of a strictly phylogenetic over a 'natural' system (i.e. one simply based on similarity) of classification, and indicates the discovery of 'phylogenetische Gesetzmäßigkeiten' as the primary task of systematics. Rainer Zangerl, one of the translators of Hennig's *Phylogenetic Systematics*, explains in the preface that 'Gesetzmäßigkeit' is 'one of the worst examples' of German words without proper English equivalents. He and D. Dwight Davis had decided to use 'conformity to law' as a translation, 'fully realizing that there is no perfect congruence between these concepts. For example, a Gesetzmässigkeit may be no more than the repetitive occurrence of a specific phenomenon'. However, as Hennig refers to 'nomothetic' sciences in the context of 'Gesetzmäßigkeiten', it is highly probable that he really assumed laws governing the course of phylogenesis.

From 1953, there is a growing proportion of phylogenetic reasoning in Hennig's publications. In Hennig et al. (1953) he introduced his famous 'argumentation scheme', clarified the relation of his concept of 'apomorphy' and the traditional homology concept, demonstrated the procedure of analysing phylogenetic relationships, and presented an 'outline of a phylogenetic tree' ('Entwurf eines Stammbaumschemas') of insects. In 1957, he explained his method by various illustrations that were later included in *Phylogenetic Systematics* and explicitly stated that his approach is linked to the 'theory of descent', i.e. Darwinian evolution. In many of his subsequent taxonomic papers, he applied his 'argumentation scheme', whereby he normally polarised characters by referring to assumed evolutionary or geological processes.

Hennig's first English publication begins with the sentence, 'Since the advent of the theory of evolution, one of the tasks of biology has been to investigate the phylogenetic relationships between species' (Hennig 1965: 97). In *Phylogenetic Systematics*, as well as in his later publications, e.g. the *Stammesgeschichte der Insekten* (1969), and even more so in the posthumously published manuscripts (for a complete list see Schmitt 2013: 193f.), Hennig nearly always inverted the description of his approach. Where he originally began with the problems of classification, he later began with revealing phylogeny. He made it unmistakeably clear that the analysis of the phylogenetic relationships is the main objective of his approach. While in his earlier publications he used this analysis as a means to construct a classification, he later treated the phylogenetic system (= classification) as a natural consequence of phylogenetic analysis.

What Hennig regarded as novel and necessary may nowadays appear self-evident: that 'relationships' or 'relatedness' in systematics must be defined in terms of genealogy, i.e. recency of ancestors in common, that the identification of monophyletic taxa is crucial (even if one finally decides to leave certain paraphyletic taxa in the classification), that a hypothesis of monophyly can only be substantiated by shared apomorph homologies (synapomorphies), and that, in short, the procedure of phylogenetics means the 'search for the sister group'. Nevertheless, even if he did not elaborate a concise recipe for polarising characters (which is arguably the core of any cladistic analysis), he laid the foundations of what is now rightly called the 'Hennigian revolution'.

References

Hennig, W. (1936a). Beiträge zur Kenntnis des Kopulationsapparates der cyclorrhaphen Dipteren. *Zeitschrift für Morphologie und Ökologie der Tiere*, 31, 328–370.

Hennig, W. (1936b). Beziehungen zwischen geographischer Verbreitung und systematischer Gliederung bei einigen Dipterenfamilien: ein Beitrag zum Problem der Gliederung systematischer Kategorien höherer Ordnung. *Zoologischer Anzeiger*, 116, 161–175.

Hennig, W. (1943). Ein Beitrag zum Problem der 'Beziehungen zwischen Larven- und Imaginalsystematik'. *Arbeiten über morphologische und taxonomische Entomologie aus Berlin-Dahlem*, 10, 138–144.

Hennig, W. (1947). Probleme der biologischen Systematik. *Forschungen und Fortschritte*, 21/23, 276–279.

Hennig, W. (1949). Zur Klärung einiger Begriffe der phylogenetischen Systematik. *Forschungen und Fortschritte*, 25, 136–138.

Hennig, W. (1950). *Grundzüge einer Theorie der phylogenetischen Systematik*. Berlin: Deutscher Zentralverlag.

Hennig, W. (1952). Autorreferat: Grundzüge einer Theorie der phylogenetischen Systematik. *Beiträge zur Entomologie*, 2, 329–331.

Hennig, W. (1955). Meinungsverschiedenheiten über das System der niederen Insekten. *Zoologischer Anzeiger*, 155, 21–30.

Hennig, W. (1957a). *Wirbellose I. Taschenbuch der Zoologie*, vol. 2, Leipzig: Thieme Verlag.

Hennig, W. (1957b). Systematik und Phylogenese. In *Bericht über die Hundertjahrfeier der Deutschen Entomologischen Gesellschaft Berlin*, ed. H. –J. Hannemann. Berlin: Akademie-Verlag, pp. 50–71.

Hennig, W. (1959). *Wirbellose II. Taschenbuch der Zoologie*, vol. 3, Leipzig: Thieme Verlag.

Hennig, W. (1965). Phylogenetic systematics. *Annual Review of Entomology*, 10, 97–116.

Hennig, W. (1966). *Phylogenetic Systematics*. Urbana, IL: University of Illinois Press.

Hennig, W. (1969). *Die Stammesgeschichte der Insekten*. Frankfurt am Main: Waldemar Kramer & Co.

Hennig, W., Bollmann, H. and Machatschke, J. (1953). Kritische Bemerkungen zum phylogenetischen System der Insekten. *Beiträge zur Entomologie*, 3 (Sonderheft), 1–85.

Kühne, W.G. (1978). Willi Hennig 1913–1976: Die Schaffung einer Wissenschaftstheorie. *Entomologica Germanica*, 4, 374–376.

Meisenheimer, J. (1921). *Geschlecht und Geschlechter im Tierreiche. 1. Die natürlichen Beziehungen*. Jena: Gustav Fischer.

Meisenheimer, J. (1930). *Geschlecht und Geschlechter im Tierreiche. 2. Die allgemeinen Probleme*. Jena: Gustav Fischer.

Schlee, D. (1978). In Memoriam Willi Hennig 1913–1976. Eine biographische Skizze. *Entomologica Germanica*, 4, 377–391.

Schmitt, M. (2002). Willi Hennig (1913–1976) als akademischer Lehrer. In *Fokus Biologiegeschichte*, ed. J. Schulz. Berlin: Akadras, pp. 53–64.

Schmitt, M. (2010). Willi Hennig, the cautious revolutioniser. *Palaeodiversity*, 3, Supplement, 3–9.

Schmitt, M. (2013). *From Taxonomy to Phylogenetics: Life and Work of Willi Hennig*. Leiden: Brill.

Senglaub, K. (2001). Erinnerungen eines ehemaligen Zoologie-Studenten an seinen akademischen Lehrer Willi Hennig im Winter-Semester 1946/47 in Leipzig. *Sitzungsberichte der Gesellschaft Naturforschender Freunde zu Berlin, Neue Folge*, (2000) 39, 165–168.

Sulloway, F.J. (1996). *Born to Rebel: Birth Order, Family Dynamics, and Creative Lives*. New York: Pantheon.

Vogel, J. and Xylander, W.E.R. (1999). Willi Hennig – Ein Oberlausitzer Naturforscher mit Weltgeltung. *Berichte der Naturforschenden Gesellschaft der Oberlausitz*, 7/8, 145–155.

3

Willi Hennig's legacy in the Nordic countries

OLE SEBERG, TORBJØRN EKREM,
JAAKKO HYVÖNEN and PER SUNDBERG

3.1 Introduction

Writing about historical events, about the major paradigm shift in systematics since Darwin, is not necessarily easy, even if the record appears to be relatively good. Evidently, it becomes even more difficult when the authors, and nearly all the sources, have been involved in the events. Some biases are easy to define, some more subtle. Going through parts of the recent literature there are obvious contradictions, omissions and errors – and there is no reason to believe that this chapter will be any different. To some extent, interpretation of events is in the eye of the beholder.

This chapter is about the introduction of phylogenetic systematics into the Nordic countries and the key persons involved. It does not, however, attempt to follow the subsequent rather rapid spread in different disciplines in each respective Nordic country. Inevitably some may perhaps understandably wonder why they are not mentioned. However, the decision not to do so is solely the responsibility of the authors.

Throughout this chapter the two terms 'phylogenetic systematics' and 'cladistics' are used interchangeably.

The Future of Phylogenetic Systematics: The Legacy of Willi Hennig, eds. D. Williams, M. Schmitt and Q. Wheeler. Published by Cambridge University Press.
© The Systematics Association 2016.

3.2 Hennig in the Nordic countries

Hennig did not travel much and rarely attended scientific meetings, which made him feel uneasy (Schmitt 2013), but he did go abroad while employed at the Staatliches Museum für Naturkunde in Stuttgart, attending the 12th International Congress of Entomology in London in 1964 and visiting major European collections in the 1960s, including those in the capitals of Denmark, Finland and Sweden (Schmitt 2013: 80).

Hennig had, however, been to Denmark much earlier. During World War II, he was enrolled in the 218th Infantry Division of the Wehrmacht and was stationed first in Vordingborg in 1941 and then moved to Haderslev. The 218th Infantry Division was a part of the occupation troops, and until August 1943 this was, to the extent possible during wartime, a relatively easy job, with good access to local commodities, which could be sent back to the family in Germany. However, Hennig did not stay long in Denmark and in January 1942 the 218th Division was rapidly deployed to Kholm in Russia to join the 39th Army, Heeresgruppe Nord, to fight the Russian counteroffensive launched to stop Operation Barbarossa. The Division was to stay around Kholm until January 1944 when it was forced to retreat through Estonia towards Riga, but Hennig was seriously wounded in his right arm and shoulder long before and sent back to Germany to recuperate (Schmitt 2013: 47–49). The recent biography of Hennig by Schmitt (2013) treats these aspects of Hennig's life and his scientific achievements in detail.

3.3 Hennig and Sweden

A key figure for spreading Hennig's ideas, not only in the Nordic countries but in general, was Lars Zakarias Brundin (1907–93). Brundin did his dissertation in 1934 on beetles from Lapland (Brundin 1934) and continued to publish on beetles until the mid-1950s. His interest in chironomid midges (Chironomidae, Diptera) developed when, in 1936, he was hired by the government's Institute for Fresh Water Research and became a key figure in the development of the application of chironomids as ecological indicators. During this period, he gradually developed a keen interest in chironomid taxonomy and eventually in biogeography of the southern continents, which, since Hooker (see e.g. Seberg 1988), is one of the truly classical problems in the field of biogeography. Brundin got a lectureship in entomology at the Swedish Museum of Natural History in Stockholm in 1952, becoming professor of entomology in 1957, a position he held until retirement in 1973. His 1966 study *Transantarctic Relationships and Their Significance, as evidenced by Chironomid midges, with a Monograph of the subfamilies Podonominae and Aphroteniinae and the Austral Heptagyinae* is a classic in modern biology, not least for its introduction (Brundin 1966). Brundin picked up Hennig's methodology during the 1950s realising that his approach and persistence in accepting only strictly monophyletic groups were key to developing solid biogeographic hypotheses.

Of course, Brundin's (1966) radical break with established tradition and its followers elicited a response (Darlington 1970), followed by his own counter-response (Brundin 1972a); Brundin's further scientific work was divided between theoretical papers on biogeography (e.g. Brundin 1988), phylogenetic systematics (e.g. Brundin 1972b) and empirical papers on chironomids (e.g. Serra-Tosio and Brundin 1990).

An event that indirectly became of crucial importance for the spread of Hennig's ideas globally was the 4th Nobel Symposium on *Current Problems of Lower Vertebrate Phylogeny* held June 1967 (Ørvig 1968, see also Schultze 2009) at the Swedish Museum of Natural History in honour of Erik Stensiö (1891–1984), who had been professor at the Department of Paleozoology from 1923 to 1959. Although the only truly phylogenetic paper presented at the symposium was given by Brundin (1968), the meeting brought together a number of scientists, such as Niels Bonde, Brian Gardiner, Peter Humphry Greenwood (1927–95), Daniel Goujet, Søren Løvtrup (1922–2002), Roger Miles, Gareth Nelson, Colin Patterson (1933–98) and Hans-Peter Schultze, who had all grasped the novelty and potential importance of Hennig's work, which had just been translated into English (Hennig 1965, 1966). However, Brundin's talk was not given at the symposium venue but at the Royal Swedish Academy of Sciences (Kungliga Vetenskapsakademien). The symposium has become a milestone in systematics; though the straightjacket of traditional thinking was still pretty obvious:

> But however primitive the acanthodians might have been, their status as fishes seems secure. And as a group of fishes they must be related to something in the Recent fauna. (Nelson 1968: 140)

Clearly, it takes time to get rid of old paradigms. Though for others present the pending paradigm shift was not emergent and Hennig's ideas remained just an extra tool in the toolbox:

> The sister group concept recently introduced by Hennig (1966; and earlier papers; see also Brundin, 1966; and this volume) is certainly a useful tool for this purpose, because it enforces a phylogenetic way of thinking and a careful consideration of the evidence, but as far as vertebrates are concerned it is as present not easy to handle. (Jarvik 1968: 522)

Not only Brundin's writings but also the opportunity to interact with him personally had a tremendous influence on the field as he obviously was a source of inspiration not least for his close Swedish colleagues, like the botanists Kåre Bremer and Hans-Erik Wanntorp, but also for visitors like Nelson, Bonde, Ole A. Sæther and perhaps Risto Tuomikoski (see below).

In the mid-1960s, Kåre Bremer (Fig 3.1) met Brundin when his school class was visiting the Swedish Museum of Natural History and Brundin gave a captivating talk

Fig 3.1 Kåre Bremer in conversation with Gareth Nelson in Miami in 1985. Photo: Birgitta Bremer.

on biogeography. Bremer started his career as a master student at the Department of Botany at Stockholm University in 1970, with the intention of becoming a systematist. Having obtained his degree, he was subsequently employed at the institute from 1972 to 1980. He left this job in 1980; first to become head curator at the Swedish Museum of Natural History and later, in 1989, professor of Systematic Botany at Uppsala University. In 2004, he returned to Stockholm to become vice chancellor of the University – a job he resigned from in 2013 to return to science.

While studying in Stockholm, Bremer met Hans-Erik Wanntorp (Fig 3.2), at that time an assistant teacher; they got to know each other and started collaborating in many botany courses during the 1970s. Wanntorp was well read in zoology and knew a lot about the emerging debate in systematics – he actually subscribed to and read every issue of *Systematic Zoology*. In the early 1970s, Wanntorp was uncontestably the only botanist in Sweden familiar with recent developments in systematics. The senior teachers and scientists in botany in Stockholm certainly did not know much, if anything, about the subject. Accordingly, it was Wanntorp who introduced Bremer to the ongoing debate and encouraged him to read *Systematic Zoology*, which he did (*pers. comm.*) and as a consequence added a cladistic analysis to his PhD thesis (Bremer 1976). At the Department of Botany in Stockholm, however, an enthusiastic group of young supporters had already formed in the mid-1970s.

Fig 3.2 Hans-Erik Wanntorp at Uppsala University in 1990. Photo: Birgitta Bremer.

Among them were Birgitta Bremer, Arne A. Anderberg and Anders Tehler. After his thesis, Kåre Bremer suggested to Wanntorp that they should publish a paper together on the significance of cladistics for botany. Initially, and typically for him, Wanntorp was sceptical about the idea, but Bremer convinced him, and together they published 'Phylogenetic systematics in botany' (Bremer and Wanntorp 1978), which had a tremendous impact. This paper was quickly followed by another concerning the problem of reticulations (e.g. hybridisation) in phylogeny (Bremer and Wanntorp 1979), and yet more were to follow (Bremer and Wanntorp 1981, 1982); even a popular text intended for a Swedish audience was published (Bremer and Wanntorp 1982).

Together with Rolf Dahlgren (see below), Bremer performed the first computer-based analysis (using PAUP) of the angiosperms (Dahlgren and Bremer 1985). However, the use of computers in cladistics was still in its infancy, and most users were on a steep learning curve, so the published analyses were largely run with the default parameter settings (including MAXTREE=100, which made the run terminate when 100 equally parsimonious trees were collected in memory instead of increasing incrementally). Hence, the analyses were nowhere near finding the shortest trees, let alone all of them. If their matrix is re-analysed with either the settings changed or using a contemporary computer program, for example Hennig86

(Farris 1988), collecting as many equally parsimonious trees as possible (it cannot run to completion), all that remains of the tree structure (Dahlgren and Bremer 1985: figs 2 and 3) is a clade consisting of Fumariaceae plus Papaveraceae. However, Hennig86 was not available to Bremer.

Bremer specialised in the sunflower family (Asteraceae), but throughout his scientific career phylogenetic systematics has been a guiding principle in his research, no matter what type of data used (morphological or molecular). He produced a large number of empirical studies and classificatory papers, and contributed to the development of historical biogeography. However, from a purely theoretical point of view his papers on 'Combinable component consensus' (semi-strict consensus; Bremer 1990) and 'Branch support and tree stability' (Bremer 1994) have become standards in the field. The latter was identified by ISI as the most highly cited paper in the Agriculture, Biology and Environmental Sciences by an author in Sweden from 1995 to 1999.

Bremer was a founding fellow of the Willi Hennig Society, which was created in 1980, and its president from 1985 to 1988. He also acted as chairman of the organising committee of the 6th Meeting of the Society in Stockholm in 1988.

Originally, Wanntorp wanted to study zoology, but due to the tough competition for positions in zoology he chose botany instead, and during his whole active career he has been employed at the Department of Botany at Stockholm University. Wanntorp was extremely well-informed and competent in many areas of organismal biology but publishing was not his strength. It is a fair guess that Wanntorp's most important papers are those he wrote with Bremer (see above) even though one of his own papers stands out on its own: 'Historical constraints in adaptation theory: traits and non-traits' (Wanntorp 1983), an important contribution to the then emerging field of historical ecology. Today, in retirement, Wanntorp has returned to his original zoological interests and spends his time with floristics and faunistics and, not least, studying beetles.

Until the mid-1980s, Stockholm was the only centre of cladistics in Sweden, but researchers from the Swedish Museum of Natural History with a cladistics background rapidly started to disperse across Sweden, primarily to the University of Gothenburg (e.g. Christer Erséus and Per Sundberg) where they heavily influenced the local developments in systematics.

Thus, cladistic analysis among the zoologists in Gothenburg was introduced by Erséus and Sundberg, both (at that time) working on marine invertebrates. Per Sundberg (Fig 3.3), with his background in mathematics and multivariate statistics, had applied various phenetic approaches to his systematic studies but became more and more interested in cladistic analyses at the beginning of the 1980s. In 1985, Sundberg even published a paper in a popular botanical journal explaining how cladistics analyses were to be conducted (Sunberg 1985). To illustrate this, he used a set of nuts, nails and bolts as 'organisms', as these familiar objects did not

Fig 3.3 Per Sundberg doing fieldwork on the beach west of Valdivia, Chile in 2009.
Photo: Pierre de Wit.

require any specialist knowledge and, thus, could easily be understood, although the
reader would have to decide what 'conditions' were apomorphic and what were plesiomorphic. This kind of dataset is still used in biological classes in Gothenburg to
illustrate the basic principles behind parsimony analysis. A similar idea was used by
Rasmussen and Seberg in Copenhagen at their phylogenetic course (see below) using
the contents of pencil cases. Sundberg and Erséus did not face any real opposition
among senior zoological systematists, even though they did not pick up the ideas.

At the end of the 1980s, Erséus moved back to the Natural History Museum in
Stockholm, taking with him the phylogenetic approach to taxonomy and systematics. He has now returned to the University of Gothenburg, again as professor in
systematics and biodiversity.

Almost at the same time, Sundberg was awarded a research council funded position in phylogenetic systematics and formed a research group focussing on phylogeny and systematics. From the beginning, this group took a keen interest in
molecular and more statistical approaches to phylogenetics. Sundberg is now professor of systematics and biodiversity at the University of Gothenburg. Members
of the group were active in the first discussions about the highly contentious
PhyloCode (e.g. De Queiroz and Gauthier 1992, 1994, De Queiroz and Cantino
2001), which met strong opposition from their botanical colleagues in Gothenburg
and which lead to heated discussions (Lidén et al. 1997).

Fig 3.4 Lennart Andersson in the field near Leticia, Colombia 1994. Photo: Claes Persson.

Some of the professors and senior researchers at the botanical department were initially strongly opposed to cladistics, whereas the younger PhD students were more interested and ready to pick up the new trends. Magnus Lidén was the first botanist in Gothenburg to publish a cladogram (Lidén 1986), which was also included in his PhD thesis. Lidén was an informal supervisor of another PhD student, Bengt Oxelman, and together they formed an active research group in cladistics. The opposition among senior researchers vanished, at least among the more active, and Lennart Andersson (1948–2005; Fig 3.4) later turned into an eager proponent of phylogenetic thinking in systematics (see e.g. Andersson and Chase 2001). Together Andersson and Sundberg started an advanced undergraduate 10-week course in systematics attracting students from both zoology and botany. Part of this course is still included in a Nordic master's programme in systematics. Andersson and Sundberg also worked together to create a common programme in systematic biology, based entirely on phylogenetic systematics both in research and teaching at all levels. Thus, when it came to a phylogenetic approach to systematics, there was widespread consensus among systematists at University of Gothenburg by the beginning of the 1990s that it was the way to go.

The transition from traditional to phylogenetic systematics was probably easier in Sweden than in most other countries. An important reason for this was that the early proponents were also active in the research councils and other scientific committees and thus could influence funding. A further reason might have been that the key players early on started to form strong research groups in an environment where taxonomists used to work in splendid isolation. These research groups were able to attract postgraduates and had a significant impact on undergraduate teaching in biology and systematics, thereby virtually training a whole generation of Swedish biologists in the field. However, the introduction of cladistics in Sweden did not take place entirely without opposition. This is evident in the scientific evaluation made by the Swedish professor Arne Strid (though employed at University of Copenhagen) of an application from one of Bremer's student to the Swedish Research Council:

> Among the many fashions taxonomy has been subject to during its long history, cladism excels in several characteristics reminiscent of religious movements: (1) One prophet (Hennig), (2) a series of articles of faith and holy texts, (3) an astonishing belief in the superiority of its own methodology, and (4) Balkanization into mutually hostile fractions. (Evaluation sent to Vetenskapsrådet; July 7, 1988)

Nonetheless, and despite these negative statements (which were made entirely out of context), the student was awarded a grant, probably because of other, more positive reviews and interventions from members of the council.

3.4 Hennig and Denmark

The first time Hennig's theoretical work (Hennig 1950) was mentioned publicly in the Nordic countries was probably in the discussion period following the French entomologist Claude Dupuis' paper on the value and taxonomic significance of characters in Tachinidae, Diptera (Dupuis 1956), given at the XIVth International Congress of Zoology in Copenhagen in 1953:

> Les critères de la plus ou moins grande valeur phylétique des caractères morphologiques sont exposés par W. Hennig dans un livre recent (1950, Berlin, *Grundzüge einer Theorie der phylogenetischen Systematik*). J'adopte en grand partie les vues de cet auteur et y renvoie. (Dupuis 1956: 476)

Whether Søren Ludvig Tuxen (1908–83; Fig 3.5), one of the very first to use Hennig's terminology outside the German-speaking world (see below), was present at this event is unknown. Tuxen was registered at the congress and, as the subject was entomological, it was very likely he was present. Hennig, however, was not in attendance.

Due to his upbringing, Tuxen was virtually bilingual and spoke fluent German (Kristensen 1983). Therefore, he had many relations to German scientists and shortly after the war he participated in the newly established 'Phylogenetisches Symposium' series started in Hamburg in 1956 (Kraus 1984, Kraus and Hoßfeld 1998) and which is still ongoing. It is certain that Tuxen gave talks at the 1959 meeting on 'Ontogenie und Phylogenie' (Tuxen 1960) and the 1961 meeting on 'Probleme der Metamerie des Kopfes' (Tuxen 1963). Though Hennig occasionally attended (he did so in 1963), there is no documentation to suggest that he and Tuxen met there, due to the scanty records of the first meetings – but it remains a possibility. Tuxen was a staff member at the Zoological

Fig 3.5 Søren L. Tuxen (1908–83) and Wolf Herre (1909–97) in conversation in front of University of Copenhagen's main building. Photo: NN.

Museum[1] in Copenhagen throughout his career, his main interest being primary wingless hexapods, Protura in particular. However, Tuxen did his Dr Phil. dissertation (then corresponding to Dr Scient., nearly equivalent to the German 'Habilitation') on a very different topic: 'The hot springs of Iceland: their communities and their zoogeographical significance' (Tuxen 1944).

At the 10th International Congress of Entomology, Montreal, August 1956, Tuxen (1958) gave a talk on the 'Relationships of Protura' in which he refers to a recent paper by Hennig (1953) – but does not mention Hennig's 1950 book:

> If we have to rank a group among other groups the decision depends on the value given to each of the characters. Hennig in a recent paper (1953) has tried to direct this decision by introducing some terms which in fact do not introduce anything new, but which may help to clear up the mind and to avoid confusion. He calls primitive characters plesiomorph and derived characters apomorph; the aim of the phylogenist then is to find synapomorphs and to rule out symplesiomorphs, i.e. to find those characters common for two or more groups that are derived and rule out those that are only common inheritance. (Tuxen 1958:493)

In contrast to Hennig (see Hennig 1966: 122), however, Tuxen was not willing to accept the idea that, unless convergence can be justified by external evidence, special similarity is a first indicator of relationship (e.g. in insects, halters = modified

---

[1] The Botanical Museum, the Botanical Garden, the Zoological Museum and the Geological Museum in Copenhagen have, since 2004, all been united in the Natural History Museum of Denmark.

hindwings or modified forewings are potential – and actual, as it turns out – synapo-morphies for Diptera and Strepsiptera, respectively). Tuxen's reluctance stems from his a priori belief that convergence was a common phenomenon, thus foreshad-owing heated discussions yet to come. Otherwise, the paper includes an emerging modern character analysis pointing out primitive and advanced characters. As indicated above, it is difficult to change old habits and their accompanying fuzzy logic, thus Tuxen also wrote (1958: 496) vague statements like: 'They [the Protura] are related [to real insects], but not very closely, to the Collembola.' However, in his 1959 paper (Tuxen 1959: 412–413) in the Snodgrass Festschrift, he presented a considerably more mature phylogenetic discussion of the relationships between Protura and the remaining insects and myriapods.

In 1963 his points of view further matured:

> it goes without saying that there here are as (nearly) everywhere, no recent group is to be directly derived from any other recent group. If ever we find a fossil Proturan – though the probability is very slight, when the minute size and the fragility of the creatures is taken into consideration – we might get little more understanding of direct evolution. Of course we would not find a creature with primitive Proturan features devoid of specialized ones; all species must have their specializations just in order to exist. (Tuxen 1963: 306–307)

However, some misunderstandings still prevailed as Tuxen – mixing character state trees with phylogeny – wrote:

> The area of 'phylogenetic trees' is now on the decline; two dimensions, or even three, are too few to illustrate what happened, already for reasons of a specialization for every species. (Tuxen 1963: 308)

Whereas Tuxen is well known in entomological circles for his studies on Protura, being a museum curator he did not teach and his influence never reached beyond his own closed spheres. His use of Hennigian methods remained hidden to most, though at a rather advanced age he used Hennig's terminology precisely (e.g. Tuxen 1971) and performed modern cladistic analyses (e.g. Tuxen 1977).

The entomologist Ole Engel Heie (Fig 3.6) is probably less widely known outside Denmark except by those who work on aphids, as he did. Heie received his degree in biology in 1951, and taught biology at widely different levels: primary schools, colleges of education and various universities. He obtained his Dr Phil. in 1967 with a dissertation on fossil aphids (Heie 1967) while employed at Skive Seminarium (Skive College of Education); he became professor at the Institute of Biology at Danmarks Lærerhøjskole (today Royal Danish School of Educational Sciences) from 1981 until he retired in 1994. He contributed extensively to the taxonomy of extinct as well as extant aphids, being the author of a series of publications on aphid taxonomy: 'The Aphioidea (Hemiptera) of Fennoscandia and Denmark, Vol. I–VI' (1980–95) in the series *Fauna Entomologica Scandinavica* and *Bladlus*, Vol. 1–2

Fig 3.6 Ole E. Heie giving a talk at the 8th International Symposium on Aphids held in Catania, Italy, June 2009. Photo: Xiaolei Huang.

(2004a, b) in the series *Danmarks Fauna*. Heie was very interested in phylogeny and became acquainted with Hennig's work (1953, 1957) while studying for his dissertation on fossil aphids (see above), which he defended in 1966. Like Tuxen, Heie initially focused on Hennig's character terminology: 'and e.g. certain characters are primitive and other derivative, plesiomorphic and apomorphic after Hennig's terminology (Hennig 1953,1957)' (Heie 1967: 226).

This also guided his discussion of characters in the 'Systematic and phylogeny' section of the paper, but nonetheless, the phylogenetic tree presented by Heie (1967: Fig 3.7) as an improvement over one published slightly earlier, remained very traditional with no obvious relationship to a cladogram. However, Heie's remarks on the use of fossil age as a key to the determination of character polarity clearly foreshadowed discussions about the 'paleontological argument' (Nelson 1978): 'In addition to that, the fossil finds tell us that numerous characters and variations of these, of which many are undoubtedly not primitive, were found already in Tertiary aphids, e.g. ...' (Heie 1967: 244).

Heie's strong interest in phylogeny is evident in his two textbooks which both have very explicit cladistic contents: A textbook on evolution (*Evolutionslære*), first published in 1969, and a textbook on zootaxonomy (*Zootaxonomi*), published in 1983. The latter textbook had a relatively narrow circulation as it was largely an internal publication, published at Heie's place of employment. However, the textbook on evolution was more widely distributed and used by primary school teachers who attended brush-up courses in biology at Danmarks Lærerhøjskole and by students at high schools. It was even translated into Swedish (1973). Several years prior to

Fig 3.7 Niels Peder Kristensen in his office in the mid-1960s. Photo: SNM; Probably Jens W. Rasmussen.

publication of the 1969 edition, xeroxed copies of *Evolutionslære* were in use by students whose main subject was biology at some colleges of education, and by a group of biology students at University of Copenhagen (*pers. comm.*). The second printing of the textbook (1976) had an even wider circulation than the first edition. However, the book had relatively little impact, perhaps because it hit the wrong audience at the wrong time. Thus, Seberg (see below) has a clear recollection of reading the book at high school, but not dwelling on its phylogenetic contents at any length. Heie also published a number of true cladistic analyses (e.g. Heie 1976).

In contrast to Heie, Niels Peder Kristensen (1943–2014; Fig 3.7), who became associate professor in 1972 and professor from 1995 to 2013 at the Zoological Museum, is probably known by all entomologists, not least for his work on Lepidoptera, his overview of insect phylogeny (e.g. Kristensen 1975) and, since 1995, his co-editorship of the 'Insecta' section of the *Handbook of Zoology/Handbuch der Zoologie*, plus more than half a dozen other editorships.

From a historical standpoint, Kristensen began using Hennig's terminology and methods in a series of papers in 1967–8 following an extended research visit to the University of Bristol in 1966–7. Kristensen visited the entomologist Howard E. Hinton's (1912–77) laboratory and during that visit he was inspired by Hinton to read Hennig. This led to discussion of several contemporary hot topics, for example, monophyly vs. non-monophyly of the Mecoptera and the possible morphological autapomorphies of the Lepidoptera, for example, 'primitive' Lepidoptera, have 'primitive' biting mouthparts, in contrast to the vast majority of Lepidoptera, which share a coilable proboscis, the 'advanced' state. Hinton advocated the creation of a separate order, Zeugloptera, for the former group. Hinton's biography in 'the Biographical Memoirs of Fellows of the Royal Society' (Salt 1978) succinctly states that Hinton 'was descriptive rather than analytical, and his achievements is characterized by range rather than by depth'. This description corresponds well with Engel and Kristensen's (2013: 596–597) description of his treatment of the above taxonomic problems; Hinton even suggested that Tuxen's pet group (see above), the Protura, was a class of its own distinct from Insecta.

However, of much greater importance for the spread of phylogenetic systematics beyond entomological circles, was Kristensen's publication in 1970 of a small, but hastily written textbook in systematic entomology (*Systematisk Entomologi*) primarily for use in a course on invertebrate zoology for undergraduate students (bachelor's degree) in biology. The primary reason for writing the textbook was that the standard textbook, *Invertebrate Zoology* (Barnes 1968), only treated insects very fragmentarily. Kristensen's book was – in fact – written and proofread so hurriedly that the important key figure in the introduction (Kristensen 1970: 11, figs A and B), which is an ultra-brief introduction to Hennigian argumentation, was wrong and corrections had to be distributed as a separate page together with the book! Kristensen, at that time just a museum curator, did not teach. This was unfortunate, as the professors and assistant teachers using Kristensen's textbook in classes, either did not realize, or at least never told their students, that the systematic principles used in Kristensen's book were radically different from those used in Barnes'. However, a few students spotted the differences and discussed them. They were attracted by the fact that in only a few other places in systematics was it so easy to diagnose taxa precisely; surely an important issue if you have to pass an examination: 'The only obvious synapomorphy common to all members of the order [Lepidoptera] is scaly wings, which is the basis of their richness in wing patterns' (translated from Kristensen 1970: 144).

A statement like this was diagnostic for all Lepidoptera and thus was unusually precise in comparison with the traditional ill-defined limits of many higher taxa in both animals and especially plants. It was fairly obvious that this could remove the more 'artistic' interpretation of systematic data, viz. changing it from an art to a science. However, the time was not yet ripe for extensive discussion of Hennig's work; that came to prevail as cladistics spread across all fields related to or using taxonomy.

It would be very difficult to overestimate Niels Bonde's (Fig 3.8) importance for the introduction of cladistics into Denmark. Bonde was hired in 1965 first as assistant professor and then, in 1966, as associate professor at the Institute of Geology at University of Copenhagen. Bonde's inspiration to take up cladistics did not come from the entomological circle in Copenhagen but from the American ichthyologist

Fig 3.8 Niels Bonde on the Jameson Land, Greenland expedition in July 2012.
Photo: The Polish–Danish Jameson Land expedition.

Gareth Nelson. Nelson was visiting the Department of Vertebrate Palaeontology at the Swedish Museum of Natural History (Naturhistoriska Riksmuseet) in Stockholm in 1966 as a postdoctoral student. He came across Brundin's just published monograph (Brundin 1966) in the department's reading room and read the introduction. Subsequently, professor Tor Ørvig (1916–94) introduced Nelson to Brundin who gave him a copy of his book without further ado. Later in 1966, Nelson decided to spend the rest of his postdoctoral time at the Natural History Museum in London, travelling to London by car, for scientific reasons, it was rather obvious to stop in Copenhagen to meet with Bonde and Eigil Nielsen (1910–68) who were both working on fossil fishes (Nelson 2014). During his short visit Nelson introduced Bonde to phylogenetic systematics (Bonde 1999, Bonde et al. 2003). A year later, during a research visit to the Department of Vertebrate Palaeontology in Stockholm, Brundin gave Bonde a copy of his *magnum opus* and Bonde had the opportunity to discuss Hennig's ideas with Brundin.

As previously mentioned, Bonde took part in the Nobel Symposium on phylogeny of the lower vertebrates. The symposium was attended by another Dane, Søren Løvtrup (1922–2002), who spent his active scientific career in Sweden where he had become Professor of Zoology at University of Umeå, following his dissertation in 1953 (Løvtrup 1953). Løvtrup is perhaps best known for his critique of neo-Darwinism (1987), but he also axiomatized the logical foundation of cladistic classifications in a series of papers of limited influence (e.g. Løvtrup 1973,

1975), and wrote a textbook on phylogeny of vertebrates (Løvtrup 1977) where this axiomatization is repeated and applied. The book is based purely on cladistics principles. Løvtrup (1973, 1975), like Hennig (1969), opted for the unviable idea of a strict relationship between the number of dichotomies in a cladogram and taxonomic rank.

In contrast to museum scientists like Tuxen and Kristensen, Bonde had to teach, which he surely enjoyed. Thus, Bonde's largest influence was related to his teaching activities. In 1975, Bonde started to teach vertebrate palaeontology, one of no less than three master level courses in palaeontology, then taught at the University of Copenhagen. Rapidly, rumours spread among the more systematically inclined students that Bonde's lectures were surprisingly interesting, and actually rather entertaining. Thus, even students for which Bonde's lectures were not part of the curriculum chose to attend his course. Here Bonde very convincingly dismantled classical textbook stories, like Simpson's famous scenario for the evolution of horses (Simpson 1951, see also Mayr and Ashlock 1991: 254–258), or some of the many, often very primitive just-so stories, explaining the phylogeny of humans and primates (see e.g. Bonde 1977). These experiences, and a course in zoological systematics at the Zoological Museum (see below), truly became eye-openers and made the phylogenetic puzzle fall into place for a number of students interested in systematics.

In addition to his teaching activities aimed at master students, Bonde appeared, mostly uninvited, and always with indefatigable energy, at PhD defences (which in Denmark are public events), guest lectures, etc., all over relevant parts of the Faculty of Natural Sciences – and even beyond. In all fairness, Bonde's in-depth knowledge of the actual subject matter of many of these defences or presentations was not necessarily great. Nonetheless, during the question sessions Bonde would ask questions, which, to the uninitiated, were no less than weird (but great fun for others): 'You have just postulated that taxon A is related to taxon C, but why? As I understood your talk, taxon A shares only a number of "primitive" characters with taxon C, but two "advanced characters" with taxon B?'; or: 'How can members of family X be equally related to members of both families Y and Z?' Out of general ignorance and because of his persistent bullying of the poor PhD student or guest lecturer, Bonde created a lot of animosity during these years, but his strategy worked! Obviously, Bonde used a lot of his time to promote cladistics in these years and only produced a smaller number of scientific papers. Most had some theoretical content and often appeared in rather obscure journals or publications with a limited circulation (e.g. Bonde 1981, 2001) and thus their general impact was limited.

Hence, from a Danish perspective, Hennig's ideas entered the scientific community by several different routes. First and foremost, they were picked up three times independently by the 'silent revolutionaries', entomologists Tuxen, Heie and Kristensen. As Tuxen and Kristensen did not teach, the message was not widely

Fig 3.9 Nils Møller Andersen and Gitte Petersen during the 2001 course on 'Phylogenetic Systematics and Historical Biogeography'. Photo: Finn N. Rasmussen.

disseminated, reaching only a very limited number of university students and professors. Basically, Heie taught, but simply taught the 'wrong audience'. Therefore, it was Bonde (a paleontologist/ichthyologist), influenced by Brundin and Nelson, who became the standard-bearer of the Cladistic Revolution in Denmark.

Another person with considerable influence on the emerging field of cladistics was Nils Møller Andersen (1940–2004; Fig 3.9). Andersen became associate professor at the Zoological Museum in 1975 and, like Kristensen, is widely known in the entomological world. Andersen was a world specialist in Hemiptera (Andersen 1982, Andersen and Weir 2004), particularly water bugs or water striders. Even though he originally started his career as a convinced pheneticist, he very quickly embraced Hennig's new method (Andersen 1977). Thus, his Dr Scient. dissertation from 1982 (Andersen 1982) is a *tour de force* in cladistic methodology. In general, Andersen had very little interaction with students and rarely taught. However in the mid-1970s, he gave a specialist course on the principles of zoological systematics primarily aimed at the museum's own (at that time rather few) masters and PhD students. However, unavoidably a few interested external students (viz. students not preparing a masters or PhD at the museum) attended, making a total audience of never more than ten persons. Andersen's belief in phenetics had probably already started to falter and his doubts were most likely enforced by the students who almost at once challenged it – most had attended Bonde's lectures and were convinced about the superiority of the cladistic approach. Passions were running high, disagreements were often profound, but all listened and learned a lot, and Andersen's faith in phenetics gradually disappeared. This was a great experience, and the majority of the participants ended as convinced cladists: Henrik Enghoff (specialist in Myriapoda, Zoological Museum), Jens Thorvald Høeg (specialist in Crustaceans, Department of Biology, University of Copenhagen), Ebbe Schmidt Nielsen (1950–2001; Lepidoptera specialist, who eventually moved to CSIRO in Canberra, and was one of the masterminds behind the Global Biodiversity Information Facility), Jens Bødtker Rasmussen (1947–2005; specialist in reptiles, particularly snakes, Zoological Museum) and Ole Seberg (specialist in angiosperms, Institute of Systematic Botany, University of Copenhagen).

Andersen quickly adapted the new methods to his research and rapidly grasped the usefulness of the emerging field of historical ecology (Spence and Andersen 1994, Andersen 1999), molecular systematics, etc. Andersen produced an impressive number of papers on extant and extinct water striders; most were empirical, but he also produced a few theoretical papers (Andersen 1978, 2001). However, as noted, his interaction with students was minimal, consequently he had a broader influence in the international entomological community rather than his local community.

In 1988, Andersen and Seberg (see below) invited Ian Kitching and Chris Humphries (1947–2009), both from the Natural History Museum in London, to teach what was maybe the first international PhD course on phylogenetics systematics and biogeography in the Nordic countries. Kitching introduced the PhD students to the computer-based cladistics program PAUP, ver. 2.4 (Swofford 1985), which was compiled on an IBM 4341 mainframe computer (the only one on campus), and Humphries gave a series of lectures on historical biogeography, etc. This course marks the start of a series of PhD courses in different settings: in 1992, 'Current approaches in systematics: theory and practice' was arranged by Seberg, Lars Werdelin (Natural History Museum of Sweden) and Wanntorp, with David Mark Hillis (University of Texas), Michael Donoghue (now Yale University) and Jonathan Coddington (Smithsonian Institution) as guest lecturers, and since 1995 the advanced PhD course 'Phylogenetic systematics', which has been running regularly every alternate year at University of Copenhagen, has had a broad range of guest lecturers including David Swofford (now Duke University), Steve Farris (emeritus at the Swedish Museum of Natural History) and Pablo Goloboff (Fundación/Istituto Miguel Lillo, Argentina). In 1994, the 13th Meeting of the Hennig Society was held at the Zoological Museum under Andersen's chairmanship.

In many ways, botanical systematics, in its broadest understanding, lagged far behind similar developments in zoology. This was true in Denmark as well. However, several of the students participating in Andersen's course in zoological systematics shared a broader interest in systematics, which clearly went beyond their chosen master's degree topic. For these students it was quite obvious that mimicking Bonde's annoying style, which when applied to the teaching of botanical systematists made the unprepared professors excellent targets.

Fig 3.10 Demonstrations/examination in the Botanical Garden in Copenhagen in 1999 lead by Signe Frederiksen. Photo: Finn N. Rasmussen.

Master students with an interest in angiosperm systematics were exposed to a series of lectures by Rolf Martin Theodor Dahlgren's (1932–87), who was professor at the Botanical Museum; his lectures were not phylogenetic in any modern sense of that word. However, Dahlgren's lectures were delivered with a rare, catching enthusiasm and despite their alternative interpretation of phylogeny, they nonetheless were a radical break from the very old-fashion systematics, which had dominated botany in Copenhagen. Undoubtedly, Dahlgren was one of the most – if not the most – influential botanist of his time, not least in monocotyledon systematics but also in angiosperm taxonomy in general. He focused heavily on character analysis (Dahlgren and Clifford 1982) and gave a lot of attention to secondary compounds and their use in plant systematics. Thus, Dahlgren did not become one of the students' targets, except when Bonde appeared at PhD defences, etc., and he would have undoubtedly embraced phylogenetic systematics had it not been for his untimely death. Dahlgren had already started, together with Finn Rasmussen (Dahlgren and Rasmussen 1983) and Kåre Bremer (Dahlgren and Bremer 1985), to embrace phylogenetic systematics when, in 1987, he died in a car accident.

Instead of being one of the University of Copenhagen's most pleasant spots, the Botanical Garden became a major battlefield. Here a well-attended supplement to Dahlgren's lectures with hands-on examinations of the specimens in the garden (Fig 3.10) was established. During these excursions, the professors were flabbergasted

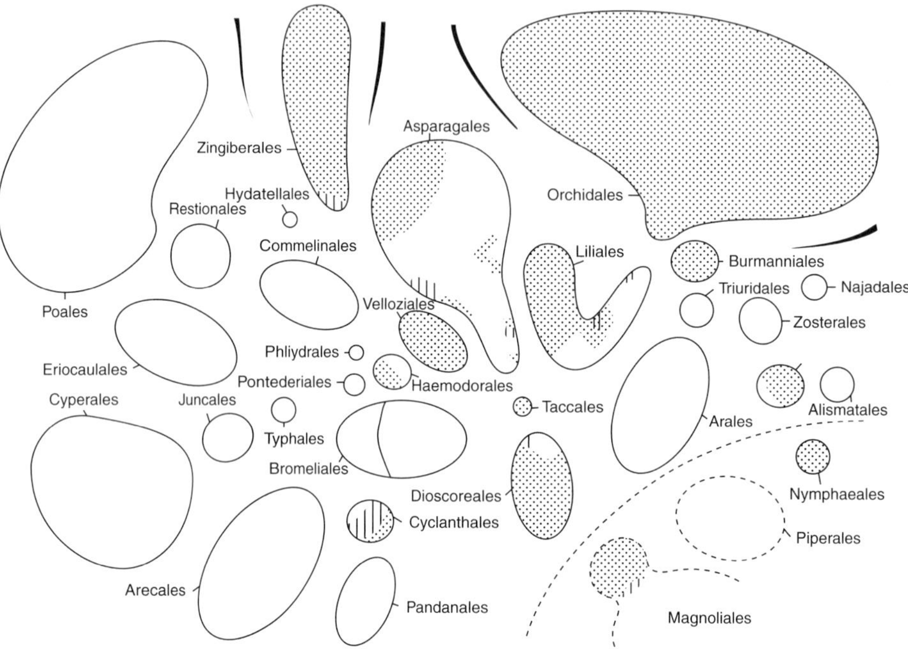

Fig 3.11 A typical dahlgrenogram or dahlgrams showing the occurrence of epigynous (dotted), half-epigynous (vertical hatching) and hypogynous flowers (unshaded) in the monocots. Note the peculiar shape of Liliales, which indicates the parts of the order that are supposed to be related to Orchidales, whereas other parts are supposed to be related to Asparagales (from Dahlgren and Clifford 1982: 111, diagram 34. Copyright Elsevier).

when their knowledge of plant relationships was challenged by the students, and often with easily accessible, hard facts and by an argumentation they were unable to defend their traditional views against. Seberg (see below) has vivid recollection of these events, with what would now be considered unusual statements: 'Most members of family A are related to members of family B, though some are more related to family C, etc.' The belief in such dual (or even more elaborate) relationships is the main reason for the strange, balloon-like shape of the figures attempting to depict relationships between higher taxonomic groups in the angiosperms; the so-called 'dahlgrenograms' or 'dahlgrams' (Fig 3.11; e.g. Dahlgren 1975, Dahlgren and Clifford 1982).

During the late afternoons and early evenings, students writing their master theses and PhD students from the Institute of Systematic Botany discussed different matters relevant to their theses including the use of cladistics. Initiated by these discussions Finn Nygaard Rasmussen (Fig 3.12), who finished his PhD in 1980 (Rasmussen 1982) and Ole Seberg, who began his PhD in 1982 (completed

Fig 3.12 Finn N. Rasmussen testing the latest Danish flora in the field in 2006. Photo: Leif Bolding.

in 1985: Seberg 1988), decided to create a course on cladistics in theory and practise, aimed at fellow students plus interested professors. The first course, which took place in 1984, was a great success, as approximately twice as many registered as the 20 that could possibly be admitted. Thus it was decided to repeat it the following year; this time for external students, too – with a comparable number of registered students and a higher rejection rate. For the first time, cladistic analyses were performed manually with strips of paper and pencils. The second time they were performed using the computer program PHYLIP (ver. 2.8, Felsenstein 1985; which was soon replaced with PAUP; see above) on a mainframe computer. For most students, this was a new experience as most had only worked with computers as word processors, if at all. Two journal clubs were created, a closed local one with a restricted number of participants, and a larger public onc, called 'Carlsberg and Cladistics', which, beside botanists, was primarily attended by entomological colleagues from the Zoological Museum.

Rasmussen's major scientific impact is in Orchidology and among his other significant contributions he has been co-editor of *Genera Orchidacearum*, a six-volume series describing and classifying all known genera of orchids in a phylogenetic context (Pridgeon et al. 1999, 2001, 2003, 2005, 2009, 2014). Additionally, Rasmussen contributed to or wrote a series of scholarly research papers on orchids as well as contributing to the discussions of cladistics in botany, most notably, perhaps, through a joint paper with Dahlgren: 'Monocotyledon evolution' (Dahlgren

Fig 3.13 Ole Seberg giving an address to the participants at 4th International Conference on the Comparative Biology of the Monocotyledons in Copenhagen City Hall in August 2008. Photo: Gitte Petersen.

and Rasmussen 1983). Though explicitly stated in the text, it is, however, often neglected that most of the paper consists of what might be called 'reverse cladistics' (Dahlgren and Rasmussen 1983: 350); the figures shown are not cladograms. The whole exercise can best be described as reverse character analysis: how many times has each character evolved if Dahlgren's ideas about the relationships between families of monocotyledons are accepted as true? In fact, the only truly phylogenetic analysis in the paper is the cladistic analysis of the Zingiberales (gingers, bananas, etc.) performed by Rasmussen (Dahlgren and Rasmussen 1983: 351–356). Perhaps, the most interesting parts of the paper are a diagrammatic representation of a 'Hennigian argumentation' – a 'how to do' cladistics with paper and pencil – and a graphic representation of the pruning or cutting rules, a simple manner to explain and identify mono-, para- and polyphyletic groups (Dahlgren and Rasmussen 1983: 256–266).

Already, during his participation in Møller Andersen's (see above) course in systematic zoology, Ole Seberg (Fig 3.13) decided to buy and read the English translation of Hennig's book (Hennig 1966), which at that time was not available in any public library in Denmark. Seberg's master thesis, which focused on the genus *Acalypha* L. (Euphorbiaceae) from the Galápagos Islands, was finished in 1982. It included a classic two-step manual phylogenetic analysis. The first step was an analysis of a large selection of mostly South American species of *Acalypha* in order to find the sister group of the Galápagos Island species such that the characters could be polarised, viz. finding the plesimorphic and apomorphic states by outgroup comparison (Watrous and Wheeler 1981). The second step was a cladistics analysis of what turned out to be only four Galápagos species. The first analysis, the 'search for the sister group', was a great challenge hampered by a lack of knowledge in dealing with ties; that is taxa sharing the same number of, but only partially overlapping, synapomorphies. However, Seberg was able to circumscribe the sister group and the second analysis was undertaken fairly easy.

However, real problems, which led to considerable delay in publication, started when Seberg decided to publish his thesis work in a local botanical journal hostile to cladistic reasoning. Being a well-behaved student, he submitted his thesis to the *Nordic Journal of Botany* and received a very harsh and inconsiderate review that clearly showed the reviewer had no understanding of cladistic methodology:

> To sum up the paper is in my opinion not acceptable for publication as it stands. Hypotheses and speculations certainly have their place in all scientific work, but the results should not be presented as facts. The simplest way of revising the paper would be to cut the phylogenetic discussions and limit it to a taxonomic revision of Galápagos *Acalypha*. (Anonymous review to the section editor of *Nordic Journal of Botany*, 3 January 1983)

Unfortunately, the section editor also lacked knowledge of cladistics and rejected the paper unless the 'hypotheses and speculations' were removed. Seberg had no other option but to appeal to the editor-in-chief arguing that if it were the policy of the *Nordic Journal of Botany* not to publish papers with phylogenetic content, it would only be acceptable provided it was clearly stated in the instruction to authors. This put the editor-in-chief on full alert and after a few deliberations the manuscript was sent to a new reviewer, and subsequently accepted, with very few changes (Seberg 1984). Since then, phylogenetic systematics has been, and continues to be, an overriding principle for Seberg's research, irrespective of the data type (morphology or sequence data) and, at the beginning, supplemented by historical biogeographic analyses. He has also, on a smaller scale, contributed theoretical papers in both areas (Seberg 1989, Humphries and Seberg 1989). He became Dr Scient. in 2005 (Seberg 2005).

From early on, Seberg has been a fellow of the Willi Hennig Society and later editor of Cladistics, the society's journal. Thus, from 1992 to 1993 he was co-editor with Randall Toby Schuh (American Museum of Natural History) and later, in 1994, with Edward Clairborne Theriot (then of the Academy of Natural Sciences, Philadelphia now at the Texas Natural Science Center and Texas Memorial Museum), and in 1995 he became managing editor. Seberg was the society's president from 1997 to 1999 and participated in organising the 13th Meeting of Hennig Society in 1994 in Copenhagen (see above).

Fig 3.14 Risto Toumikoski in the Oulanka National Park, Finland in 1969 during a mire vegetation course. Photo: Rauno Ruuhijärvi.

Thus, there can be no reasonable doubt that Rasmussen and Seberg can take credit for the introduction of cladistics in Danish botany.

3.5 Hennig and Finland

As in Denmark, cladistics was introduced to Finland by several routes. While some persons can be pointed out as central figures, it seems that Hennig's ideas were adopted independently by several biologists. Risto Tuomikoski (1911–89; Fig 3.14), who was extraordinary professor at the University of Helsinki and a very versatile scientist with broad interests in bryology, entomology, mycology, vegetation ecology and linguistics (Koponen 2013a), knew Brundin's and Hennig's work through his entomological research. Like Brundin, Tuomikoski's main research focus was on Diptera, producing more than 30 papers in this field (Vilkamaa 2013) – but never produced any phylogenetic analyses. Late in his career, Tuomikoski wrote a single

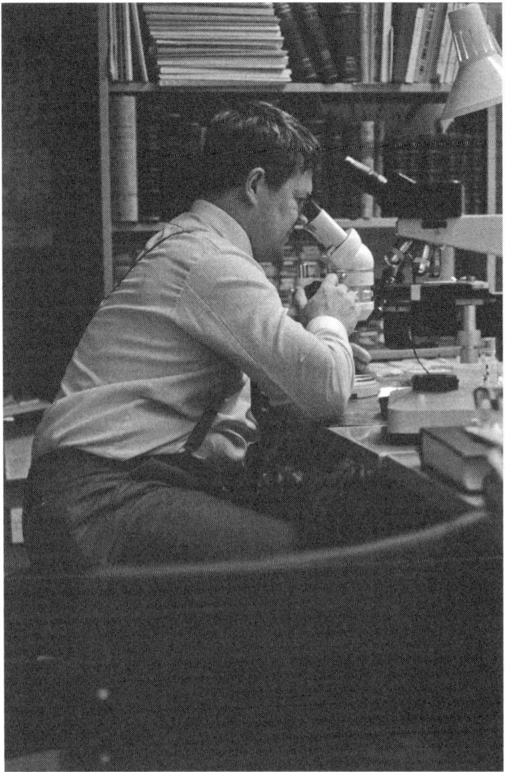

Fig 3.15 Timo Koponen working at the microscope in his office in the cryptogam collections of the Botanical Museum, University of Helsinki in the mid-1980s. Photo: Jaakkoo Hyvönen.

theoretical paper dealing with some principles of cladistics, his contribution having an astonishingly clear idea of what was about to come:

> [I]f the recent works of Hennig and Brundin are carefully studied by the systematic monographers and biogeographers and strict methodical reasoning in phylogenetic matters is accepted and put into practice, a new area of fruitful research will dawn, greatly contributing to a better understanding of the history of life on earth. (Tuomikoski 1967: 147)

Strangely, however, Tuomikoski (1967: 145–147) favoured paraphyletic groups in classifications. Tuomikoski had only two PhD students; the mycologist Harri Harmaja and the bryologist Timo Koponen. It was Tuomikoski, who urged Timo Koponen (Fig 3.15) to read Brundin's (1966) and Hennig's (1950, 1966) books together with his own 1967 paper, and advised Koponen to apply phylogenetic systematics in his thesis (Koponen 1968). Even though Koponen's thesis was the first botanical paper to employ cladistics, it was overlooked by Bremer and Wanntorp in their early publications (Koponen 2013b).

Surprisingly, there were no discussions about phylogenetics at Koponen's PhD defence. Surprising, because of the paper's minimalistic discussion of methodology and its use of terms like plesiomorphous and apomorphous, which ought to have been an excellent opportunity for debate. Even the choice of outgroup was only very briefly touched upon: 'some comparative studies were carried out on the related families Rhizogoniaceae and Bryaceae' (Koponen 1968: 134). It is worth mentioning Koponen's rationale for choosing Hennig's 'cladistic-phyletic' method (apart from being urged to do so by Tuomikoski):

> Any numerical-phenetic classification is out of the question in the present case because the number of characters needed is high, according to Michener and Sokal (1957) 60 at least. The organization of mosses is rather simple and, accordingly, the number of useful characters is low especially when, as in the present study, attention is mainly limited to taxa above the specific rank. Biochemical data may, however, later raise the number of characters … so that a numerical classification may be attempted. The low number of characters also hinders classification on the cladistic-phyletic basis. (Koponen 1968: 133)

However, Koponen continued to use Hennig's methodology (Koponen 1973, 1980).

Meanwhile Finnish entomologists became interested in phylogenetic systematics independently of Tuomikoski. Thus, when Lic. Phil. Mauri Hirvenoja visited the Hydrobiologische Station zu Plön in Schleswig-Holstein (currently Max-Planck-Institut für Evolutionsbiologie) in 1962 to deepen his knowledge of aquatic fauna, he became acquainted with Dieter Schlee[2] who in 1967 became Willi Hennig's assistant in the Staatliches Museum für Naturkunde, Stuttgart (Schmitt 2013). Hirvenoja and Schlee shared an interest in chironomids, and in 1968 Hirvenoja received Schlee's PhD thesis (Schlee 1968), the nexus of which was a Hennigian treatment of phylogeny. Hirvenoja thought he did not need to do such an analysis in his own study. However, Ernst Josef Fittkau (1927–2012) and Friedrich Reiss (1937–99), colleagues in Plön, offered their assistance in revising the German language of Hirvenoja's thesis and Fittkau proposed to add a cladogram, which Hirvenoja did. Hirvenoja even added symbols to show instances of parallelisms. He did not, however, revise the text and the addition was only to present the results of a detailed comparative analysis of the anagenesis of the organs of all developmental stages whenever possible. Hirvenoja completed his thesis in 1968 but was unable to defend it until 5 years later (Hirvenoja 1973). Brundin acted as an opponent even though Hirvenoja would have preferred Tuomikoski as this would have enabled ample discussion in his native Finnish. At the disputation dinner, Tuomikoski was

---

[2] Strangely, Schlee appears to have left his job at Staatliches Museum für Naturkunde in Stuttgart around 1995–6 without leaving any trace of his later whereabouts.

seated beside Brundin; what followed was a somber discussion. After the dinner one of the guests told Hirvenoja that Tuomikoski was not fascinated by Brundin's biogeographical work considering it to be exaggerated so as to draw conclusions on the distribution of midges and their relationship to plate tectonics.

Other contemporary entomological theses such as those of Samuel Panelius (1965) and Martin Meinander (1972; 1940–2004) also included discussions on phylogenetic relationships. However, they did not grasp Hennigian methodology but instead tried to merge it with traditional eclectic or phenetic views.

By the early 1980s, monographs using Hennigian methodology were being published, for example, by Rauno Väisänen (1984), who is currently Head of Nature Conservation Research Unit and National Board of Waters. His work treated fungus gnats, a group Tuomikoski also studied; Väisänen's work was not supervised by Tuomikoski although he met him on several occasions and was familiar with Koponen's PhD thesis. Väisänen considered lengthy discussions with Mauri Hirvenoja about phylogenetic analyses as very influential for this part of his work. By that time phylogenetic textbooks in English were appearing, Väisänen was able to cite Wiley (1981). The publication of Väisänen´s work was delayed partly due to lack of funds to print such a voluminous monograph and because by the time of his defense Väisänen was about to leave the university world to make a career in Finnish environmental agencies, frustrated by the prospect of very long delays in the publication of further papers (the manuscripts were filed and forgotten; they were partly published later: Väisänen 1996, 2013a, b, 2014).

Jyrki Muona, who is currently professor and senior curator at the Finnish Museum of Natural History and a specialist in beetles, undertook his PhD at the University of California, Davis, in 1981. In his thesis he did not use cladistic methods but instead compatibility analysis as developed by George Estabrook, Kent Fiala and others (see below). Muona took up cladistic methodology when he returned to Finland to the University of Oulu and during his postdoctoral period in Australia and the Western Pacific in 1985–86. In Australia, he became acquainted with Ebbe Schmidt Nielsen and together they had, sometimes heated, debates about systematics and its theory with local pheneticists. In 1988, Muona attended the 7th meeting of the Willi Hennig Society in Stockholm and presented a paper on his studies on biogeography of Eucnemidae of Southeast Asia and the Western Pacific (Muona 1991) and has since then been deeply involved in cladistics. He has been editor of Cladistics from 1998 to 2001 and twice president of the Willi Hennig Society (2008–10, 2014–).

3.6 Hennig and Norway

After having received his first degree from University of Oslo, Ole Anton Sæther (1936–2013, Fig 3.16) was hired in its Department of Limnology from which he got

Fig 3.16 Ole A. Sæther at the 17th International Symposium on Chironomidae in Tianjin, China in 2009. Photo: Elisabeth Stur.

his cand. real. in 1963 (Sæther 1963). Shortly afterwards, he was promoted to lecturer and remained there until 1969, when he was hired at the Freshwater Institute of the Fisheries Research Board of Canada following a short interlude as a visiting scientist at the institute in 1967. Sæther returned to Norway in 1977 to become professor at the Museum of Zoology in Bergen. He was a specialist in chironomids and had almost surely read and been inspired by Brundin's 1966-work, but he was apparently also acquainted with Hennig's method by reading papers of the German entomologists Karl Strenzke (1917–61) of Forschungsinstitut und Naturmuseum Senckenberg, Frankfurt am Main, and Dieter Schlee, both publishing on chironomid systematics. Sæther was a prolific writer and was hugely productive in descriptive and analytical systematics; he was also interested in environmental monitoring. During his career he published more than 40 phylogenetic analyses (Ekrem and Andersen 2007, Andersen 2013), the first, in 1971 (Sæther 1971), while still in Canada. When Sæther returned to Norway he became the focal point for the introduction of cladistics into Norway and changed the curriculum for students to include both an introduction to phylogenetic systematics and historical biogeography. He supervised a number of master and PhD students in chironomid systematics and insisted that their projects should include phylogenetic and biogeographic analyses. Sæther persistently argued for the use of a Hennigian argumentation scheme even some time after the introduction of computer programs in cladistics (Sæther 1990) and was perhaps never completely convinced that modern methods in parsimony performed better in reconstructing chironomid phylogenies. However, he turned to the use of PAUP and

MacClade as these became available and was always quite explicit in his description of characters, states and weights. Sæther is perhaps most widely known for his promotion of and persistent use of the concept of 'underlying synapomorphies' (Sæther 1979a, 1986) or 'canalized evolutionary potential' (Sæther 1983) – a postulated tendency to develop similar apomorphic traits in different lineages – in phylogenetic reconstruction, an advocacy that even reached the popular scientific literature (Sæther 1979b). The concept of 'underlying synapomorphies' was shared with both Tuomikoski (1967: 141) and Brundin (1968: 479–480; 1972b: 111) and can be traced back at least to the Swedish professor of geology at the University of Uppsala, Gunnar Säve-Söderbergh (1910–47; 1934: 10–11) and perhaps even earlier: 'Thus we find that the existence of a definite evolutionary trend will have to be used as a systematical character in the larger classification' (Säve-Söderbergh 1934: 11).

Sæther also developed several indices, now mostly forgotten, such as 'the adjusted evolutionary index' (Sæther 1970), and separated out what he called 'objective' and 'subjective apomorphies' (Sæther 1983, 1986). But it was from Sæther that Norwegian students first learnt about Hennig's methods.

The first Norwegian botanist to adopt cladistics was Inger Nordal, who started her career as assistant professor in botany at Institute of Biology, University of Oslo. Nordal received her PhD from the University of Uppsala in 1977 (Nordal 1977) and became Professor in 1987 at University of Oslo. Her research focused on evolution, phylogeny, population biology and taxonomy of angiosperms, primarily from Africa south of the Sahara and in arctic-alpine areas. Nordal's first paper reflecting a truly phylogenetic point of view was written in collaboration with Thomas Duncan of the University of California in Berkeley (Nordal and Duncan 1984). The paper is an analysis of *Haemanthus* L. and *Scadoxus* Raf. (Amaryllidaceae) and includes a very thorough character analysis. The characters were polarized using outgroup comparison but their justification was sought in a paper by Crisci and Stuessy (1980), which had since been demonstrated as wrong (Watrous and Wheeler 1981). The data were analysed on a mainframe computer in California using both compatibility and parsimony algorithms and represents a very early use of computer-based phylogenetic analyses. Comparison and discussion of methodologies, and results in the paper are, with hindsight, imprecise and unfocused, even to the extent that a classical eclectic phylogeny in an earlier paper (Björnstad and Friis 1971: 202–204) is called a 'tentative cladogram' (Nordal and Duncan 1984: 145). It would not have been predicted at that time that compatibility analysis (Meacham and Estabrook 1985) would be a methodology of only passing interest, very rarely used today. Perhaps more interestingly, Nordal and Duncan's paper has a short discussion of the correspondence between cladograms and taxonomy. In a later theoretical paper, Nordal (1987) offered a belated defence of compatibility analysis made with reference to Brundin's and Sæther's concept of underlying synapomophies (see above). Additionally, Dahlgren and

Rasmussen's (1983) rather imprecise distinction between parsimony and character compatibility is cited in favour of the latter, which may be stretching their statements too far.

3.7 Conclusion

Following a brief lag period and a universally slow beginning, the introduction of cladistic methodology in the Nordic countries was relatively fast, though, as elsewhere, marked by traditionally isolated and dogged pockets of resistance – or simply met by neglect and sheer ignorance. A primary source of inspiration, influencing not only his Swedish colleagues but also scientists in other Nordic countries, such as Risto Tuomikoski in Finland, Ole A. Sæther in Norway, and Niels Bonde in Denmark, was beyond doubt the entomologist Lars Brundin. That Brundin's influence reached beyond Scandinavia has been repeatedly emphasized (Hull 1988, Wanntorp 1983, Bonde et al. 2003, Nelson 2014). However, there were three different and unrelated entomological introductions of Hennig's idea into the scientific community of Denmark: Søren Tuxen who most likely picked up on the ideas more or less directly from Hennig; Niels P. Kristensen who was influenced by Hinton while visiting University of Bristol; and Ole E. Heie who was inspired by his fundamental interest in phylogenetic methods and found Hennig's early works on his own. For Seberg, it is tempting to draw a direct line from the influence of Heie's textbook (at high school) to Kristensen's textbook in systematic entomology (at the bachelor level), finally via Bonde's and Møller Andersen's teaching (at the master level), as being instrumental in shaping his own introduction to phylogenetic systematics – but this may be due to the benefit of hindsight.

The transfer of Hennig's ideas from zoology (or, more precisely, entomology) to botany was rather direct and seamless in Sweden: from Brundin to Kåre Bremer and Hans-Erik Wanntorp. In Denmark, the botanical resistance at University of Copenhagen towards Hennig's methodology was more profound, and the introduction primarily took place through Finn N. Rasmussen and Ole Seberg. In Norway, it appears that Inger Nordal was the first to adopt cladistics reasoning, apparently independent of Sæther, but it never played a central role in her research profile. In Finland, inspiration came partly through Tuomikoski, whose taxonomic expertise was exceptionally broad (Koponen 2013a), but also through others entomologists. The very first use of Hennigian argumentation in botany was most likely that made by Tuomikoski's student, the Finnish bryologist, Timo Koponen (1968). Koponen's thesis was reprinted in Duncan and Stuessy (1985).

One of the biggest problems for many traditional taxonomists – zoologists, palaeontologists, botanists alike – has been the requirement that only strictly monophyletic groups are acceptable in classifications. Oddly, recent years have seen an astonishing number of publications, not least botanical papers, related to paraphyletic taxa (Brummitt and Sosef 1998, Brummitt 2002, Freudenstein 1998, Nelson et al. 2003, Ebach et al. 2006, Schmidt-Lebuhn 2012, Stuessy and Hörandl 2014, and references therein). The most recent and bizarre entry in this debate was a coordinated effort made by two Norwegian botanists for the acceptance of paraphyletic taxa: 'Paraphyletic taxa should be accepted' (Nordal and Stedje 2005); these authors were basically advocating a return to eclectic classifications. Interestingly, this links directly back to Nordal's interest in character compatibility (see also Estabrook 2008: 125–126); nevertheless, no less than 150 scientists signed this petition. Their main arguments are stability in naming and the apparent havoc many name changes cause for our understanding of processes like species extinction. Actually, the latter argument is a misrepresentation of E.O. Wilson's contribution to the symposium *Linnaean Taxonomy in the 21th Century* in 2001(see e.g. Williams et al. 2005) and if stability is obtained by conveying only partial, erroneous information about relationships to the knowledgeable, and information that can only be interpreted outright erroneously by the uninitiated, its value is indeed a moot point (see e.g. Dias et al. 2005, Rieppel 2005, Williams et al. 2005, Zachos 2014). Thus, maybe most biologists will know that families Hominidae, a monophyletic group, and Pongidae, a paraphyletic group, are not sister groups; but many non-biologists will not. That the chimpanzee (or alternatively the gorilla), one of three extant members of the Pongidae, is the extant sister group of *Homo sapiens* and thus more closely related to *Homo* than the two other extant member of Pongidae, is totally intractable from this classification.

However, in passing it is worthwhile mentioning that Hennig's strong emphasis on monophyly may have deeper roots than is usually understood. Consider briefly the following statement made by Säve-Söderbergh (see above):

> And the aim of classification in Zoology is to express, in the form of a grouping into larger units of the individuals, the natural phyletic relations of the animals. For this purpose, in each group one has to include animals more closely related to each other than to animals of any parallel group. Each group must include the descendants of one ancestral form – the group has to represent a monophyletic assemblage. (Säve-Söderbergh 1934: 2)

Acknowledgements

We are very grateful to a number of colleagues that have helped us compile this account, most notably Niels Bonde (University of Copenhagen), Kåre Bremer (University of Stockholm), Ole E. Heie (Holte, Denmark), Jens T. Høeg (University of Copenhagen), Mauri Hirvenoja (Vantaa, Finland), Timo Koponen (University of Helsinki), Niels P. Kristensen (University of Copenhagen), Jyrki Muona (University of Helsinki), Thomas Pape (University of Copenhagen), Gitte Petersen (University of Copenhagen), Finn N. Rasmussen (University of Copenhagen) and Rauno Väisänen (Vantaa, Finland). Henrik Enghoff (University of Copenhagen), Gitte Petersen (University of Copenhagen) and Dennis W. Stevenson (New York Botanical Garden) kindly read an earlier draft of the manuscript.

References

Andersen, N.M. (1977). A new and primitive genus and species of Hydrometridae (Hemiptera, Gerromorpha) with a cladistics analysis of relationships within the family. *Entomologica Scandinavica*, 8, 301–316.

Andersen, N.M. (1978). Some principles and methods of cladistics analysis with notes on the uses of cladistics in classification and biogeography. *Zeitschrift für Zoologische Systematik und Evolutionsforschung*, 16, 242–255.

Andersen, N.M. (1982). The semiaquatic bugs (Hemiptera, Gerromorpha). Phylogeny, adaptations, biogeography, and classification. *Entomograph*, 3, 1–455.

Andersen, N.M. (1999). The evolution of marine insects: phylogenetic, ecological and geographical aspect of species diversity in marine water striders. *Ecography*, 22, 98–111.

Andersen, N.M. (2001). The impact of W. Hennig's "phylogenetic systematics" on contemporary entomology. *European Journal of Entomology*, 98, 133–150.

Andersen, N.M. and Weir, T.A. (2004). Australian water bugs. Their biology and identification (Hemiptera-Heteroptera, Gerromorpha and Neomorpha). *Entomograph*, 14, 1–344.

Andersen, T. (2013). In memory of Ole Anton Sæther (1936–2013). *Chironomus Newsletter on Chironomidae Research*, 26, 16–19.

Andersson, L. and Chase, M.W. (2001). Phylogeny and classification of Marantaceae. *Botanical Journal of the Linnean Society*, 135, 275–287.

Barnes, R.D. (1968). *Invertebrate Zoology.* 2nd edition. Philadelphia, PA: W.B. Saunders Company.

Björnstad, I.N. and Friis, I. (1971). Studies on the genus *Haemanthus* L. (Amaryllidaceae) I. The infrageneric taxonomy. *Norwegian Journal of Botany*, 19, 187–206.

Bonde, N. (1977). Cladistic classification as applied to vertebrates. In *Major Patterns in Vertebrate Evolution* (NATO Advanced Study Institutes Series, vol. 14), ed. M.K. Hecht, P.C. Goody and B.M. Hecht. New York, London: Plenum Press, pp. 741–804.

Bonde, N. (1981). Problems of species concepts in palaeontology. In *International Symposium on "Concept and Method in Palaeontology"*, Contributed Papers, ed. J. Martinell. Barcelona, Spain: Department de Paleontologia, Universitat de Barcelona, pp. 19–34.

Bonde, N. (1999). Colin Patterson (1933–1998): a major vertebrate palaeontologist of the century. *Geologie en Mijnbouw*, 78, 255–260.

Bonde, N. (2001). L'Espéce et la dimension du temps. *Biosystema*, 19, 29–62.

Bonde, N., Høeg, J.T. and Seberg, O. (2003). Introduktion af fylogenetisk Systematik. Kladismens fremme i Danmark og andre land. *Dansk Naturhistorisk Forenings Årsskrift*, 13, 8–34.

Bremer, K. (1976). The genus *Relhanina* (Compositae). *Opera Botanica*, 40, 1–85.

Bremer, K. (1990). Combinable component consensus. *Cladistics*, 6, 369–372.

Bremer, K. (1994). Branch support and tree stability. *Cladistics*, 10, 295–304.

Bremer, K. and Wanntorp, H.-E. (1978). Phylogenetic systematics in botany. *Taxon*, 27, 317–329.

Bremer, K. and Wanntorp, H.-E. (1979). Hierarchy and reticulations in systematics. *Systematic Zoology*, 28, 624–627.

Bremer, K. and Wanntorp, H.-E. (1981). The cladistic approach to plant classification. In *Advances in Cladistics, Proceedings of the First Meeting of the Willi Henning Society*, ed. V.A. Funk and D.R. Brooks. New York: New York Botanical Garden, Bronx, pp. 87–94.

Bremer, K. and Wanntorp, H.-E. (1982). Fylogenetik systematic [Phylogenetic systematics]. *Svensk Botanisk Tidsskrift*, 76, 177–183.

Brummitt, R.K. (2002). How to chop up a tree. *Taxon*, 51, 31–41.

Brummitt, R.K. and Sosef, M.S.M. (1998). Paraphyletic taxa are inherent in Linnean classification: a reply to Freudenstein. *Taxon*, 47, 411–412.

Brundin, L. (1934). *Die Coleopteren des Torneträskgebietes. Beitrag zur Ökologie und Geschichte der Käfer-walt in Schwedisch-Lappland*. Lund, Sweden: Carl Bloms Boktryckei.

Brundin, L. (1966). Transarctic relationships and their significance, as evidenced by chironomid midges. With a monograph of the subfamilies Podonominae and Aphroteniinae and the Austral Heptagyiae. *Kungliga Svenska Vetenskapsakademiens Handlingar, Fjärde series*, 11, 1–472.

Brundin, L. (1968). Application of phylogenetic principles in systematics and evolutionary theory. In *Current Problems of Lower Vertebrate Phylogeny, Proceedings of the 4th Nobel Symposium held in June 1967 at the Swedish Museum of Natural History (Naturhistoriska riksmuseet) in Stockholm*, ed. T. Ørvig. Stockholm: Almqvist and Wiksell, pp. 473–495.

Brundin, L. (1972a). Phylogenetics and biogeography. *Systematic Zoology*, 21, 69–79.

Brundin, L. (1972b). Evolution, causal biology, and classification. *Zoologica Scripta*, 1, 107–120.

Brundin, L. (1988). Phylogenetic Biogeography. In *Analytical Biogeography. An Integrated Approach to the Study of Animal and Plant Distributions*, ed. A.A. Myers and P.S. Giller, New York, London: Chapman and Hall, pp. 343–239.

Brundin, L. (1993). From Grimsgöl to Gondwanaland: half a century with chironomids. *Cladistics*, 9, 358–367.

Crisci, J.V. and Stuessy, T.F. (1980). Determining primitive character states for phylogenetic construction. *Systematic Botany*, 5, 112–135.

Dahlgren, R. (1975). A system of classification of the angiosperms to be used to demonstrate the distribution of characters. *Botaniska Notiser*, 128, 119–147.

Dahlgren, R.M.T. and Clifford, H.T. (1982). *The Monocotyledons: A Comparative Study*. London: Academic Press.

Dahlgren, R. and Bremer, K. (1985). Major clades of the Angiosperms. *Cladistics*, 1, 349–368.

Dahlgren, R. and Rasmussen, F.N. (1983). Monocotyledon evolution. Characters and phylogenetic estimation. *Evolutionary Biology*, 16, 255–395.

Darlington P.J., Jr. (1970). A practical criticism of Hennig–Brundin 'Phylogentic [sic] Systematics' and Antarctic biogeography. *Systematic Zoology*, 19, 1–18.

De Queiroz, K. and Gauthier, J. (1992). Phylogenetic taxonomy. *Annual Review of Ecology and Systematics*, 23, 449–480.

De Queiroz, K. and Gauthier, J. (1994). Towards a phylogenetic system of biological nomenclature. *Trends in Ecology and Evolution*, 9, 27–31.

De Queiroz, K. and Cantino, P.D. (2001). Phylogenetic nomenclature and the PhyloCode. *Bulletin of Zoological Nomenclature*, 58, 254–271.

Dias, P., Assis, L.C.S. and Udulutsch, R.G. (2005). Monophyly vs. paraphyly in plant systematics. *Taxon*, 54, 1039–1040.

Duncan, T. and Stuessy, T.F. (1985). *Cladistic Theory and Methodology*. New York: Van Nostrand Reinhold Co.

Dupuis, C. (1956). Variations convergentes ou comparables de certains caractères des Tachinaires, notamment des Phasiinae (Dipt. Larvaevoridae); leur signification taxonomique différente selon les lignées. In *Proceedings of the XIVth International Congress of Zoology, Copenhagen 5–12. August 1953*. Copenhagen: Danish Science Press, Ltd., pp. 474–476.

Ebach, M.C., Williams, D.M. and Morrone, J.J. (2006). Paraphyly is bad taxonomy. *Taxon*, 55, 831–832.

Ekrem, T. and Andersen, T. (2007). Professor Ole Anton Sæther 70 years: Four decades of chironomid research. In *Contribution to the Systematics and Ecology of Aquatic Diptera: A Tribute to Ole A. Sæther*, ed. T. Andersen. Colombus, OH: The Caddis Press, pp. 1–15.

Engel, M.S. and Kristensen, N.P. (2013). A history of entomological classification. *Annual Review of Entomology*, 58, 585–607.

Estabrook, G.F. (2008). Fifty years of character compatibility concepts at work. *Journal of Systematics and Evolution*, 46, 109–129.

Farris, J.S. (1988). Hennig86, ver. 1.5. Computer program and manual, published by the author.

Felsenstein, J. (1985). PHYLIP, version 2.8. Tape distributed by the author.

Freudenstein, J.V. (1998). Paraphyly, ancestors, and classification: a response to Sosef and Brummitt. *Taxon*, 47, 95–104.

Heie, O.E. (1967). Studies on fossil aphids: (Homoptera: Aphidoidea): Especially in the Copenhagen collection of fossils in Baltic amber. *Spolia Zoologica Musei Hauniensis*, 26, 1–273.

Heie, O.E. (1969). *Evolutionslære*. Copenhagen: Aschehoug.

Heie, O.E. (1976). *Evolutionslære*. 2. opl. Copenhagen: Aschehoug.

Heie, O.E. (1973). *Evolutionslära*. Stockholm: Natur och Kultur Förlag.

Heie, O.E. (1976). Taxonomy and phylogeny of the fossil Elektraphididae Steffan, 1968 (Homoptera: Aphidoidea). *Entomologica Scandinavica*, 7, 53–58.

Heie, O.E. (1980). The Aphidioidea (Hemiptera) of Fennoscandia and Denmark. I. General part. The families Mindaridae, Hormaphididae, Thelaxidae, Anoeciidae, and Pemphigidae. *Fauna Entomoligica Scandinavica*, 9, 1–236.

Heie, O.E. (1982). The Aphidioidea (Hemiptera) of Fennoscandia and Denmark. II. *The family Drepanosipidae Fauna Entomoligica Scandinavica*, 11, 1–176.

Heie, O.E. (1983). *Zootaxonomi*. Emdrup. Denmark: Danmarks Lærerhøjskole.

Heie, O.E. (1986). The Aphidioidea (Hemiptera) of Fennoscandia and

Denmark. III. The family Aphididae: subfamilies Pterocommatinae and tribe Aphidini of subfamily Aphidinae. *Fauna Entomoligica Scandinavica*, 17, 1–314.

Heie, O.E. (1992). The Aphidioidea (Hemiptera) of Fennoscandia and Denmark. IV. The family Aphididae: Part 1 of tribe Macrosiphini of subfamily Aphidinae. *Fauna Entomoligica Scandinavica*, 25, 1–189.

Heie, O.E. (1994). The Aphidioidea (Hemiptera) of Fennoscandia and Denmark. V. The family Aphididae: Part 2 of tribe Macrosiphini of subfamily Aphidinae. *Fauna Entomoligica Scandinavica*, 28, 1–242.

Heie, O.E. (1995). The Aphidioidea (Hemiptera) of Fennoscandia and Denmark. IV. The family Aphididae: Part 3 of tribe Macrosiphini of subfamily Aphidinae, and family Lachnidae. *Fauna Entomoligica Scandinavica*, 31, 1–217.

Heie, O.E. (2004a). Bladlus 1. *Danmarks Fauna*, 87, 1–428.

Heie, O.E. (2004b). Bladlus 2. *Danmarks Fauna*, 88, 428–864.

Hennig, W. (1950). *Grundzüge einer Theorie der phylogenetischen Systematik.* Berlin: Deutscher Zentralverlag.

Hennig, W. (1953). Kritische Bermerkungen zum phylogenetischen System der Insekten. *Beiträge zur Entomologie*, 3, 1–85.

Hennig, W. (1957). Systematik und Phylogenese. *Bericht Hundertjahrfeier Deutsche Entomologische Gesellschaft, Berlin*, 1956, 51–71.

Hennig, W. (1965). Phylogenetic systematics. *Annual Review of Entomology*, 10, 97–116.

Hennig, W. (1966). *Phylogenetic Systematics.* Urbana, IL: University of Illinois Press.

Hennig, W. (1969). *Die Stammgeschichte der Insekten.* Frankfurt am Main: W. Kramer.

Hirvenoja, M. (1973). Revision der Gattung Cricotopus van der Wulp und Ihrer Verwandten (Diptera, Chironomidae). *Annales Zoologici Fennici*, 10, 1–363.

Hull, D.L. (1988). *Science as a Process.* Chicago, IL: University of Chicago Press.

Humphries, C.J. and O. Seberg (1989). Graphs and generalized tracks: some comments on method. *Systematic Zoology*, 38, 69–76.

Jarvik, E. (1968). Aspects of vertebrate phylogeny. In *Current Problems of Lower Vertebrate Phylogeny. Proceedings of the 4th Nobel Symposium held in June 1967 at the Swedish Museum of Natural History (Naturhistoriska riksmuseet) in Stockholm*, ed. T. Ørvig. Stockholm: Almqvist and Wiksell, pp. 496—527.

Koponen, T. (1968). Generic revision of Mniaceae Mitt. (Bryophyta). *Annales Botanici Fennici*, 5, 117–151.

Koponen, T. (1973). Rhizomnium (Mniaceae) in North America. *Annales Botanici Fennici*, 10, 1–26.

Koponen, T. (1980). A synopsis of Mniaceae (Bryophya) II. *Ortomnion. Annales Botanici Fennici*, 17, 35–55.

Koponen, T. (2013a). The biography of professor Risto Tuomikoski: bryologist, entomologist, linguist, mycologist, vegetation scientist and ecologist. *Bryobrotheria*, 11, 9–80.

Koponen, T. (2013b). Risto Tuomikoski as a bryologist. *Bryobrotheria*, 11, 110–139.

Kraus, O. (1984). Die Veranstaltung "Phylogenetisches Symposion": Rückblick auf 25 Tagungen (1955–1982). *Verhandlungen des NaturwissenschaftlichenVereins in Hamburg. Neues Folge*, 27, 277–289.

Kraus, O. and Hoßfeld, U. (1998). 40 Jahre "Phylogenetisches Symposion" (1957–1997): eine Übersicht, Anfänge, Entwicklung, Dokumentation und

Wirkung. *Jahrbuch für Geschichte und Theorie der Biologie*, 5, 157–186.

Kristensen, N.P. (1970). *Systematik Entomologi*. Copenhagen: Munksgaards Förlag.

Kristensen, N.P. (1975). The phylogeny of hexapod 'orders'. A critical review of recent accounts. *Zeitschrift für Zoologische Systematik und Evolutionsforschung*, 13, 1–44.

Kristensen, N.P. (1983). S. L. Tuxen, 8. August 1908–15. juni 1983. *Videnskabelige Meddelelser fra dansk naturhistorisk Forening*, 144, 157–170.

Lidén, M. (1986). Synopsis of Fumarioideae with a monograph of the tribe Fumarieae. *Opera Botanica*, 88, 1–136.

Lidén, M., Oxelman, B., Backlund, A., et al. (1997). Charlie is our darling. *Taxon*, 46, 735–738.

Løvtrup, S. (1953). Studies on amphibian embryogenesis. *Comptes Rendus des Travaux du Laboratoire Carlsberg, Série Chimie*, 28, 371–466.

Løvtrup, S. (1973). Classification, convention and logic. *Zoologica Scripta*, 2, 49–61.

Løvtrup, S. (1975). On the phylogenetic classification. *Acta Zoologia Cracoviensia*, 20, 500–523.

Løvtrup, S. (1977). *The Phylogeny of Vertebrata*. London: John Wiley and Sons.

Løvtrup, S. (1987). *Darwinism: The Refutation of a Myth*. London: Croom Helm Ltd.

Mayr, E. and Ashlock, P.D. (1991). *Principles of Systematic Zoology*. 2nd edition. New York: McGraw-Hill.

Meacham, C.A. and Estabrook, G.F. (1985). Compatibility methods in systematics. *Annual Review of Ecology and Systematics*, 16, 431–446.

Meinander, M. (1972). A revision of the family Coniopterygidae (Planipennia). *Acta Zoologica Fennica*, 136, 1–357.

Michener C.D. and Sokal, R.R. (1957). A quantitative approach to a problem in classification. *Evolution*, 11, 130–162.

Muona, J. (1991). The Eucnemidae of South-East Asia and the Western Pacific: a biogeographical study. *Australian Systematic Botany*, 4, 165–182.

Nelson, G.J. (1968). Gill-arch structure in Acanthodes. In *Current Problems of Lower Vertebrate Phylogeny. Proceedings of the 4th Nobel Symposium held in June 1967 at the Swedish Museum of Natural History (Naturhistoriska riksmuseet) in Stockholm*, ed. T. Ørvig. Stockholm: Almqvist and Wiksell, pp. 129–143.

Nelson, G. (1978). Ontogeny, phylogeny, paleontology and the biogenetic law. *Systematic Zoology*, 27, 324–345.

Nelson, G. (2014). Cladistics at an earlier time. In *The Evolution of Phylogenetic Systematics*, ed. A. Hamilton. Berkeley, CA: University of California Press, pp. 139–149.

Nelson, G., Murphy, D.J. and Ladiges, P.Y. (2003). Brummitt on paraphyly: a response. *Taxon*, 52, 295–298.

Nordal, I. (1977). Taxonomic studies on African Amaryllidaceae. *Acta Universitatis Upsalensis*, 437, 1–16.

Nordal, I. (1987). Cladistics and character weighting: a contribution to the compatibility versus parsimony discussion. *Taxon*, 36, 59–60.

Nordal, I. and Duncan, T. (1984). A cladistic analysis of Haemanthus and Scadoxus. *Nordic Journal of Botany*, 4, 145–153.

Nordal, I. and Stedje, B. (2005). Paraphyletic should be accepted. *Taxon*, 54, 5–8.

Ørvig, T. (ed.) (1968). *Current Problems of Lower Vertebrate Phylogeny. Proceedings of the 4th Nobel Symposium held in June 1967 at the Swedish Museum of Natural History*

(Naturhistoriska riksmuseet) in Stockholm. Stockholm: Almqvist and Wiksell.

Panelius, S. (1965). A revision of the European gall midges of the subfamily Porricondylinae (Diptera: Itonididae). *Acta Zoologica Fennica*, 113, 1–157.

Pridgeon, A.M., Cribb, P, J., Chase, M.W. and Rasmussen, F. (eds.) (1999). *Genera Orchidacearum. Volume 1: Apostasioideae and Cypripedioideae.* Oxford: Oxford University Press.

Pridgeon, A.M., Cribb, P, J., Chase, M.W. and Rasmussen, F. (eds.) (2001). *Genera Orchidacearum. Volume 2. Orchidoideae (Part 1).* Oxford: Oxford University Press.

Pridgeon, A.M., Cribb, P, J., Chase, M.W. and Rasmussen, F. (eds.) (2003). *Genera Orchidacearum. Volume 3. Orchidoideae (Part 2) Vanilloideae.* Oxford: Oxford University Press.

Pridgeon, A.M., Cribb, P, J., Chase, M.W. and Rasmussen, F. (eds.) (2005). *Genera Orchidacearum. Volume 4. Epidendroideae (Part 1).* Oxford: Oxford University Press.

Pridgeon, A.M., Cribb, P, J., Chase, M.W. and Rasmussen, F. (eds.) (2009). *Genera Orchidacearum. Volume 5. Epidendroideae (Part 2).* Oxford: Oxford University Press.

Pridgeon, A.M., Cribb, P, J., Chase, M.W. and Rasmussen, F. (eds.) (2014). *Genera Orchidacearum. Volume 6. Epidendroideae (Part 3).* Oxford: Oxford University Press.

Rasmussen, F.N. (1982). The gynostemium of the neottiod orchids. *Opera Botanica*, 65, 1–96.

Rieppel, O. (2005). Proper names in twin worlds: monophyly, paraphyly, and the world around us. *Organisms, Diversity and Evolution*, 5, 89–100.

Salt, G. (1978). Howard Everest Hinton. 24. August 1912–2. August 1977.

Biographical Memoirs of Fellows of the Royal Society, 24, 151–182.

Schlee, D. (1968). Vergleichende Merkmalsanalyse zur Morphologie und Phylogenie der Corynoneura-Gruppe (Diptera, Chironomidae). Zugleich eine Allgemeine Morphologie der Chironomiden-Imago (♂). *Stuttgarter Beiträge zur Naturkunde*, 180, 1–150.

Schmidt-Lebuhn, N. (2012). Fallacies and false premises: a critical assessment of the arguments for recognition of paraphyletic taxa in botany. *Cladistics*, 28, 174–187.

Schmitt, M. (2013). *From Taxonomy to Phylogenetics: Life and Work of Willi Hennig.* Leiden, Brill.

Schultze, H.-P. (2009). The international influence of the Stockholm School. *Acta Zoologica (Stockholm)*, 90 *(Suppl. 1)*, 22–37.

Seberg, O. (1984). Taxonomy and phylogeny of the genus *Acalypha* (Euphorbiaceae) in the Galápagos Archipelago. *Nordic Journal of Botany*, 4, 159–190.

Seberg, O. (1988). Taxonomy, phylogeny, and biogeography of the genus *Oreobolus* R. Br., with comments on the biogeography of the South Pacific continents. *Botanical Journal of the Linnean Society*, 96, 119–195.

Seberg, O. (1989). Genome analysis, phylogeny, and classification. *Plant Systematics and Evolution*, 166, 159–171.

Seberg, O. (2005). *The Triticeae: Branch in the Tree of Life.* Copenhagen: SNM, Botanisk Centralbibliotek.

Serra-Tosio, B. and Brundin, L. (1990). Redescription of the male of *Microzetia mirabilis* Seguy, 1965, an endemic Chironomid Midge from the Crozet Island (Diptera, Chironomidae). *Annales de la Société entomologique de France*, 26, 411–419.

Simpson, G.G. 1951. *Horses.* New York: Oxford University Press.

Spence, J.R. and Andersen, N.M. (1994). Biology of water striders: interaction between systematics and ecology. *Annual Review of Entomology*, 39, 101–128.

Stuessy, T.F. and Hörandl, E. (2014). The importance of comprehensive phylogenetic (evolutionary) classification: a response to Schmidt-Lebuhn's commentary on paraphyletic taxa. *Cladistics*, 30, 291–293.

Sundberg, P. (1985). Numerisk kladistik [Numerical cladistics]. *Svensk Botanisk Tidskrift*, 79, 205–217.

Swofford, D. (1985). PAUP, Ver. 2.4. Tape distributed by the author.

Sæther, O.A. (1963). *Østensjøvann. Biologi og miljøfaktorer i en grunn, kulturpåvirket sjø. (Østensjøvann. Biology and environmental factors in a shallow, eutrophied lake).* Unpublished thesis. University of Oslo, Norway.

Sæther, O.A. (1970). Nearctic and Palaeartic *Chaoborus* (Diptera: Chaoboridae). *Bulletin of the Fisheries Research Board of Canada*, 174, 1–57.

Sæther, O.A. (1971). Nomenclature and phylogeny of the genus *Harnischia* (Diptera: Chironomidae). *Canadian Entomologist*, 103, 347–362.

Sæther, O.A. (1979a). Underlying synapomorphies and anagenetic analyses. *Zoologia Scripta*, 8, 305–312.

Sæther, O.A. (1979b). Underliggende synapomorfi og enestående innvendig paralellisme belyst ved eksemplar fra Chironomidae og Chaoboridae. *Entomologisk Tidsskrift*, 100, 173–180.

Sæther, O.A. (1983). The canalized evolutionary potential: Inconsistencies in phylogenetic reasoning. *Systematic Zoology*, 32, 343–359.

Sæther, O.A. (1986). The Myth of Objectivity – Post-Hennigian Deviations. *Cladistics*, 2, 1–13.

Sæther, O.A. (1990). Midges and the electronic Ouija board. The phylogeny of the Hydrobaenus group (Chironomidae, Diptera) revised. *Zeitschrift für Zoologische Systematik und Evolutionsforschung*, 28, 107–136.

Säve-Söderbergh, G. (1934). Some points of view concerning the evolution of the Vertebrates and the Classification of this group. *Arkiv för Zoologi*, 26 A (17), 1–20.

Tuomikoski, R. (1967). Notes on some principles of phylogenetic systematics. *Annals Entomologici Fennici*, 33, 137–147.

Tuxen, S.L. (1944). The hot springs of Iceland, their animal communities and their zoogeographical significance. *The Zoology of Iceland* 1, 11, 1–216.

Tuxen, S.L. (1958). Relationships of Protura. In *Proceedings of the 10th International Congress of Entomology, Montreal, August 17–25, 1956, 1.* Ottawa: Mortimer Ltd, pp. 493–497.

Tuxen, S.L. (1959). The phylogenetic significance of entognathy in entognathous apterygotes. *Smithsoniam Miscellanous Collections*, 137, 379–416.

Tuxen, S.L. (1960). Ontogenie und Phylogenie bezogen auf die Apterygoten Insekten. *Zoologischer Anzeiger*, 164, 362–363.

Tuxen, S.L. (1963). Phylogenetical trends in the Protura as shown by relationship between recent genera. *Zeitschrift für Zoologische Systematik und Evolutionsforschung*, 1, 277–310.

Tuxen, S.L. (1971). All Apterygota are true insects with no symphylan connection. *Proceedings of the XIIIth International Congress of Entomology, Moscow* 1968, I, 311–312.

Tuxen, S.L. (1977). The genus *Berberentulus* (Insecta, Protura) with key and phylogenetical considerations. *Revue d'Écologie et de Biologie du Sol*, 14, 597–611.

Vilkamaa, P. (2013). Risto Tuomikoski as an entomologist. *Bryobrothera*, 11, 158–174.

Väisänen, R. (1984). A monograph of the genus *Mycomya* Rondani in the Holarctic region (Diptera, Mycetophilidae). *Acta Zoologia Fennica*, 177, 1–346.

Väisänen, R (1996). New *Mycomya* species from the Himalayas (Diptera, Mycetophilidae). 1. *Subgenus Mycomya s. str. Entomologica Fennica*, 7, 99–132.

Väisänen, R. (2013a). New *Mycomya* species from the Himalayas (Diptera, Mycetophilidae). 2. Subgenera Calomycomya, Cymomya, Neomycomya and Pavomya subg. n. *Zootaxa*, 3666, 301–318.

Väisänen, R. (2013b). New *Mycomya* species from the Himalayas (Diptera, Mycetophilidae). 3. Subgenera Cesamya and Mycomyopsis. *Zootaxa*, 3737, 129–153.

Väisänen, R. (2014). New *Mycomya* species from South-East Asia (Diptera, Mycetophilidae). *Zootaxa*, 3815, 526–540.

Wanntorp, H.-E. (1983). Historical constraints in adaptation theory: traits and non-traits. *Oikos*, 4, 157–160.

Wanntorp, H.-E. (1993). Lars Brundin 30 May 1907–17 November 1993. *Cladistics*, 9, 357–367.

Watrous, L.E. and Wheeler, Q.D. (1981). The outgroup comparison method of character analysis. *Systematic Zoology*, 30, 1–11.

Wiley, E.O. (1981). *Phylogenetics. The Theory and Practice of Phylogenetic Systematics.* New York: John Wiley and Sons.

Williams, D.M., Ebach, M.C. and Wheeler, Q. (2005). 150 reasons for paraphyly: a response. *Taxon*, 54, 858.

Zachos, F.E. (2014). Paraphyly – again? A plea against dissociation of taxonomy and phylogenetics. *Zootaxa*, 3764, 594–596.

Hennigian systematics in France, a historical approach with glimpses of sociology

Pascal Tassy

Auprès de mon arbre,
Je vivais heureux
J'aurais jamais dû m'éloigner d'mon arbre
Auprès de mon arbre,
Je vivais heureux
J'aurais jamais dû le quitter des yeux.

Georges Brassens, *Auprès de mon arbre* (1955)

4.1 Introduction

Hennigian systematics, that is, phylogenetic systematics, developed in France (as elsewhere) with a slow tempo impossible to imagine today when instantaneous communication is the rule. Although phylogenetics prevailed in the 1970s–1980s, the mode of its historical ascension bears strong sociological aspects. The central idea of this presentation is that the early events of the history are explained not only by debates on the nature of phylogenetics in the neo-Darwinian paradigm but also by the academic positions of both actors and disciplines. After decades, the

The Future of Phylogenetic Systematics: The Legacy of Willi Hennig, eds. D. Williams, M. Schmitt and Q. Wheeler. Published by Cambridge University Press.

relatively quiet nature of the debate among French process and pattern cladists, explains, at least partly, the present development in France of LisBeth, an original three-item analysis (3ia) program.

4.2 Historical path

Entomology circles

The first chapter in the development of Hennigian systematics has already been well documented by Dupuis (1979), as a long, laborious, obscure movement constrained by the place of entomology among the scientific disciplines in the French academic system.

Even in these entomological circles the earliest references to Hennig went unnoticed (Gouin 1950, Dupuis 1956); as Dupuis (1979: 26) wrote, they were 'sans lendemain'. French entomologists did not read articles from Belgian authors either (Kiriakoff 1952, 1955) or, if they read them, they did not benefit from them.

Nevertheless, the first Hennigian scheme of argumentation published in France came from entomology for chironomids by Serra-Tosio (Fig 4.1; Serra-Tosio 1968: 161, pl. 12). The world of entomology is so large and subdivided into tight communities, that French entomologists, even dipteran specialists, underestimated or simply ignored this article (Matile, *pers. comm.* 1987). Neither did Serra-Tosio's work reach other scientific communities. As a matter of fact, Serra-Tosio met Lars Brundin (1907–93; Serra-Tosio's 'intellectual father', Serra-Tosio, *pers. comm.*, 2002) during a symposium on chironomids at Plön in Germany in 1964. It was illuminating. Back in his laboratory of zoology and hydrobiology at the University of Grenoble, he taught himself Hennigian systematics in splendid isolation. Serra-Tosio (1968) published his work on the phylogeny of Diamesini chironomids in the local journal, *Travaux du Laboratoire d'Hydrobiologie et de Pisciculture, Grenoble*, a journal of narrow circulation. If you were not interested in chironomids or hydrobiology, there was little opportunity to read this article, and, consequently, to learn something about the implied methodology. Three years later, Serra-Tosio's (1971) thesis included a chapter on methodology (Fig 4.2). Again, this work did not influence colleagues: 'My colleagues were competent … open to discussions … but none adhered …. With time, I noticed that indifference turned into somewhat sceptical interest, then embarrassed questions, and finally late concerns, as if my colleagues felt having missed the point at one time' (Serra-Tosio, *pers. comm.*, 2002).

Palaeontology circles

The same might be said about the French palaeontological circles in the early stages – but this ended differently (Goujet 2000).

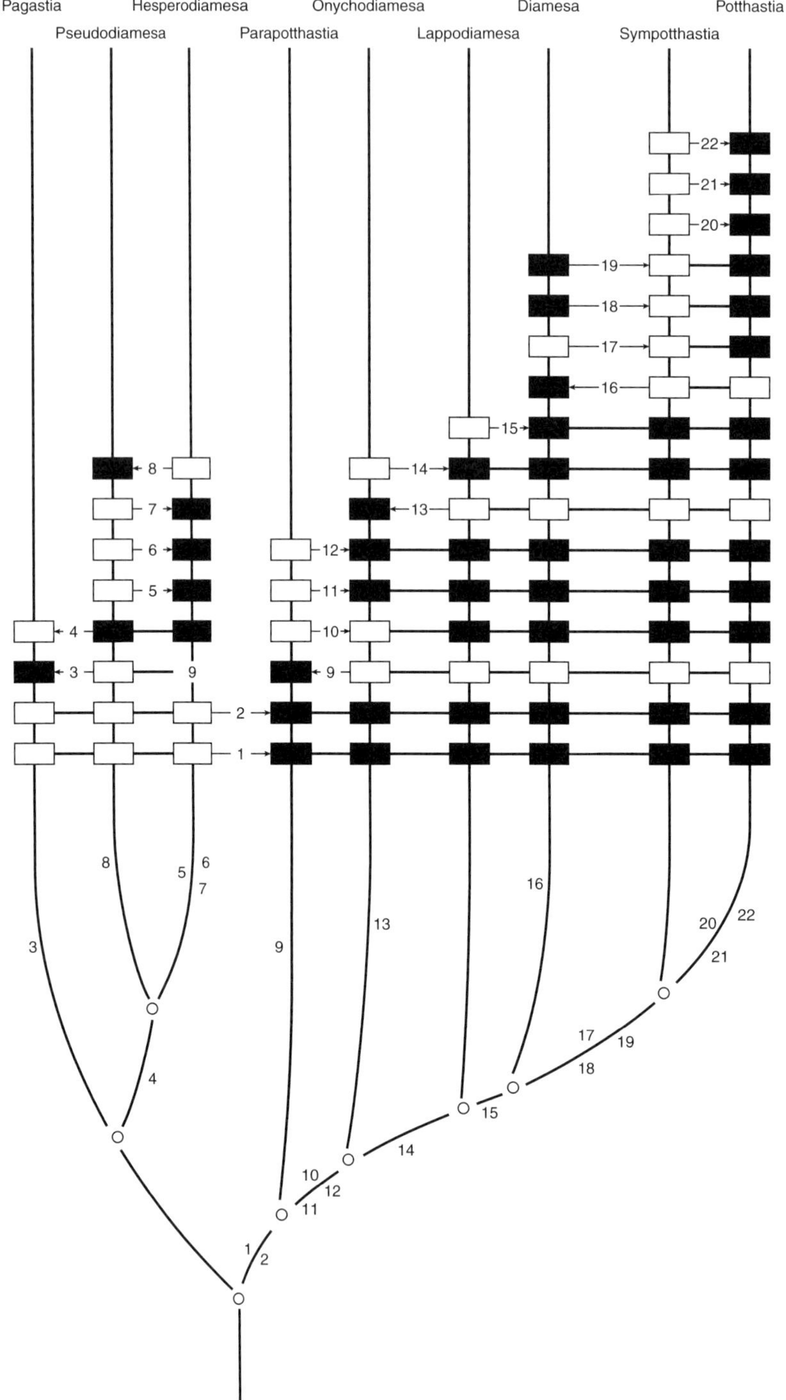

Fig 4.1 First argumentation scheme of phylogenetic systematics in the French literature: phylogenetic interrelationships of the Diamesini. Reproduced from Serra-Tosio 1968: 161, pl. 12, courtesy of Université de Grenoble and B. Serra-Tosio.

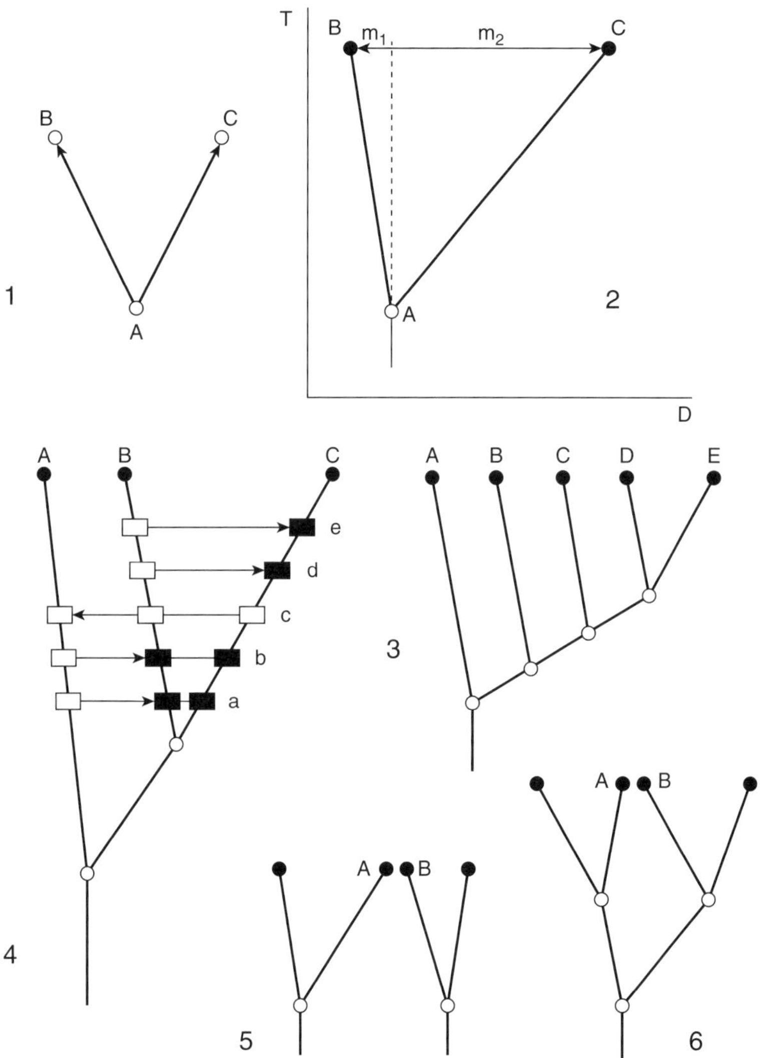

Fig 4.2 Early methodological phylogenetic schemes: 1, speciation; 2, deviation rule; 3, successive events of vicariance; 4, Hennig's principle; 5, convergence; 6, parallelism. Reproduced from Serra-Tosio 1971: plate 2, courtesy of B. Serra-Tosio.

Sigal (1966) – a micropalaeontologist – mentions Hennig's (1950) work but did not discuss his ideas. Moreover Sigal's theoretical work was not well received: for many a palaeontologist, micropalaeontology should serve as a tool for stratigraphy, and not for a theory of systematics. As Sigal (*pers. comm.*, 1992) declared: 'it was still too early and things were not ripe enough to reach the few palaeontologists interested in this domain.'

Again, symposia functioned as starting events. The first one was the Nobel symposium in Sweden on early vertebrates, held in 1967 (Ørvig 1968) and again Brundin (1968) played a decisive role. Ichthyologists and palaeoichthyologists left that symposium with Hennig's ideas in their heads (Bonde 2000). In France, Daniel Goujet was the go-between for Sweden and France, especially for the Muséum national d'Histoire naturelle. In this latter institution the book edited by Ørvig (1968) was of primary importance: there the Laboratoire de Paléontologie was well known to be largely devoted to the study of the differentiation of vertebrates and, under Jean-Pierre Lehman's direction, palaeoichtyology was especially active. The same situation was found at the Naturhistoriska Riksmuseet in Stockholm. During the 1960s and 1970s, palaeontologists from both institutions met often. Contacts between Sweden and France were important and exchanges on phylogenetic systematics through its impact on palaeoichthyology were numerous.

Yet the man who had a decisive influence was the palaeomammalogist Robert Hoffstetter (1908–99). In 1972, he exposed Hennigian concepts at the 17th International Zoological Congress in Monaco (Hoffstetter 1973).

What failed among entomologists, prevailed among palaeontologists. Hoffstetter was professor at the University Pierre-et-Marie-Curie (then named Paris VI). His lectures on palaeontology and evolution included phylogenetic systematics. Armand De Ricqlès, who worked at the University Paris VII (future University Denis-Diderot) on the same campus, also included phylogenetic systematics in his teaching courses. His stay, in 1973, in Tor Ørvig's laboratory in Stockholm was crucial. There, he inevitably met Lars Brundin (De Ricqlès, *pers. comm.*). In the semi-popular science journal *La Recherche*, then somewhat comparable to *Scientific American*, he anonymously reviewed the book on interrelationships of fishes edited by Greenwood et al. (1973) (Tassy 2011). In his review, De Ricqlès (Anonymous 1975), quoted the French philosopher Pascal, referring to the rise of Hennig's method as the 'triumph of the spirit of geometry over the spirit of finesse'. Seminars on phylogenetics were organized at Universities Paris VI (palaeontology), Paris VII (comparative biology) and at the Museum (palaeontology). Various communities, not only palaeontologists, attended these seminars. At one of them Søren Løvtrup, an embryologist from Umeå (Sweden), invited by Armand De Ricqlès, explained his axiomatic approach to Hennig's principles which were detailed in his book on the phylogeny of vertebrates (Løvtrup 1977) to a stunned audience composed of both 'traditionalists' (such as Jean-Pierre Lehman at the Museum and Charles Devillers at the University Paris VII) and 'Hennigians' (such as Hoffstetter). A different context indeed, compared to what happened in the Laboratoire d'Entomologie of the Muséum national d'Histoire naturelle. There, when Hennig himself visited the laboratory to study the collection of dipterans, he discussed his ideas only with his colleagues, and then in strict secrecy (Matile, *pers. comm.*, 1978).

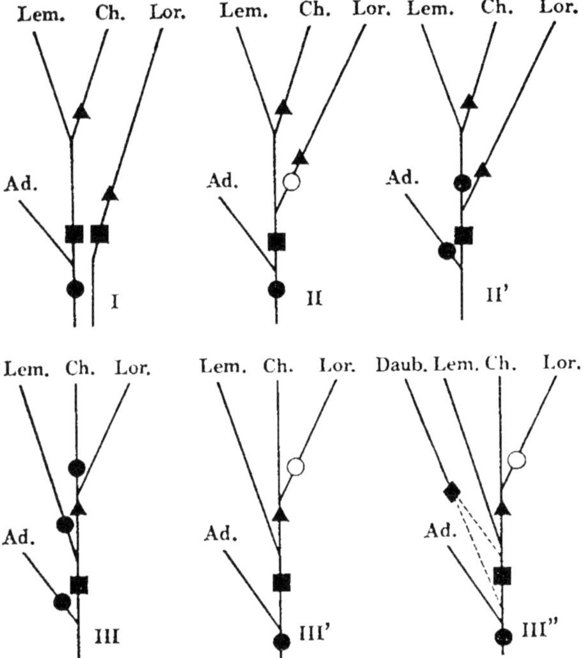

Fig 4.3 Competitive hypotheses about the phylogeny of Stresirhini and the use of parsimony as a means to select the tree (III″). After Hoffstetter (1974: 334, Fig 2), courtesy of *Journal of Human Evolution.*

In 1973, a congress on vertebrate palaeontology organized by J.-P. Lehman (Anonymous 1975) in Paris at the Muséum national d'Histoire naturelle was of major importance. Besides lively discussions derived from the use of Hennigian methods, several young French palaeontologists were especially impressed by the impact created by the vista of others such as Niels Bonde (1975) and Roger Miles (1975).

In the 1970s, Hoffstetter's research played a leading role (Hoffstetter 1974) (Fig 4.3), but his teaching was the major influence on students and young colleagues (Hublin 1978, Muizon 1978, Eisenmann 1979, Janvier and Blieck 1979, Tassy 1979): palaeontology was going to play the leading role in disseminating cladistics in France (Fig 4.4) and methodological papers soon appeared (Janvier 1981, Tassy 1981). These early publications insisted on the fact that fossils should be analysed like extant taxa and that time does not provide information about relationships. These two aspects made the so-called Cladistic Revolution especially evident in palaeontological circles. But some palaeontologists abandoned this approach, such as Vera Eisenmann, whose first cladogram was her last (Eisenmann 1979).

A story about a pioneering paper by Janvier (1978) published by *Annales de Paléontologie* provides a summary of the academic situation in the late 1970s.

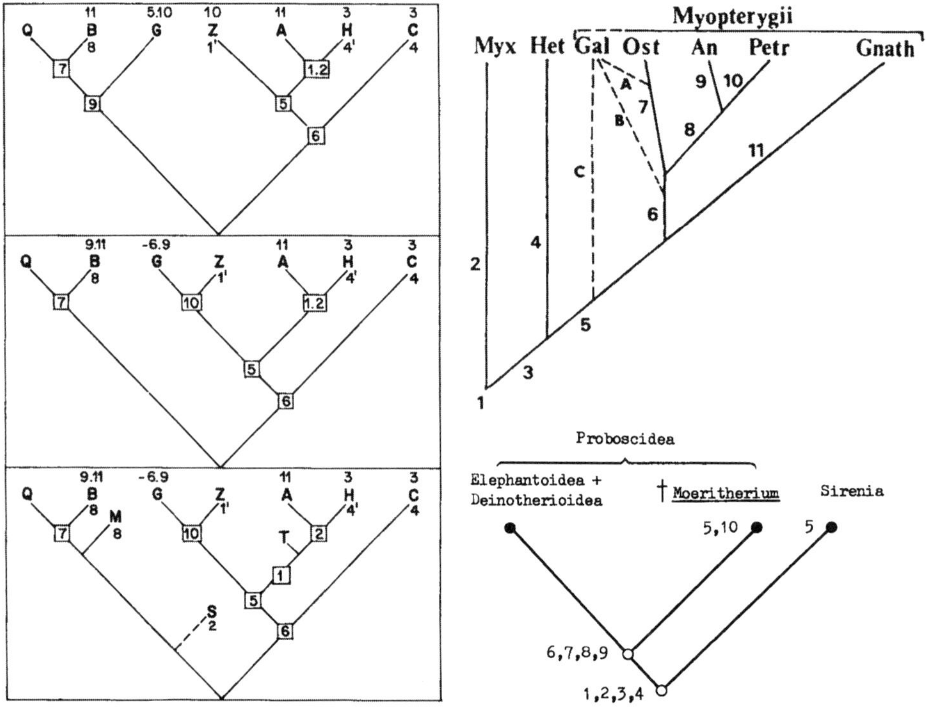

Fig 4.4 Early cladograms from French palaeontology. (a) Three hypotheses concerning the phylogeny of horses (after Eisenmann 1979, courtesy of Académie des Sciences – A: asinus horses, B: *E. burchelli*, C: caballus horses, G: *E. grevyi*, H: onager horses, Q: *E. quagga*, Z: *E. zebra*); (b) interrelationships of Craniota (after Janvier and Blieck 1979: 294, Fig 8 courtesy of *Zoologica Scripta*); (c) relationships of *Moeritherium* (after Tassy 1979: 86, courtesy of Académie des Sciences).

Janvier's paper showed the paraphyly of Agnatha and somehow foreshadowed the current conception of the ostracoderms as stem gnathostomes. The editor (J.-P. Lehman) was embarrassed for two reasons: (1) a young researcher should not publish on such deep relationships; and (2) the cladogram is too straightforward a picture. The paper was finally published with the same phylogenetic results but with a 'soft cladogram' (Fig 4.5), a tree with elegant curved lines in place of geometrical branches (Janvier, *pers. comm.*).

The year 1978 was a crucial one at the Muséum national d'Histoire naturelle. Goujet and Matile (1978) – a palaeontologist and an entomologist – organized a meeting on phylogenetic systematics and distributed a compendium of French translations from several articles along with excerpts of papers by Hennig (including the entire 1965 article; Hennig 1965), Brundin, Schaeffer et al. and Wiley, with a preface by Hoffstetter (Fig 4.6). The same year in the same place, the Société

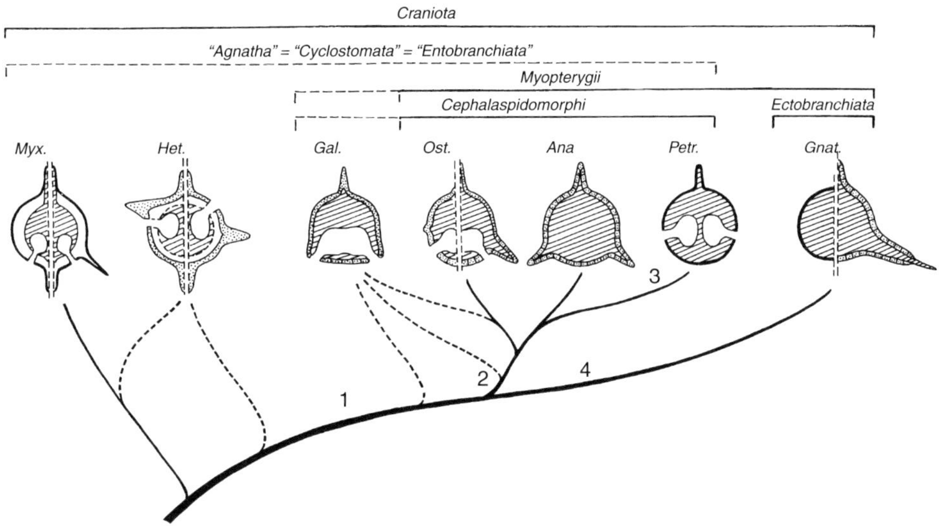

Fig 4.5 'Soft cladogram' of agnathan craniotes. After Janvier 1978: [27], Fig 14, courtesy of *Annales de Paléontologie*.

Zoologique de France organized a meeting on classification with a few cladistics orientated publications (Gasc 1979, Hoffstetter 1979, Renous 1979).

In 1980, cladistics was popularized for a more general audience with an article written by three young palaeontologists (Janvier et al. 1980) published in *La Recherche*. In this article Hennig was compared to Darwin, no less. It also included the already famous story of the cow, salmon and dipnoan relationships that resulted from discussions at the annual symposium of vertebrate palaeontology and comparative anatomy meeting at Reading in 1978; the article was nicely illustrated by Philippe Janvier (Fig 4.7). The staff of *La Recherche* were both excited to publish the new method and afraid of the reactions from the establishment. But their excitement was stronger. The result was immediate: any scientist could learn of the existence of this new method of phylogeny reconstruction and classification.

Yet, we have to be modest. Twelve years later, the 11th Annual Meeting of the Willi Hennig Society (1992) was organized in Paris with the support (somewhat reluctant in the beginning) of the CNRS (Earth Science Division). The meeting was held at the headquarters of the CNRS. During a coffee break, the director of the CNRS, Claude Paoletti (a molecular biologist) turned up to discuss the meeting with Philippe Janvier, one of the organizers, and to make sure that he was not hosting a peculiar sect lead by an unknown guru (Janvier, *pers. mea culpa comm.*).

In 2002, again in Paris, a workshop was organized at the Institut Pasteur on the topic 'Dynamics of the biological disciplines: classification, embryology, genetics'

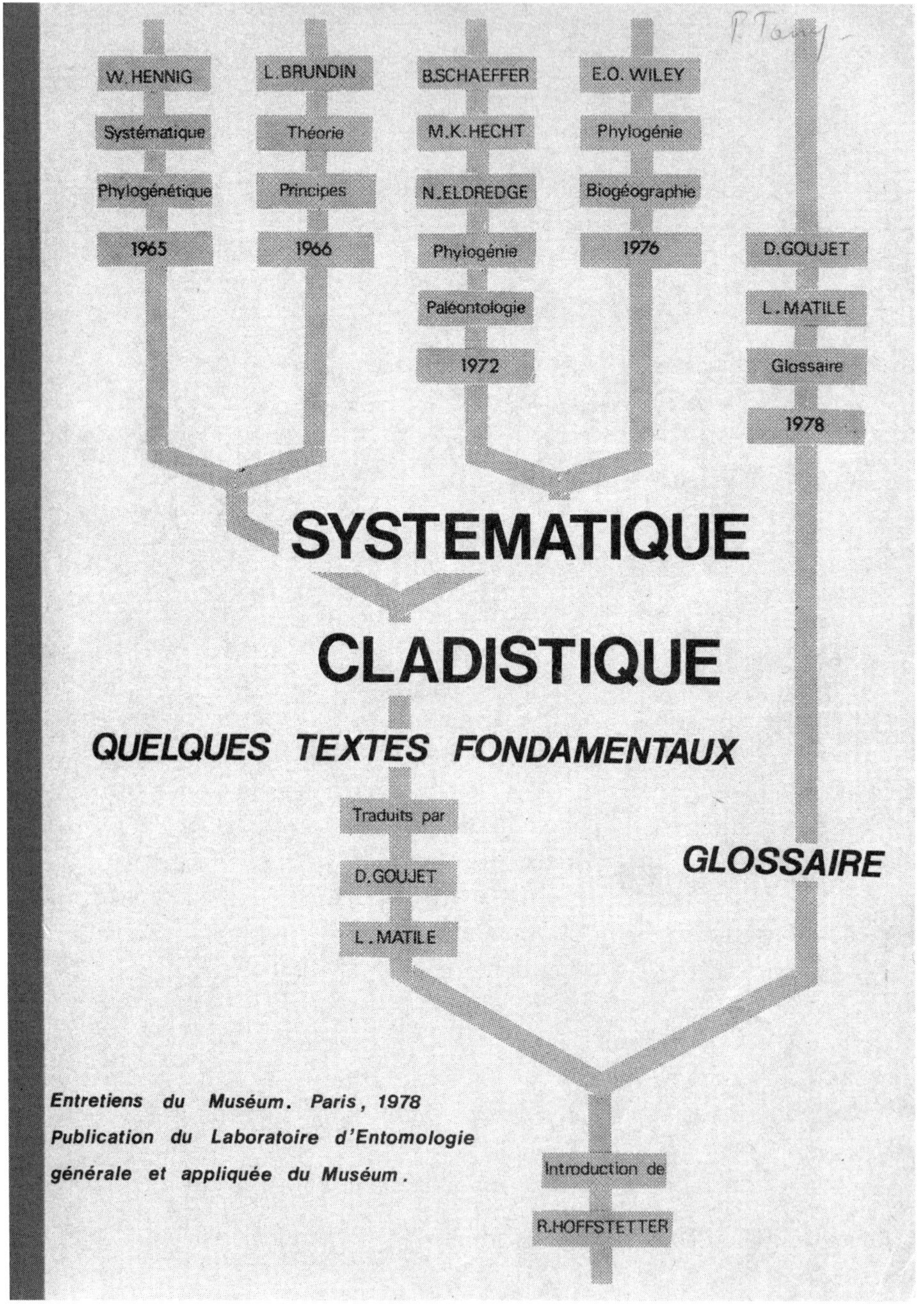

Fig 4.6 Cover of *Systématique cladistique. Quelques textes fondamentaux* (Goujet and Matile 1978).

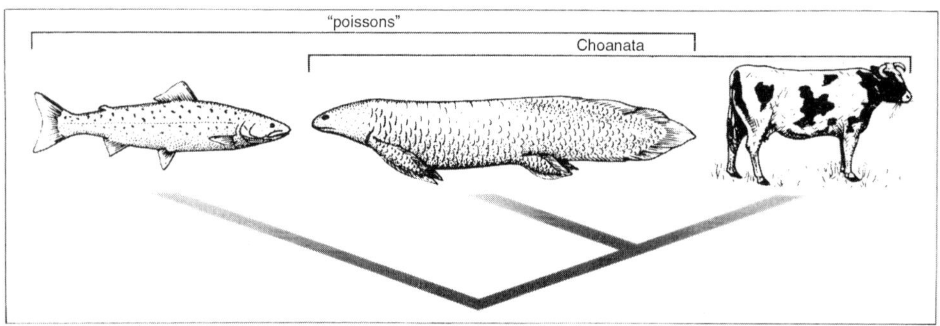

Fig 4.7 The story of the cow, salmon and dipnoan. After Janvier et al. (1980: 1399, Fig 4); drawing by P. Janvier. Reproduced courtesy of P. Janvier and *La Recherche.*

attended by biologists of various fields. This event is also a way to measure the impact of systematics in the early twenty-first century. During his talk on nomenclature, Marc Ereshefsky asked the audience, 'How many of you are familiar with cladistics?' Very few arms rose!

Universities

Hoffstetter's and De Ricqlès' university courses gave a formidable impetus to cladistics. Hennig's work was discussed again and again, endlessly. In the 1980s few French palaeontologists had not heard about cladistics. Most invertebrate palaeontologists, more inclined toward stratigraphy than to phylogeny, were largely reluctant. However, vertebrate palaeontologists who were familiar to comparative anatomy and character analysis generally practiced cladistic analyses and most students were readily convinced of the logical basis of cladistics.

Palaeontology – not entomology – played a major role in France, because it was an active domain in universities. Universities are the place where disciplines are branches of instruction, not just of research. The centralised French system had been lethal to natural history in biology circles. Botany and zoology, and consequently entomology, were underrated disciplines and marginalized when compared to genetics. Almost nobody cared for systematics in French biological research. Entomologists were mostly working in museums, especially the Muséum national d'Histoire naturelle in Paris. There, they had far less opportunities to convey the new methodology to students. Moreover, most biologists in universities and at the CNRS considered entomology to be an old-fashioned discipline. The people working on *Drosophila melanogaster* were the least interested in the systematics of dipterans. In such a system, a discipline thought by the majority to be obsolete cannot yield and spread innovations. Despite his pioneering role, Serra-Tosio could not include cladistics in his teaching program at the University of Grenoble before 1992: 'The moral of the story is that the early cladists were just a few, and their activities scattered. Then, the palaeontologists came and heavily changed the situation' (Serra-Tosio, *pers. comm.*, 1993). In contrast, palaeontologists, especially vertebrate palaeontologists, were part of the French educational system, especially at the universities Paris VI and VII. Palaeontologists were academically subordinated to geology. Yet geologists let palaeontologists organise their teaching as they wished. Probably for the worst reasons: geologists did not care much about what was going on in palaeontology. Palaeontologists, although active, formed a small community which could not pretend to lead earth sciences. In the late 1960s, a clash occurred in France between geology and geophysics over plate tectonics. In the late 1970s–1980s, no such a clash ever occurred between geologists and palaeontologists about cladistics. Difficulties occurred within palaeontological circles when some realised that the new generation of French palaeontologists was rather numerous and claimed that phylogenetics should become a prominent scientific activity. The 'gradual schools' of evolutionary palaeontology tried to stop the trend at both the universities of Montpellier (vertebrate palaeontology) and Dijon (invertebrate palaeontology), respectively, led by Louis Thaler and Henri Tintant. Stratigraphic palaeontology was undergoing difficult times. In other words: if time does not provide information about relationships, then what is left for palaeontologists to do? In 1981, Jean Chaline organized in Dijon a CNRS-funded colloquium on evolution, mainly process-orientated (Chaline 1983). Cladists, at first moved aside, then were finally tolerated (Goujet et al. 1983). But we know the end of the story: cladistics was there to stay. In 1982, Hoffstetter's festschrift (Buffetaut et al. 1982) aptly displayed the state of vertebrate palaeontology and the emergence of cladistics in the early 1980s.

Pluralism, a characteristic of the French regeneration of systematics, can be illustrated in two ways. In 1984, the Société Française de Systématique was founded by both neontologists and palaeontologists to support systematics as a fundamental science. Since the beginning, it serves as a forum for methodological issues. This society was not devoted to cladistics (even if cladistics turned to be the most active and attractive topic) but was pluralistic. It remains as such today (http://sfs.snv .jussieu.fr/). A second example is given by a French textbook on phylogenetic reconstruction (Darlu and Tassy 1993). This textbook, for undergraduate and graduate students, describes every method that was used in the early 1990s as objectively as possible (even if it is easy to understand the preferences of the authors). The chapter on likelihood methods, for example, does not focus on the one case of long-branch attraction but includes a thoughtful study of the two zones where parsimony and likelihood are doomed to yield conflicting solutions.

In conclusion, museums and meetings played a major role in the spread of Hennigian systematics in France as elsewhere. But the specificity of the French story is the role of universities and of palaeontology in a pluralistic context.

Unnoticed conflict between Hennigian and pattern cladistics in the 1980s

Another French characteristic is the underestimation of the emergence of pattern cladistics. What I mean here is that during the 1980s pioneering books by Eldredge and Cracraft (1981), Wiley (1981) and Nelson and Platnick (1981) were read as variations on the same theme. New definitions of schemes of phylogenetic argumentation (cladograms) and of plesiomorphy and apomorphy by Nelson and Platnick (1981) were not understood as the profound changes they turned out to be. As is clear, computers stole the show, and cladistics and parsimony were understood as one monolithic entity. Only careful readers could find a few papers on the debate written in French (Janvier 1983, Hull 1986, Tassy 1987), and one can read the transcription of a roundtable on several questions including pattern cladistics in Tassy (1986).

In France, during the last 20 years, the primary conflict in systematics was between cladistics (of every kind) on the one hand and proponents of likelihood on the other (Tassy 2011). That is, there were two ways of seeing phylogenetics: as systematics or as statistics. Debates about process and pattern in systematics were more gentlemanly in France than in the United States and in the United Kingdom (e.g. Cibois et al. 2004, Deleporte and Lecointre 2005). It is probably because French cladists were not such strong characters as Colin Patterson, Gareth Nelson and Steve Farris. This statement does not mean that process cladists had accepted pattern approaches. Deleporte (2005), for instance, turns a constant deaf ear to pattern arguments. De Ricqlès (2005) recognizes the distinction between pattern and process but minimizes the contrast between the two. Hence, on a more general level, the underrating of the independence of pattern cladistics and the weight put on pragmatism when using parsimony analyses, are probably what characterises the rise of cladistics in France.

4.3 Present cladistic debates

The debate between three-taxon analysis (3ta), or three-item analysis (3ia), and parsimony have, for more than 20 years, been vivid, to say the least. Nelson (2004) is right when he claims that the revolution is not finished: the evolution of systematics did not stop with Kluge and Farris' (1969) article. Yet, year after year the arguments went pretty far (compare summer exchanges of 2013 in the journal *Cladistics* between Ebach and Williams (2013) and Farris (2013)). Is it sound to qualify Hennig's transformation series a 'myth', as do Williams and Ebach (2008: 259)? Is it sound to qualify 3ta as 'creationism' or 'fraudulent propaganda' as does Farris (2013: 228)? Rhetoric and flashy style are no help, I fear.

I view this debate in the following manner. How can we apply Hennig's concept of congruence so that a computer can read the data? The first answer is the search for minimal length trees, that is, for maximal congruence (or maximal number of synapomorphies). The second answer, provided by 3ia (e.g. Nelson and Platnick 1991), is the maximum congruence of hypotheses of relationships of the form A(BC). In France, pattern cladists developed a program called LisBeth (formerly Nelson05) based on 3ia (Cao et al. 2007, Zaragüeta Bagils et al. 2012). This program has been used in many different fields such as palaeobotany (Corvez 2012), biogeography (Ung 2013) and ontologies (Grand 2013). It is still in progress but in terms of performances, as implemented in LisBeth, it exemplifies perfectly the virtues and weaknesses of the method: its virtues are the necessity to look at and carefully and clearly code characters so that quality prevails over quantity, a view that need not be further addressed. But weaknesses do exist: the somewhat interrogative statement by Nelson (1994: 137) on ancestral states followed by the positive conclusions on 'homologues' (Williams 2004) or 'synapomorphic homologues' (Nelson 2011: 139) – to cite only a few papers, but notwithstanding criticisms of 'matrix entries' and 'integrity of evidence' (Farris 2012, Williams and Ebach 2012) – turns indeed into a rigid concept of primary homology connected to the hierarchical representation and treatment of hypotheses of homology. It implies a rejection of reversals and losses as informative shared traits (synapomorphies). We know that this is exactly the basis for the present battle between cladists. This aspect of the congruence of hypotheses of relationships can be viewed in terms of constraints applied to character evolution. The old debates on constraints applied to initial hypotheses of homology did not reach so much excessive wording. For instance, Dollo's parsimony (as implemented in routine parsimony programs), based on the uniquely derived character of Le Quesne (1972) (and its emphasis on reversals – a nightmare for aficionados of 3ia), caused no dramatic exchanges (Farris 1977). Beside discordant concepts of transformation, this can be simply due to the fact that Dollo's parsimony was never intended to be the general reference system. As for LisBeth, I confess that up to the present I have not read – and much has been written about it – any argument that finally makes me give up the idea (and I have been close to doing so, Tassy 2005) that in our analyses characters should be free to evolve in any manner. Authors who believe, and argue, that systematics must be phylogenetic, connected to the concepts of transformation series, descent with modification (I see no difference between transformation and modification) and propinquity of descent are legion. Although I would surely not conclude that 'everything goes', I am convinced, along with Feyerabend (1975: 32), that 'all methodologies, even the most obvious ones, have their limits', and less obvious ones as well. Yet discussing these limits should be welcomed and not discarded.

4.4 Conclusion

Hennigian systematics developed in France through a precocious reconciliation of process cladistics and pattern cladistics mainly due to initial misunderstanding, or even ignorance, as to what pattern cladistics is. This is, indeed, not the best way to proceed. Yet one positive consequence resulted in a long but relatively gentle series of discussions that have occurred since the beginning. In Paris, one can walk on the wild side on both sides of Rue Cuvier: the Muséum national d'Histoire naturelle and the University Pierre-et-Marie-Curie have long been a common nest for palaeontologists and phylogeneticists for discussions and debates, and still are. I am wondering if there is any other place today where master degree students could learn phylogenetics by manipulating data with parsimony, likelihood/Bayesian or 3ia programs (even Neighbor Joining, if they insist), as they do in Paris in the masters course called 'Systématique Evolution Paléontologie' (see http://lis-upmc.snv.jussieu.fr/sep/index.html). Pluralism is the key. As France proudly thinks of herself as a rational country, where the Cartesian spirit is supposed to inspire every citizen, the balance in understanding the strengths and weaknesses of the various methodologies is left to the intelligence of our informed students. I cannot convince myself that this is a risky bet.

Acknowledgements

I thank David Williams, Michael Schmitt, Quentin Wheeler and Ian Kitching for their invitation to participate in the celebration of the 100th birthday of Willi Hennig in London, 27 November 2013. David Williams greatly helped to improve the English. Armand De Ricqlès, Philippe Janvier and René Zaragüeta Bagils read early versions of this chapter. I largely benefited from their comments. Yet, mistakes and biased memories and opinions remain mine.

References

Anonymous (1975). Analyse de 'Interrelationships of fishes', P.H. Greenwood, R.S. Miles and C. Patterson (ed.). *La Recherche*, 53, 190.

Anonymous (ed.) (1975). *Problèmes actuels de paléontologie – Evolution des vertébrés*. Colloques Internationaux du Centre national de la recherche scientifique n° 218 (Paris 4–9 juin 1973), 2 volumes, Paris: Editions du CNRS.

Bonde, N. (1975). Origin of 'higher groups': viewpoints of phylogenetic systematics. In *Problèmes actuels de paléontologie – Evolution des vertébrés*, ed. Anonymous. Colloques Internationaux du Centre national de la recherche scientifique n° 218 (Paris 4–9 juin 1973), Paris: Editions du CNRS vol. 1, pp. 293–324.

Bonde, N. (2000). Colin Patterson: the greatest fish palaeobiologist of the twentieth century. In *Colin Patterson (1933–1998): A Celebration of his Life*, ed. P.L. Forey, B.G. Gardiner and C.J. Humphries. London: The Linnean Society of London, Special Issue 2, pp. 33–38.

Brundin, L. (1968). Application of phylogenetic principles in systematics and evolutionary theory. In *Current Problems of Lower Vertebrate Phylogeny, Nobel Symposium 4*, ed. T. Ørvig. Stockholm: Almqvist and Wiksell, pp. 473–495.

Buffetaut, E., Janvier P., Rage J.-C. and Tassy P. (ed.) (1982). Phylogénie et paléobiogéographie. Livre jubilaire en l'honneur de Robert Hoffstetter. *Geobios*, 15 (supplement 1; Mémoire special 6), Lyon.

Cao N., Zaragüeta Bagils R. and Vignes-Lebbe R. (2007). Hierarchical representation of the hypotheses of homology. *Geodiversitas*, 29, 5–15.

Chaline, J. (ed.) (1983). *Modalités, rythmes, mécanismes de l'évolution biologique. Gradualisme phylétique ou équilibres ponctués?* Paris: Colloques Internationaux du Centre national de la recherche scientifique n° 330, Editions du CNRS.

Cibois, A., Bourgoin T. and Silvain J.-F. (ed.) (2004). Avenir et pertinence des méthodes d'analyse en phylogénie moléculaire. *Biosystema*, 22. Paris: Société Française de Systématique.

Corvez, A. (2012) *L'origine de la mégaphylle chez les monilophytes*. Paris: Unpublished Thèse de Doctorat du Muséum national d'Histoire naturelle.

Darlu, P. and Tassy, P. (1993). *La reconstruction phylogénétique. Concepts et méthodes*. Paris: Masson.

Deleporte, P. (2005). Taxons, classifications: de la théorie au cahier des charges. *Biosystema*, 24, Paris: Société Française de Systématique. pp. 43–52.

Deleporte, P. and Lecointre, G. (2005). Philosophie de la systématique. *Biosystema*, 24, Paris: Société Française de Systématique.

De Ricqlès, A. (2005). La distinction entre «patterns» et «processes» est-elle désuète en systématique? *Biosystema*, Paris: Société Française de Systématique 24, pp. 33–41.

Dupuis, C. (1956). Variations convergentes ou comparables de certains caractères des Tachinaires, notamment des Phastinae (Dipt. Laravaevoridae); leur signification taxonomique différente selon les lignées. Copenhagen: *Proceedings of the 14th International Zoological Congress (Copenhagen 1953)*, 474–476.

Dupuis, C. (1979). Permanence et actualité de la systématique. La «systématique phylogénétique» de W. Hennig (historique, discussion, choix de références). *Cahiers des Naturalistes*, 34, 1–69.

Ebach, M.C. and Williams, D.M. (2013). E quindi uscimmo a riveder le stelle. *Cladistics*, 29, 227.

Eisenmann, V. (1979). Caractères évolutifs et phylogénie du genre *Equus* (Mammalia, Perissodactyla). *Comptes rendus de l'Académie des Sciences, Paris*, ser. D 288, 497–500.

Eldredge, N. and Cracraft, J. (1981). *Phylogenetic Patterns and the Evolutionary Process*. New York: Columbia University Press.

Farris, J.S. (1977). Phylogenetic analysis under Dollo's law. *Systematic Zoology*, 26, 77–88.

Farris, J.S. (2012). Fudged 'phenetics'. *Cladistics*, 28, 231–233.

Farris, J.S. (2013). Pattern taxonomy. *Cladistics*, 29, 228–229.

Feyerabend, P. (1975). *Against the Method. Outline of an Anarchistic Theory of Knowledge*. London: New Left Books.

Gasc, J.-P. (1979). Relations entre la phylogénie et la classification: évocation des débats actuels entre phénéticiens et cladistes. *Bulletin de la Société Zoologique de France*, 103 (1978), 167–178.

Gouin, F. (1950). Analyse de: Hennig (1948), Die Larvenformen der Dipteren. 1.Teil. *L'année biologique* (1950), 26, 515–516.

Goujet, D. (2000). The French connection. In: *Colin Patterson (1933–1998): A Celebration of his Life*, ed. P.L. Forey, B.G. Gardiner and C.J. Humphries. London: The Linnean Society of London, Special Issue 2, 79–81.

Goujet, D., Janvier P., Rage J-C. and Tassy P. (1983). Structure ou modalité de l'évolution: point de vue sur l'apport de la paléontologie. In *Modalités, rythmes, mécanismes de l'évolution biologique. Gradualisme phylétique ou équilibres ponctués?*, ed. J. Chaline. Paris: Colloques Internationaux du Centre national de la recherche scientifique n° 330, Editions du CNRS, pp. 137–143.

Goujet, D. and Matile, L. (ed.) (1978). *Entretiens du Muséum. Systématique cladistique. Quelques textes fondamentaux. Glossaire.* Paris: Muséum national d'Histoire naturelle, Laboratoire d'Entomologie Générale et Appliquée.

Grand, A. (2013). *Représentation sémantique des phénotypes: métamodèle et ontologies pour les caractères taxonomiques et phylogénétiques.* Paris: Unpublished Thèse de Doctorat du Muséum national d'Histoire naturelle.

Greenwood, P.H., Miles, R.S. and Patterson, C. (ed.) (1973). *Interrelationships of Fishes.* London: Academic Press.

Hennig, W. (1950). *Grundzüge einer Theorie der phylogenetischen Systematik.* Berlin: Deutscher Zentral Verlag.

Hennig, W. (1965). Phylogenetic Systematics. *Annual Review of Entomology* 10, 97–116.

Hennig, W. (1966). *Phylogenetic Systematics.* Urbana, IL: University of Illinois Press.

Hoffstetter, R. (1973). Origine, compréhension et signification des taxons de rang supérieur: quelques exemples tirés de l'histoire des mammifères. *Annales de Paléontologie,* 59, 137–169.

Hoffstetter, R. (1974). Phylogeny and geographical deployment of the Primates. *Journal of Human Evolution,* 3, 327–350.

Hoffstetter, R. (1979). Phylogénie et classification: l'exemple des Primates. *Bulletin de la Société Zoologique de France,* 103 (1978), 183–188.

Hublin, J.-J. (1978). *Le torus occipital transverse et les structures associées: évolution dans le genre Homo.* Paris: Unpublished Thèse de Doctorat de 3e cycle, Université Paris VI.

Hull, D. (1986). Les fondements épistémologiques de la classification biologique. In *L'Ordre et la diversité du vivant,* ed. P. Tassy. Paris: Fayard -Fondation Diderot, pp. 163–203.

Janvier, P. (1978). Les nageoires paires des Ostéostracés et la position systématique des Céphalaspidomorphes. *Annales de Paléontologie,* 64, 113–142.

Janvier, P. (1981). The phylogeny of Craniata, with particular reference to the significance of fossil 'agnathans'. *Journal of Vertebrate Paleontology,* 1, 121–151.

Janvier, P. (1983). Le divorce de l'oiseau et du crocodile. *La Recherche,* 149, 1430–1432.

Janvier, P. and Blieck, A. (1979). New data on the internal anatomy of the Heterostraci (Agnatha), with general remarks on the phylogeny of the Craniata. *Zoologica Scripta,* 8, 287–296.

Janvier, P., Tassy, P. and Thomas, H. (1980). Le cladisme. *La Recherche,* 117, 1396–1406.

Kiriakoff, S. (1952). Analyse de: W. Hennig; *Grundzüge einer Theorie der phylogenetischen Systematik. Bulletin et Annales de la Société royale d'entomologie de Belgique*, 88, 294.

Kiriakoff, S. (1955). Le système phylogénétique: principes et méthodes. *Bulletin et Annales de la Société royale d'entomologie de Belgique*, 91, 147–158.

Kluge, A. and Farris, J.S. (1969). Quantitative phyletics and the evolution of anurans. *Systematic Zoology*, 18, 1–32.

Le Quesne, W. (1972). Further studies based on the uniquely derived character concept. *Systematic Zoology*, 21, 281–288.

Løvtrup, S. (1977). *The Phylogeny of Vertebrata*. New York: John Wiley and Sons.

Miles, R.S (1975). The relationships of the Dipnoi. In *Problèmes actuels de paléontologie: Evolution des vertébrés*, ed. Anonymous. Paris: Colloques Internationaux du Centre national de la recherche scientifique n° 218 (Paris 4–9 June 1973), Editions du CNRS Vol. 1, 133–148.

Muizon, C. de (1978). *Contribution à l'étude systématique, anatomique et fonctionnelle des Phocidae pliocènes de la Formation Pisco (Pérou). Considérations générales sur la phylogénie et la paléobiogéographie des Phocidae*. Paris: Thèse de Doctorat de 3e cycle, Université Paris VI.

Nelson, G. (1994). Homology and systematics. In *Homology: The Hierarchical Basis of Comparative Biology*, ed. B.K. Hall. San Diego, CA: Academic Press, pp. 101–149.

Nelson, G. (2004). Cladistics: its arrested development. In *Milestones in Systematics*, ed. D.M. Williams and P.L. Forey. Boca Raton, FL: The Systematics Association-CRC Press, pp. 127–147.

Nelson, G. (2011). Resemblance as evidence of ancestry. *Zootoxa*, 2946, 137–141.

Nelson, G. and Platnick, N.I. (1981). *Systematics and Biogeography. Cladistics and Vicariance.* New York: Columbia University Press.

Nelson, G. and Platnick, N.I. (1991). Three-taxon statement: a more precise use of parsimony? *Cladistics*, 7, 351–366.

Ørvig, T. (ed.) (1968). *Current Problems of Lower Vertebrate Phylogeny, Nobel Symposium 4*. Stockholm: Almqvist and Wiksell.

Renous, S. (1979). Confrontation entre certaines données morphologiques et la répartition géographique des formes actuelles de sauriens: interprétation phylogénétique et hypothèses paléogéographiques. *Bulletin de la Société Zoologique de France*, 103 (1978), 219–224.

Serra-Tosio, B. (1968). Taxonomie phylogénétique des Diamesini: les genres *Potthastia* Kieffer, *Sympotthastia* Pagast, *Parapotthastia* n.g. et *Lappodiamesa* n.g. (Diptera, Chironomidae). *Travaux du Laboratoire d'Hydrobiologie de Grenoble*, 59–60, 117–164.

Serra-Tosio, B. (1971). *Contribution à l'étude taxonomique, phylogénétique, biogéographique et écologique des Diamesini (Diptera, Chironomidae).* Thèse Doctorat ès-Sciences, Université de Grenoble.

Sigal, J. (1966). Le concept taxinomique de spectre. Exemples d'application chez les foraminifères. Proposition de règles de nomenclature. *Mémoire Hors-Série de la Société Géologique de France*, 3, 1–126.

Tassy, P. (1979). Relations phylogénétiques du genre *Moeritherium* Andrews, 1901 (Mammalia). *Comptes rendus de l'Académie des Sciences, Paris*, ser. D 289, 85–88.

Tassy, P. (1981). Phylogeny as the history of evolution and phylogeny as a result of scientific investigation: the significance of paleontological data. In *International Symposium on «Concepts and*

Methods in Paleontology» Contributed Papers, ed. J. Martinell. Barcelona, Spain: Departament de Paleontologia, Universitat de Barcelona, pp. 65–73.

Tassy, P. (ed.) (1986). *L'ordre et la diversité du vivant*. Paris: Fayard-Fondation Diderot.

Tassy, P. (1987). *Pour la bibliothèque – Mark Ridley: Evolution and Classification. The Reformation of Cladism*. Longman, London et New York, 201 p., 1986. *Annales de Paléontologie*, **73**, 235–239.

Tassy, P. (2005). Fait et théorie: quelle connaissance de base pour la cladistique structurale? *Biosystema*, 24, Paris: Société Française de Systématique pp. 63–74.

Tassy, P. (2011). Une histoire de géométrie et de finesse (ou: comment parler de phylogénétique?). *Comptes Rendus Palevol*, 10, 341–346.

Ung, V. (2013). *Nouvelles inférences cladistiques sur la biogéographie historique du Sud-Est asiatique et de la région de l'Ouest du Pacifique. Réflexion sur une approche intégrative de la dimension temporelle en biogéographie historique*. Thèse de Doctorat du Muséum national d'Histoire naturelle. Paris.

Wiley, E.O. (1981). *Phylogenetics. The Theory and Practice of Phylogenetic Systematics*. New York: John Wiley and Sons.

Williams, D.M. (2004). Homologues and homology, phenetics and cladistics: 150 years of progress In *Milestones in Systematics*, ed. D.M. Williams and P.L. Forey. Boca Raton, FL: CRC Press pp. 181–224.

Williams, D.M. and Ebach, M.C. (2008). *Foundations of Systematics and Biogeography*. New York: Springer Science and Business Media.

Williams, D.M. and Ebach, M.C. (2012). Phenetics and its application. *Cladistics*, 28, 229–230.

Zaragüeta Bagils, R., Ung, V., Grand, A., et al. (2012). LisBeth: new cladistics for phylogenetics and biogeography. *Comptes Rendus Palevol*, 11, 563–566.

5

Are we all cladists?

ANDREW V.Z. BROWER

> As though it mattered for the definition of the concept 'truth' that we cannot recognize truth itself and everywhere in science are limited to hypotheses concerning truth. (Hennig 1966: 94)
>
> We can never know of a theory that it is true, even if it is true. (Popper 1983: 79)
>
> You wouldn't know the truth if it came up and bit you in the ass. (Darrel Frost, *pers. comm.*, September 1996)

5.1 Introduction

This chapter addresses a fundamental philosophical question in systematics: how should we understand the relationship between an empirical phylogenetic hypothesis and the "true tree"? I argue that cladists (starting with Willi Hennig), unlike other kinds of systematists, hold a philosophical perspective that embraces the unknowability of the actual historical course of evolutionary diversification. This means that the justification for cladistic methods and results must be sought not in their capacity for accurately representing the hierarchy of life, but elsewhere, by reliance on epistemological principles such as parsimony. At least from the cladist perspective, claims of the purported greater "accuracy" of model-based phylogenetic methods are not only unsubstantiated but meaningless.

The Future of Phylogenetic Systematics: The Legacy of Willi Hennig, eds. D. Williams, M. Schmitt and Q. Wheeler. Published by Cambridge University Press.
© The Systematics Association 2016.

5.2 What does it mean to be a "Hennigian"?

When we contemplate the work of Willi Hennig, we are impressed that the ideas of a person of humble origins and modest professional standing can nevertheless influence a scientific discipline in permanent and fundamental ways. Although Hennig was a man of his time, influenced in his consideration of species by Ernst Mayr and other "new taxonomists" of the mid-twentieth century, and perhaps neither a profound philosopher nor a great communicator, he nevertheless established an epistemological framework for the practice of systematic biology that endures and inspires to the present day. In this way, Hennig stands out as one of the few individuals, along with Linnaeus and Darwin, whose single-handed contributions to our understanding of the methods and goals of systematic biology represent deserved eponymous Kuhnian paradigm shifts (Kuhn 1970).

The title of this chapter is based on oddly inclusive circumscriptions of what it means to be a "Hennigian" expressed recently by stalwarts of the Willi Hennig Society in an article and two textbooks:[1]

> Why is it that phylogenetic classification carried the day, and virtually all systematists (at least the younger generations) are now Hennigian phylogenetic systematists? (Mishler 2009: 63)

> Relationship still means genealogical relationship, synapomorphy is still the mark of common ancestry, and monophyletic groups are the only natural groups regardless of whether one uses a parsimony algorithm or a likelihood algorithm to analyse one's data. That makes us all phyogeneticists, and if you wish to use a label, it makes us all Hennigians. (Wiley and Lieberman 2011: xiii)

> In contemporary systematics, several methods are used to make these judgments based on Ockham's razor (parsimony) or stochastic evolutionary models (likelihood and Bayesian techniques). Although they differ in their criteria, they all agree that groups must be monophyletic in the Hennigian sense, that classifications must match genealogy exactly, and that evidence must rely on special similarity (if differently weighted). All systematists today, whether they like it or not, are Hennigian cladists. (Wheeler 2012: 18–19)

While the requirement of monophyly as discovered via synapomorphy and the explanation of these patterns by historical bifurcation of ancestral lineages are certainly central to Hennig's approach and championed by contemporary cladists, it seems to me that there are additional philosophical and methodological issues that set at odds the schools of systematics cozily intermingled by these statements. In

---

[1] As noted by Williams et al. (2010), Mishler, Wiley and Lieberman, and Wheeler are not alone in their assertions of phylogenetic comity: Felsenstein (2004: 146) made the following claim: "both [phenetic and cladistic] methods can be considered to be statistical methods, making their estimates in slightly different ways."

particular, I question assertions that model-based phylogenetic inference is either "Hennigian" or "cladistic" in the modern sense. The aim of this chapter is to trace the development of philosophical perspectives and identify critical differences showing that not "all systematists today" are cladists (see also Brower 2000a).

5.3 Theories of truth

A fundamental split in the way systematic biologists (and other scientists) view their understanding of natural phenomena is the opposing perspectives of the so-called correspondence and coherence theories of truth (Oldroyd 1986, Rieppel 2005a, Assis 2009), which I briefly contrast/caricature here. No one who thinks about these ideas has an absolutely polarized perspective, but it is instructive to describe the ends of the continuum before discussing nuances in the middle. Those who entertain a correspondence theory of truth believe that natural entities and processes exist and do things independent of our knowledge of them (cf. Ghiselin 1997), and that the aim of science is therefore to develop explanatory theories that conform as closely as possible to the actual nature of these things. This is sometimes also called realism or foundationalism. Alternatively, adherents of a coherence theory of truth consider the "true nature of things" to be inscrutable or inaccessible to direct empirical investigation, and rely instead upon the congruence of evidence in support of one explanatory theory or another that ultimately is grounded in what might be called common sense (intersubjective corroboration of uncontroversial low-level observations), rather than absolute certainty (Wägele 2005). I refer to those with coherentist sympathies as "empiricists" (cf. Van Fraassen 1980, Sober 2008).

If you are shooting arrows at a target, the closer to the center your arrow hits, the more accurate is your shot. The bull's eye is the "truth" to which the accuracy of arrows corresponds to a greater or lesser degree. If you do not know where the bull's eye is, all you can hope to do is to shoot your arrows in a tight cluster that suggests that your archery is precise. That is coherence.

Karl Popper claimed to "believe in metaphysical realism" (Popper 1983: 80), and yet he did not believe that scientific knowledge was absolute:

> The empirical basis of objective science has ... nothing "absolute" about it. Science does not rest upon rock bottom. The bold structure of its theories rises, as it were, above a swamp. It is like a building erected on piles. The piles are driven down from above into the swamp, but not down to any natural or 'given' base; and when we cease our attempts to drive our piles into a deeper layer, it is not because we have reached firm ground. We simply stop when we are satisfied that they are firm enough to carry the structure, at least for the time being. (Popper 1965: 111)

At first blush it might seem counterproductive and anti scientific to question the premise that systematic biology is concerned with "reality." Indeed, many realists (including cladists) evidently feel so justified in their metaphysical beliefs to that effect that they see no need to articulate or defend their positions (cf. Kluge 2001, Lee 2002). Instead, they take an authoritarian stance to criticize the lack of "justification" for contrasting views by labeling them variously as operationalist, instrumentalist, phenomenologist, or creationist (more about these below). But as Popper said, "[t]he way in which knowledge progresses, and especially our scientific knowledge, is by unjustified (and unjustifiable) anticipations, by guesses, by tentative solutions to our problems, by *conjectures*." (Popper 1965: vii), "[a] scientific result cannot be justified. It can only be criticized, and tested" (Popper 1979: 265), and "[t]here is, again, no attempt on my part to *justify* positively, or establish, in the traditional sense, that a preference for one theory rather than another one is the correct one" (Popper 1983: 24). When there is no rock bottom, there can be no certain ground for justification of empirical theories (see also Rieppel 2005b).

In this way, it is evident that many concepts in systematics, such as common ancestry of higher taxa and descent with modification as the causal agent of organismal diversity, are theories that are not straightforwardly "justifiable," or accessible via observation. The more circumspect empirical approach recognizes that the critical scientific questions are not the ontological "what do we know," but the epistemological "how do we know what we know." How do we know there were common ancestors? How do we know that descent with modification took place? The evidence supporting these explanatory historical process theories comes entirely from syntheses of empirical systematic comparisons of features of extant and extinct organisms. Again, the issue is not the rejection of any and all "theories," but the distinction of which theories are premises of systematics, and which are conclusions inferred from systematics.

An illustrative example of differing realist/empiricist worldviews is the long-standing problem of the nature of the "natural system," which in the post-Darwinian era was rebranded as the "problem of phylogeny." In the eighteenth and early nineteenth centuries, Linnaeus and others believed that there was a real natural system of relationships among taxa (thus the title of Linnaeus 1758), which conformed to the Plan of Creation, in contrast to various "methods" of classification which were acknowledged to be artificial contrivances of convenience for identification and other "special purposes" (Bicheno 1827, Macleay 1830). Even in those times, thoughtful systematists (e.g. Strickland 1841) saw that the natural system, as manifest from "natural affinities" was an irregularly bifurcating hierarchy (see discussion in Schuh and Brower 2009). As shown by the American philosopher Ronald H. Brady (1985), the empiricist Darwin (1859) clearly recognized that his theory of descent with modification provided a naturalistic mechanism to explain the pattern of groups nested within groups discovered from comparative anatomy by his pre-evolutionary systematist forerunners. After Darwin, the theory of descent with modification came to be seen, through the voluminous efforts of Whig historians/

evolutionary taxonomists like Mayr (1982), not as an explanation of evidence but as a reified causal process, and therefore shifted the "problem of phylogeny" from the epistemological realm of empiricism to the ontological realm of belief (Brower 2000a, Winsor 2009). Concomitantly, the conception of the natural system changed from the grand empirical pattern of nature to a metaphysical outcome of phylogenetic divergence, and systematics became subsumed within evolutionary biology (Wheeler 1995).

The broader culture of modern systematics is so steeped in this post-Darwinian perspective that it penetrates even to the definitions of fundamental terms. For example, the modern standard homology definition is "similarity due to common ancestry" (Haas and Simpson 1946, Mayr 1982, Futuyma 1986, Nixon and Carpenter 2012a), which entails that systematics cannot provide evidence to support the theory of evolution, because the theory is an a priori assumption of systematics (Brady 1985). By contrast the subtly different definition of Brower and de Pinna (2012; "homology is the relationship among parts of organisms that provides evidence for common ancestry") keeps the horse before the cart.

Another good example of post-Darwinian anti-empiricism is the naïve reification, via "tree thinking," of phylogenetic hypotheses as pictures of the evolutionary process in action: "trees communicate the evolutionary relationships among the elements, such as genes, or species, that connect a sample of branch tips. Under this interpretation, the nodes on a tree are taken to correspond to actual biological entities that existed in the past: ancestral populations or ancestral genes," while methods and evidence are no longer of any consequence whatsoever: " tree thinking does not necessarily entail knowing how phylogenies are inferred by practicing systematists" (Baum et al. 2005: 979).

5.4 Hennig: realist or empiricist?

[P]hylogenetic analysis is most certainly empirical, for in applying the parsimony criterion, it chooses among alternative hypotheses of relationship on nothing other than their explanatory power. (Farris 1983: 36)

Willi Hennig died in 1976, and the scope of systematic theories and methods subsequently advanced under his aegis is certainly larger and more elaborate than anything he himself wrote or conceived. Nor was he infallible in his views. Nevertheless, Hennig (1966) has been widely quoted in support of diverse arguments, and many of his statements are somewhat overdetermined, which is to say they may be (and have been) interpreted to mean different things by authors with different agendas. Given this background of Hennigian scholasticism, the reader should consider arguments presented here with a degree of circumspection: who knows what Hennig would have thought of computer-generated phylogenetic trees, let alone trees generated by fundamentally different "Hennigian" methods?[2] Notwithstanding, it is interesting to contemplate what his attitude about such things might be if he were extant today.

Received wisdom suggests that "a scientific realist stance is the only metaphysical position that genuinely corresponds to the intentions behind the original Hennigian phylogenetics systematics project" (Vergara-Silva 2009: 278). I am not certain that this is correct, at least in the foundationalist sense, and in this section I attempt to reveal Hennig as empiricist, rather than realist, through his own words.

At the outset of his Chapter 2, "Tasks and methods of taxonomy," Hennig (1966) drew a distinction between "tasks," which represent his efforts at providing causal ontological grounding (Rieppel 2005c, Rieppel and Kearney 2007) for the raw materials of phylogenetic systematics (taxa and characters) in such processes as ontogeny and tokogeny, and "methods," which are more concerned with the epistemological treatment of empirical data to assess relationships among taxa. Thus, Hennig said,

> The phylogenetic system is a task whose goal is as unattainable as that of any other science. What we tentatively call the 'phylogenetic system' of a group of organisms can consequently never be anything final. But we are justified in calling such a provisional system a 'phylogenetic system' (in distinction to other possible systems) only if, with the aid of the presently known facts and methods, it can be made probable that it represents phylogenetic relationships more accurately than any other system. (Hennig 1966: 29)

And,

> The task of the phylogenetic system is not to present the result of evolution, but only to present the phylogenetic relationships of species and species groups on the basis of the temporal sequence of origin of sister groups. (Hennig 1966: 194)

And,

> The supposition that two or more species are more closely related to one another than to any other species and that, together they form a monophyletic group, can only be confirmed by demonstrating their common possession of derivative characters ('synapomorphy'). (Hennig 1965: 114)

In other words, according to Hennig, the "natural system" as a topic of scientific inquiry is not a mirror of evolution (the "temporal" aspect of cladograms relates

---

[2] In a critique of Sneath and Sokal (1973), Hennig's acolyte Dieter Schlee (1975) argued that "numerical phyletic" methods (including those of Farris 1969, 1970) were "defective" and "impracticable," but this appears, at least in the case of Farris, to be based on Schlee's incorrect assumption that Farris believed that symplesiomorphies provide evidence of grouping. On the other hand, Kluge (1976: 69) explicitly rejected an empirically supported monophyletic classification of pygopodid lizards on the grounds that symplesiomorphic taxa were "relatively little patristically different" from one another – a clear example of anti-Hennigian argumentation. Perhaps Schlee's criticisms played a role in purging such syncretic proclivities from early practitioners of the cladistic approach.

only to relative order of branching events and not to absolute time), and our know-ledge that there even *is* a "natural system" is provided by empirical data interpreted through well-reasoned methodologies, not by a priori belief in a causal mechanism. These sentiments have been echoed elsewhere:

> Phylogeneticists hold that the study of phylogeny ought to be an empirical science, that putative synapomorphies provide evidence on genealogical relationship, and that (aside possibly from direct observation of descent) those synapomorphies con-stitute the only available evidence on genealogy. (Farris 1983: 7)

Such statements would seem to place Hennigian phylogenetic systematics, aka "cla-distics" unequivocally in the realm of empiricism. Likewise, discussing homoiology, Hennig (1966: 117) refers to "true homologies" as "character correspondences that were actually taken over from the common ancestor" – a seemingly metaphysical claim. But of course Hennig recognized that "true homologies" cannot be observed de facto:

> The question of how parallelisms in the narrow or broad sense are to be explained genetically is not of such great importance to phylogenetic systematics, for which it is a question of finding criteria that make it possible to decide whether or not the occurrence of identical characters or whole character complexes in different spe-cies is based on the fact that these were taken over from one stem species that is common only to these species (synapomorphy).[3] (Hennig 1966: 119)

… and must be inferred on the grounds of character congruence: "The common occurrence of parallelisms and homoiologies … indicates the necessity for phylo-genetic systematics to take into account as many characters as possible in deciding kinship relationships" (Hennig 1966: 121). Failing corroboration by character con-gruence, Hennig resorted to his auxiliary principle –which as Farris (1983) pointed out is not a claim about the world, but a methodological rule equivalent to parsi-mony. This argument, as well as his subsequent discussion, provide clear evidence that Hennig took a coherentist approach to assessing "truth":

> For phylogenetic systematics this means that the reliability of an assumption concerning the phylogenetic relationships of systematic groups increases with the number of individual characters that can be fitted into transformation series. (Hennig 1966: 132)

Others (e.g. Rieppel 2007a; see below) have labeled Hennig a realist. To be sure, in his "concluding remarks," Hennig (1966: 234) urged systematists to "recognize that evolution is a fact" – a seemingly realist exhortation that provides expedient

---

[3] Note that, at least in this part of his book, Hennig's definition of "true homology" ("character cor-respondences that were actually taken over from the common ancestor," Hennig 1966: 117) is the same as his definition of "synapomorphy" (characters "taken over from one stem species that is common only to these species," Hennig 1966: 119).

causal grounding for phylogenetic systematics to the exclusion of various "typologi-cal" or hybrid schemes. But as Hennig argued earlier (Hennig 1966: 14; see below), this "fact" is a result of systematics, intersubjectively corroborated by overwhelm-ing empirical evidence, and not an a priori foundationalist claim. Surely Hennig recognized that or he would not have made this statement at the conclusion of his book, after elaborating epistemological caveats about the limits of knowledge such as those quoted above. We can therefore view this recommendation as Hennig's way of allowing his readers to skip all the philosophical argumentation and cut to the chase of his method, albeit at the expense of the carefully reasoned empiricist perspective.[4] As we have seen, though, that leap of faith is one which Hennig himself did not make. I identify this leap aversion as a critical component of the demar-cation of cladistics from evolutionary taxonomy in all its historical and modern manifestations.

Another aspect of Hennig's writing that could lead readers to characterize him as a realist appears in his distinction between phylogenetic systematics and what he variously called idealistic-morphological, typological, Aristotelian, evolutionary or "adaptiogenetic" systematics (cf. Hennig 1966, 1975). Statements such as,

> [t]he idea … that phylogenetic systematics is based logically and/or historically on purely morphological or at least nonphylogenetic systems, and the view often derived from this idea that a pure (idealistic) morphological or at least nonphylo-genetic system therefore merits precedence over the phylogenetic one because it stands closer to the natural facts and contains fewer hypothetical elements, is absolutely wrong. (Hennig 1966: 11)

are easy to interpret as invocations of evolutionary background knowledge and seem to directly contradict my thesis. However, closer reading shows that Hennig's argument was epistemological and not ontological: as he said,

> [w]ith such an attitude, the varying distinctness with which the relations between natural things are conspicuously visible, changing from case to case, rather causes a continuous changing of viewpoint, resulting in a 'system' in which the relations between natural things appear to be projected through one another in such a tan-gled way that this picture is completely useless for solving problems, for gaining an understanding of the nature of the diversity of nature. (Hennig 1966: 12).

This is plainly an epistemological statement about the systematist's confusion result-ing from viewing any kind of resemblance as evidence of grouping, rather than an ontological statement about a proper evolutionary mindset: the argument is about which methods and evidence are appropriate for inference of relationships, not

---

[4] Or perhaps, as Platnick (1979) suggested, Hennig deliberately obscured his empiricism, in order to make his methodology less unpalatable to the ontologically committed systematic "leaders" of his day (e.g. Ernst Mayr and George Gaylord Simpson).

about metaphysical views regarding whether or not evolution took place. Hennig was trying to articulate his view, in contrast to that of those he was critiquing, that only certain kinds of similarity are pertinent to the inference of phylogenetic relationships (i.e. synapomorphies, and not symplesiomorphies or convergent features) – distinctions first noticed long before, by Macleay (1819). Going on, Hennig (1966: 12) pointed out, "it is not individuals, but semaphoronts, i.e., individuals in a definite, very brief interval of their life span that must be regarded as the elements of systematics." Again, this is clearly a statement about how characters are observed, not what they "mean" from an ontological perspective. Hennig (1966: 14) concluded his demarcation of phylogenetic from idealistic-morphological, typological, Aristotelian, evolutionary or "adaptiogenetic" systematics:

> The theory of descent, that is the perception that the existing diversity of life on the earth arose historically from an earlier simpler condition, and that the semaphoronts – the elements of all systematic efforts in biology – must be regarded among other things as members of a community of descent, is thus derivable from biological systematics (not "morphological systematics"). This is, in fact, its most important result to date.

Thus, Hennig's argument is that the "theory of descent" is a *result* of phylogenetic systematics, not part of its background knowledge. We will return to this below.

5.5 The inconsistency of parsimony: what would Willi do?

As far as may be determined, the only articulated "justification" that has ever been offered for abandoning parsimony in favor of maximum likelihood or other model-based approaches to phylogenetic inference is the 35-year-old observation that under particular hypothetical circumstances, parsimony might be statistically inconsistent (Fleckenstein, 1978) – that is, as more data are gathered, the most parsimonious tree deviates topologically from the "true tree" with an increasing degree of support or certainty. A commonly invoked version of this phenomenon has come to be known as "long-branch attraction" (cf. Hendy and Penny 1989, Bergsten 2005).[5]

Fig 5.1 shows an unrooted topology for four taxa (the simplest case preferred by modelers such as Huelsenbeck and Hillis (1993), Huelsenbeck (1995), Swofford et al. (2001)). On the left is the "true tree," in which character 3 unites taxa A and

---

[5] Another common argument is that "parsimony" is really a hodgepodge of different character state transformation "models" (e.g. Dollo, Camin-Sokal, etc.), and to draw a sympathetic parallel between these and the multiplicity of alternative models used in statistical approaches (cf. Swofford et al. 1996, Felsenstein 2004). See critique of this stratagem in Schuh and Brower (2009).

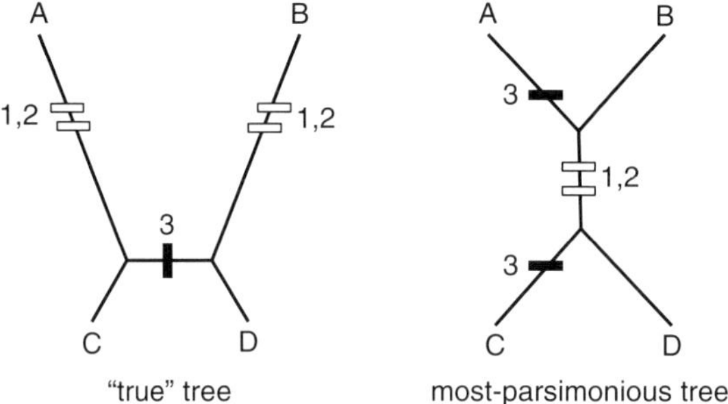

Fig 5.1 Long-branch attraction. See discussion in text.

C (or taxa B and D), and characters 1 and 2 undergo independent transformations on the branches leading to both A and B. The length of this tree is five steps. On the right is the most parsimonious tree for the same data. Here, A and B (or C and D) are united by characters 1 and 2, and character 3 undergoes homoplastic transformations on the branches leading to A and C. This tree is four steps long.

It is immediately apparent that when we speak of the "true tree" we have abandoned the realm of empiricism and have become preoccupied with whether or not the evidence conforms to some independently conceived reality, something which, as we have seen, most philosophers view as outside the realm of science (but which is embraced by systematists who test models with simulations). Given the argument in the previous section that Hennig was an empiricist, it is interesting to examine this four-taxon problem from a "Hennigian" perspective.

As Hennig (1966: 94) said in the epigram at the top of this chapter, "we cannot recognize truth itself and everywhere in science are limited to hypotheses concerning truth." The implication of that statement seems clear: Hennig would not have been concerned with the tree on the left and would select as his preferred hypothesis of relationships the most parsimonious tree on the right. As Hennig's "auxiliary principle" states,

> the presence of apomorphous characters in different species "is always reason for suspecting kinship [i. e., that species belong to a monophyletic group], and that their origin by convergence should not be assumed a priori" (Hennig 1953). This was based on the conviction that "phylogenetic systematics would lose all the ground on which it stands" if the presence of apomorphous characters in different species were considered first of all as convergences (or parallelisms), with proof to the contrary required in each case. Rather the burden of proof must be placed on the contention that "in individual cases the possession of common apomorphous characters may be based only on convergence (or parallelism)." (Hennig 1966: 121–122)

As noted above, Farris (1983) pointed out that Hennig's auxiliary principle is an invocation of parsimony. And as Rieppel (2007a: 180) said, "parsimony in phylogeny reconstruction is coherence-theoretically motivated." Thus, it seems clear that Hennig would have chosen the right-hand tree, unless he had compelling evidence to do otherwise: "where convergences occur or are to be suspected, only the most subtle distinction of the individual characters and the most discriminating evaluation will protect us from false conclusions" (Hennig 1966: 103). In Hennigian systematics, homoplastic characters are identified one at a time, not as classes of evidence (e.g. "third positions" in protein-coding sequence data) that are believed a priori to be collectively misleading.

5.6 The realist philosophy of model-based approaches

> It is regrettable that even the most recent authors present and recommend their own methods but never explain why their method deserves preference over those of their predecessors. (Hennig, 1966: 85)

To provide a contrast to the Hennigian empiricist position outlined above, I will briefly describe the world view of evolutionary taxonomists who prefer deterministic models of evolution. Generally speaking, Maximum Likelihood and Bayesian methods are so complicated and difficult to explain that their advocates merely allude to the inconsistency problem to highlight the weakness of parsimony, and then set about describing the large array of parameter choices, algorithmic alternatives and selection criteria that comprise these methods. The rhetorical trick, established by Felsenstein (1978) but employed by many others, is to say that parsimony could be misleading, and then to bury deep in the statistical details the admission that unless the assumed evolutionary model is "true" that the ML (or Bayesian) tree also might be inconsistent. Thus, the justification of evolutionary models necessary for these methods is dependent upon an assertion that the models are more realistic than the "model" of parsimony (an appeal to a correspondence theory of truth), even though ultimately we have no idea of whether a given model actually corresponds to the true course of evolution or not. This is a fundamental flaw.

Swofford et al. drew a number of sharp distinctions between their own views and those of cladists:

> An alternative position is that parsimony is required as a method of scientific inquiry regardless of any considerations about whether it is more or less likely to recover the true phylogeny than other methods. Some proponents of this view hold that since the truth is essentially unknowable, we should abandon the search for it and simply choose the most parsimonious solution for its own sake. We do not subscribe to this position. Although the true phylogeny may be "unknowable," it can nevertheless be estimated, and we view phylogenetic methods as means to that end rather than an end in themselves. (Swofford et al. 1996: 426, footnote)

This statement conflates the principle of parsimony as a philosophical perspective and parsimony as a systematic method, implies that advocates of parsimony are operationalists (see below), and (apparently unintentionally) poses the conundrum of how we "estimate" the "unknowable" in a realist context. Swofford et al. proceeded to describe many methods that attempt this, but ultimately were stymied: "Farris' (1983, 1986) point that a maximum likelihood method will guarantee consistency only if evolution proceeds according to the assumed model is of course true" (Swofford et al. 1996: 427, footnote); and, "(i)f the model of evolution used to evaluate the likelihood of given trees does not reflect the actual evolutionary processes, then maximum likelihood analyses will be subject to systematic error" (Swofford et al. 1996: 495) ... just like parsimony! (See Farris 1999, 2008, and Schuh and Brower 2009 for further discussion.)

Bayesian phylogenetic analysis, dependent on the same models as priors, fares no better. In addition, most Bayesian analyses assume uninformative priors for topology, which systematists have shown behave unpredictably (Pickett and Randle 2005) and statisticians say "cannot be uncritically accepted" (Efron 2013).

But if model-based efforts to avoid statistical inconsistency are red herrings, then what are these model-based approaches really about? The reinsertion of autapomorphy and complementary symplesiomorphy, in the form of "information about branch lengths," as evidence in phylogenetic inference – in other words, the resurrection of idealistic-morphological, typological, Aristotelian, evolutionary or "adaptiogenetic" systematics based on preconceived notions about how evolution works – exactly the perspective to which Hennig was vehemently opposed.

5.7 What is branch length?

> [O]nly synapomorphy justifies the presumption of monophyly a group of species. (Hennig 1966: 93)
>
> Phylogeneticists group by synapomorphy, others by all similarities, and this distinction has been well understood since Hennig (1966) called attention to it. (Kluge and Farris 1999: 210)[6]

As is well known, the cladistic optimality criterion is minimization of tree length (given a defined set of character and state-transformation weights), which can be unambiguously compared among alternative topologies for a given data set. Because they contribute to tree length, both synapomorphies and autapomorphies are clearly part of this calculus (although cladists do not consider autapomorphies to be relevant as evidence because, as features unique to a single terminal taxon, their contribution to tree length does not vary among alternative topologies). However,

---

[6] But see footnote 2.

even when a given data set yields a single most parsimonious tree, state transformations of homoplastic characters may be hypothesized to have taken place on various alternative branches. As unobservable events with no effect on tree length, these are immaterial to parsimony, but for model-based methods they represent variations in "branch length" that are used to evaluate likelihood.

In order for long-branch attraction to occur, as in the Felsenstein zone example shown in Fig 5.1, there must be a value, "branch length," that may be measured or estimated in some way. The concept is not straightforward and confusion about its precise meaning abounds. Felsenstein (2004: 70) said, "[b]ranch lengths are numbers that are supposed to indicate for a given branch how many changes of state have occurred in the branch" (a measurement). Swofford et al. (1996: 428–429) equated branch length with "amounts of evolutionary time" and Wiley and Lieberman (2011: 212) defined short branch lengths as "relatively short periods of time or slow rates of mutation and fixation." (Note that if branch length were equivalent to time, then sister taxa would always have equal branch lengths. This is false, so branch length cannot equal time.) By contrast, Swofford et al. (1996: 440) also said, "the length of a branch then represents the expected number of substitutions per site along that branch, with no implication as to the actual amount of evolutionary time it represents" (an expectation). Page and Holmes (1998: 186) made this dichotomy explicit: "branch lengths can represent two quite different quantities: the *expected* amount of evolutionary change and the *actual* or realized amount of evolutionary change."

Typically, ML advocates criticize parsimony-based phylogenetic hypotheses for failing to account for differences in branch length and thus ignoring evidence that, when incorporated, they think better corresponds to reality.[7] Under the ML perspective, it seems not only that long branches need to be shortened: "[w]hereas parsimony ignores information on branch lengths when evaluating a tree, maximum likelihood considers that changes are more likely along long branches than short ones, and the estimation of branch lengths is an important component of the method" (Swofford et al. 1996: 429); but also that short branches need to be lengthened: "[w]e must correct the branch lengths that are reconstructed by parsimony to allow additional events" (Felsenstein 2004: 72). It is these corrections by addition of unobserved events and/or discounting of putative homology as homoplasy (contra Hennig's Auxiliary Principle), based on a deterministic a priori model of evolution, that are asserted to be the fix that allows ML to select the "true" tree when parsimony erroneously selects the tree with the minimum number of observed character state transformations (Fig 5.1). "However, estimates of branch lengths can be greatly affected by the choice of model, hence for questions involving rates of evolutionary change, having a good model is very important" (Page and Holmes 1998: 198).

---

[7] It is quite remarkable how this mirrors old criticisms of cladistics by Mayr: "the method ignores the fact that phylogeny has two components, the splitting of evolutionary lines and the subsequent evolutionary changes of the split lines." (Mayr 1982: 230).

Indeed, even with model parameters held constant, branch lengths of taxa may change when other taxa are added or removed from the analysis – the so-called "node density effect" (Sanderson 1990, Hugall and Lee 2007); thus, branch length is a relative concept that depends not only on evolution supposed to have occurred on the branch itself, but also elsewhere in the tree.

The problem of branch length estimation is obviously another empiricist/realist issue: the empiricist takes the evidence (the synapomorphies, at least) at face value; the realist worries that the empirical data might be misleading in one way or another and therefore feels the need to adjust them so they correspond to the truth. As Felsenstein's (2004) book enumerates, there are a great variety of methods and models by which branch lengths may be estimated (see also discussion in Farris 1983). The lack of agreement on what "branch length" is, how it might be estimated or measured, and whether it represents a meaningful measure of any pertinent quantity to begin with (cf. Farris et al. 2001) would certainly appear to represent a challenge for the realist perspective. "And as long as there is no objective and generally binding standard of measurement by which we may measure the 'amount and nature of evolutionary change' or 'magnitude of anagenetic steps' and thereby demonstrate that one of the proposed systems is more correct than the others, there can and will be no generally accepted Aristotelian ('evolutionary') system" (Hennig 1975: 253).

5.8 Parsimony, simplicity, models and coherence

Much has been written (cf. Farris 1982, Sober 1988) about the meaning of "simplicity" and its relationship to the concept of parsimony. A perennial evolutionary taxonomists' objection to cladistic parsimony, that evolution does not necessarily proceed parsimoniously (evolution is not simple) and therefore parsimony is invalid, was refuted by Farris, who noted that this criticism misinterprets cladistic parsimony as a metaphysical statement when it is in fact an epistemological rule: "that reasoning presumes a general connection between minimization and the supposition of minimality, but it is now plain that no such general connection exists" (Farris 1983: 13). Cladistic parsimony does not assume that there is little or no homoplasy, it simply minimizes ad hoc hypotheses of homoplasy for a given data set implied among alternative hypotheses of relationship. Nor does it assert that the most parsimonious tree is "true." Again, the contrast between empiricist and realist views is stark.

It is apparent that, despite likelihoodist claims of the superior realism of "biologically-inspired phylogenetic parameterizations, such as the GTR + Γ model" (Huelsenbeck et al. 2011), the realist appeal is a smokescreen. Felsenstein (2004: 219) himself admitted, for example, that "[t]here is nothing about the gamma distribution that makes it more biologically realistic than any other distribution, such as the lognormal. It is used because of its mathematical tractability." Nevertheless, while realists reject cladistic parsimony as a criterion for tree selection because it could be wrong (while ignoring the fact that their own methods could also be wrong), they still invoke the parsimony principle when selecting models for their ML or Bayesian analyses. Likelihood ratio tests compare the likelihoods of nested sets of models and reject more heavily parameterized models if simpler models explain the data equally well. The Akaike Information Criterion (Akaike 1974) and Bayesian Information Criterion (Schwarz 1978) penalize more complex models so that all things being equal, simpler models are selected (Sullivan and Joyce 2005). These are not "reality-based" criteria, but are instead based on epistemological considerations (Schuh and Brower 2009). Indeed, Sober (2009) considered the entire model selection process to embody an instrumentalist epistemology (see below). Not only that, but alternate model selection criteria do not always concur in their selection of models, and there does not seem to be an agreed upon meta-criterion for model selection alternatives (Ripplinger and Sullivan 2008). Such methodological ambiguities reveal the model-based evolutionary taxonomic endeavor to be entirely indifferent to realist considerations.

In order to assess whether or not the commonly stated claim that "model-based methods outperform parsimony" is true, Rindal and Brower (2011) performed a meta-analysis of 1000 empirical studies of phylogenetic relationships published in the journal *Molecular Phylogenetics and Evolution*. In the empirical world, of course, there is no way to know whether one hypothesis of relationships is more "accurate" (metaphysically true) than another, and so the only resort is a coherence test: how often do the methods give different results, offering a contrast that could be interpreted as one method potentially performing "better" than the other? The answer: of 504 articles comparing parsimony and one or more model-based methods, only two produced strongly supported, incongruent trees by the methods compared. This evidence quite conclusively demonstrated that model-based methods almost never "outperform parsimony" (and conversely, it must be admitted, parsimony almost never "outperforms" model-based methods). The result has since been corroborated (Smith 2013, Brower and Rindal 2013). Thus, whether one considers phylogenetic parsimony to be a "simple model" (cf. Steel 2002, Goloboff 2003) or just employs it because of an empiricist frame of mind (cf. Goloboff and Pol 2005), it is to be preferred on epistemological grounds, as more "big P" parsimonious, as an explanatory approach less dependent on auxiliary hypotheses.

5.9 The Hennigian aesthetic of heuristic accessibility

A further contrast between empiricist and realist approaches to phylogenetic inference is the latter's apparent preoccupation with "sophistication." As we have seen, the increased complexity of model-based methods cannot achieve its stated aim of greater conformity to reality, nor from a coherentist perspective does it appear to produce results that differ appreciably from those of parsimony. Thus, I can only interpret the widespread fascination with these methods as a Kuhnian sociological phenomenon in which novelty and complexity themselves are viewed as virtues (cf. Anisimova et al. 2013). In this light, the rise of model-based phylogenetic methods could be considered as an example of disruptive innovation (Christenson 1997) in which newer technologies replace older ones, with or without a rational basis. In such a narrative, it might be argued that parsimony analyses often result in multiple equally parsimonious trees (bad), while ML trees are usually fully resolved (good). Bayesian analyses can assess larger datasets in shorter times than ML analyses. But these ostensibly practical advantages are somewhat illusory: sometimes the evidence is not sufficient to provide resolution, and an unresolved parsimony tree may simply be illustrating that fact; modern parsimony algorithms (e.g. TNT; Goloboff et al. 2008) are at least as fast as most model-based programs. Critics of disruptive innovation (e.g. Lepore 2014) have pointed out that the phenomenon often replaces a superior product with an inferior one: consider the sound quality of 8-track tapes compared to LPs, or MP3s to CDs. Enthusiasm for novelty and complexity is not, in itself, a philosophical justification for discarding one set of theories for another.

In contrast to this technophilic exuberance, Hennig (1965: 97) said, "[I]nvestigation of the phylogenetic relationship between all existing species and the expression of the results of this research in a form which cannot be misunderstood, is the task of phylogenetic systematics." This would seem to express a rather different sentiment: that clarity, rather than sophistication, is the desideratum. As Farris (1986: 26) said, "Hennig brought phylogenetic systematics to preeminence by his unwavering insistence that systematics be based on clear – and clearly stated – scientific principles." The heuristic accessibility of cladistics may be understood as a sort of aesthetic parsimony, of the same sort that mathematicians strive for when they seek elegant proofs. In this vein we may appreciate that a cladogram is a graphical representation of implied character state transformations among terminal taxa, and conversely, that characters mapped on a cladogram may be directly converted back into a data matrix. Try that with a Bayesian tree! Indeed, with "sophisticated" model-based approaches, one wonders, sensu Baum et al. (2005), if even the "practicing systematists" actually know how their phylogenies are inferred. In Riepellian terms (e.g. Kearney and Rieppel 2006; see below), we might consider the inter-translatability of cladograms and data matrices to be a secondary level of "causal grounding," inasmuch as the evidence underlying the phylogenetic hypothesis is scrutable from the results of the analysis.

5.10 -isms

Kluge (2001: 203) said that the goal of the "philosophy of operationalism … is to maintain a logical gap between theory and evidence (Hull 1968)"[8] and,

> Brower's [2000b] point is that operational parsimony is justified because it makes no background assumptions, whereas phylogenetic parsimony cannot be justified that well, because it assumes "descent, with modification." This leaves unanswered what guiding theory is used to collect data, and, presuming that there must be one, the results of an analysis of those data cannot be theory free. (Kluge 2001: 203)

Yet Kluge had previously argued the opposite (Siddall and Kluge 1997: 320): "if we discover tomorrow that all life is the product of product only of special creation, we can still do cladistics, operationally, in terms of summarizing observed character generalities." Furthermore, Brower (2000b) said neither that the pattern cladistic approach he advocated "makes no background assumptions": "[s]ome background knowledge is required to justify the raw data, its tabulation into a matrix, the hierarchical pattern, and the grouping algorithm" (Brower 2000b: 146), nor that the cladistic method is "theory free": "[a]n observation is an existential statement that relies on theory" (Brower 2000b: 146). Nevertheless, Kluge's (2001) criticisms of Brower (2000b) provides a good starting point to discuss ideas underlying the variety of labels applied by some authors to ideas they don't like and to the people who espouse them.

No thoughtful systematist believes that observations are theory free, but some assume more or different theories than others (Schuh and Brower 2009, Wilkins and Ebach 2014). Sober (2008: 131–132) made a useful distinction between statements of "absolute" and "relative theory-neutrality," as follows:

> (EA) There exists a set of observation statements that presuppose no theories whatever, and these can be used to evaluate any theories we wish to consider.
>
> (AE) For any set of competing theories, there exists a set of observation statements that presuppose none of the theories under test, and these can be used to evaluate those theories.
>
> The statement (EA) characterizes *absolute* theory-neutrality, while (AE) defines *relative* theory-neutrality. The claim that all statements are theory laden impugns (EA), but leaves (AE) untouched, (AE) expresses the important point that observation

---

[8] This is not the original operationalism described by Bridgeman (1927: 5): "we mean by any concept nothing more than a set of operations; the concept is synonymous with the corresponding set of operations." Bridgeman was interested in empirical measurement of physical variables such as extreme pressure and temperature, in circumstances outside the range within which traditional instruments functioned, and his philosophical writings investigated the epistemology of such concepts. See Chang (2009) for an accessible historical summary.

> statements need to be *epistemically independent* of the hypothesis they are used
> to test. [Italics in original.]

A caricature operationalist sensu Hull (1968) or Kluge (2001) might hold an (EA) point of view, but I think pattern cladists' views are more like (AE). This seems to be in agreement both with Brady's (1985) argument for the independence of systematics and Brower's (2000b) perspective that evolution need not be an assumption of cladistics, and to directly contradict Kluge's (1997, 2001) repeated assertions that the strong ontological claim of "descent, with modification" is necessary background knowledge for "phylogenetic parsimony."[9]

Related to operationalism is a philosophical perspective called instrumentalism, originally conceived by medieval philosophers as a way to avoid being burned at the stake for proposing naturalistic explanations for phenomena, in defiance of church doctrine. Instrumentalists are coherentists who hold that scientific hypotheses are tested and corroborated or refuted in a more-or-less closed system that is not concerned with "truth" (Blackburn 1994). Rieppel (2007a, b) labeled a variety of systematic practices as instrumentalist: total evidence, DNA barcoding, dynamic homology, genomics and even the cladistic parsimony criterion itself, in which individual hypotheses of homology are tested by character congruence. In his view (see also Kearney and Rieppel 2006), each of these endeavors to a greater or lesser degree suffers from a lack of appropriate consideration for character coding and causal grounding of homology statements.[10]

Does the need for this sort of causal grounding imply that a realist/correspondence viewpoint is ultimately necessary with regards to inference of the true course of phylogenetic diversification, and does it require an assumption of "descent, with modification"? In short, I think that if we are able to resort back to the sort of argumentation implied by Popper's pilings, quoted above, then no, or at least not to a foundationalist sort of rock bottom. Obviously, as Hennig (1966) argued, ontogeny and tokogeny are observable phenomena, so the "descent" aspect is easily encompassed within the empirical sphere. The "modification" side is less immediately accessible, and Hennig relied on the argument of "reciprocal illumination" to avoid the accusation of a *petitio principii*:

> The empirico-critically constructed 'natural system,' as an encaptic system of graduated diversity of form, served as proof for the theory of descent. And only the

---

[9] Of course, Sober (1988) has also argued that cladistic parsimony does make specific, if hazily articulated, assumptions about evolution. But then he is not a cladist (or even an empirical scientist), and he is also a tireless advocate of an argument he calls the "likelihood justification for parsimony."

[10] At the same time, because of the overwhelming amount of data that pertains to systematic problems, Rieppel (2005d: 148) has acknowledged that there seems "to be no choice, but to accept at least some degree if instrumentalism in phylogeny reconstruction, and this is another reason for being frugal with ontological commitments."

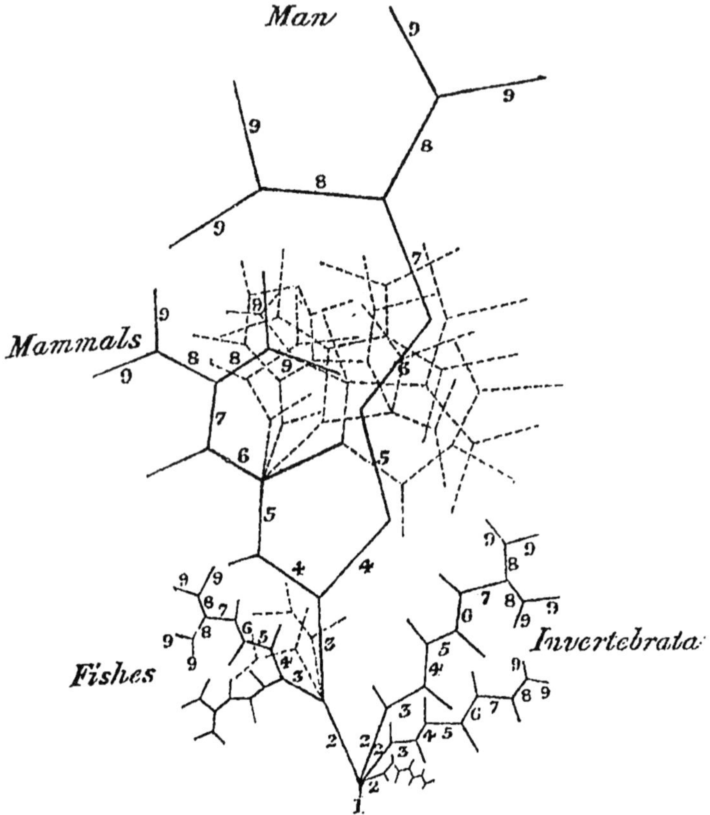

Fig 5.2 A hierarchical representation of the developmental diversification patterns of animals from Barry (1837; image in public domain). As Amundson (2005: 70) said, "it is impossible for a modern reader to resist seeing this as a phylogenetic tree."

theory of descent, in turn, uses scientific, empirico-critically based ideas to explain the encaptic diversity of form expressed by the 'natural system.' (Hennig 1966: 22, quoting Günther 1956)

A fortiori, history shows that an assumption of material evolutionary modification is not needed to perceive a pattern of groups nested within groups based on empirical evidence: von Baer's (1828) Law implies a hierarchy of ontogenetic diversification that is strongly suggestive of a cladogram (Fig 5.2), and nested taxonomic groups were scrutable to Linnaeus (1758), among others, from study of the hierarchical patterns of characters they exhibited. It is clear from pre-Darwinian endeavors such as these that knowledge of or credence in the material modification of parts by evolution is unnecessary to conceive and elaborate a fairly detailed and not entirely fanciful "natural system." It has been suggested (Brady 1985) that the natural hierarchy is an empirical fact, or an epistemological axiom (Brower 2000b).[11]

As we have seen, such considerations do not contradict Hennig's (1966: 11) denial that "morphological systematics" had logical precedence over "phylogenetic systematics," but rather support his position that the theory of descent is a result of the latter. Thus, I am sticking to my guns that "descent, with modification" as a material process explains the results of systematics and is not part of its background knowledge (Brower 2000b, Brower and de Pinna 2014). Indeed, I consider the exclusion of "descent with modification" and concomitant concern with particular processes and amounts of evolution as background knowledge to be another fundamental aspect of the demarcation of Hennigian cladistics from evolutionary taxonomy. Remember: Hennig, like Darwin before him, saw the theory of evolution as a result of empirical systematic inference, and an explanation of the natural system.

A further realist criticism of the empiricist perspective advocated in this chapter is the unfortunately frequent insinuation that if a systematist does not think cladistics requires evolutionary assumptions then s/he does not believe in evolution, and/or is a creationist (e.g. Farris 2011, 2012, 2014, Nixon and Carpenter 2012a). There once were and may still be systematists who do not believe in evolution, but there are others (and if the argument if this chapter is correct, we may count Willi Hennig among them) who believe that their concept of evolution stems in large part from the results of systematics – the explanans/explanandum distinction pointed out by Brady (1985). The problem, if there is any substance to it at all, seems to turn upon how empiricist/coherentist systematists achieve causal grounding for their phylogenetic schemes: "If one does not have a question or hypothesis, this is not science, it is simple description, and it is uncertain what is being described" (Nixon and Carpenter 2012b). This seems a bit harsh: were Linnaeus, Cuvier and Owen not doing "science" in their comparative anatomical investigations of the Natural System because they did not "believe in evolution"? Or are they given a bye because they had a "hypothesis" (relationships among living things represent the Plan of Creation as a manifestation of the wisdom of God), even though we recognize now that it was metaphysical nonsense? Clearly, discovering the phenomenon of relationship and then explaining the causes of relationship are two different things. The former is the realm of systematics, the latter is the realm of evolutionary biology.

I might also observe that a "question or hypothesis" is not the same thing as an assumption of background knowledge. The former are then subject of test within the investigation at hand, while the latter is not. You can believe that having a rabbit's foot in your pocket will bring you luck (an assumption), or you can do an experiment

---

[11] Some iconoclastic provocateurs have suggested that "the Tree" is a poor or false metaphor/model, and that life has evolved as more of a network (e.g. Doolittle 2009, Morrison 2014). That is a paradigm shift I am not ready to accept.

to test whether you are luckier with or without a rabbit's foot in your pocket. It might not be quite so easy to corroborate or falsify the hypothesis that "descent with modification" caused the diversity of some group of taxa (Brower 2002), which is why Popper (1974) called Darwinism a "metaphysical research programme."

Nor does the privileged role of parsimony as a phylogenetic method have anything to do with evolution, for as argued above and recognized by Hennig and Farris, parsimony is an epistemological rule, not a claim about how the world is. Indeed, it is the realist paranoia that the evolution is not parsimonious that opens the Pandora's box of "more realistic" models that is the main target of this chapter.

5.11 Ragnarøkkr

> There is agreement among the editors that parsimony alone is not sufficient and that the dataset should be subjected to partitioned likelihood analyses as well.[12]

There are two narratives about the development of systematics. The received history is a linear, progressive transformation series of disruptive innovations, from a Simpsonian "art," to phenetics, to cladistics, to "modern statistical approaches" – each representing a technological advance upon the preceding one, and with "progress" driven largely by computers (cf. Baum and Smith 2013). Incongruously, although there is a strong realist undercurrent to this worldview, the post-cladistic methodological fashions seem to evolve purely by pragmatism, or attraction to novelty driven by a combination of poor training and technophilia. Unfortunately, the wishful admonitions of Mishler, Wiley, Lieberman and Wheeler quoted in the introduction appeal to this historical view.

The alternative history, argued here, is that there was a philosophical fork in the road 50 years ago, when Hennig and subsequent cladists recognized that only shared, derived character states represent evidence of phylogenetic grouping. Pheneticists did not recognize this then, and evolutionary taxonomists do not recognize it now. Cladists have a coherent and epistemologically defensible disciplinary philosophy; modelers do not, and whatever "philosophy" they may hold, if they have any at all, is antithetical to ours. Thus, not all practitioners of phylogenetics are Hennigian cladists.

A final observation is that, sadly, those who actually are cladists seem to spend more time disagreeing among ourselves than we do combatting the more pressing issue of the metastasis of evolutionary taxonomy in the form of ML, Bayesian phylogenetics and phylogenomics that is currently overwhelming twenty-first-century systematics. Perhaps the philosophical gulf between "us" and "them" is so vast that it seems futile

---

[12] From a December 2013 letter by an editor of a systematic entomology journal, summarily rejecting a manuscript containing a cladistic analysis of molecular and morphological data for a group of butterflies.

to try to engage the modelers: even the most potent rhetoric is wasted on someone who does not understand or is not willing to listen to what you are saying. But, as the epigram at the head of this section exemplifies, a lot of people in influential positions now view cladistics as an obsolete or "not sufficient" phylogenetic approach.

Brady (1985: 117) said, "the province of science is not a democracy." Perhaps, but it is a social realm, and research paradigms rise and fall because of the successful communication of ideas to the rest of the world (Kuhn 1970). Despite its manifest merits, cladistics has a public relations problem, and it seems to me that cladists' efforts are better devoted to proselytizing unbelievers than to excommunicating our fellows for relatively insignificant heresies. If cladists want to preserve Willi Hennig's legacy and to ensure that systematics survives as a coherent and intelligible empirical discipline, we need to work together to present a united front against these assaults on reason, and to not grind our axes to the point where they no longer have the mass to do their work.

Acknowledgments

I would like to thank my friend and colleague David Williams for inviting me to participate in the Hennig Centennial Symposium at the Linnean Society, and to contribute to this book. We disagree about some things, but we agree about many more. I thank him, Malte Ebach, Ivonne Garzón-Orduña, Gareth Nelson, and Jon Todd for comments on the manuscript. Funds to support my participation were provided by a collaborative grant, *Dimensions US-Biota-São Paulo: Assembly and evolution of the Amazon biota and its environment: an integrated approach*, supported by the US National Science Foundation (NSF DEB 1241056), National Aeronautics and Space Administration (NASA), and the Fundação de Amparo à Pesquisa do Estado de São Paulo (FAPESP Grant 2012/50260–6), by the MTSU Biology Department and Molecular Biosciences PhD program, the Systematics Association and the Linnean Society of London. I dedicate this chapter to everyone who calls him or herself a cladist, and particularly to those brave souls who have admitted to being a "pattern cladist."

This chapter is an expanded version of a talk presented at the celebration of Willi Hennig's centennial at the Linnean Society, November 27, 2013.

References

Akaiki, H. (1974). Information theory as an extension of the maximum likelihood principle. In *Second International Symposium on Information Theory*, ed. B. Petrov and F. Csaki. Budapest: Akademiai Kiado, pp. 267–281.

Amundson, R. (2005). *The Changing Role of the Embryo in Evolutionary Thought: Roots of Evo-Devo*. Cambridge: Cambridge University Press.

Anisimova, M., Liberles, D.A., Philippe, H., et al. (2013). State of the art methodologies dictate new standards for phylogenetic analysis. *BMC Evolutionary Biology*, 13, 161.

Assis, L.C.S. (2009). Coherence, correspondence, and the renaissance of morphology in phylogenetic systematics. *Cladistics*, 25, 528–544.

Barry, M. (1837). Further observations on the unity of structure in the animal kingdom, and on congenital anomalies, including "hermaphrodites"; with some remarks on embryology, as facilitating animal nomenclature, classification, and the study of comparative anatomy. *Edinburgh New Philosophical Journal*, 22, 345–364.

Baum, D.A. and Smith, S.D. (2013). *Tree Thinking: an Introduction to Phylogenetic Biology*. Greenwood Village, CO: Roberts and Co.

Baum, D.A., Smith, S.D. and Donovan, S.S.S. (2005). The tree-thinking challenge. *Science*, 310, 979–980.

Bergsten, J. (2005). A review of long-branch attraction. *Cladistics*, 21, 163–193.

Bicheno, J.E. (1827). On systems and methods. *Transactions of the Linnean Society of London*, 15, 479–496.

Blackburn, S. (1994). *The Oxford Dictionary of Philosophy*. Oxford: Oxford University Press.

Brady, R.H. (1985). On the independence of systematics. *Cladistics*, 1, 113–126.

Bridgman, P.W. (1927). *The Logic of Modern Physics*. New York: Macmillan.

Brower, A.V.Z. (2000a). Homology and the inference of systematic relationships: some historical and philosophical perspectives. In *Homology and Systematics: Coding Characters for Phylogenetic Analysis*, ed. R.W. Scotland and R.T. Pennington. London: Taylor and Francis, pp. 10–21.

Brower, A.V.Z. (2000b). Evolution is not an assumption of cladistics. *Cladistics*, 16, 143–154.

Brower, A.V.Z. (2002). Cladistics, phylogeny, evidence and explanation – a reply to Lee. *Zoologica Scripta*, 31, 221–223.

Brower, A.V.Z. and de Pinna, M.C.C. (2012). Homology and errors. *Cladistics*, 28, 529–538.

Brower, A.V.Z. and de Pinna, M.C.C. (2014). About nothing. *Cladistics*, 30, 330–336.

Brower, A.V.Z. and Rindal, E. (2013). Reality check: a reply to Smith. *Cladistics*, 29, 464–465.

Chang, H. (2009). Operationalism. In *The Stanford Encyclopedia of Philosophy* (Fall 2009 Edition), ed. E.N. Zalta, http://plato.stanford.edu/archives/fall2009/entries/operationalism/.

Christenson, C.M. (1997). *The Innovator's Dilemma*. Boston, MA: Harvard Business Press.

Darwin, C. (1859). *On the Origin of Species*. London: John Murray.

Doolittle, W.F. (2009). The practice of classification and the theory of evolution, and what the demise of Charles Darwin's tree of life hypothesis means for both of them. *Philosophical Transactions- Royal Society of London, Series B Biological Sciences*, 364, 2221–2228.

Efron, B. (2013). Bayes' Theorem in the 21st Century. *Science*, 340, 1177–1178.

Farris, J.S. (1969). A successive approximations approach to character weighting. *Systematic Zoology*, 18, 374–385.

Farris, J.S. (1970). Methods for computing Wagner trees. *Systematic Zoology*, 19, 83–92.

Farris, J.S. (1982). Simplicity and informativeness in systematics and phylogeny. *Systematic Zoology*, 31, 413–444.

Farris, J.S. (1983). The logical basis of phylogenetic analysis. In *Advances in Cladistics*, 2, ed. N.I. Platnick and V.A.

Funk. New York: Columbia University Press, pp. 7–36.

Farris, J.S. (1986). On the boundaries of phylogenetic systematics. *Cladistics,* 2, 14–27.

Farris, J.S. (1999). Likelihood and inconsistency. *Cladistics,* 15, 199–204.

Farris, J.S. (2008). Parsimony and explanatory power. *Cladistics,* 24, 825–847.

Farris, J.S. (2011). Systemic foundering. *Cladistics,* 27, 207–221.

Farris, J.S. (2012). Counterfeit cladistics. *Cladistics,* 28, 227–228.

Farris, J.S. (2014). "Taxic homology" is neither. *Cladistics,* 30, 113–115.

Farris, J.S., Källersjö, M. and De Laet, J.E. (2001). Branch lengths do not indicate support – even in maximum likelihood. *Cladistics,* 17, 298–299.

Felsenstein, J. (1978). Cases in which parsimony or compatibility methods will be positively misleading. *Systematic Zoology,* 27, 401–410.

Felsenstein, J. (2004). *Inferring Phylogenies.* Sunderland, MA: Sinauer Associates.

Futuyma, D.J. (1986). *Evolutionary Biology.* Sunderland, MA: Sinauer Associates.

Ghiselin, M.T. (1997). *Metaphysics and the Origin of Species.* Albany, NY: State University of New York Press.

Goloboff, P.A. (2003). Parsimony, likelihood, and simplicity. *Cladistics,* 19, 91–103.

Goloboff, P.A., Farris, J.S. and Nixon, K.C. (2008). TNT a free program for phylogenetic analysis. *Cladistics,* 24, 774–786.

Goloboff, P.A. and Pol, D. (2005). Parsimony and Bayesian phylogenetics. In *Parsimony, Phylogeny, and Genomics,* ed. V.A. Albert. Oxford: Oxford University Press, pp. 148–159.

Günther, K. (1956). Systematik und Stammesgeschichte der Tiere 1939–1953. *Fortschrift Zoology, (N. F.)* 10, 33–278.

Haas, O. and Simpson, G.G. (1946). Analysis of some phylogenetic terms, with attempts at redefinition. *Proceedings of the American Philosophical Society,* 90, 319–349.

Hendy, M.D. and Penny, D. (1989). A framework for the quantitative study of evolutionary trees. *Systematic Zoology,* 38, 297–309.

Hennig, W. (1953). Kritische Bemerkungen zum phylogenetischen System der Insekten. *Beiträge zur Entomologie,* 3, 1–85.

Hennig, W. (1965). Phylogenetic systematics. *Annual Reviews of Entomology,* 10, 97–116.

Hennig, W. (1966). *Phylogenetic Systematics.* Urbana, IL: University of Illinois Press.

Hennig, W. (1975). "Cladistic analysis or cladistic classification?": a reply to Ernst Mayr. *Systematic Zoology,* 24, 244–256.

Huelsenbeck, J.P. (1995). Performance of phylogenetic methods in simulation. *Systematic Biology,* 44, 17–48.

Huelsenbeck, J.P., Alfaro, M.E. and Suchard, M.A. (2011). Biologically inspired phylogenetic models strongly outperform the no common mechanism model. *Systematic Biology,* 60, 225–233.

Huelsenbeck, J.P. and Hillis, D.M. (1993). Success of phylogenetic methods in the four-taxon case. *Systematic Biology,* 42, 247–264.

Hugall, A.F. and Lee, M.S.Y. (2007). The likelihood node density effect and consequences for evolutionary studies of molecular evolution. *Evolution,* 61, 2293–2307.

Hull, D.L. (1968). The operational imperative – sense and nonsense in operationalism. *Systematic Zoology,* 17, 438–457.

Kearney, M. and Rieppel, O. (2006). Rejecting the "given" in systematics. *Cladistics*, 22, 369–377.

Kluge, A.G. (1976). Phylogenetic relationships in the lizard family Pygopodidae: an evaluation of theory, method and data. *Miscellaneous Publications of the Museum of Zoology, University of Michigan*, 152, 1–72.

Kluge, A.G. (1997). Testability and the refutation and corroboration of cladistic hypotheses. *Cladistics*, 13, 81–96.

Kluge, A.G. (2001). Parsimony with and without scientific justification. *Cladistics*, 17, 199–210.

Kluge, A.G. and Farris, J.S. (1999). Taxic homology = overall similarity. *Cladistics*, 15, 205–212.

Kuhn, T.S. (1970). *The Structure of Scientific Revolutions*. Chicago, IL: University of Chicago Press.

Lee, M.S.Y. (2002). Divergent evolution, hierarchy, and cladistics. *Zoologica Scripta*, 31, 217–219.

Lepore, J. (2014). The disruption machine. *The New Yorker*, 23 June.

Linnaeus, C. (1758). *Systema Naturae*, (10th edition, facsimile reprint, 1956). London: British Museum (Natural History).

MacLeay, W.S. (1819). *Horae Entomologicae: or Essays on the Annulose Animals*. London: S. Bagster.

MacLeay, W.S. (1830). On the dying struggle of the dichotomous system. *Philosophical Magazine*, 7, 431–445; 8, 53–57, 134–140, 200–207.

Mayr, E. (1982). *The Growth of Biological Thought*. Cambridge, MA: Belknap Press.

Mishler, B.D. (2009). Three centuries of paradigm changes in biological classification: Is the end in sight? *Taxon*, 58, 61–67.

Morrison, D.A. (2014). Is the Tree of Life the best metaphor, model, or heuristic for phylogenetics? *Systematic Biology*, 63, 628–638.

Nixon, K.C. and Carpenter, J.M. (2012a). On homology. *Cladistics*, 28, 160–169.

Nixon, K.C. and Carpenter, J.M. (2012b). More on errors. *Cladistics*, 28, 539–544.

Oldroyd, D. (1986). *The Arch of Knowledge: an Introductory Study of the History of the Philosophy and Methodology of Science*. New York: Methuen.

Page, R.D.M. and Holmes, E.C. (1998). *Molecular Evolution: A Phylogenetic Approach*. Oxford: Blackwell Science.

Pickett, K.M. and Randle, C.P. (2005). Strange Bayes indeed: uniform topological priors imply non-uniform clade priors. *Molecular Phylogenetics and Evolution*, 34, 203–211.

Platnick, N.I. (1979). Philosophy and the transformation of cladistics. *Systematic Zoology*, 28, 537–546.

Popper, K.R. (1965). *Conjectures and Refutations: the Growth of Scientific Knowledge*. New York: Harper Torchbooks (1968 reissue).

Popper, K.R. (1974). Darwinism as a metaphysical research programme. In *The philosophy of Karl Popper*, ed. P.A. Schilpp. La Salle, IL: Open Court, pp. 133–143.

Popper, K.R. (1979). *Objective Knowledge: An Evolutionary Approach*. Oxford: Clarendon Press.

Popper, K.R. (1983). *Realism and the Aim of Science*. New York: Routledge.

Rieppel, O. (2005a). The philosophy of total evidence and its relevance for phylogenetic influence. *Papeis Avulsos de Zoologia*, 45, 77–89.

Rieppel, O. (2005b). A skeptical look at justification. *Cladistics*, 21, 203–207.

Rieppel, O. (2005c). Monophyly, paraphyly, and natural kinds. *Biology and Philosophy*, 20, 465–487.

Rieppel, O. (2005d). A note on reality, ontology, and pattern cladism. *Neues Jahrbuch für Geologie und Palaontologie-Monatshefte*, 3, 142–150.

Rieppel, O. (2007a). The nature of parsimony and instrumentalism in systematics. *Journal of Zoological Systematics and Evolutionary Research*, 45, 177–183.

Rieppel, O. (2007b). Parsimony, likelihood, and instrumentalism in systematics. *Biology and Philosophy*, 22, 141–144.

Rieppel, O. and Kearney, M. (2007). The poverty of taxonomic characters. *Biology and Philosophy*, 22, 95–113.

Rindal, E. and Brower, A.V.Z. (2011). Do model-based phylogenetic analyses perform better than parsimony? A test with empirical data. *Cladistics*, 27, 331–334.

Ripplinger, J. and Sullivan, J. (2008). Does choice in model selection affect maximum likelihood analysis? *Systematic Biology*, 57, 76–85.

Sanderson, M.J. (1990). Estimating rates of speciation and evolution: a bias due to homoplasy. *Cladistics*, 6, 387–391.

Schlee, D. (1975). Numerical phyletics: an analysis from the viewpoint of phylogenetic systematics. *Entomologica Scandinavica*, 6, 193–208.

Schuh, R.T. and Brower, A.V.Z. (2009). *Biological Systematics: Principles and Applications*, 2nd edition. Ithaca, NY: Cornell University Press.

Schwarz, G. (1978). Estimating the dimensions of a model. *Annals of Statistics*, 6, 461–464.

Siddall, M.E. and Kluge, A.G. (1997). Probabilism and phylogenetic inference. *Cladistics*, 13, 313–336.

Smith, M.R. (2013). Likelihood and parsimony diverge at high taxonomic levels. *Cladistics*, 29, 463.

Sneath, P.H.A. and Sokal, R.R. (1973). *Numerical Taxonomy*. San Francisco: W.H. Freeman.

Sober, E. (1988). *Reconstructing the Past: Parsimony, Evolution and Inference*. Cambridge, MA: MIT Press.

Sober, E. (2008). Empiricism. In *The Routledge Companion to the Philosophy of Science*, ed. S. Psillos and M. Curd. London: Routledge, pp. 129–138.

Sober, E. (2009). Parsimony and models of animal minds. In *The Philosophy of Animal Minds*, ed. R. Lurz. Cambridge: Cambridge University Press, pp. 237–257.

Steel, M. (2002). Some statistical aspects of the maximum parsimony method. In *Molecular Systematics and Evolution: Theory and Practice*, ed. R. DeSalle, G. Giribet and W. Wheeler. Basel, Switzerland: Birkhäuser Verlag, pp. 125–139.

Strickland, H.E. (1841). On the true method of discovering the natural system in zoology and botany. *Annals and Magazine of Natural History, (1st ser.)* 6, 184–194.

Sullivan, J. and Joyce, P. (2005). Model selection in phylogenetics. *Annual Reviews in Ecology and Systematics*, 36, 445–466.

Swofford, D.L., Olsen, G.J., Waddell, P.J. and Hillis, D.M. (1996). Phylogenetic inference. In *Molecular Systematics*, ed. D.M. Hillis, B.K. Mable and C. Moritz. Sunderland, MA: Sinauer Associates, pp. 407–514.

Swofford, D.L., Waddell, P.J., Huelsenbeck, J.P., et al. (2001). Bias in phylogenetic estimation and its relevance to choice between parsimony and likelihood methods. *Systematic Biology*, 50, 525–539.

Van Fraassen, B.C. (1980). *The Scientific Image*. New York: Oxford University Press.

Vergara-Silva, F. (2009). Pattern cladistics and the 'realism-antirealism debate' in the philosophy of biology. *Acta Biotheoretica*, 57, 269–294.

Von Baer, K.E. (1828). *Über Entwickelungsgeschiichte der Thiere: Beobachtung und Reflexion.* Königsberg: Börntrager.

Wägele, J.-W. (2005). *Foundations of Phylogenetic Systematics.* Munich: Verlag Dr. Friedrich Pfeil.

Wheeler, Q.D. (1995). The "old systematics": classification and phylogeny. In *Biology, Phylogeny and Classification of Coleoptera: Papers Celebrating the 80th Birthday of Roy A. Crowson*, ed. J. Pakaluk and S.A. Slipinski. Warszawa: Muzeum i Instytut Zoologii PAN, pp. 31–62.

Wheeler, W.C. (2012). *Systematics: A Course of Lectures.* Chichester: Wiley-Blackwell.

Wiley, E.O. and Lieberman, B.S. (2011). *Phylogenetics: Theory and Practice of Phylogenetic Systematics.* Hoboken, NJ: John A. Wiley and Sons.

Wilkins, J.S. and Ebach, M.C. (2014). *The Nature of Classification: Relationships and Kinds in the Natural Sciences.* Basingstoke, New York: Palgrave Macmillan.

Williams, D.M., Ebach, M.C. and Wheeler, Q.D. (2010). Beyond belief: the steady resurrection of phenetics. In *Beyond Cladistics: the Branching of a Paradigm*, ed. D.M. Williams and S. Knapp. Berkeley, CA: University of California Press, pp. 169–195.

Winsor, M.P. (2009). Taxonomy was the foundation of Darwin's evolution. *Taxon*, 58, 43–49.

6

How much of Hennig is in present day cladistics?

MICHAEL SCHMITT

6.1 The 'father of cladistics'

The publication of Willi Hennig's *Grundzüge einer Theorie der phylogenetischen Systematik* (*Fundamentals of a Theory of Phylogenetic Systematics*) in 1950 marks the beginning of the so-called Hennigian (Dupuis 1990, Mishler 2000, Wheeler 2008) or Cladistic Revolution (Engel and Kristensen 2013). Naturally, Hennig's approach did not remain as it was originally conceived, with Hennig himself, as well as others, correcting, adding and modifying this approach in the years following this first publication. Hennig only participated in the discussion until 1976, the year he died, so it is no longer certain that subsequent modifications of phylogenetics reflect his original ideas. It is possible that some cladistic paths have deviated from Hennig's 'Phylogenetic Systematics' to such a considerable degree that perhaps they even contradict his key positions, as previously noted by Hull (1989: 12: 'Thus it comes as a surprise to discover Platnick …, declaring that anyone who 'argues for the primacy of evolutionary theorizing over systematics has the cart before the horse both historically and logically' …, the very view that Hennig … dismissed as 'durchaus irrig' [completely wrong]').

The Future of Phylogenetic Systematics: The Legacy of Willi Hennig, eds. D. Williams, M. Schmitt and Q. Wheeler. Published by Cambridge University Press.
© The Systematics Association 2016.

In this chapter, I aim to reveal how accurately modern cladistics follows Willi Hennig's original intentions and where his original claims are possibly distorted or ignored.

6.2 Sources for comparison

In his banquet speech to the 32nd Willi Hennig Society meeting on August 6, 2013, held in Rostock, Rudolf Meier posed the question: 'What would Willi Hennig say?' It is certainly an interesting question but only in certain cases possible to answer. To compile a picture of Hennig's way of thinking and of his arguments as complete as possible, I used his publications, mainly his detailed books of 1950, 1966, and 1984. In addition, I have based my decisions on numerous letters by him, kept in the archive of the Staatliches Museum für Naturkunde, Stuttgart (SMNS), by his sons, and those in the bequest of Klaus Günther (now held in the Staatsbibliothek Berlin).

What 'present-day cladistics' or 'modern cladistics' means is similarly difficult to outline. Whoever looks at the history of post-Hennigian cladistics must realise that there are different lines of reasoning, different schools, even heavily fighting camps. Carpenter gave an entertaining account of the 'cladistics of cladists' (Carpenter 1987), which should not hide the profound theoretical differences between several of the 'modern cladists.' Ebach et al. (2008) presented a more elaborated 'new cladistics of cladists.' Although I have some reservation as to the inclusion of pre-Hennigian authors like Adolf Naef and Johann Wolfgang von Goethe in the 'coterie of cladists,' I refer to this chapter for an overview of the most prominent branches of present-day cladistics.

I tried to find a way to condense different cladistic approaches without ignoring or concealing the differences by using some major publications: Farris (1983), Kitching et al. (1998), Wägele (2000), Williams and Forey (2004), Schuh and Brower (2009), Wiley and Lieberman (2011), and the numerous articles published in the journal *Cladistics*, especially those in the discussion forum 'Letters to the editor.' From these sources, I take the statements on the respective aspects, use them as they were in case they concur with each other or treat them separately when they differ.

6.3 Aspects compared

The selection of aspects of Willi Hennig's 'Phylogenetic Systematics' I compare to the respective parts of 'modern cladistics' is, of course, arbitrary to a certain degree. I used Hennig's 'self-review' of 1952 as a guideline to infer what he had regarded most important of his new approach. In this peculiar paper, Willi Hennig reviewed his own 1950 book. Thus, he could freely emphasise what he regarded important and omit what he felt was less relevant.

6.4 Concept of 'relationship'

Hennig left no doubt that 'relationship' in his terminology exclusively means 'genealogical relationship.' This is exactly the wording in Kitching et al. (1998: 2) and Wiley and Lieberman (2011: 11). Farris (1983) also implicitly applied this concept of 'relationship.' Consequently, we can conclude that the usage of 'phylogenetic relationship' in modern cladistics entirely matches that of Hennig. Nevertheless, there is a caveat. Where Hennig uses the phrase 'phylogenetische Beziehungen' in the manuscript that D. Dwight Davis and Rainer Zangerl translated (e.g. Hennig 1982: 26), it is translated as 'phylogenetic relationships' in Hennig (1966: 20). However, Hennig also wrote 'phylogenetische Verwandtschaft' (e.g. Hennig 1982: 27 and 78), which likewise is translated as 'phylogenetic relationship' in Hennig (1966: 20 and 74) where a more appropriate translation would be 'relatedness.' The difference might not appear dramatic but could be responsible for a hidden difference in attitude among traditional (i.e. mostly German-speaking) phylogeneticists and modern cladists as in German 'phylogenetische Beziehung' always means 'phylogenetic relatedness.' For native English speakers, 'phylogenetic relationships' might not necessarily have the same connotation.

6.5 Concept of 'monophyly'

Hennig's original definition, or rather paraphrase, of 'monophyly' (1950: 307f.) is so clumsy that it did not find its way into general use. A literal translation of his definition reads:

> Therefore only those groups of species – and these are all groups of higher rank – can be regarded monophyletically evolved that finally can be traced back to a common stem species. In a phylogenetic system are, consequently, only those group formations justified, which must be considered monophyletic in this sense. We have to add to this definition, as we have described above in detail, that a monophyletic group must not only comprise species stemming from a common stem species, but it must, moreover, comprise all species which descended from this stem species. (translation M.S.)

Later (Hennig 1966: 73 = Hennig 1982: 77), he gave a more concise definition, however still longer than actually needed. Interestingly, he defined 'monophyletic' in 1966 only by mentioning the descendants of a stem species ('comprising all descendants of a single stem species'). Only incidentally he stated that also the stem species is to be included in the monophylum comprising all its descendants.

Therefore, the essence of Hennig's concept of 'monophyly' is 'comprising a stem species and all of its descendants.' Again, this fully agrees with the definitions given by Kitching et al. (1998: 10f) and Wiley and Lieberman (2011: 9).

Vanderlaan et al. (2013) have pointed out that a series of authors has coined different definitions of the term 'monophyly,' some of which do not or not exactly correspond to Hennig's definition(s). Admittedly, Hennig did not explicitly refer to different epistemological levels when he found different wordings for the different definitions he gave in different publications. Nevertheless, I dare to state that he would have emphasised that *defining* monophyly is different from *recognising* it empirically. As far as I see, much of the debate summarised by Vanderlaan et al. is caused by confusing these two levels. Hennig *defined* 'monophyly' in terms of genealogy, i.e. ancestry. But he was, of course, aware of the fact that we cannot empirically observe phylogeny but have to infer it from empirical data. Thus, his – somewhat clumsy – definitions pertain to genealogy, whereas his statement (Hennig 1974) that hypotheses on monophyly of taxa are based on the recognition of synapomorphies, those on paraphyly on symplesiomorphies, and those on polyphyly on convergencies (= independently acquired similarities).

Thus, we can safely assume that Willi Hennig would agree with the usage of this term in present-day cladistics. Actually, all the representatives of 'modern cladistics' cited above apply the same concept of 'monophyly.' It is less certain that he would have accepted the phrase 'reciprocal monophyly' (e.g. in Flot et al. 2010), or 'mutual monophyly' (as in Sharma et al. 2014) since 'monophyly' sensu Hennig refers to a property of a taxon, not to a relation between two or more taxa. 'Reciprocal monophyly' originally meant (when introduced by Neigel and Avise 1986) 'a group of individuals labelled as a species that are more closely related genetically to one another than to individuals labelled as other species, in regard to their maternally transmitted mitochondrial DNA.' However, 'like monophyly, the term reciprocal monophyly has also changed definition over time' (Vanderlaan et al. 2013).

6.6 Concept of 'apomorphy'

In 1950, Hennig equated 'apomorph' with 'derived' (Hennig 1950: 106, 144). In his posthumously published book he specified 'transformed in comparison with its homologous partner character' (Hennig 1984: 40). This is more or less what we find in the introductions to modern cladistics, for example in the glossary of Kitching et al. (1998: 200). However, Wiley and Lieberman (2011: 13f.) distinguish 'evolutionary novelty' – what Hennig would have equated with 'apomorphy' –from a 'character state that is shared by members of a more restricted monophyletic group nested within the larger group' and call this an 'apomorphy.' Hennig would certainly not have accepted a concept of 'apomorphy' without reference to evolutionary transformation, thus making 'synapomorph' nothing else than 'shared exclusively by' a number of taxa (e.g. Platnick 1979: 540). The latter phrase would fit into a traditional Hennigian framework only if we consider it to be an empirical tool for assessing a state to a character, but not a definition proper.

6.7 Principles of grouping

In Hennig's works it is unequivocally clear that a 'phylogenetic system,' which might more appropriately be called a 'phylogenetic classification,' even if Hennig stated (1974: 282) that he had 'mostly' ('meist') avoided this term, must only contain monophyletic taxa (e.g. Hennig 1950: 307, Hennig 1966: 207 = Hennig 1982: 200). Implicitly, it is also clear that species can be terminal taxa in a cladogram but not be monophyletic (since 'monophyletic' is defined as 'comprising a stem species and all of its descendants'). Consequently, all taxa in a strictly phylogenetic classification must be either monophyla or single species. There is no disagreement in this respect between Hennig and modern cladists, as demonstrated by the respective quote from Wiley and Lieberman (2011: 233): 'The constituents of phylogenetic classifications are taxa: species and monophyletic taxa.'

6.8 Method of polarising characters

Hennig did not use, perhaps did not even know, the term 'polarisation' of character states. Nevertheless, he determined what he called the 'Merkmalsphylogenie' (Hennig 1982: 98) or 'character phylogeny' (Hennig 1966: 95), later known as 'Lesrichtung der Merkmale' (direction in which to read a transformation series; Wägele 2000: 169). Hennig gave four 'criteria' for determining the direction of transformation of a series of character states: geological character precedence, chorological progression, ontogenetic character precedence, and correlation of series of transformations. None of these 'criteria' provides a general tool for determining character polarity. However, as several authors have emphasised already, Hennig applied implicitly what Kluge and Farris (1969), Wiley (1981: 139), and Watrous and Wheeler (1981) devised as the 'outgroup comparison method' (a more detailed history is given by Nixon and Carpenter 1993), under the label 'distribution of the characters among the taxa' (in Hennig and Schlee 1978). In practice, Hennig discussed and pondered on the polarity of each character in question independently. In many cases using ad hoc arguments, often applying intuitively the outgroup comparison, and in some cases (Hennig 1983: especially pp. 26f.) he relied fully on the authority of other authors. Quite regularly, he applied adaptiogenetic reasoning; he would start from an evolutionary process, for example a transition from one mode of life to another, to infer on the direction of a character transformation.

Kitching et al. (1998), Schuh and Brower (2009), as well as Wiley and Lieberman (2011), correctly describe Hennig's way of character polarisation as 'a priori' polarisation and treated his approach fairly. However, all three sets of authors emphasise the epistemological and practical problems. Instead, they – as probably do most modern cladists – clearly prefer polarisation by 'a posteriori rooting,' a method called 'outgroup addition' by Wägele (2000: 175f.).

It is evident that Hennig's approach of discussing each character separately against plausible evolutionary scenarios is not feasible for molecular characters. Therefore, it is futile to ask if Hennig would have insisted on independent polarisation of transformation series when facing the problem of polarising molecular characters. Nevertheless, judging from his own practice of determination of polarity, we can fairly conclude that he would have disliked the method of 'a posteriori rooting,' as this circumvents the separate evaluation of each character in question (see also Wägele 2004).

6.9 Weighting characters

Hennig did not explicitly apply a concept of weighting characters differently, but it is evident that he considered more complex characters to be more relevant (Hennig 1950: 175). According to Wägele (2004: 101), the traditional Hennigian method 'implies a priori weighting.' Kitching et al. (1998: 99 ff.), as well as Wiley and Lieberman (2011: 196 ff.), describe a priori weighting as a legitimate way to reduce homoplasy. However, they clearly regard a posteriori or 'successive approximations' (Farris 1969) as less arbitrary. Schuh and Brower (2009: 125ff.) offer a balanced discussion of procedures and problems but clearly state that ' 'goodness' of characters as indicators of relationships cannot be judged a priori.'

6.10 Concept of 'homology'

In the *Grundzüge*, the term 'homology' appears only once (Hennig 1950: 176) and briefly, to separate 'true homologies' from 'homoiologies.' Initially, Hennig left the epistemic relation between his concept of 'apomorphy' (or 'synapomorphy') versus 'homology' open. Only in 1953 did he clarify that apomorph and plesiomorph states of a character are homologous (Hennig et al. 1953). Already in this paper, but more extensively in his 1966 book (Hennig 1966: especially pp. 93ff.), he explained the relation between 'apomorph,' 'plesiomorph,' and 'homologous,' and referred to Remane's (1952) so-called 'homology criteria' (they are rather empirical indications: 'sameness of position in comparable fabric systems,' 'linkage of intermediate forms,' 'special quality of the structures'). Hennig regarded Remane's 'principal criteria of homology' as only accessory ones because he regarded 'belonging of the characters to a phylogenetic transformation series' the 'real principal criterion.'

During the past few years, there has been a lively discussion on the epistemic status of 'homology' in cladistics, building on Nelson and Platnick's (1981) and Patterson's (1982) exclusion of 'symplesiomorphy' from 'homology.' These authors equated 'homology' and 'synapomorphy.' Farris (2014) gave a concise summary of this discussion and made clear that this position is 'anti-Hennigian.' However, despite this dissent, the use of 'homology' in modern cladistics does not differ significantly from Hennig.

6.11 Role of ancestors

Hennig emphasised that his approach aims at a 'phylogenetic system' rather than a 'phylogenetic classification' (Hennig 1984: 17). The final goal of his method is the elaboration of a general reference system in biology. This system comprises exclusively 'closed communities of descent' (geschlossene Abstammungsgemeinschaften), i.e. monophyla and (terminal) species. Consequently, the crucial step of constructing such a system is the establishment of sister group relationships. This means that ancestors are, in Hennig's original approach, (1) hypothetical and (2) only means to an end, namely the determination of character polarity. As Hennig held the opinion that it was impossible to provide general methods for polarising transformation series (Hennig 1984: 46f.), he used in many cases hypotheses on the evolutionary transformation from an ancestor to its descendants. Also, as Hull (1989: 137) pointed out, the graphical representations of phylogenetic relationships – cladograms – only show sister group relations and not ancestor-descendant relations.

How to deal with ancestors is a greatly debated topic in modern cladistics, for example Kitching et al. (1998: 13f.), Schuh and Brower (2009: 105f.), and Wiley and Lieberman (2011: 241ff). However, these discussions pertain mostly to the inclusion of ancestral species in a phylogenetic classification, less so the principal relevance of ancestors in a phylogenetic analysis and methodology.

6.12 Concept of 'species as individuals'

Whether or not species should be considered as individuals has been the subject of extensive and vivid debates in the literature (for references see e.g. Hull 1976, Ghiselin 1987, Rieppel 2010, 2011). In spite of the controversies, there is hardly any disagreement between Hennig's view (e.g. Hennig 1950: 116ff., 312) and Wiley and Lieberman (2011: 27). Comparison with Kitching et al. (1998) is not possible because they neither provide a definition of 'species' nor treat the question of individuality. Schuh and Brower (2009: 45) only shortly comment on that point, finally arguing that in cladistics 'define' is used in an alternative way, i.e. not in the logical sense of defining a 'class,' but providing the empirical basis for the circumscription of taxa. In any case, for the praxis of cladistic analysis this question is irrelevant.

6.13 Use of the term 'phylogeny'

Hennig left no doubt in his writing that he started from 'phylogeny' (synonymous with 'Phylogenese') as a unique historical process. He expressed this view explicitly in 1957 (Hennig 1957). Consequently, he would certainly not have accepted the usage of 'a phylogeny' in the sense of 'a phylogenetic hypothesis' as, for example, in Dumont et al. (2010). Kitching et al. (1998) and Wiley and Lieberman (2011) strictly use 'hypothesis of relationships' or something similar. Amazingly, Schuh and Brower (2009: 44) explain clearly that 'phylogeny (or phylogenesis) refers to a process of branching diversification of taxa.' However, many modern authors of original papers on cladistic analyses of certain taxa seemingly have not taken notice of this clarification, as we find examples of 'phylogenies' frequently, especially in molecular analyses.

6.14 Graphical representations and their meaning

Hennig presented his hypotheses on phylogenetic relationships from time to time as graphs, which he called 'trees' ('Stammbäume'). After he had developed the 'argumentation scheme,' first published in 1957 (Hennig 1957), he used this type of graph also. However, in his famous reply to Ernst Mayr's 1974 paper 'cladistic analysis or cladistic classification,' he used 'Stammbaum' (phylogenetic tree), 'cladogramm' (his quotation marks), and 'phylogenetisches System' interchangeably (Hennig 1974). Hennig put 'cladogramm' in quotation marks because he disliked the term 'cladistics' as a characterisation of his approach. First because it was invented by his opponent Ernst Mayr, but second because he emphasised that 'phylogenesis' means more than 'cladogenesis' (Schmitt 2013: 147f.).

From this, it is clear that he would never ever have accepted the statement that cladograms and trees were not the same (Kitching et al. 1998: 15f.,Wiley and Lieberman 2011: 101f.). Schuh and Brower (2009: 106) only shortly treat the relation between 'cladogram' and 'tree' without explicitly pondering on arguments. Platnick (1977) originally suggested that any given cladogram could be converted into several trees, as the cladogram representing the phylogenetic hypothesis (A(B(CD))) could, among other possibilities, result in a 'tree' representing the ancestor–descendant relationships A→B→C→D. Since within a strict phylogenetic framework there cannot be supraspecific ancestors, this transformation would only be acceptable as long as A, B, and C represent single species. But even more important is that within a Hennigian framework each accepted taxon – including the terminal ones – has to be demonstrably monophyletic, and it is logically impossible that any terminal taxon could 'in reality' be ancestor of any other terminal taxon. An ancestor cannot possess a character in an apomorph state as compared to its descendant. Consequently, the cladogram (Fig 1.11 in Kitching et al. 1998: 16) (Lamprey(Shark(Salmon Lizard))) would, in a Hennigian framework, exactly and exclusively mean that the salmon and the lizard share an ancestor that is not an ancestor of any other taxa in question, the salmon and the lizard plus the Shark, and all four have one ancestor exclusively in common. A 'tree' Lamprey→Shark→Salmon→Lizard would be completely nonsensical.

Hennig was certainly aware that a 'tree' does hardly represent an 'argumentation scheme,' but he would definitely have insisted that any 'argumentation scheme' represents one and only one possible tree.

6.15 Conclusion

The results of these comparisons are summarised in Table 6.1. In my opinion, there are only two areas in which traditional Hennigian phylogenetics and modern cladistics differ crucially: First, while Hennig discussed and determined the polarity of each character separately and would most probably insist that this must be done so, modern cladists first calculate an 'unrooted tree' and polarise the character states by rooting that tree. The second main difference lies in the non-equivalence of cladograms and trees in modern cladistics, two terms that Hennig definitely used interchangeably. I regard the remaining differences as less relevant and suppose that Hennig would have found a way to cope with them.

I am certain that Hennig would have insisted on a priori weighting of characters, as he so often pondered on polarity with reference to evolutionary transformations. Of such transformations, some certainly had a higher impact on the holomorph of the organism in question than others, and therefore it seems natural to regard some character transformations more important than others. I had in mind the fact that modern cladists vigorously refuse a priori weighting, and I have met with this attitude in discussions. However, the examined textbooks (Kitching et al. 1998, Schuh and Brower 2009, Wiley and Lieberman 2011) are much less strict than assumed. Both books treat a priori weighting fairly but critically and do not characterise it as non-scientific.

With all necessary caveats, I dare to state that Willi Hennig would never have accepted the idea that his approach ('phylogenetic systematics') could pursued without recourse to the theory of evolution. However, it might well be that he would have agreed with Brower's (2000) statement that 'evolution is not a necessary

Table 6.1 Summary of the comparison of basic concepts in traditional Hennigian phylogenetics and in modern cladistics.

Topic	Hennig	Cladistics
Concept of 'relationship'	Strictly genealogic	Strictly genealogic
Concept of 'monophyly'	Stem species + all descendants	Stem species + all descendants
Concept of 'apomorphy'	Transformed relative to the ancestor/sister	Transformed relative to the ancestor/sister (really?)
Grouping principle	Monophyla exclusively	Monophyla exclusively
Polarising characters	For each character individually, by intuitive outgroup comparison, by reference to authorities	Outgroup comparison exclusively, by 'outgroup addition' (collectively
Weighting	Regularly, intuitively, not especially deliberated	Critically debated, preferably a posteriori
Homology	Not especially defined, implicitly as 'transformation series', i.e. corresponding plesiomorph and apomorph character states	Sometimes 'homologous' = 'synapomorph', however disputed and mostly as Hennig
Ancestors	Nearly neglected	Rather a problem
Species as individuals	Yes	Yes/no
'Phylogeny'	The real and unique process of descent	A hypothesis on phylogenetic relationship
Graphical representation	Tree = cladogram = argumentation scheme	Tree ≠ cladogram

☐ : complete agreement
▨ : agreement with restrictions
▨ : wide disagreement
■ : complete disagreement

assumption of cladistics.' Possibly this is the crucial difference between 'phylogenetic systematics' and 'cladistics' (see also Panchen 1992: 7).

That Hennig did not use cladistic computer programmes is trivial: he died in 1976, 9 years before PAUP (Swofford 1985) and 12 years before Hennig86 (Farris 1988) became available. Nevertheless, since he had seriously discussed the use of non-morphological characters in phylogenetics (Hennig 1950: 161–166, Hennig 1966: 103–107) and was interested in the possible exploitation of molecular data (Schmitt 2013: 106), it is certain that he would have accepted the use of computers. In his own analyses, he regularly applied the principle of parsimony, as convincingly demonstrated by Farris (1983). Thus, it is reasonable to speculate that he would have appreciated parsimony-based algorithms. As explained under 'Method of polarising characters,' he would certainly have drawn a line separating his method from 'modern cladistics.' Nevertheless, the basic principles are the same.

Acknowledgements

My sincere thanks go to the Linnean Society of London for the chance to present the talk on 27 November 2013, on which this chapter is based, to Gabriele Uhl (Greifswald, Germany) for critically reading the manuscript, and to David M. Williams and an anonymous reviewer for numerous hints that helped to improve my original text linguistically and scientifically.

References

Brower, A.V.Z. (2000). Evolution is not a necessary assumption of cladistics. *Cladistics*, 19, 143–154.

Carpenter, J.M. (1987). Cladistics of cladists. *Cladistics*, 3, 363–375.

Dumont, H.J., Vierstraete, A. and Vanfleteren, J.R. (2010). A molecular phylogeny of the Odonata (Insecta). *Systematic Entomology*, 35, 6–18.

Dupuis, C. (1990). Hennig, Emil Hans Willi. In *Dictionary of Scientific Biography*, Volume 17 (Supplement 2), ed. F.L. Holmes. New York: Charles Scribner's Sons, pp. 407–410.

Ebach, M.C., Morrone, J.J. and Williams, D.M. (2008). A new cladistics of cladists. *Biology and Philosophy*, 23, 153–156.

Engel, M.S. and Kristensen, N.P. (2013). A history of entomological classification. *Annual Review of Entomology*, 58, 585–607.

Farris, J.S. (1969). A successive approximation approach to character weighting. *Systematic Zoology*, 18, 374–385.

Farris, J.S. (1983). The logical basis of phylogenetic analysis. In *Advances in Cladistics* 2, Proceedings of the Second Meeting of the Willi Hennig Society, ed. N.I. Platnick and V.A. Funk. New York: Columbia University Press, pp. 7–36, 701–702.

Farris, J.S. (1988). Hennig86 version 1.5, Computer programme and manual, published by the author.

Farris, J.S. (2014). Homology and misdirection. *Cladistics*, 30, 555–561.

Flot, J.-F., Couloux, A. and Tillier, S. (2010). Haplowebs as a graphical tool for delimiting species: a revival of Doyle's "field for recombination" approach and its application to the coral genus Pocillopora in Clipperton. BMC Evolutionary Biology 10:372. DOI 10.1186/1471-2148-10-372.

Ghiselin, M.T. (1987). Species concepts, individuality, and objectivity. *Biology and Philosophy*, 2, 127–143.

Hennig, W. (1950). *Grundzüge einer Theorie der phylogenetischen Systematik*. Berlin: Deutscher Zentralverlag.

Hennig, W. (1952). Autorreferat: Grundzüge einer Theorie der phylogenetischen Systematik. *Beiträge zur Entomologie*, 2, 329–331.

Hennig, W. (1957). Systematik und Phylogenese. In *Bericht über die Hundertjahrfeier der Deutschen Entomologischen Gesellschaft Berlin*, ed. H.-J. Hannemann. Berlin: Akademie-Verlag, pp. 50–71.

Hennig, W. (1966). *Phylogenetic Systematics*. Urbana: University of Illinois Press.

Hennig, W. (1974). Kritische Bemerkungen zur Frage 'Cladistic analysis or cladistic classification?'. *Zeitschrift für Zoologische Systematik und Evolutionsforschung*, 12, 279–294.

Hennig, W. (1982). *Phylogenetische Systematik*. Berlin-Hamburg: Paul Parey.

Hennig, W. (1983). *Stammesgeschichte der Chordaten* (Fortschritte der Zoologischen Systematik und Evolutionsforschung 2). Berlin-Hamburg: Paul Parey.

Hennig, W. (1984). *Aufgaben und Probleme stammesgeschichtlicher Forschung*. Berlin-Hamburg: Paul Parey.

Hennig, W. and Schlee, D. (1978). Abriß der phylogenetischen Systematik. *Stuttgarter Beiträge zur Naturkunde*, Serie A, Nr. 319, 1–11.

Hennig, W., Bollmann, H. and Machatschke, J. (1953). Kritische Bemerkungen zum phylogenetischen System der Insekten. *Beiträge zur Entomologie*, 3 (Sonderheft), 1–85.

Hull, D.L. (1976). Are species really individuals? *Systematic Zoology*, 25, 174–191.

Hull, D.L. (1989). The evolution of phylogenetic systematics. In *The Hierarchy of Life*, ed. B. Fernholm, B., K. Bremer and H. Jörnvall. Amsterdam: Elsevier (Excerpta Medica), pp. 3–15.

Kitching, I.J., Forey, P.L., Humphries, C.J. and Williams, D.M. (1998). *Cladistics. The Theory and Practice of Parsimony Analysis*, 2nd edition. Oxford: Oxford University Press.

Kluge, A. and Farris, J.S. (1969). Quantitative phyletics and the evolution of the anurans. *Systematic Zoology*, 18, 1–32.

Mayr, E. (1974). Cladistic analysis or cladistic classification. *Zeitschrift für Zoologische Systematik und Evolutionsforschung*, 12, 94–128.

Mishler, B.D. (2000). Deep phylogenetic relationships among 'plants' and their implications for classification. *Taxon*, 49, 661–683.

Neigel, J.E. and Avise, J.C. (1986). Phylogenetic relationships of mitochondrial DNA under various demographic models of speciation. In *Evolutionary Processes and Theory*, ed. E. Nevo and S. Karlin. New York: Academic Press, pp. 515–534.

Nelson, G.J. and Platnick, N.I. (1981). *Systematics and Biogeography, Cladistic and Vicariance*. New York: Columbia University Press.

Nixon, K.C. and Carpenter, J.M. (1993). On outgroups. *Cladistics*, 9, 413–426.

Panchen, A.L. (1992). Classification, Evolution, and the Nature of Biology. Cambridge-New York-Oakleigh: Cambridge University Press.

Patterson, C. (1982). Morphological characters and homology. In *Problems of Phylogenetic Reconstruction*, ed. K.A. Joysey and A.E. Friday. New York: Academic Press, pp. 21–74.

Platnick, N.I. (1977). Cladograms, phylogenetic trees, and hypothesis testing. *Systematic Zoology*, 26, 438–442.

Platnick, N.I. (1979). Philosophy and the transformation of cladistics. *Systematic Zoology*, 28, 537–546.

Remane, A. (1952). *Die Grundlagen des natürlichen Systems, der vergleichenden Anatomie und der Phylogenetik. Theoretische Morphologie und Systematik I*. Leipzig: Akademische Verlagsgesellschaft Geest and Portig.

Rieppel, O. (2010). Reydon on species, individuals and kinds: a reply. *Cladistics*, 26, 341–343.

Rieppel, O. (2011). Species are individuals – the German tradition. *Cladistics*, 27, 629–645.

Schmitt, M. (2013). *From Taxonomy to Phylogenetics – Life and Work of Willi Hennig.* Leiden – Boston: Brill.

Schuh, R.T. and Brower, A.V.Z. (2009). *Biological Systematics – Principles and Applications,* 2nd edition. Ithaca, London: Comstock Publishing Associates.

Sharma, P.P., Kaluziak, S.T., Pérez-Porro, A.R., et al. (2014). Phylogenomic interrogation of Arachnida reveals systemic conflicts in phylogenetic signal. *Molecular Biology and Evolution.* DOI 10.1093/molbev/msu235.

Swofford, D.L. (1985). *PAUP: Phylogenetic Analysis Using Parsimony.* User's manual. Champaign, IL: Illinois Natural History Survey.

Vanderlaan, T.A., Ebach, M.C., Williams, D.M. and Wilkins, J.S. (2013). Defining and redefining monophyly: Haeckel, Hennig, Ashlock, Nelson and the proliferation of definitions. *Australian Systematic Botany,* 26, 347–355.

Wägele, J.-W. (2000). *Grundlagen der Phylogenetischen Systematik.* München: Dr. Friedrich Pfeil.

Wägele, J.-W. (2004). Hennig's Phylogenetic Systematics brought up to date. In *Milestones in Systematics,* ed. D.M. Williams and P.L. Forey. London: CRC Press, pp. 101–125.

Watrous, L.E. and Wheeler, Q.D. (1981). The out-group comparison method of character analysis. *Systematic Zoology,* 30, 1–11.

Wheeler, Q.D. (2008). Undisciplined thinking: morphology and Hennig's unfinished revolution. *Systematic Entomology,* 33, 2–7.

Wiley, E.O. (1981). *Phylogenetics. The Theory and Practice of Phylogenetic Systematics.* New York: Wiley.

Wiley, E.O. and Lieberman, B.S. (2011). *Phylogenetics – Theory and Practice of Phylogenetic Systematics.* Hoboken, NJ: Wiley-Blackwell.

Williams, D.M. and Forey, P.L. (ed.) (2004). *Milestones in Systematics.* Boca Raton, FL: CRC Press.

7

The evolution of Willi Hennig's phylogenetic considerations

RAINER WILLMANN

In 1937 the German botanist Walter Zimmermann (1892–1980) stated that biological classification or systematics only makes sense if the order of organisms reflects their phylogenetic relationships (Zimmermann 1937). This was not representative of the prevailing views at that time. One reason was the conviction that the theory of descent goes beyond experience, led only by theory, by ideas. In addition, there were attempts to group according to practical requirements ("Zweckgruppierungen" in Zimmermann 1937: 944), which Willi Hennig complained about 13 years later (Hennig 1950a: 11). Biological classification was generally understood as ordering according to similarity; of course, with a consideration of the evolutionary background. However, how much this evolutionary background was taken into consideration depended on the views (or even preferences) of any particular author. This is the reason why Simpson (1961: 110), for example, could point out that systematics is a useful art. Moreover, many biologists doubted that one could ever get a true impression of phylogenetic relationships or of the true course of evolution (Hennig 1950a: 188, 192). Even Simpson (1961: 107) underlined "[t]hat classification can or should express phylogeny is an evident error." It was in such an intellectual milieu that Hennig spent the first decades of his scientific career, during which he shaped a coherent theory and methodology for biological systematics.

The Future of Phylogenetic Systematics: The Legacy of Willi Hennig, eds. D. Williams, M. Schmitt and Q. Wheeler. Published by Cambridge University Press.
© The Systematics Association 2016.

7.1 Walter Zimmermann and phylogenetic systematics

Zimmermann's two treatments entitled *Die Methoden der Phylogenetik* (*Methods in Phylogenetics*; Zimmermann 1937, 1943), one of them of book length, were among the most influential for Hennig (Hennig 1950a: 2). Many other ideas important to Hennig, however, came from Adolf Naef (1855–1914), the concept of monophyly and the identification of groups that were later called paraphyletic, for example. Of course, both Zimmermann and Hennig developed their thoughts based on a phylogenetic tradition that for German-speaking scientists began with Ernst Haeckel (1834–1919; Willmann 2003; for a discussion of Zimmermann's contribution to the theory of phylogenetic systematics, with respect to his 1937 publication only, see Donoghue and Kadereit 1992). Zimmermann had clearly defined "phylogenetic relationship," what it is based upon, and that the degree of relationship is determined by the relative recency of common ancestry; common ancestry meaning that two or more species share an immediate past phylogenetic splitting (speciation) event.

Zimmermann formulated a paragraph concerning the degree of relationship that may sound familiar to systematists as Hennig had used similar phrases in some of his publications. Stating that relationships between three taxa are the basic scheme of any phylogenetic reasoning, Zimmermann wrote:

> In phylogenetic ordering we focus on relative groupings or the degree of relationship. We say: the plants or organs B and C are more closely related to each other than to another plant or organ A. The common ancestor or the primordial form of B and C (X_2) lived not as long ago as the ancestor common to all three plants or characters (X_1)... A statement about phylogenetic relationship which cannot be represented in a basic scheme such as that of Fig 172b [see Fig 7.1 for a similar graph reproduced from Zimmermann 1943: figs 2–7] does not exist ... Relative age of the ancestors X_1 and X_2 is the only direct measure of phylogenetic relationship. (Zimmermann 1937: 989–990)

Zimmermann later clarified the principles of phylogenetic work when he pointed out that grouping according to relationship means to demonstrate that two of three organisms are more closely related to each other than they are to the third (Zimmermann 1943: 38). At the same time, he stated that with three taxa there are exactly three possibilities for their possible phylogenetic relations (Zimmermann 1943: figs 2–7, reproduced as Fig 7.1).

The degree of "similarity," which was an immediate indicator of relationship in non-phylogenetic systems, is merely a preliminary measure of degree of relationship in phylogenetics. The two are not directly interconnected, as character change is not proportional to time in different evolutionary lineages ("Stammlinien") (Zimmermann 1937: 994–995, Zimmermann 1943: 40). Furthermore, convergent evolution may be deceptive in attempts to recognize relationships. While the difficulties arising from convergences were well known to biologists, this was not

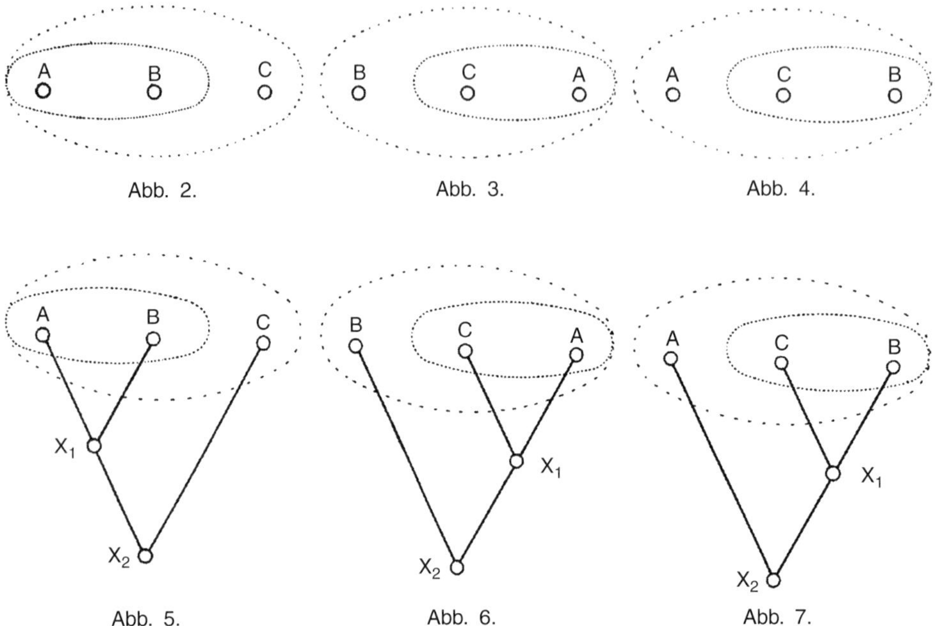

Fig 7.1 The three possible phylogenetic relationships between three taxa A–C, represented in two different ways. After Zimmermann (1943: figs 2–7).

the case with the dangers arising from unequal evolutionary rates. Zimmermann pointed out the following example (Zimmermann 1937: 997–998): if there are three species or groups A, B, and C, of which B and C are closest relatives, the following problem may arise. If C has departed considerably with respect to its characters from B, while B and A have changed only little in comparison to the ancestor of A, B and C, then one might place A with B and achieve the wrong stem tree, if the degree of similarity or dissimilarity was decisive. However, in phylogenetics the degree of relationship, as defined by Zimmermann, is the sole criterion for establishing a stem tree (Zimmermann 1937: figs 174–176). Another difficulty in recognizing true relationships may be caused by recurrent evolution (Zimmermann 1937: 998).

In order to avoid confusion that might arise from different meanings of the term "relation," Zimmermann (1937: 996, 998, 999) consistently used the terms genealogical relation, phylogenetically related, degree of phylogenetic relation-ship, etc.

Zimmermann also addressed the structure of a stem tree. The reasons being are some authors have asserted that phylogenetic trees must be net-like. Zimmermann responded that net-like relationships can only be expected within species (Zimmermann 1937: 1044). Hybrids between distantly related forms that might cause a net-like arrangement cannot be expected. This allowed the derivation of a hierarchy of categorical ranks for the groups included in a tree. Zimmermann wrote that two most closely related groups, such as B and C in Fig 7.1 (Abb. 4), may correspond to a family and A, B and C in Fig 7.1 (Abb. 7) may correspond to an order (Zimmermann 1937: 989).

Zimmermann pointed out that in phylogenetics distinguishing primitive from derived characters is required and he provided a few hints on how to do so (Zimmermann 1937: 1001, 1943: 42–46; intuitive, idealistic, and, hence, typological methods do not allow for conclusions concerning "primitive" versus "derived" ("ursprünglich → abgeleitet") characters, Zimmermann 1937: 973). In phylogenetics, biological objects are combined according to their relative degree of relationship (Zimmermann 1937: 989). At this point, I will enumerate some of them because Hennig (1950a) gave a similar list. For example, Zimmermann assumed that primordial characters are generally more widespread than derived characters, but this would require knowledge of relationships. Another method is to prove that two characters evolved in correlation with each other, as this means that if one of them is derived (or not), the other one is also derived (or not). Still another method is to use early ontogenetic stages and to apply the biogenetic law, which is, however, said to lead to erroneous conclusions in about 20% of the study cases. Possibly, Zimmermann cautiously wrote, atavisms may provide evidence as to the direction of character change. Finally, Zimmermann (1937: 1002, 1037, 1943: 42–43) stated that fossils are best suited to this purpose (Zimmermann 1937: 1002–1005, 1016, 1032; 1943: 42–44).

Perhaps one of the most interesting steps in Zimmermann's train of thought is his recognition that presenting phylogenetically subordinate groups as correlated in a phylogenetic system is not allowed (Zimmermann 1943: 52). As an example, he mentioned that angiosperms are subordinate to gymnosperms, while in almost all systems the two are coordinated and given the same categorical ranks.

What Zimmermann did not realize is:

1. Shared derived characters only are indicative of closest relationship; he mainly relied on the value of the number of characters used in phylogenetics and thus was close to phenetics in this respect.
2. Characters must be valued before phylogenetic conclusions can be drawn, which means that decisions must be made as to whether they are primitive or derived.

Furthermore, Zimmermann held the view that species-rich groups (and even extant groups) may be the ancestors of other such groups. In Hennig's terms, Zimmermann accepted paraphyletic groups and did so despite his revealing comments that phylogenetically subordinate groups must not be presented as coordinate.

7.2 The positioning of Willi Hennig as a theoretician

By the end of 1943 Willi Hennig had already published 66 articles. In a paper on the agamid genus *Draco* the question arose as to whether the color patterns may have evolved independently. An answer to this required an analysis of relationships and patterns of geographic distribution (Hennig 1936). Again, in 1939 Hennig had also assembled most of the material for his extensive work on the dipteran larvae in order to clarify the phylogenetic systematics of Diptera (see the introduction to Hennig 1952a). Therefore, Hennig began to gather various views on the methods, and the importance and goals of systematics from the contemporary literature. Hennig's first book on the theory of phylogenetic systematics, which finally appeared in 1950, dealt with the distinction between confused ideas and phylogenetically usable ones and was meant to condense the latter and include ideas of his own to form a sound theory and methodology (Hennig 1950a).

Before 1943, Hennig brought together many of his arguments. Where his thinking was going can be derived from his forceful reply to criticism of one of his papers by Krüger and the famous limnologist August Thienemann (Thienemann and Krüger 1937). This publication shows how certain Hennig already was with respect to his views on systematics. He stressed that the animal system is at the same time a theory of its phylogenetic relationships (Hennig 1943: 143). At the beginning of this paper, Hennig (1943: 139) denied that relationships based on similarity are the sole criterion of a "phylogenetic" systematics. The phrase "the more similar two forms or groups of forms the closer their phylogenetic relationship" is not valid (cf. Zimmermann 1937, 1943; Hennig 1950a: Fig 38, reproduced here as Fig 7.2). Hennig explained that there are two possible conclusions: either one denies that systematics should express phylogenetic relationships and group animals according to their degree of similarity, or systematics uses tools that are different from those applied for determining quantitative relationships of similarity. In the first case, Hennig (1943: 140) continued, different systems for larvae and imagoes (in entomology) would be justified, just as egg systematics or pupae systematics, and in many instances special systematics for the two sexes. It would not be expected that the different systems would be congruent. This situation is one of the reasons why the genuine task of systematics should be to determine phylogenetic relationships. As phylogenetic relationships cannot be derived from the degree of (morphological) similarity, it is necessary to evaluate and discern similarity relationships that indicate closer phylogenetic relationship from those that indicate distant relationship and those that do not indicate phylogenetic relationship at all. Hennig (1943: 143) pointed out that there cannot be different phylogenetic systems based on the larvae or the imagoes, etc., because systematics is based on individuals and groups of individuals. Only the relationships of similarity can be different in larvae, on the one hand, and imagoes, on the other.

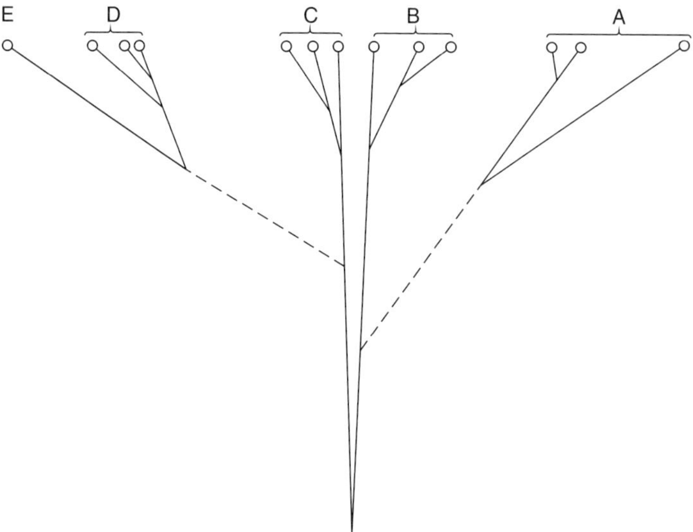

Fig 7.2 Similarity in form is not essential to phylogenetic systematics; degree of phylogenetic relationship is. Groups B and C have not deviated much from their most recent common ancestor and remained relatively similar to each other. Nevertheless, B is most closely related to A, and C is most closely related to E plus D. After Hennig (1950a: Fig 38).

The subsequent sections are intended to show how Hennig's ideas developed. With the exception of publications that were translated into English before Hennig (1966b, 1966c, 1975, 1981), the translations herein are my own. I have been careful to use wording that reflects Hennig's usage of language in the respective periods of his scientific career. By focusing on what Hennig himself wrote I hope to provide readers with a glimpse of his way of thinking. I will not discuss in detail who was influenced by Hennig directly and who was not as there were different paths that phylogenetic systematics took in different countries during his lifetime (for a recent review of the development in the UK and USA, see Rieppel 2014). Hennig influenced mid-European zoologists in the 1960s not only via his book *Phylogenetic Systematics* (Hennig 1966b), but especially via his reviews on insect systematics published in 1953 and 1962 (Hennig 1953, 1962), his paper on the biogeography of the Diptera of New Zealand (Hennig 1960, for an English translation see Hennig 1966c) and finally his *Stammesgeschichte der Insekten* (Hennig 1969, title *Insect Phylogeny* in the English translation, Hennig 1981). His theory and methods were discussed at various congresses and symposia. For example, Stammer focused on them in his introductory talk at the 11th International Congress of Entomology in Vienna (Stammer 1960) and several authors published papers in order to demonstrate the advantages of "Hennig's method" (e.g. Illies 1967: 126, 127). While it is true that the message "phylogenetics is the search for the sister group" was at the heart of Brundin's (1968) paper and may have been the stimulus to pursue strict phylogenetics among some British and American systematists (Rieppel 2014: 119), it must be remembered that the very idea of "the search for the sister group" was at the heart of the methodological considerations of Hennig himself (e.g. Hennig 1953: 10, Hennig 1955: 24, Hennig 1960: 224; see also Kiriakoff 1955: 153 and Günther 1956: 49). On the other hand, several influential German authors remained ignorant of Hennig's work.

Hennig's first book, which appeared in 1950 (Hennig 1950a), was largely completed in 1946. As there was little chance of getting the book printed due to a paper shortage in Germany after World War II (Hennig 1947: 276, 1948: 2), Hennig decided to publish two precursory essays addressing the major topics. These were to appear in 1947 and 1949 (Hennig 1947, 1949), but the introduction to the first volume of the *Larvenformen der Dipteren* (Hennig 1948) is of particular importance, as it succinctly summarizes his views on the "Theorie der Zoologischen Systematik" in 19 pages. None of these works (the 1950a book included) can be used as an introduction to the methods of phylogeny reconstruction. These came later and were developed in the course of Hennig's practical work. Again, neither of the two essays provides the reader with new insight when compared to the 1950a book. Nevertheless, extracts from it show what Hennig himself considered to be of particular importance.

Königsmann (1975) presented definitions and explanations of terms relevant in phylogenetic systematics, a survey that was checked by Hennig himself. Dupuis reviewed the works of Hennig, the literature which dealt with Hennig's ideas between 1948 and 1966 and summarized "les discussions 'cladistes' depuis 1966" (Dupuis 1978). Richter and Meier (1994) presented a sketch of the development of Hennig's phylogenetic concepts, Tassy (1999) reviewed Hennig's treatment of fossil taxa and Rieppel (2006) highlighted at great length the philosophical background of Zimmermann's and Hennig's chain of thought. A sketch about Hennig as a philosopher was also given by Schmitt (2013) who, in his book about the life and work of Willi Hennig, included a detailed chapter on Hennig's systematic thoughts.

7.3 Hennig's early theoretical papers: 1947–1950

Hennig's starting point was clear: the process of phylogenesis produces a system of organisms by the successive splitting of reproductive communities. Hence, every organism is phylogenetically related to a fixed degree with every other organism. The system can be reconstructed and the classification of organisms must reflect these relationships unambiguously. As the phylogenetic system is only one of many possible groupings of organisms that is based upon natural material interconnections between all existing species and that expresses these interconnections at the same time; it is only this that can serve as a general reference system. Hennig did not summarize his point of view in sentences such as these until the early 1950s, but his ideas are clear even from his earliest theoretical writings. It appears that the goal of phylogenetic systematics was clear to him, but that he did not really know how to achieve it until 1952.

Convinced of the central position of systematics in biology, Hennig (1947) complained that the general importance of this field appeared to be regarded as inferior by other scientific branches for three main reasons: systematics had been unable to develop a wide array of theories; some biologists dealing with systematics belittled the tasks and importance of their own research; and some authors even stated that systematists do not have to think about the basis for the order of organisms they have to classify. Hennig argued that in principle biological systematics is the ordering of organisms according to any of their similarities or dissimilarities, and thus it is theoretically possible to do systematics according to almost innumerable points of views (Hennig 1947: 276). Therefore, one has to decide whether there is one system which biologists should prefer over all others (see also Hennig 1948: 3). Hennig's solution was that genetic (and only genetic) relationships are inherent in all other natural organismic relationships, therefore the phylogenetic system is the general reference system that needs to be established and biological systematics is the branch of biology that must discover it (Hennig 1947: 277, Hennig 1948: 3).

The units of practical work in systematics. Hennig (1947: 276) stated that it is not individuals but character bearers ("Merkmalsträger") that are the true units of systematic work. A character bearer is not an individual during its entire ontogeny, but the individual during a period of its life that is short enough to appear as if it does not change. The background for this viewpoint is that many organisms may change their "gestalt" or their spatial occurrence considerably during their lifetime, for example metamorphic insects.

However, relationships between ontogenetic stages are only some of the so-called "hologenetic relationships" ("Hologenie," a term coined by Zimmermann 1934: 159) that connect all living and former character bearers (Hennig 1947). The theory of descent implies that the entirety of hologenetic relationships has to be considered, including those which today are called phylogenetic relationships. While it is true that in many cases the morphological similarities among organisms were simply re-interpreted in terms of genetic relations (with which Hennig meant evolutionary relationships), genetic systematics is neither logically nor methodologically dependent on morphological systematics (see also Hennig 1948: 3). The reason is simple: genetic relationships are reflected in all organismic connections. This means that an arrangement of organisms that is assumed to reflect the genetic relationships correctly will also show the natural relationships not considered when establishing that order. On the other hand, one may be convinced of the correctness of phylogenetic assumptions derived from morphological similarities, if they show an evolutionarily plausible pattern in the geographical or ecological distribution of the respective organisms (Hennig 1947: 277). Hennig added that while phylogenesis is the source of all kinds of relationships among organisms, all kinds of similarities can be used for detecting phylogenetic relationships and for establishing the phylogenetic system (Hennig 1948: 3). With these remarks Hennig made clear that knowing the phylogenetic relationships is the presupposition for establishing the phylogenetic system.

However, as Hennig (1948: 3–4) emphasized, in phylogenetic systematics it is not the degree of similarities we wish to know. A quantitative analysis of characters does not resolve the genetic relationships because there is no strict correlation between degree of similarity ("Ähnlichkeitsgröße") and degree of relationship ("Verwandtschaftsgrad") (Hennig 1948: 9–10; see also Fig 7.2). In phylogenetic systematics, the different traits in the baupläne of organisms are *valued* with respect to their significance for indicating genetic relationships (Hennig 1947: 277, 1948: 4b; my emphasis). One has to determine which characters indicate close phylogenetic relationship from those that reflect a lesser degree of relationship. The metamorphosis of individual organisms is of particular interest in this context because one can use different ontogenetic stages as if they were independent organisms. Thus, Hennig linked the notion of the character bearer to phylogenetics. Of course, the phylogenetic relationships of different metamorphic stages of one and the same individual are identical. Should different systems be the result of research based on different metamorphic stages, one has to go on searching for other grouping principles until one system results.

The method of phylogenetic systematics is to utilize the results of relationships gained from straightforward characters to clarify the genetic relationships from other characters, such as morphological, geographical, ecological similarities, etc. Hennig (1947: 277) pointed out that this is not a case of circular reasoning (as one might think due to the fact that one uses the results of systematic work as presuppositions for further studies), but it is the method of reciprocal illumination.

Hennig (1947) explained that usually the scope (Geltungsbereich) within which biological observations are considered to be valid is the species. The usual definition of species is a group of individuals which belong to a natural reproductive community or to a complex of vicarious reproductive communities (Hennig 1947: 277–278). This implies that the limits of species are basically determined by the reproductive relationships among the individuals and that species are natural groups (Hennig 1949: 136). However, in many instances biological laws apply to systematic groups that comprise more than one species, while in others for subgroups of only one species; every biological study is dependent on the systematic ordering of the organisms (Hennig 1947: 277).

Hennig (1947: 278) emphasized that the category 'species' occupies a preferential position. The reason for this is because species are in a particular equilibrium. On the one hand, they tend to differentiate; on the other, there is a conserving principle, expressed in the bisexual mode of reproduction. It is in the nature of life that the pressure for differentiation or pressure for diversification ("Vermannigfaltigungsdruck") is generally so strong that it breaks the forces that tend to conserve the species limits. Therefore, species split and new equilibria and species borders are formed. As this is a relatively long-lasting process, one may sometimes differ in the opinion as to whether or not there is still one species, or a complex of most closely related species, whenever the old equilibrium is broken while the new one has not yet been stabilized (Hennig 1947: 278).

One of the most concise and clearest summaries of what Hennig understood as species is given in the introduction to his *Larvenformen der Dipteren*:

> In systematics, natural reproductive communities are called species. "However, this definition requires an amendment: Geographically, genuine reproductive communities are often restricted to extremely small areas. However, if one takes members of such a reproductive community into another (neighbouring) area, they integrate themselves into a similar reproductive community without difficulties. This means that one has to extend the term "species" such that "species" is defined as a reproductive community or a complex of reproductive communities which substitute one another in space and which originated obviously through the disruption (Zerfall) of a formerly truly unified reproductive community" (Hennig 1948: 7).

Hennig added that "space" does not necessarily mean geographic space but the multidimensional living space (see also Hennig 1950a: 64). Hennig continued by mentioning that quite often:

> [S]ome of the reproductive communities substituting one another in space are characterized by the possession of particular traits. These smaller reproductive communities or even still smaller subunits of these are usually called races (or better subspecies) ... As there are all degrees of transitions of isolation between subspecies and independent species and because the reproductive relationships existing in nature between different forms of animals can only rarely be detected by simple means ... it is useful to conceive the species in a wide sense and to speak of species whenever it appears to be probable that the animal forms (organisms) under consideration really belong to a homogenous reproductive community ... or to a clearly circumscribed complex of reproductive communities replacing each other in space. (Hennig 1948: 7)

> It must be pointed out, however, that there is an essential difference between the simple morphological species concept of the old imprint and the species concept of phylogenetic systematics ... despite the fact that the latter uses often morphological criteria in practice. (Hennig 1948: 7–8)

In 1950 (Hennig 1950a: 38), Hennig cited Naef (1919), who stated that: "A reproductive community is called a species." However, Hennig criticized the term "reproductive community" as it excludes those individuals (or, better, semaphoronts) that do not participate in reproduction. Therefore, Hennig (1950a: 39) suggested a new formulation: a species is "the highest of those group categories of the hierarchical phylogenetic system, in which all semaphoronts belonging to it are connected by autogenetic relationships." (Among the autogenetic relationships Hennig mentioned, for example, the ontogenetic ones which connect the different semaphoronts of one and the same individual.) Later he added that a species consists of a number of vicarious reproductive communities, which led him to the following definition: "A species [is] a complex of reproductive communities which substitute each other in space, or … a complex of vicarious reproductive communities" (Hennig 1950a: 64, translated).

Hennig (1947: 278) pointed out that every higher category [taxon] comprises several lower species groups which are believed to have originated through the decay ("Zerfall") of a stem species from which no other species can be derived. Two years later, Hennig proposed that the general phylogenetic pattern results from what we know about the origin of species: new species arise through the splitting ("Zerfall") of existing species into several daughter species (Hennig 1949: 136). That species have a real existence is certain.

This led Hennig to the question of whether groups of higher rank in the phylogenetic system also are real entities. Very often, he wrote, one has considered them to be merely mental constructs. However, as a stem species continues to exist in its entirety via the descendant species, one also must assign them with reality and individuality ("in gewissem Sinne Realität und Individualität"). This is comparable to a protozoan that divides and survives as the entirety of its descendants (Hennig 1947: 279). Hennig repeatedly stressed that *all* descendants are thus involved. In his 1950 book, he devoted almost 20 pages to this topic (Hennig 1950a: 111–130).

Hennig saw as an important future task in systematics to objectify its group categories:

> If it is not possible today to say in which respect groups from different areas of the system, for example from plants and different 'stems' of animals which are assigned the same categorical ranks (e.g. families, orders etc.) are comparable entities, then this is a deficit which hampers starting work on the mentioned tasks of phylogenetic systematics most. (Hennig 1947: 279, translated)

With these words Hennig opened his nearly 20 year long discussion on how to deal with categorical ranks (see below).

Hennig (1949) found that ambiguous definitions of the terms used by him are inevitable. He stressed that because morphological relationships sometimes have been called phylogenetic relationships, this caused confusion, as many authors did not realize the difference. Above all, the term "relationship" had been used in different ways. Following Zimmermann (1937, 1943), Hennig (1949: 136–137, 1950a: 107) proposed to restrict "relationship" in biology to "phylogenetic relationship." For relationship in form the term similarity ("Ähnlichkeit") should be used.

Hennig (1947: 278–279) concluded that a hierarchy is the only possible way to reflect thc genetic relationships of species. The (phylogenetic) system displays a hierarchy which is based upon division and which has species as the dividing entity. As a result, he emphasized that in contrast to morphological relationships, phylogenetic relationships can always be presented in a correct and exact way in one dimension (Hennig 1948: 8, g 1949: 136, 1950a: 105). In another form of representation the phylogenetic relationships can be arranged according to a hierarchical or encaptic type of system.

What are the criteria for assembling species into more comprehensive groups? Hennig (1948: 8) explained that the hierarchy is determined by the period of time that has passed since the disruption of the stem species of extant species groups. Thus, Hennig considered time to be the criterion for relative closeness of relationships, not characters. However, while only characters can serve to detect relationships, he could not offer a practical solution: in 1947, Hennig had emphasized that one has to tell similarities indicating close relationships of descent from similarities indicating distant relationships, which means that characters must be evaluated. However, the main problem is that without a perfect fossil record to hand, the evaluation of characters is not possible by criteria which can be directly observed (Hennig 1948: 8). Having said this, Hennig could not present any ideas on how to recognize close relationships between groups.

This viewpoint began to change with volume 2 of the *Larvenformen der Dipteren* (Hennig 1950b). From then on he used the terms *apomorphic* and *plesiomorphic* (and also *apoecous* and *plesioecous*) (Hennig 1950b: vi). Obviously, however, he was able to insert these terms only just before the volume went to press, so that he could not expressly refer to shared derived characters as evidence for sister group relationships (although the characters of many groups he mentioned are uniquely derived ones). Only in later publications the mention of these terms and the use of synapomorphies as the foundation for phylogenetic hypotheses became a well-established part of his phylogenetic reasoning.

A major problem arises from the fact that detecting phylogenetic relationships is largely dependent on the interpretation of similarity:

> Often, systematics relies credulously on the equation of community of similarities and community of descent which means that systematics assumes that species and species groups which are most similar to each other are phylogenetically closer related than each of them is related to other species or species groups which do not belong in this community of similarity. (Hennig 1949: 137, translated)

However, often the degree of morphological similarity does not correspond to the degree of phylogenetic relationship (see also Hennig 1948: 4; cf. Fig 7.2). Species-poor groups often exhibit numerous or obvious special characters which are important for judging similarities.

> In the system, these groups appear coordinated to species-rich neighbouring groups. However, close examination may show that the phylogenetic relationships do not point to a coordinated neighbouring group but to a subgroup of the latter so that the species-poor, morphologically well characterized group is to be considered, in a genealogic-phylogenetic sense, only as a subgroup of the group which was hitherto coordinated as its species-rich neighbouring group. (Hennig 1949: 137, translated)

Hennig added that in current systematics one often finds no difficulty in removing morphologically well-characterized subgroups from species-rich units and to coordinate them to the entirety of all other subgroups, disregarding the degree of phylogenetic relationships. However, phylogenetic systematics as a science must not tolerate a distortion of its principles.

Hennig (1949: 137) proposed that groups which departed from their closest relatives and which, in comparison with their sister groups (here, the term "Schwestergruppen" appears for the first time in his writings), have been assigned a categorical rank which does not correspond to their phylogenetic position shall be called "apomorphic." A species-poor unit placed next to a species-rich one which is its sister group, Hennig called "stenomeric." As a typically apomorphic group, Hennig (1949: 137) mentioned man as a subgroup of catarrhine primates, which are placed opposite to "animals" in naïve systems. Among other examples, Hennig pointed to the Acrania with its very few species as an example for stenomeric organisms, as it is the sister group of the Craniota with more than 60 000 extant species. Stenomery is opposed to eurymery (species richness).

In 1949, Hennig also introduced the term plesiomorphy (Hennig 1949: 138). "Apomorphy" and "plesiomorphy" describe the morphological position of a *group* (my emphasis) in relation to its stem form. Plesiomorphy is the term describing the position of a group near that of its stem species. Vernacular equivalents would be, for example, "primitive" and "derived." Replacing them with scientific terms was, according to Hennig, justified because vernacular terms may lead to misunderstandings. For example, many plesiomorphous groups are not morphologically primitive and the term "derived" is often mixed up with the notions like "specialized," "more highly evolved," or "better adapted" (Hennig 1949: 138).

Hennig (1949: 138) wrote that in many instances there can be no doubt which group of any two is apomorphous and which is plesiomorphous. Often, however, groups in their entirety cannot be called apomorphous or plesiomorphous, because many groups will be plesiomorphous in certain respects and apomorphous in others ("Spezialisationskreuzungen" is the term Hennig used to denote a mosaic of plesiomorphic and apomorphic features). In such cases, the proposed terms do not apply to entire groups but were to be used for single traits instead (Hennig 1949: 138). So, Hennig used the terms plesiomorphic and apomorphic for both, taxa and characters. Only later did his usage shift towards a preferential connection with characters until he finally restricted it entirely to characters (e.g. Hennig 1980: 16, 1984: 41) stressing that "it is fundamentally wrong to speak of plesiomorphic and apomorphic groups or the plesiomorphic and apomorphic sister group" (Hennig 1984: 41; see also Richter and Meier 1994: 215). Only the application of "apomorphic" and "plesiomorphic" to characters allowed Hennig to conclude that the two terms are an important amendment of "homology," as this latter term is entirely devoid of any indication of the phylogenetic direction (see also Königsmann 1975: 107). Finally, Hennig (1980: 18–19) even questioned the usefulness of "homology" (see below).

As new terms corresponding to apomorphy and plesiomorphy, Hennig (1949: 138) also coined the words *apecy* versus *plesiecy* and *apochory* versus *plesiochory*, both referring to a mode of life or range of distribution which either deviates considerably from the characteristics of the stem form or are quite similar to it.

The deviation rule and its significance for systematics. Hennig emphasized that the two alternatives, plesiomorphic groups vs. apomorphic groups, reflect an evolutionary law that he called:

> the "rule of the uneven deviation of the daughter species (in short: deviation rule)".
>
> It is a matter of fact that, when species split, the daughter species depart to different degrees from the morphological position of their common stem species. Above that it seems often as if, morphologically speaking, the stem species coexisted along with its daughter species, which had 'split off' from it ('problem of the surviving stem species'). (Hennig 1949: 138; see Fig 7.3)

In the introduction to his *Larvenformen der Dipteren*, Hennig explained that this phenomenon may lead to the situation that phylogenetic relationships within a group of species cannot be resolved:

> the disruption of a reproductive community into several ones does not necessarily lead to differences in particular body structures in the daughter communities, or the resulting morphological differences may be so small that they become entirely veiled by characters which evolve during the disruption processes of succeeding daughter communities. Therefore systematics cannot recognize all communities of descent as such … This implies that in such a case systematics is not capable of determining, by its means, the degree of relationship among the species of a systematic group. (Hennig 1948: 10)

In practical systematic work consideration of the deviation rule is a presupposition for the recognition that one group of two or more is plesiomorphic relative to the others, be it in single characters or in its entire organization (Hennig 1953: 13–14).

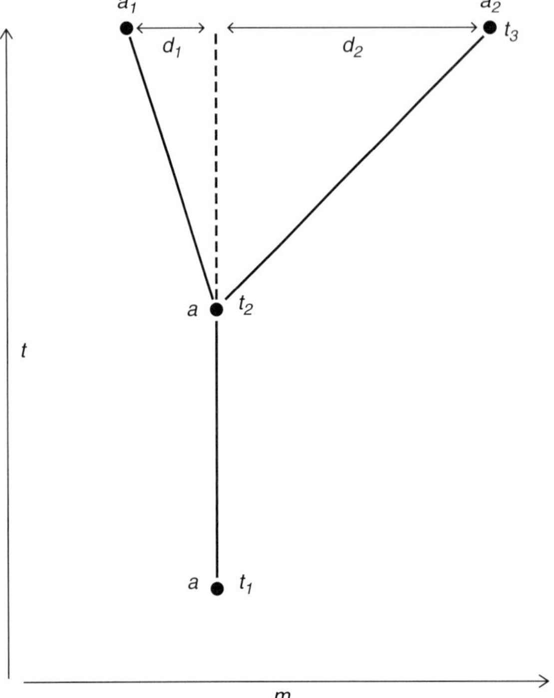

Fig 7.3 Hennig's graph explaining the deviation rule (Hennig 1950a: Fig 25) as redrawn by Günther (1956: Fig 2). Species a_1 and a_2 deviate to different degrees from their stem species (a). The stem species ceases to exist at the point in time t_2 which is the time of its splitting. d_1, d_2 are morphological distance of species a_1 and a_2 from their stem species.

7.4 The recognition of paraphyletic groups

Traditionally, the choice of taxonomic categorical rank, such as genus, family, order, etc., depending on the number and peculiarity of the characters of a group, or the morphological distance from the characters of other groups. However, there is no simple, close correlation between the number and size of similarities and the degree of phylogenetic relationship (Hennig 1948: 10). In phylogenetic systematics, a group contains only species which form a community of descent ("Abstammungsgemeinschaft") and, at the same time, the group must contain all species belonging to that community. Hennig related these statements to the use of categorical ranks. If, he wrote, one uses the qualitative method for the determination of the absolute categorical rank, a certain group A with many or peculiar morphological characters will be assigned a high rank. The remaining species, i.e. those that do not differ much (in contrast to A) from the primordial bauplan, would be united in another systematic group, say B. This will also be the case, if A is more closely related to a subgroup of B. In this case, B is not a group of most closely related species. Hennig called the species remaining in B after exclusion of A a "rest body" ("Restkörper"). Hennig (1948: 12) chose the term because it reflects a common result of practical taxonomy.

It is clear, Hennig (1948: 11) summarized, that the resulting groups are not in accordance with the principles of phylogenetic systematics (see also Hennig 1949: 137, 1950b: 57) and of course the "Restkörper" are the paraphyletic groups of his later papers. However, he invented the term "paraphyletic" only when he was writing the manuscript that became *Phylogenetic Systematics* (Hennig 1966b). Nevertheless, the word was first published in his paper on the phylogenetic system of insects (Hennig 1962: 35).

7.5 The fate of taxonomic ranks in Hennig's writings

Hennig (1948: 13–14) pointed out that detecting degrees of relationships among species united in groups of higher systematic levels is impossible in practice. As this would be one prerequisite for determining absolute taxonomic ranks, he concluded that relating the degree of relationship to ranking would have to be abandoned. As a consequence, Hennig envisaged another possibility:

> The question whether a certain group of high level is to be called a genus, family, order etc. could be answered using the age of the respective group, i.e. by reference to that point in time of earth history when the stem form of the species united in that group separated from the stem form of its nearest relative(s). (Hennig 1948: 13–14, translated)

With this Hennig was able to decouple ranks and ranking from the phylogenetic system of organisms which he wanted to help achieve. However, instead of ignoring the topic, it became a major part of his 1950 book (more accurately, it was the reverse: the introduction to his 1948 publication was largely based on the draft of the 1950 book).

Hennig (1950a) repeated at length that many problems were related to the fact that phylogenetic relationships are often insufficiently known. This precludes determination of the absolute taxonomic ranks of any subgroups. If, however, systematics could resolve the phylogenetic relations between all species completely, one could assign groups of a certain degree of phylogenetic relatedness a particular taxonomic rank. For example, one could recognize groups that are related to each other in the first degree as "species groups," groups related to each other in the second degree subgenera, those related to each other in the third degree as genera, and so on (Hennig 1950a: 203). Although Hennig regarded this as impossible a number of other difficulties would remain.

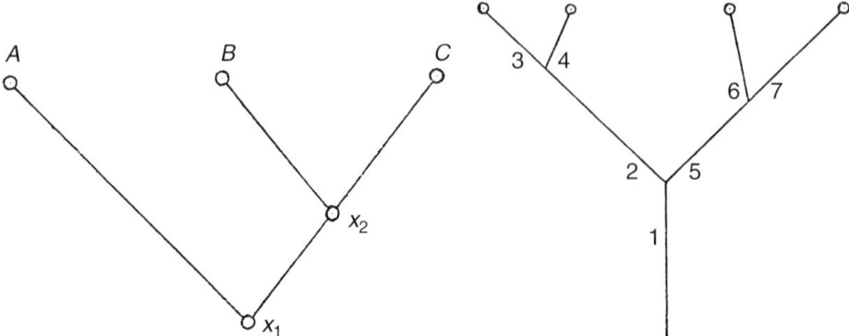

Fig 7.4 The meaning of "phylogenetic relationship" (left) and the principle of assigning absolute categorical ranks. On the diagram 1 is a family, 2 and 5 are genera, 3, 4, 6, 7 are species. After Hennig (1950: Fig 23).

The problem can be stated thus: the rank of a group will depend on the rank of the group to which it is phylogenetically coordinated; in today's terminology, its sister group (Fig 7.4). On the other hand, the rank of the latter group depends on the rank of its subunits and their rank, once again, is dependent on their immediate coordinates. To make things simple, the rank of any group is indirectly dependent on the ranks of all other taxonomic groups of the biological system (Hennig 1950a: 214, also Hennig 1965c: 84). But this is not the whole story. If fossil forms are considered, or new species are added, it becomes obvious that the ranks that were once agreed upon are not final absolute decisions. Thus, Hennig concluded that the difficulties in determining the absolute ranks of taxonomic groups according to their degree of phylogenetic relationships cannot be solved (Hennig 1950a: 215). However, Hennig suggested another way of finding appropriate ranks.

For Hennig (1950a) there were two possibilities. Either one merely estimates the degrees of relationships that are unknown and assesses ranks pending on this estimation, or the absolute rank is assigned without considering degrees of relationships at all, but (and this is difficult to understand) without ignoring the fundamentals of phylogenetic systematics (Hennig 1950a: 205). In current systematics, Hennig said, absolute ranks are almost always assigned according to morphological differences, but a determination of the absolute ranks of taxonomic categories [taxa] by the degrees of morphological differences will be largely subjective (Hennig 1950a: 206). At the same time, assessment of the rank according to degree of morphological similarity almost always leads to incongruity with phylogenetic relationships. Hennig explained that there are many authors who state that whenever one group deviates extremely from its closest relatives within a larger group of organisms, it must be assigned a rank which is the same as that of the remaining relatives (Hennig 1950a: 209). A well-known example from the insects is the Strepsiptera:

> According to their phylogenetic relationships this small and unique group of insects belongs in the malacoderm Coleoptera. However, in the current system they are placed as a group of equal rank alongside the order Coleoptera as a separate order. (Hennig 1950a: 210, translated)

As the hierarchy of the biological system mirrors the relative time of their origin, the search for absolute ranks of higher taxonomic groups must rely on their age. Whether two

> coordinated units are recognized as subfamilies, families, orders or categories of any other rank would therefore … depend … on defining their time of origin as the origination time of subfamilies, families, orders or any other rank of order. (Hennig 1950a: 215–216, translated)

Hennig emphasized that paleogeographic events which caused today's subdivision of the Earth's surface are responsible for the vicariance patterns and the origin of new organismic groups (Hennig 1950a: 215–220, also Hennig 1948).

He saw a great advantage in using the age of groups as criterion for their rank. The relatively large independence from exact knowledge of phylogenetic relationships would allow ranks to be assigned even for groups where their relationships are not well known. Furthermore, Hennig thought that while there is nothing in current systematics which might justify a comparison of, say, families of certain plants with those of vertebrates, molluscs, or insects, for example, his proposal at least would make the groups objectively comparable (Hennig 1950a: 245).

The vicariance patterns which Hennig envisaged as starting points are set in the period from the upper Mesozoic to recent times. As Hennig's biogeographical discussion focused on insects, he was inclined to assign groups which originated during this time span ranks covering the entire spectrum from superfamily to genus and species group. Groups stemming from the Triassic would be assigned ordinal rank, while Carboniferous groups would be ranked as classes (e.g. Hennig 1950a: 257). Here, the paleontological method comes into play which consists of referring to dated and phylogenetically correctly interpreted fossils (Hennig 1950a: 255).

The logical consequence is that species described from former epochs which have left no extant descendants must be assigned the same ranks as that of the existing groups which originated during those periods of time. For example, if extant groups which appeared first in the Carboniferous were ranked as classes, this rank would apply equally to all other Carboniferous species. Hennig finally concluded that fossil and Recent organisms cannot be united in the same system but, for example, in a stem tree (Hennig 1950a: 258–259).

In addition to biogeographical and paleontological dating methods, Hennig discussed yet another possibility for assigning groups absolute categorical ranks: the parasitological method. If the absolute rank of one member of a host and parasite pair was determined, then an appropriate rank could also be assigned to the other member. The theoretical background relies on the general presupposition that host and parasite largely evolve together in parallel.

After 1950, Hennig often repeated that the age of groups would be a good measure for categorical ranks. Obviously, he had lost sight of the background of this idea as its development was based on the assumption that sister group relationships will never be solved satisfactorily across the entire organismic world. Soon after 1950, Hennig could have envisaged the phylogenetic system as the sole criterion for assigning categorical ranks, but until 1965 he did not.

Instead, Hennig delved even deeper into the matter as set out in 1948 and 1950 (Hennig 1948, 1950a). In 1966 he had, once again, with a different train of thought, decoupled the assignment of categorical ranks from the phylogenetic system, writing:

> In principle, there is nothing against determining the absolute rank of taxa on the basis of degree of morphological divergence, provided this is done within the limits set by phylogenetic systematics – that sister groups be coordinate, and thus have the same absolute rank. (Hennig 1966b: 155)

In effect, this implied two orders of organisms: the phylogenetic system, which would indicate degrees of relationship, and a classification based on similarity (or geological age, the two would not necessarily exclude one another) – a combination of both would lead to confusion.

One advantage of categorical ranks is, Hennig repeated (1965c: 88), that if they are used properly, it might allow comparisons between distantly related groups. For this purpose the taxa themselves, i.e. groups of particular ranks must be comparable. Comparability would be given if sister groups are assigned the same rank as both age and morphological starting point of sister groups would be identical.

The idea to relate categorical ranks to epochs in Earth history, still defended in *Phylogenetic Systematics* (Hennig 1966b), decoupled ranking from the phylogenetic system and was therefore not appropriate. Griffiths did not accept Hennig's attempt to redefine Linnaean categories as age classes, as it could not be recommended to "attempt to apply the Linnaean category names in a way which has nothing to do with their original meaning" (Griffiths 1976: 170). Again, clinging to categorical ranks was proving to be a mere burden.

The fact that sister groups deserve the same rank made Stammer stress that there must be as many categories as there are sequential phylogenetic splits: "Here, one demands simply too much from taxonomy" (Stammer 1960: 5), he complained. Cracraft (1974) dealt at length with how to order ranked taxa according to the degrees of phylogenetic relationship (his solution: sequencing according to strict rules), Farris (1976) proposed eight prefixes for category names, as commonly used (e.g. *Hyper*classis, *Pico*ordo, *Hyperpico*ordo) and Van Valen (1978: 287) pointed to formal difficulties with ranking fossils. He explained that a single short-lived species with no descendants must be given a rank equal to its sister group which may consist of innumerable species. Hence the discussion began to eliminate Linnean categories altogether. Ax (1984: 243–257, 1987: 235–249) reviewed the issue *in extenso*.

In 1965, Hennig introduced the new idea that categorical ranks can be substituted by a numbered ordering (Hennig 1965a: 98). The result is a hierarchical arrangement of monophyletic groups according to their exact degree of phylogenetic relationship. Hennig pointed out that "it needs little reflection" to see that classical categorical ranks are incompatible with the theoretical foundations of phylogenetic systematics (Hennig 1965a: 115). "This should already have been shown by the fact that sister groups must have the same rank in a phylogenetic system ... for sister groups can, of course, have morphologically unfolded ... with completely different rates of evolution" (Hennig 1965a: 115). In agreement with this, Hennig (1968: 19) presented preliminary results of his work on the phylogenetic system of the Diptera as a written system:

> A. Polyneura (= Tipulomorpha)
> B. "Oligoneura" (preliminary name as used in the text)
>   B.1 Psychodomorpha
>   B.2 unnamed group
>   B.2a Culicomorpha
>   B.2b unnamed group
>   B. 2b1 Bibionomorpha
>   B. 2b2 Brachycera

When Hennig dealt with all fossils and Recent insects, he gave individual groups long sequences of numbers from which their position can immediately be seen:

> The sister group relationship can always be deduced from the last figure in the series. For example, I consider the Ectognatha (2) to be the sister group of the Entognatha (1), the Coccina (2.2.2.2..3.2..2.2.2.1.2) the sister group of the Aphidina (...1.) and so on. (Hennig 1969: 10, translation from Hennig 1981: xviii)

Interestingly, Hennig had not used categories in his review of insect phylogeny and systematics from 1953 (Hennig 1953), but it appears that by that time he had not realized that this was indeed the practice in general, i.e. at all levels of the phylogenetic system. With respect to his textbook on invertebrates ("Wirbellose"), it is likely Hennig would have abandoned categorical ranks entirely if he had had time to finalize these works (Hennig 1980, 1986, the manuscript being finalized in 1975–76). One can arrive at this conclusion from the fact that only remnants of ranking appear in part 2 where one categorical rank is used twice – and only twice (Hennig 1986). In his *Stammesgeschichte der Chordaten*, Hennig made no use of categorical ranks at all (Hennig 1983[1]).

---

[1] That Hennig used categorical ranks in his treatment of the Diptera in the *Handbuch der Zoologie* (1973) appears to be due to conditions laid down by the editors.

It was left to Griffiths to point out where the true discrepancies between the usage of categorical ranks and phylogenetic systematics lay:

> The aim of systematics is to represent the systematic structure (order) of the real world ... Classification is the name of a logical activity, that of ordering concepts into classes ... Ordering according to element/system relations may be called systematic ordering or systematization. (Griffiths 1974: 87, 90)

Griffiths pointed out that individuals, species and monophyletic groups are systems and, hence, are real (Griffiths 1974: 99, 103): "The traditional names (genus, family, etc.) were intended by Linnaeus as terms of Aristotelian logic. They are inappropriate because of the inherent confusion between systems and classes in this logic" (Griffiths 1974: 118). Griffiths further stated that "there is no logical reason why taxa (monophyletic groups) *must* be classified (that is, ordered into classes or categories)," thus "it is possible to avoid the difficulties inherent in defining categories (classes) of taxa by presenting an unclassified hierarchy of taxa" ... "The nomenclature of Hennig's (1969) review of fossil insects is an important advance" (Griffiths 1976: 168).

7.6 Hennig's first phylogeny book (Hennig 1950a)

In the second half of 1950, Hennig's book *Grundzüge einer Theorie der phylogenetischen Systematik* appeared. However, although many German biologists dealt with theoretical issues in systematics, it was not widely read. Thematically simple sections were difficult to conceive because in order to be exact, Hennig explained every little detail. Many sentences were extremely convoluted – Hennig tended to write in long sentences. Again, he used a multitude of little-known or newly introduced biological terms and sometimes this made it even more difficult to grasp the content of his sentences (for more details, see Schmitt 2013: 66).

Many pages of his book deal with categorical ranks. As these chapters are interconnected with many other of his papers, I have treated this topic above.

As in his two precursory papers (Hennig 1947: 277 and Hennig 1948: 3), Hennig (1950a: 29–30) summarized the primary justification for the phylogenetic system being *the* general reference system of organisms: the phylogenetic system is that from which all relationships in every other system in biology can be easily deduced. This is because the historical development of all organisms is necessarily mirrored in these relationships. At the same time, Hennig underlined the relationships between the hierarchical structure of the system of organisms and the theory of descent (Hennig 1950a: 22). He pointed out that as early as 1844 Darwin had written that the possibility of erecting a hierarchical system of organisms can only be explained by the theory of phylogenetic relationships among species. Hennig then showed that only a hierarchical structure of the system is feasible and that there are two ways of presenting it: the written hierarchical system, and the phylogenetic (stem tree), which is a graphic representation of the former (Hennig 1950a: 22–23, citing Hertwig 1914: 82).

In addition, Hennig offered several subsidiary reasons in favor of the phylogenetic system, in particular relationships in the phylogenetic system can be measured in a clearly defined manner, which is not possible in morphological systems. It is true that erecting the phylogenetic system in practice usually starts with a system of morphological similarities, but phylogenetic systematics must go beyond a mere re-interpretation of morphological comparative results. Similarities reflect the phylogenetic relationships of the organisms to a certain degree. In phylogenetic systematics, these similarities are then subjected to the method of reciprocal illumination (Hennig 1950a: 26–27).

The semaphoront. Both Hennig (1950a: 9) and Zimmermann (1937) pointed out that the ultimate element in systematics is not the individual organism (or the species) but the individual during a short time span of its life, which is the character bearer or semaphoront. In insects these are, for example, eggs, larvae, pupae, and adults. A semaphoront is more or less constant and its dimension for practical work is determined by the speed with which the characters of an individual change (Hennig 1950a: 9). An individual embraces all its semaphoronts which are connected by ontogenetic relationships (Hennig 1950a: 94). Hennig then discussed what he called the science of holomorphology. The term describes studies on the morphological relationships between several semaphoronts and individuals, among them metamorphic stages, polymorphism, and space as an important element in the evolution of species. Holomorphology is important, but still only a subsidiary science for taxonomy, Hennig underlined.

Subspecies and species. Hennig suggested that the most derived populations of any species can be found at the periphery of their distributional area, which is obvious as many of these populations were described as independent species (Hennig 1950a: 74–76). These populations are best-called subspecies rather than races, as often happens. This is because the latter term has been used with different meanings (Hennig 1950a: 78); Hennig himself continued to use the term "race" (Hennig 1950a: 311, Rasse, Rassenkreise, etc.).

Pending character(s) chosen for distinguishing groups of individuals, one may reach different conclusions as to the limits of subspecies. As the same difficulties arise in distinguishing species, Hennig said that many systematists became desperate with respect to ever finding a general species definition (Hennig 1950a: 87–91). The error here is that these systematists rely on a morphological understanding of species. However,

> for intrinsic reasons the phylogenetic system (and hence the genetical species concept) is to be preferred over all other possible systems. The mistake is to have

not realized this and to have overlooked that all investigations of holomorphologi-
cal characters and differences in systematics only assist in detecting the genetical
relationships. (Hennig 1950: 93)

In phylogenetics, we really are interested in the reproductive community and in the splitting process (Hennig 1950a: 86). A species is a group of vicarious groups of individuals and thus encompasses all stages of the splitting process, the formation of new species (Hennig 1950a: 101). In other words (Hennig 1950a: 102), new species originate whenever the tokogenetic relationships among the members of a species cease to exist Hennig introduced the term *tokogenetic* for the relationships between conspecifics: "Tokogenetic relationships connect the groups of semaphoronts that are called individuals within a species" (Hennig 1950a: 39–40). These considerations led Hennig to the definition of the word "phylogenesis" (Phylogenese) thus whenever sexual contact between individuals living in distant areas ceases to occur and subspecies are formed, the genetic relationships change. This is the process of phylogenesis: "By discussing the subspecific differentiation of a species we have, hence, arrived at the field of phylogenetic relationships" (Hennig 1950a: 93–94).

Hennig also dealt with species hybridization. The occurrence of hybrids would lead to a reticulate phylogenetic pattern. Of course, the question of how often "species hybrids" occur in plants and animals largely depends on the definition of species. Almost all cases of reported species hybridization in fact concern subspecies of a species, as defined by Hennig (1950a: 100–101). Referring to Zimmermann (1931), Hennig added that a similar statement applies to plants as well, which means that phylogenetic net-like relationships (in contrast to relationships of form) can be assumed to occur also in plants within species only. This is followed by one of the most fundamental paragraphs of phylogenetic systematics:

If a species splits into two because some of the tokogenetic relationships among its individuals cease to exist, the species itself ceases to exist. It is the stem species common to its two daughter species. Both of the daughter species are phylogenetically related to each other in the first degree. The two together are a group category of higher level, which we will subsequently, for the sake of convenience, call a genus. If each of the two daughter species itself splits into two succeeding species, these four species are phylogenetically related in the second degree, because they descended from the stem species common to all of them by two splitting steps ... This scheme reflects the structure of the entire system of the group categories of higher level. Thus we accept without any reservations the standpoint of Zimmermann (1931, p. 989–990): 'The relative age relationship of the ancestors x_1 and x_2 is the only direct measure of phylogenetic relationship.' A statement about phylogenetic relationship which cannot be expressed in the basic scheme of Fig. 23a does not exist. (Hennig 1950a: 102–103, translated)

However, Hennig added,

> in Zimmermann's definition the words "relative age of ancestors x_1 and x_2 must be replaced by 'the number of splitting steps which have led from the ancestors x1 and x2 to the recent stage" (Hennig 1950a: 103, translated)

The first five sentences of the preceding paragraph imply that a species is delimited in time by two successive splitting events (cf. Hennig 1966b).

Brundin was one of the first authors to follow Hennig without compromise on this point, ideas considered too provocative by many systematists. Brundin wrote that

> "the speciation process must be looked upon as a splitting of an ancestral species into daughter species, and not as a branching off of daughter species from a persisting ancestral species...We have to hold to this interpretation even if the ancestral species...should remain practically unchanged. That is, after all, a logical consequence of the definition of phylogenetic relationship, and that definition is self-evident. (Brundin 1966: 23)

Tasks and methods of phylogenetic systematics. With few words, Hennig (1950a: 30, 34) merged taxonomy and systematics together by stating that the tasks of special systematics (taxonomy) are the same as that of special phylogenetics. He added that phylogenetic systematics will have solved its basic task when the correct correlation and subordination of the units has been determined (Hennig 1950a: 106). Hennig (1950a: 34) cited Paramonow (1935) who coined the term biosystematics for this branch of biology.

However, will it ever be possible to achieve, in practice, a fully resolved phylogenetic system? Hennig (1950a: 129) repeatedly stressed that the degree of phylogenetic relationship between species depends on the number of splitting events which separate them from their common stem species. However, in most instances it is not possible to determine all phylogenetic relations. Partial solutions are very often all one can expect.

Hennig then switched to the methods of detecting phylogenetic relationships. A chapter on the paleontological method is of particular importance because here he treated the significance of special characters and the structure of monophyletic groups.

Although fossil organisms are only partially preserved, paleontology can help to determine the direction of character change (Hennig 1950a: 145). However, it does not offer direct insight into phylogenetic relationships. First of all, fossils are generally not preserved as elements in more or less continuous evolutionary lineages (Hennig 1950a: 134, citing Naef 1919). But if a fossil species can be assigned to an extant group, paleontology can determine the minimum age of that group (Hennig 1950a: 135). Here, however, a major problem arises and once again Hennig needed to explain the importance of the deviation rule: if a stem species splits into two and

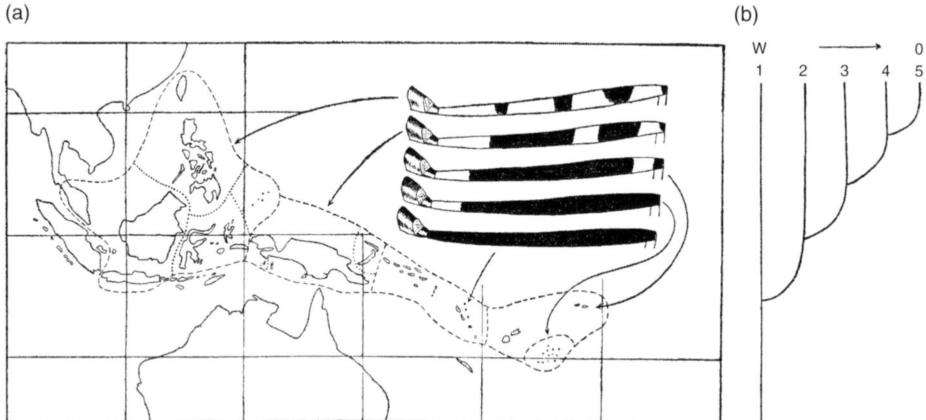

Fig 7.5 Phylogenetic relationships and degrees of difference. (a) Coloration of the hind femurs of the subspecies of *Mimegralla albimana* (Diptera, Tylidae) that exhibits an increasing degree of apomorphy geographically from West to East. Eastern forms always are derived when compared to its neighboring western subspecies. (b) The explanation is that with every split one of the two originating subspecies has remained morphologically unchanged while only the respective Eastern descendant underwent morphological change. In the case of species this would mean that a fossil with the characters of subgroup 1 could be a stem form of all subgroups and possibly a stem form of any other subgroup. After Hennig (1950a: 137–138 and Figs 27 and 28, re-published by Hennig 1966b: 135–136, Figs 39 and 40).

if one of its descendants remains morphologically plesiomorphic, it may appear that one species has split off from the other. In other words, the impression left in such cases is that one species has continued to exist relatively unchanged while the other has branched off. If this happens several times in a group of most closely related species, a fossil may look like a representative of a (such as group 1 of Fig 28 in Hennig 1950a, reproduced here as Fig 7.5) but it may in fact belong to any of the groups 1–5 or even to the stem lineage of any of the subgroups (Hennig 1950a: 138).

For Hennig, it was self-evident that the phylogenetic relationships of fossils can only be determined once the relationships among recent organisms have been clarified. However, many paleontologists held (and still hold) the view that the determination of phylogenetic relationships is their field of research, as they alone study a genuine historical branch of the biological sciences. Reasons for Hennig's view were, of course, that he was a neontologist and an entomologist, and that the characters preserved in fossil insects are usually only a fraction of what one can investigate in Recent specimens.

Despite the limited availability of characters in fossils and the limited results of their examination, Hennig noted that to reveal the phylogenetic position of any fossil requires a morphological analysis exactly as undertaken in Recent organisms. Attempts to integrate fossils in the phylogenetic system may result in consequences of general theoretical interest. If there are only primitive characters, it would not be possible to determine its exact phylogenetic relationship. Hennig used a fossil fly assigned to the genus *Syrphus* as an example. The extant representatives of *Syrphus* have the most primitive type of wing venation that occurs in the Syrphidae, which implies that all other syrphids that do not belong to *Syrphus* have ancestors whose wing venation was identical to that of *Syrphus*. Therefore, it is not possible to decide whether the fossil belongs to the lineage of the ancestors ("Ahnenreihe") of *Syrphus* or whether it represents an early stage of other derived genera (Hennig 1950a: 140). If, however, there are characters which have occurred in a certain species and have been preserved in its descendants without further transformation, a decision can be made whether or not a certain fossil belongs to a particular group of species (Hennig 1950a: 140–141). These types of characters were called special characters by Hennig ("Sondermerkmale," Hennig 1950a: 141). In order to explain this, Hennig used another example of fossil flies:

> The grouping of fossils from the Baltic Amber…with the extant rhachicerines can be founded on the existence of constantly present special characters. At the same time this excludes the possibility that the fossils belong to the stem lineage of other … groups. (Hennig 1950a: 141, translated)

The wealth of characters of a species makes it likely that even species which are in many ways similar to their ancestors and which are, hence, largely plesiomorphous, would still exhibit a number of apomorphic characters. These allow a decision to be made as to whether or not a fossil belongs in the lineage of ancestors of certain recent organisms. Hennig (1950a: 142) mentioned one fundamental principle only in passing: ancestors of a particular group must be more primitive in all of their characters than the recent representatives of that group.

When Hennig (1950a: 140–147) discussed these matters, he did not highlight the more important issues. Furthermore, he interrupted his train of thought by presenting examples. Finally, he explained how to determine the phylogenetic position of a species in a paleontological context where only few readers expected explanations of core concepts. For all these reasons, the presentation of some of the most important principles of phylogenetic systematics went largely unnoticed by readers of his book, for example, the fact that apomorphic characters reveal a particular organisms' phylogenetic position, while plesiomorphic characters are useless. Hennig did not use the terms apomorphic and plesiomorphic when explaining the method of determining relationships (Hennig 1950a: 140–141), but only in the context of comments on the deviation rule (Hennig 1950a: 142). I am inclined to believe that Hennig had not realized how far his theory and methodology of phylogenetic systematics had evolved at that time.

simpler characters in common (criterion of the complexity of characters; similarity in simple characters can be more easily explained by convergence or homoiology).

4. Characters which are not recognizably linked to certain modes of life or developmental tendencies of their bearers are better indicators of phylogenetic relationship than characters having such a dependency or which are known to be subject to certain developmental tendencies (criterion of coincident special characters). However, the problem lies in how to recognize such characters (Hennig 1950a: 178–179).

While the validity of these rules for recognizing phylogenetically useful features is, in part, not given or questionable, what was lacking were rules for recognizing derived characters. As described below, Hennig published a clear statement that apomorphic characters are decisive for the determination of phylogenetic relationships only in 1953 (Hennig 1953: 16), but Hennig (1953) presented no rule as to how to tell plesiomorphic from apomorphic character states at that time. Even as late as 1966, Hennig presented only an enumeration of the indications he published earlier (Hennig 1966b). As a result, much later, he reasoned that it is "not possible to specify unfailing rules and criteria for deciding which homologous characters are apomorphic and which plesiomorphic" (Hennig 1984: 47). This is somewhat surprising because throughout his career Hennig had to decide in hundreds of cases whether or not a certain character was derived in comparison to another state of the same character – and he did so successfully. In order to reach a decision, he compared the object he was working on with other species, using all his knowledge of animal systematics and morphology, but he failed to recognize this course of action as a sound method. Since 1969, the method is well known as "outgroup comparison" (based on the terms "exgroup" and "outside group" of Sturtevant 1942 and Throckmorton 1968: 357, respectively) and Wiley (1981: 139–146) introduced the term "outgroup rule."

However, Hennig (1950a: 192) stated that mostly, if not always, the dissolution of a species into several reproductive communities is tightly connected with its spatial dispersal. This implies that the distribution of closely related reproductive communities can be used as a criterion to determine their relationships (criterion of vicariance). And if the dissolution of a formerly united group into vicarious subgroups can be shown to be related to a certain paleogeographic event, one may well conclude that the groups that have the same type of vicariance are of the same geological age.

The deviation rule and the reality of "higher" groups. Hennig left no doubt that species are comparable to individuals ("haben individuenartigen Charakter," Hennig 1950a: 295). He held the view that a species ceases to exist when it splits into two daughter species. This statement, he admitted, did not correspond to the prevailing view according to which a species may give rise to a daughter species without ceasing to exist. An extreme view is that even recent species may be the stem forms of other recent species (Hennig 1950a: 110–111; reproduced as Fig 7.3). According to the deviation rule, daughter species (a1, a2) will deviate to different degrees from the common stem form (a). If one of two daughter species (say a1) has remained morphologically almost identical to its stem form, one might even say that the stem species (a) has remained unchanged while a daughter species (a2) had budded off from it and that the two (stem species a and daughter species a2) are finally contemporaneous. However, Hennig's own view was completely different and it is of utmost importance regarding the question of whether taxonomic groups are real entities.

A basic question must be first answered: what does it means to be an individual? Or, in other words, what is individuality (Hennig 1950a: 114)? The solution is that continuity in time between different states accounts for individuals. Hennig had mentioned unicellular organisms dividing into daughter individuals as an example (Hennig 1950a: 114, see also Hennig 1947). This standpoint makes it possible to regard the mother individual and its daughters as identical, because the mother continues to exist in her daughter cells. Any physical or spatial connection is merely a superficial criterion of identity, individuality and reality and thus also a superficial criterion for the notion of what a thing is. Hennig (1950a: 114–115) agreed with N. Hartmann that the true characters of real things are their restriction in time (origination and decay) which are also the criteria of individuality. These examples demonstrate that in phylogenetic systematics two species which have descended from a common stem species and which form a group category [taxon] of higher level must be considered as identical to its stem species. Together they form an individual of higher order and have real existence. It is common descent of the individuals belonging to particular groups which confers individuality on these groups (Hennig 1950a: 117), they are not mere abstractions as "time creates the systematic categories, not the arbitrariness of man" (Hennig 1950a: 117, referring to Brauer 1885). However, the view that a stem species continues to exist in both of its daughter species and that one does not adhere to the idea that one species may split off from another, is a presupposition. Each group has an individual history that ends only with the death of the last individual that derives from their stem species.

The view that supraspecific groups are individuals shows that the phylogenetic system is to be preferred over all others. In other systems the criterion of individual-like reality for groups of organisms does not apply (Hennig 1950a: 120) – for example, if single elements are removed from the community to which they belong and placed as equivalent units with respect to the rest of this community (Hennig 1950a: 210–211). In modern terminology: paraphyletic groups are not individuals.

As a result Hennig proposed that the taxonomy of higher categories (taxa) self-evidentially use the species as its basic unit. The task is to order species according to their degree of phylogenetic relationship into groups of successive higher levels. The degree of phylogenetic relationship mirrors the number of splitting steps by which species united in a group are separated from their common stem species. The morphological method is hereby decisive for detecting the relationships, while the degree of morphological similarity (if measurable at all) does not necessarily correspond to the degree of phylogenetic relationship. As all characters of an organism are functionally more or less tightly correlated, it is theoretically justified to consider the determination of phylogenetic relationships by using only a selection of the organism's structures. Advances in phylogenetic systematics do not rely on considering more and more characters, but on optimizing its methodology (developing further biological rules, etc.). A valuable source for checking systematic results is geographical distribution. Conversely, distribution patterns can be used to recognize phylogenetic relationships by the principle of reciprocal illumination (Hennig 1950a: 199–202).

Referring to Zimmermann, Hennig wrote that phylogeny is the sum change of ontogeny. As ontogenetic changes result in differences in the gestalt of individuals, "change of the gestalt" is the most general element in evolution (Hennig 1950a: 293–294). Ontogenetic change in individuals does not necessarily lead to the internal subdivision of a group. Such a process Hennig called "holomorphogenesis" or "holomorphogeny." Individuals' changes leading to differences in the subdivision of species were named "speciogenesis" or "speciogeny" by Hennig (1950a: 294). With the subdivision into species, evolution becomes phylogeny (Hennig 1950a: 295).

The term phylogeny, or phylogenesis, relates to the entire development of individual natural elements (Hennig 1950a: 295). Its usage is in general restricted insofar as the term is connected to the process of the subdivision of individualized elements. Such a subdivision may occur in the case of the splitting of species and species are individual-like. Strictly speaking, one can therefore speak of the evolution of a single character, for example of the wing venation of insects, but not of its phylogenesis. The reason for this is that a character is not individual-like and that it is only the property of an individual (Hennig 1950a: 295).

Monophyly and polyphyly. Redefinition of the term "monophyly" appears late in Hennig's work. Nevertheless, in his monumental study on the dipteran larvae he discussed several times whether or not certain groups are taxa that we would call monophyletic today. The term was not yet available in a Hennigian sense but the concept (developed by Naef 1917, 1919) was clear. As early as 1947, Hennig wrote that a group of higher rank, which is not a mental construct but something real and individual-like, consists of a stem species and the entirety of its descendants (Hennig 1947: 279). As a result of the non-availability of the term monophyly sensu Hennig, he had to circumscribe the nature of such groups using other terms such as "a true phylogenetic unit," a "natural group," and a "closed unit of related group" ("natürliche Verwandtschaftsgruppe," "geschlossene Verwandtschaftsgruppe"

Hennig 1950b: 419, 252, Hennig 1948: 114). This shows that Hennig actually took a suitable term that had been coined a long time before for a certain structure of an assembly of species. For the first time he did so in his book from 1950, but the newly defined term could not enter other contemporaneous works of his.

Hennig wrote:

> The result of taxonomic work is a hierarchic system of groups which are absolutely unambiguously determined by the demand that they must comprise only such species which can be derived from a common stem species but which must also embrace all extant species which are descendants of this stem species. (Hennig 1950a: 276, and 209, translated)

This comes close to the formulation of Naef:

> A systematic category [taxon] is the entirety of species which are believed to have developed from one stem form. (Naef 1917: 46, translated)

Naef was even more precise by insisting that the stem form of any systematic category [taxon] should also be a member of that category. In 1919 Naef equated "stem form" and "stem species" (see Willmann 2003 for details).

Later, Hennig demonstrated that evolutionary views depend greatly on what is understood as monophyly (Hennig 1950a: 307). It is here where Hennig for the first time linked this term to his definition of a "closed phylogenetic group." Hennig explained that many authors of his time thought that the monophyletic origin of a group is given if it descended from one pair of parents, while others called a group monophyletic if it descended from a group of the same systematic rank. However, Hennig underlined that monophyly is determined by the course of evolution. This means that the term monophyly is indisputably predefined: a monophyletic group contains all species that derive from a stem species (Hennig 1950a: 307–308; cf. Hennig 1966b: 207). Only such groups are justified in phylogenetic systematics. In 1953, Hennig proposed the terms monophylum or homophylum for the monophyletic group, but he himself always preferred the usage of the term "monophyletic group" (Hennig 1953: 9).

Hennig pointed out that it may well happen that the monophyletic origin of groups introduced in systematics cannot be demonstrated in practice (Hennig 1950a: 308).

Are polyphyletic species the source of polyphyletic groups? In *Euphrasia* there are two species of different evolutionary origin (*E. glabra* and *E. borealis*) and as they cannot be discriminated by morphological means, they have been subsumed under the same species name. This means that it is a polyphyletic species. Hennig (1950a: 309) commented on this case by writing that it is decisive whether the two alleged species combined form one reproductive community. Hennig mentioned that there are many such cases in botany, but until now no hybridization of two species (the foundation for polyphyletic species) is known in zoology. Nevertheless, higher ranked groups that originated by the splitting of a species which itself is of polyphyletic origin are not to be regarded as polyphyletic units because the criterion of monophyly, i.e. descent from just one stem species, is met (Hennig 1950a: 309–310).

The presentation of phylogenetic relationships. An important problem which is related to a truly phylogenetic system is the dichotomy of the stem tree. Many groups consist of two coordinated subgroups, for example Cyclostomata and Gnathostomata, Acrania and Craniota, and Adephaga and Polyphaga. Hennig asked whether this pattern is due to phylogeny or if it was artificial. He came to the clear conclusion that (paleo-) geographic events have influenced the systematic subdivision of many groups, so that a dichotomous splitting pattern has a natural foundation. However, the origination of species in a narrow space and under the influence of ecological mechanisms may not result in dichotomies. Especially in these cases several subsequent species splits may occur in a short period of time and then one can expect that only minor differences of the gestalt have occurred. This implies that the exact phylogenetic relationships between the groups descending from them are hardly detectable (Hennig 1950a: 332–333).

A written presentation of the phylogenetic system can only use one dimension, which is why it is necessary to deal with coordinate groups one after another. ("Coordinate groups," a term much used by Zimmermann, are sister groups in Hennig's later terminology. Hennig had used the term "Schwestergruppen" in Hennig (1949: 137) for the first time, but only once and in passing.) Which group comes first is of no importance in principle. However, very often the group that has remained more similar to the common stem species and thus appears to be more primitive than the other has been dealt with first in the literature (Hennig 1950a: 106–107, 276–277).

7.7 Developing the theory of phylogenetic systematics: Hennig's further thoughts from 1953 to 1955

With the publication of *Grundzüge einer Theorie der phylogenetischen Systematik*, Hennig far had not completed his theoretical work. He had made clear where the deficits of systematics had been, he had pointed out what the tasks of a truly phylogenetic systematics are and he had introduced and redefined a number of terms. However, "[a]lthough Hennig's clearly circumscribed goal in the *Grundzüge einer Theorie der phylogenetischen Systematik* in 1950 was the reconstruction of phylogenetic relationships, his methods were at that time not capable of achieving this goal" (Richter and Meier 1994: 217).

One important item was missing: he had not yet stated clearly how the graded (or better, hierarchically structured) degrees of phylogenetic relationships between several groups can be detected. Indeed, he had thought this was not possible, an opinion first expressed in 1948 (Hennig 1948: 20). The most important contribution at that time was the characterization of certain phylogenetic units.

This soon changed. In 1953 he reviewed the classifications of insects, as laid down in contemporary textbooks and articles (Hennig 1953). He took the opportunity to clarify a number of issues that were of importance for establishing the phylogenetic system of organisms in general. Furthermore, Hennig was preparing a major paper on dipteran phylogeny, based on wing venation fossils. His argumentation concerning the relationship of certain groups was clear and he presented diagrams that come close to his later "argumentation schemes" and to character tables. In short, he applied modern phylogenetics in these works.

Groups as individuals and names. In addition to his earlier statements that organismic groups are individuals, Hennig (1953: 3) concluded that scientific names, such as *Musca, Passer, Felis*, and *Helix*, are proper names ("Eigennamen"), at least in the phylogenetic system. These groups and their names cannot be defined, which means that the determination of the limits of a group cannot be derived from axiomatic principles. Their recognition is a matter of empirical study. On the other hand, Hennig wrote, names for the categorical ranks (genus, family, order, class, etc.) are generalized terms ("echte Allgemeinbegriffe"). They cannot but be defined. In 1955, Hennig added that proper names may well be coined with reference to certain characters of the respective group, but as proper names they have nothing to do with a morphological definition of a group (Hennig 1955: 27). Furthermore, proper names are not necessarily created with reference to derived characters: for example, the Ectognatha as a valid group (insects with primarily external mouth parts) was founded on the detection of several synapomorphies. However, ectognathous mouth parts are not among them: the name Ectognatha refers to a character that is plesiomorphic for the group (Hennig 1955: 27–28).

Note that at this time Hennig was still speaking of the synapomorphies of one monophyletic group (Ectognatha), while he had already introduced the term "autapomorphy" (Hennig 1953: 15). The incorrect application of "synapomorphy" to just one taxon may still be encountered today.

The search for the sister group. In 1953, for the first time Hennig formulated the method of phylogenetic systematics:

> As far as categories above the species level are concerned one might say, the task consists in the 'search for the sister group'. No animal group that has been shown to be monophyletic can be assigned its place in the system as long as its sister group is unknown. (Hennig 1953: 10)

Hence, a phylogenetic branching event is the formation of a pair of sister groups.

At the same time, Hennig refined the term "sister group" and its usage. While he was using the term in its plural form (a group has several sister groups) in 1949 (see above), his wording in the 1953 paper and even more so in his study of the Diptera in

1954 made clear that a monophyletic group has only one sister group (Hennig 1953, 1954). "Die *Bibioniformia ...* stehen wohl in einem Schwestergruppenverhältnis zu den *Fungivoriformia*"(Hennig 1954: 294, and many other examples).

This is also self-evident from Hennig's notion that phylogenetic work consists of the "search for the sister group." Nevertheless, from time to time, he still used the term sister group in its plural form in 1954. This indicates that his phylogenetic terminology was not yet fully settled. But note that he did not categorically exclude the possibility of a tri- or polytomy as a speciation event that may lead to more than just two descendants of a species.

A successful application of the search for the sister group (without having mentioned this term) is evident in some early papers: "A critically evaluating investigation of both larvae and imaginal forms appears to leave no doubts that the Cyclorrhapha have to be derived from a stem form which was also the stem form of the Empidiformia" (Hennig 1948: 58). Hennig's words reflect an immediate switch from practical work ("investigation") to theory (sharing a stem form), but at that time (1948) he had obviously not yet realized that shared derived characters indicate closest relationship. He found the solution only step by step and circumstantial formulations from the time in between were a result.

To decide which place a certain group must be allocated to in the system means finding out whether a group B is more closely related to a group C than to group A. The first step of this procedure is to detect similarities and differences between the groups under consideration. The second step is to decide which similarities or differences are relevant for recognizing the degree of relationship and which are not:

> Here it is decisive to state whether the similarities are plesiomorphic or apomorphic characters. Only similarities in apomorphic characters (synapomorphies) indicate closer relationship of two groups while similarities in plesiomorphic characters (or traits of the bauplan), the symplesiomorphies, are irrelevant in this context. (Hennig 1953: 14, translated)

Hennig illustrated this with an example from insect systematics: are the Mecoptera more closely related to Diptera or to Neuroptera (Hennig 1953: 14–16)? He pointed to an important methodological train of thought: if one intends to compare several groups, this must not be done with derived species as examples, but one must try to reconstruct the ground plans of the respective groups first and then compare these instead. Hennig went on to explain that there is another group of characters, which is of no value for detecting phylogenetic relationships: apomorphic characters possessed by one group only. For these he introduced the term "autapomorphy" (Hennig 1953: 15), which he first used frequently in 1954, in his review of the evolution of the dipteran wing venation (Hennig 1954).

Especially in this work, Hennig's terminology had become much clearer. For example, he wrote: "the species of the Erinnidae possess a number of characteristic synapomorphic concordances which unambiguously indicate that they are a mono-phyletic group" (Hennig 1954: 331).

In a response to a comment on his paper from 1953, Hennig added that the core of establishing phylogenetic relationships is "the *interpretation* of characters or the states of characters as either plesiomorphic or apomorphic (derived)" (Hennig 1955: 22, my emphasis).

All further steps for erecting the system are comparable to applying a "calculat-ing rule." To illustrate this, Hennig referred to the relationships among the primar-ily wingless insects and used a drawing that almost looks like a character matrix, as we understand today (Hennig 1955: Fig 1, reproduced here as Fig 7.6; for simi-lar illustrations, see Hennig 1954). The far left column shows a list of characters, while the headings for the rows lines show the groups under consideration. For each group the state of characters relevant for establishing the relationships, either lack-ing a symbol or present with a black circle], is given. This served to demonstrate that the wingless Lepismatidae (Zygentoma) are the nearest relatives of the Pterygota (winged insects), that the Archaeognatha are next to the Lepismatidae and Pterygota combined, and so on. (At that time the problem was that many authors combined the primarily wingless insects under the name "Apterygota.")

Homology, synapomorphy, and symplesiomorphy. Hennig explained that the starting point of any character comparison is to make sure that the characters selected for detecting relationships are homologous, for which there are certain criteria (Hennig 1953: 16). However, the terms homology and synapomorphy do not mean the same thing, as both synapomorphies and symplesiomorphies are homologs. Hennig added that in contrast to "homology" there is an indication of the evolutionary direction inherent in the terms symplesiomorphy and synapomorphy (Hennig 1954: 248; and: "The statement that the wing of birds is homologous to the forelimbs of the primates is as correct as the same statement in reverse."; Hennig 1980: 38). As a result, Hennig identified the homology concept as being insufficient as a fundament of a theory of the phylogenetic system (Hennig 1953: 11): the dis-tinction between plesiomorphous and apomorphous characters is decisive (Hennig 1953: 13).

Zimmermann (1937: 992) wrote that "homologous organs are organs of the same phylogenetic origin," while they can be *recognized* as such, for example, by the same position in the bauplan. How to identify homologies was discussed at length by Remane (1952), but because "we can never directly observe the phylo-genetic transformation of a character … in determining homologies we are limited to erecting hypotheses" (Hennig 1966b: 93, 94). A statement about homology, as well as a statement about phylogeny, can be disproven but never verified, as Bock (1974: 389) underlined. Both are scientific theories and so Popper's terminology can be applied, such that these statements can acquire a high degree of corroboration at best.

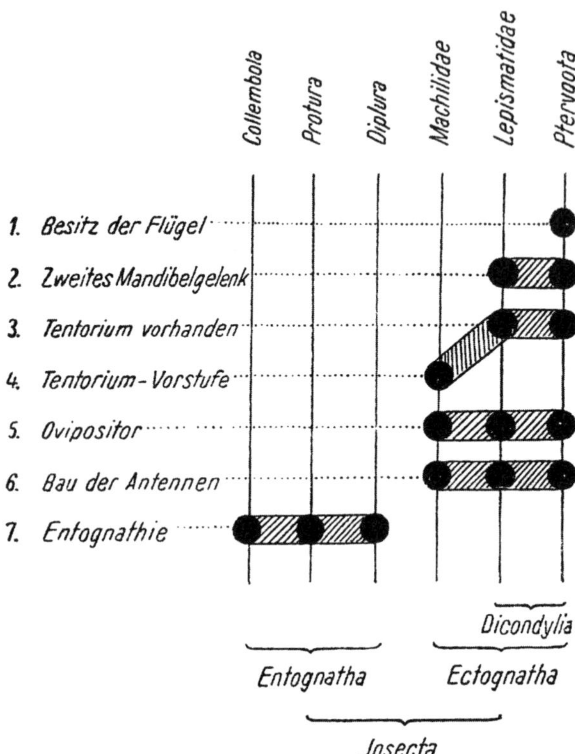

Fig 7.6 The phylogenetic relationships among the primarily wingless insects and their relation to Pterygota (winged insects). Characters are listed on the far left (1–7), the six insect groups in question are listed along the top; below are the monophyletic groups, the relationships are based on the synapomorphies (black circles). The monophyletic groups are connected by shaded areas. After Hennig (1955: 22, Fig 1).

Hennig defined "homology" as follows: "Different characters that are to be regarded as transformation stages of the same original character are generally called homologous" (Hennig 1966b: 93). Later, Hennig defined homology by reference to elements of the phylogenetic system: "Homologous are such characters that have been carried over from common ancestors of their bearers" (Hennig 1980: 18, see also Hennig 1984: 37). As Hennig (1953: 16) realized that a synapomorphy may consist in a negative or regressive feature, he concluded that "the concepts of symplesiomorphy and synapomorphy go somewhat beyond the range of what are ordinarily called 'homologous characters'" (Hennig 1966b: 94, 95).

In 1984 Hennig explained the obvious limitations of the term homology in some detail: "Without any bending of its meaning, the term homology is applicable only to structures that are actually present in an organism ('positive characters')" (Hennig 1984: 38, translated).

However, Hennig added, by definition a character can also be the absence of a structure, thus "the term symplesiomorphy expands [sprengt] the term homology" (Hennig 1984: 43). He recognized that the term homology must be broadened and that additional terms are required that are related to the change of homologous characters (including negative characters) (Hennig 1984: 38, 39). This requirement is fulfilled by the terms apomorphous and plesiomorphous.

A partial solution also came in the introduction to the fourth edition of his book *Wirbellose I* (Hennig 1980), where Hennig made the statement that symplesiomorphy and synapomorphy denominate characters that have been inherited from the common ancestor of their bearers (Hennig 1980: 17). This allows for the inclusion of absence by loss to the "homology" concept. Hennig used an example which does not allow the application of the term "homology" (Hennig 1980: 16). He explained that in phylogenetics one always deals with pairs of characters and that within such pairs the plesiomorphic and apomorphic state can be determined (Hennig 1980: 16). Then he proposed a fundamental critique of the homology concept (Hennig 1980: 18–19). He began by explaining that in insects there are character states "wings absent" (negative character) – "wings present" (positive character). Wings may be (a) primarily absent, which is a plesiomorphous trait, or (b) they may be secondarily absent, which is apomorphic. If the term "homology" only is applied to structures that are present, difficulties arise in such a case. So, Hennig developed the idea that the definition of homology includes the absence of structures, as long as this character evolved in some ancestor of its bearer. However, certainly no one will call the primary absence of wings in different insect groups a homology as it is a symplesiomorphy. Hennig concluded that under the presupposition that homologous characters include negative characters and as long as these originated in some ancestor (but not characters that are primarily absent), then "synapomorphy is always related to homologous characters, convergence never, symplesiomorphy sometimes and sometimes not. However, even under such a view it is sometimes questionable whether the term homology does not play only a theoretical, in practice hardly fruitful role" (Hennig 1980: 19).

This means that the shared symplesiomorphy in primary winglessness of, say, the Diplura and Zygentoma, cannot be addressed by using the term homology because the primary absence of a structure cannot be homologous to another case of primary absence of a structure.

In 1953 Hennig had stated that the starting point of any phylogenetic study is based on characters that are homologous. In contrast to this, he later stressed that this assertion must be treated with caution (Hennig 1980: 18). He wrote that it is often said that in phylogenetic research the first step is to discern homologous and analogous characters, and only then can it be determined whether the conformity in homologous characters is to be interpreted as a symplesiomorphy or a synapomorphy. However, Hennig now concluded, this is only correct to a limited degree.

The usage of different characters may lead to conflicting hypotheses about phylogenetic relationships. Therefore, phylogenetic hypotheses and character interpretations are interconnected, but in many cases character weighing may assist in reaching a conclusion. Bock later pointed out that "circular reasoning must be avoided in assigning relative weight" and that therefore "features must be weighed before the taxa are recognized or the classification deduced," adding that "[p]erhaps the best and most widely used [criterion for judging relative value of characters] is the complexity of the feature" (Bock 1974: 390; see above for Hennig's criteria for selecting characters in phylogenetic systematics). Furthermore, the "evolutionary origin of a new feature or major change in an existing feature is usually given greater weight than the loss of a feature ... because the chances of independent loss of a feature is greater than the independent origin of similar features"(Bock 1974: 390).

Convergence. Under the topic "convergence" one might demonstrate how Hennig's train of thought evolved – or at least how he developed his terminology. Hennig (1953) had first pointed out that apomorphies may arise independently. Hennig (1953: 17) then used the term "synapomorphy" for both derived characters that taxa share from a common ancestor and for characters that are convergencies. A year later he refined the terminology by clarifying that homologous characters may either be plesiomorphies or synapomorphies and distinguished between convergencies and synapomorphies (Hennig 1954: 248–250). Some of Hennig's explanations from 1954 reflect the process of improving the terminology. In order to stress that "apomorphy" and "convergence" are not necessarily the same, he wrote that in certain cases one still has to decide "whether the similarity is due to convergence or *true* synapomorphy" (Hennig 1954: 372; my emphasis). From then on he kept "synapomorphy" and "convergence" apart (e.g. Hennig 1955: 25). For example, while discussing possible derived similarities shared by the Zygentoma and the Pterygota, he stated "as long as it cannot be demonstrated that convergence is probable, I have to agree ... with Snodgrass that the mentioned similarities are to be viewed as "synapomorphies" in fact" (Hennig 1955: 25).

Hennig highlighted an additional criterion that may serve to establish relationships: synapomorphies of two or more groups should in the first instance be regarded as evidence of close relationship and should not be interpreted as of convergent origin (Hennig 1953: 17 and Hennig 1954: 257). Some authors, Hennig noted, had dealt with them in the opposite way. Ten years later, Hennig wrote that "it is not a sound methodological principle to favour the assumption that the presence of similar derived characters is due to convergent evolution" (Hennig 1964a: 63).

Dollo's rule. Like Zimmermann (1937), Hennig (1953: 16–17) discussed the possibility that the evolution of characters can take a reversed course. This would mean that apparent symplesiomorphies have evolved convergently. In 1954, he gave an instructive example: some Nemestrinidae (Diptera) exhibit a reticulate venation pattern in their wings and some authors believed that this is a very ancient pattern. Hennig replied that this is not correct and that "it is a revival of old structures at best" (Hennig 1954: 345). However, Hennig thought that for systematic work the effect of such reversals is limited because in general, any reversed evolution will not go so far that no synapomorphies of closely related groups are left (Hennig 1953: 17). A character that cannot be identified as a plesiomorphy or as an apomorphy that is a reversal, Hennig called a pseudoplesiomorphy (Hennig 1984: 43).

7.8 Finalizing the procedures of phylogenetic systematics (Hennig 1957)

In 1957, Hennig summarized the achievements with respect to ordering organic diversity according to the principles of phylogenetic systematics. Of course, much of what he wrote was repetitions from previous work hence these are not tackled here. However, in Hennig (1957) some of his remarks shed new light on some of his older ideas and some methodological issues became clarified for many of his fellow zoologists.

The significance of species characters. Hennig's introductory remarks deal with species concepts. He criticized the widespread view of that time that one should not describe too many species because systematics would then become a too difficult field (Hennig 1957: 51–53). Individuals do not belong to one and the same species just because they have certain characters in common. On the contrary, certain characters must be interpreted as species-specific when one can assume that they are evidence ("Steckbriefmerkmale") for a reproductive community that is separate from other such groups. The main idea is that morphological characters have no value in themselves in systematics, but are mere indicators for something decisive (Hennig 1957: 54–55).

Phylogenetic relationships and the system of organisms. Hennig also pointed out that one cannot freely decide how to order the animal species. On the contrary, after the detection of the principle of how species and species groups originate (namely by successive splitting events), ordering them has to be in strict accordance with this principle. The question then is whether or not there are objectively existing relationships between these species – independent of whether or not there is someone who can recognize them (Hennig 1957: 57). According to Hennig, in order to answer this question one must refer to the species problem. The notion that a species is a reproductive community satisfies the core of what a species is quite well. The origin of barriers within reproductive communities corresponds to the evolution of new gaps in the stream of genealogic relationships meaning that new species originate. The structure of phylogenetic relationships resulting from this process is hierarchical.

This implies that the representation of phylogenetic relationships is *necessarily* hierarchical (Hennig 1957: 58–59).

Hennig added that no one has ever shown that there are other than phylogenetic relationships between species which are hierarchical (Hennig 1957: 59, 61). This implies that systematists who do not accept the principles of phylogenetic systematics, but use the hierarchical type of system, do this without any foundation and cause confusion.

The fact that the hierarchical structure of the phylogenetic system is expressed in Hennig's (in fact, Zimmermann's, see above) definition of the term "phylogenetic relationship" (and only in this definition) also becomes obvious when the position of the species in the phylogenetic system is considered. Every higher category [taxon] corresponds to a species, Hennig wrote, namely the stem species of all those species (and only those species) which are united in this taxon (Hennig 1957: 62). This implies that a speciation process belongs to each category [taxon], irrespective of its rank. However, this only applies to the correctly constructed phylogenetic system. This system provides us with an exact list of all speciation events that play a role in the origin of the animal world as it currently is.

Methods. As in earlier publications (Hennig 1943: 139, 1948: 3–4, 1950a: Fig 38), Hennig stated that the hypothesis that the more obvious similarities are between organisms, the closer their phylogenetic relationship is, is misleading but not entirely wrong (Hennig 1957: 64). This is the reason why many groups from pre-phylogenetic times are valid when examined via phylogenetic systematics. In order to substitute this hypothesis by a better one, Hennig (1957: 65) enumerated a few simple facts. Characters are always present in different states. For example, the dipteran character "halteres" is a different state of the character "well developed hind wings." The two are homologous organs. If there is a bristle present in one species, there is the alternative of "no bristle present" in other species. These character states occur in a relationship, so that one state can be called "primitive" and the other state "derived." Thus, nothing is more obvious than the conclusion that two species that share the derived state of a character have inherited it from a stem species which is common to them only (Hennig 1957: 66). This implies that the two species are both more closely related to each other than to species which do not have this character state. This shows that the relational directions between features which are valued as the relatively primitive and derived states of one and the same character are identical with the direction of phylogenetic relationships (Hennig 1957: 67). Every group of organisms that shall be accepted as a group of most closely allied species, which is a monophyletic group, must have such derived characters (Hennig 1957: 67).

Graphic representations of relationships. In 1953, Hennig presented a diagram of relationships within the Lepidoptera (Hennig 1953: 48). By studying the literature Hennig realized that none of the classifications reflected the true phylogenetic relationships because both plesiomorphic and apomorphic characters had been used. By basing the analysis only on apomorphies he was able to confirm that the Micropterigidae are Lepidoptera and not the nearest relatives of the Trichoptera. Furthermore, it became clear that groups of higher Lepidoptera originated in a sequence of splitting events. Hennig's illustration demonstrating these discoveries is a milestone in phylogenetic argumentation. In a simple diagram (here Fig 7.7a–b), Hennig showed the occurrence of a number of characters and their states (plesiomorphic *versus* apomorphic) in various groups, using black dots for apomorphies and bars for plesiomorphies. The distributions of apomorphic characters (only a few synapomorphies are shared by the most derived Lepidoptera and, stepwise, more and more are common to more inclusive taxa) reflect the hierarchy of the phylogenetic system and thus degrees of relationship. Note that only 5 years earlier Hennig (1948) thought that such a detailed, clear and transparent demonstration of relationships would never be achievable.

The next step came four years later, when Hennig presented an even more comprehensive illustration of phylogenetic relationships (Hennig 1957: 67–68, Figs 8 and 9; his Fig 8 is reproduced here as Fig 7.8). It shows derived characters as black squares, while plesiomorphic character states are visualized by white squares. Apomorphies shared by more than one species are called synapomorphies, a term introduced by Hennig in 1953 (Hennig 1953: 14). Hennig called this graph an "Argumentierungsschema" (argumentation scheme), a word he also used in his work on the phylogenetic relationships between families of the Diptera Schizophora (Hennig 1958: 508, as an example). This expression was later refined as "Argumentationsschema" (Günther 1962: Fig 3, Illies 1967: 126; the word also appeared in Hennig's 1961 German manuscript that became *Phylogenetic Systematics*, Hennig 1966b). In English it became "argumentation plan" (Hennig 1965: 105, 106), scheme of argumentation (Hennig 1966b: 90, 91), argumentation scheme (Wiley 1981: 139) etc. Hennig viewed such a graph as a means of forcing systematists to visualize the foundation of his phylogenetic system. Later, schemes of this kind have been simplified to exclude the plesiomorphic states altogether, while the black squares symbolizing apomorphies were replaced by simple bars.

7.9 Hennig's 1966 Book

Perhaps the most important nonavailable book during the past two decades for English-speaking systematists was Hennig's *"Grundzüge einer Theorie der phylogenetischen Systematik"* ... The value of an English edition of Hennig's work is immense, but the price for it is an old book. (Bock 1968: 646)

Indeed, the manuscript that became Hennig's 1966 book was completed in 1961 and, after its delivery, Hennig had no opportunity to correct or amend any part.

(a)

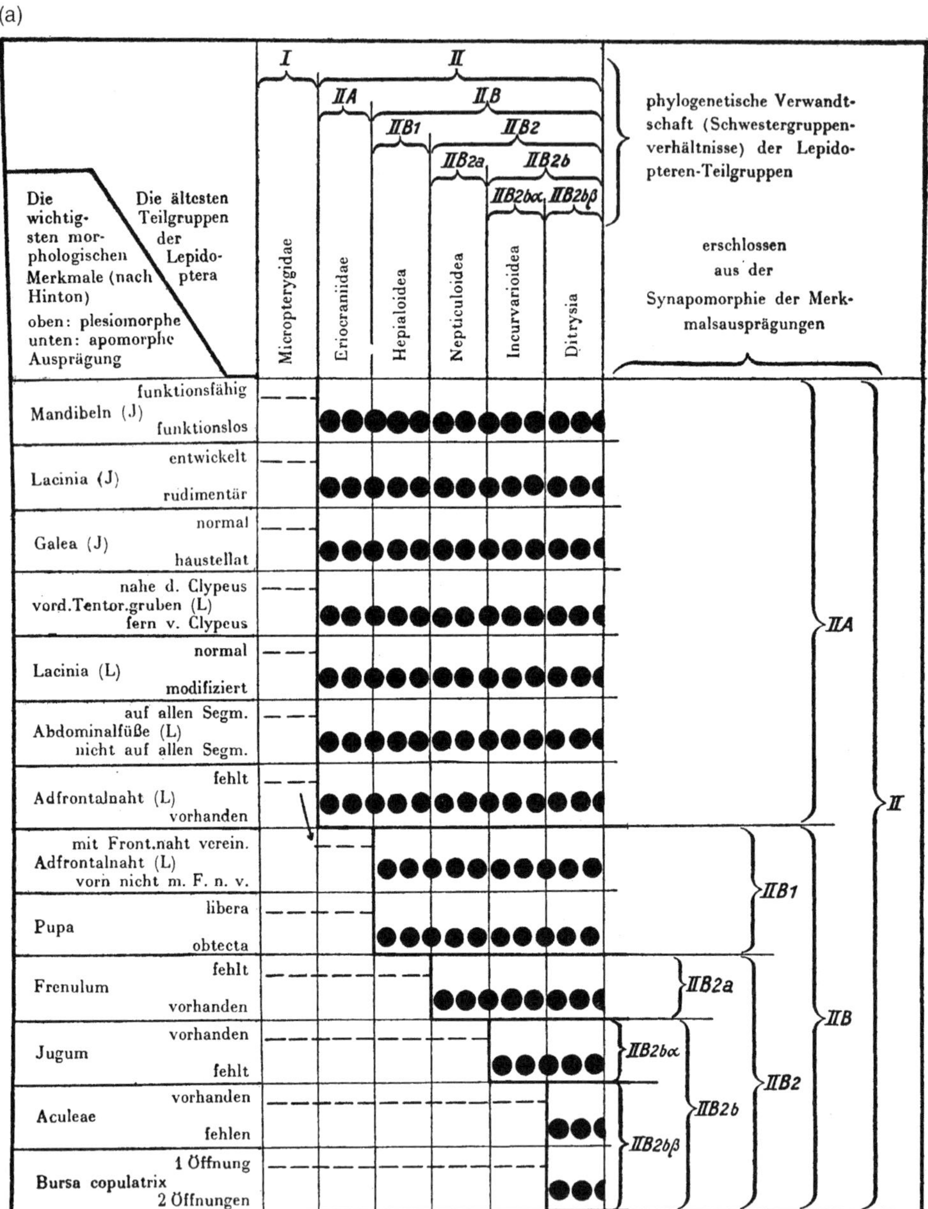

Fig 7.7 Relationships among the Lepidoptera. (a) Apomorphic characters (black dots) unite subgroups into ever smaller taxa. The Micropterygidae (I) are the sister group of the remaining Lepidoptera (II). Sequences of numbers and letters reappear in (b). Analysis of the distribution of apomorphies results in the phylogenetic scheme in (b). After Hennig (1953: Figs 8 and 9).

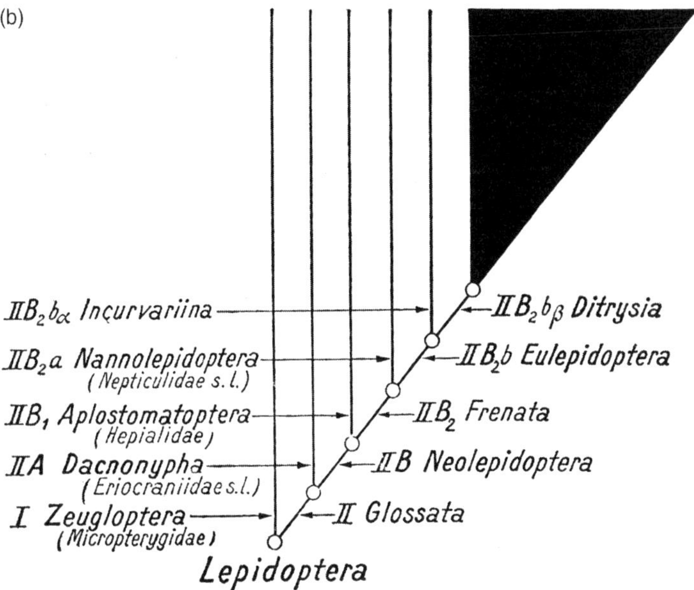

Fig 7.7 (cont)

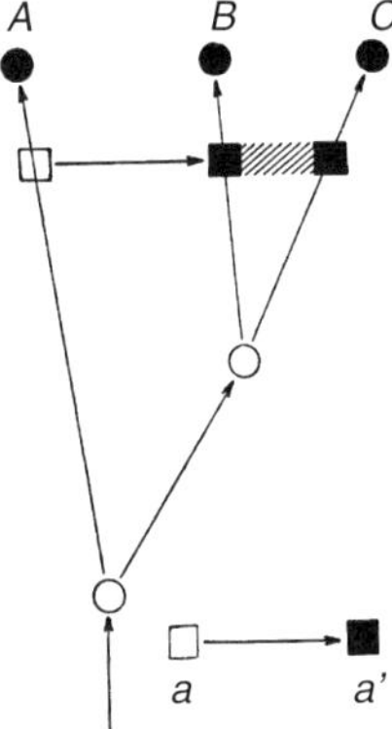

Fig 7.8 Hennig's (1957) version of the argumentation scheme for phylogenetic considerations. Apomorphic states are represented by the black squares; the plesiomorphic state of the same character is represented by the white square. The shaded area between B and C indicates their union by synapomorphy (which excludes interpretation of the character as a convergent similarity). After Hennig (1957: Fig 8).

This is regrettable as the English translation lacks a few parts and contains certain formulations that Hennig could not have accepted (Wolfgang Hennig 1982: 5; but see Schmitt 2013: 138–143, who deals with this topic at length). With respect to its content, the main difference between Hennig's 1950 book and the 1966 book was that the latter included a clear presentation of the methods of phylogenetic systematics as first described in Hennig (1957). Therefore, apart from a few sections on zoogeography there was nothing new when compared to Hennig's publications until 1957. An analysis of the figures in *Phylogenetic Systematics* (Hennig 1966b) reflects the mixture of the older elements with a few new developments. Of the 69 figures, 20 were taken from the 1950 book (with a few slightly redrawn), 9 were taken from Hennig (1957), and 7 had been published in previous publications from 1954 to 1962 (for example, the figure of the insect relationships from 1962 was derived from a 1961 presentation; Hennig 1962: 31). Some of the new figures in Hennig (1966b) represent ideas that were published in similar form in 1950 and are thus cannot be considered new. However, there are two illustrations explaining Hennig's view of the delimitation of species in time which have no earlier counterpart.

Many topics in Hennig (1966b) were laid out in much shorter sections and therefore better arranged than in Hennig (1950a). With respect to the species concept, Hennig became clearer in several important topics. First, he pointed out that in hermaphroditic and self-fertilizing species (such as parasitic flatworms):

> the absence of dioeciousness is a derived exceptional condition… In the light of this fact the question of the species concept in organisms without bisexual reproduction is no more than a relatively subordinate special problem of systematics. (Hennig 1966b: 44)

Hennig added but he did not offer a solution and did not consider the primary absence of dioeciousness in groups outside the metazoans. Second, while Hennig (1950a) had made clear where the limits of a species in time are, this was explained in much more detail in Hennig (1966b).

In accordance with the conclusion "that a hierarchic system is the appropriate representational form of phylogenetic systematics," Hennig (1966b: 58) wrote a sentence that has often been cited and discussed:

> The limits of the species in a longitudinal section through time would consequently be determined by two processes of speciation: the one through which it arose as an independent reproductive community, and the other through which the descendants of this initial population ceased to exist as a homogenous reproductive community. (Hennig 1966b: 58, here Fig 7.9, cf. Hennig 1950a: 102, see above)

A splitting of the initial population into two populations of almost equal size and equally strongly different successor populations would present no theoretical or practical difficulties for the determination of species limits. However, Hennig added that perhaps "more commonly only a small partial population splits off from the parental population and becomes a new species" (Hennig 1966b: 58). By referring to a few insect examples and paleontology, Hennig described other patterns

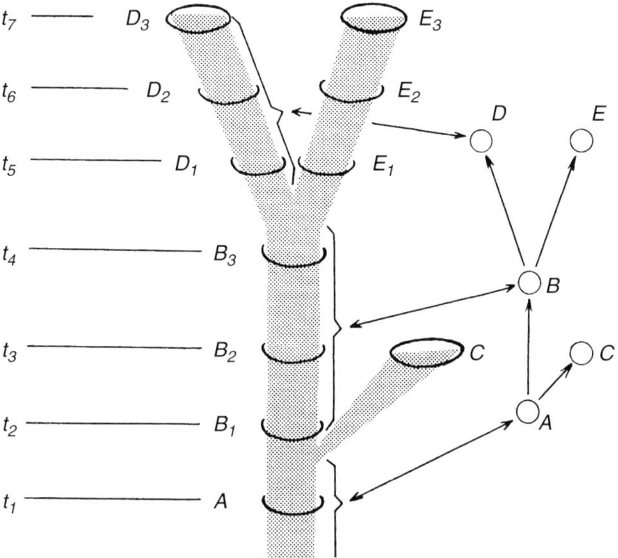

Fig 7.9 Delimitation of five species (*A–E*) in time. Stem species *A* splits into daughter species *B* and *C*; stem species *B* splits into daughter species *D* and *E*. The temporal extensions of species *A*, *B* and *D* are highlighted by brackets. B_1–B_3, D_1–D_3, etc. are temporally successive stages of each respective species. After Hennig (1966b: Fig 14).

that may occur when a species splits. If individuals from two successive populations A and B are almost identical, paleontology would be unable to conclude that these individuals belong to different species, although there was a speciation event between A and B, leading to a contemporary population of B, namely population C.

Hennig continued:

> On the other hand, there is the possibility that the descendants of a population may change without a cleavage having taken place. [In such a case] paleontology ... could not determine on the basis of the ideas we have developed so far whether it was dealing with "one and the same species" ... Apparently, there are two ways of escaping this dilemma, one is to incorporate into the genetical species concept a morphological principle criterion for the time dimension ... This course has been followed in the proposal of the concepts of "paleospecies" and "chronospecies", which are purely morphologically defined concepts. (Hennig 1966b: 62)

Hennig (1966b: 63) judged these species concepts – "paleospecies" and "chronospecies" – as the result of an unhappy attempt to determine species morphologically and returned to delimiting species by two successive processes of speciation. Hennig explained the conceptual difficulties in paleontology by referring to Simpson (1951) who once noted that it is "not useful taxonomy to classify as the same thing throughout" a species that has changed between two processes of speciation. Hennig opposed this view:

> We can point to the phenomenon of metamorphosis; different metamorphic stages of the same individual are not "one and the same thing" in this sense either. It is not clear where the difference is supposed to lie when we speak analogously of transformation stages of one and the same species. The idea that under certain circumstances populations … are to be regarded as "different species" although they are not at all "different" morphologically presents great difficulties to many authors … A narrowness of outlook in the logical interpretation of phylogenetic systematics is expressed in discussions of this question. (Hennig 1966b: 63–64; ellipses represent omitted figures and references)

and

> the problem of the so-called "living stem species" in neozoology is an artefact of a morphological or partly morphological species concept. (Hennig 1966b: 72)

Hennig noted a fossil form that is identical with a recent species (Hennig 1965b: 24). If it is also the stem species of other recent species, the two cannot be subsumed under the same species name, as this is not in accord with the theory of phylogenetic systematics. To do so would be to fall back onto a purely morphological species definition. In conclusion, Hennig further developed the biological species concept (see Willmann 1985 for more details).

In an account showing how practical work on species should be, Hennig pointed out that the description of a new species is a hypothesis, "the hypothesis that the specimens he is describing belong to a separate, previously unknown, reproductive community" (Hennig 1966b: 67).

7.10 Fossils and phylogeny reconstruction

While many paleontologists were searching for ancestors, Hennig suggested that the question "of whether certain fossils belong to the actual line of ancestors of recent species groups can only be answered when this possibility can be ruled out. On the other hand, the assertion that particular fossils *must* have recent descendents can never be proved" (Hennig 1981: 31, Hennig 1969: 35).

The methods for clarifying relationships of fossils are the same as with recent organisms: only synapomorphies indicate close relationships.

In 1954, Hennig's consideration of fossil species led him to consider the temporal dimension of phylogenetic relationships (Hennig 1954). He distinguished between several points in time: that which marks the origin of a group and the origin of its subgroups – the "age of origin" ("Entstehungsalter") and the "age of subdivision" ("Gliederungsalter"). Obviously, the age of origin of any group precedes the age of

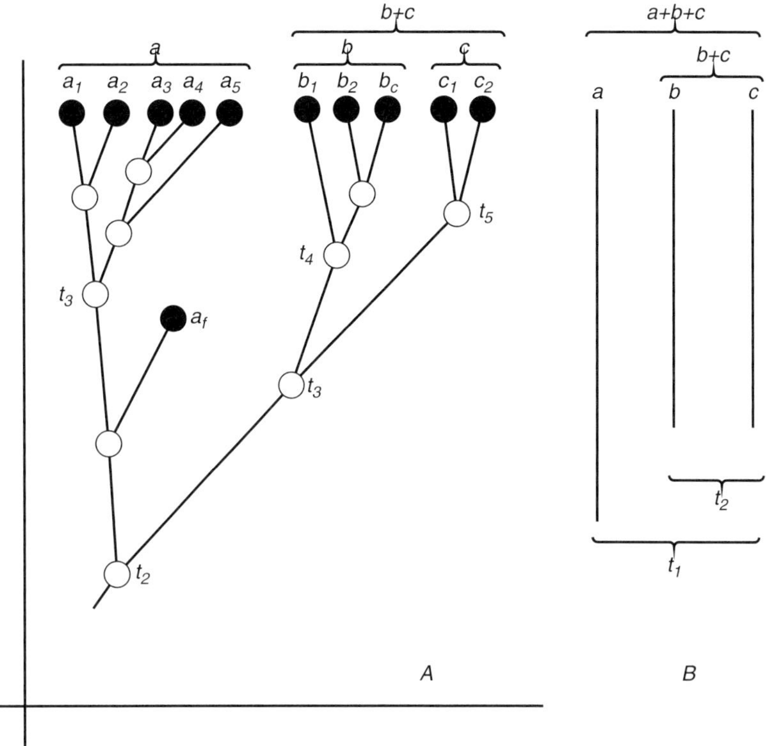

Fig 7.10 Age of origin and age of subdivision of organismic groups, beginning with t_1 (origin of the two groups a and $b + c$). The fossil form a_1 that belongs to the intersection between the age of origin and age of subdivision of a group (in this case of group a) is not a member of a higher group of its own but a member of group a. After Hennig (1954: Fig 268); from 1965 on by Hennig specified as a member of the stem group of a recent group.

subdivision (Hennig 1954: 379, 380, here as Fig 7.10). These thoughts led Hennig to the concept of stem groups.

Stem groups. The distinction between age of origin and age of subdivision of a group has significant consequences for understanding the systematic position of fossil organisms (Hennig 1954: 380). A fossil that belongs to the intersection between the age of origin and the age of subdivision is often regarded as a member of its own group. This leads to misunderstandings as the age of origin is assumed to be younger than it is because the fossil is not included as an early member of the group: "This is a consequence of morphological-typological thinking and contrasts with phylogenetic thinking" (Hennig 1954: 380, translated).

To avoid making such mistakes, Hennig recognized these fossils as a member of the recent group. Knowledge of the true systematic position of fossils allows the determination of the *terminus post quem non* of the origin of a group (Hennig

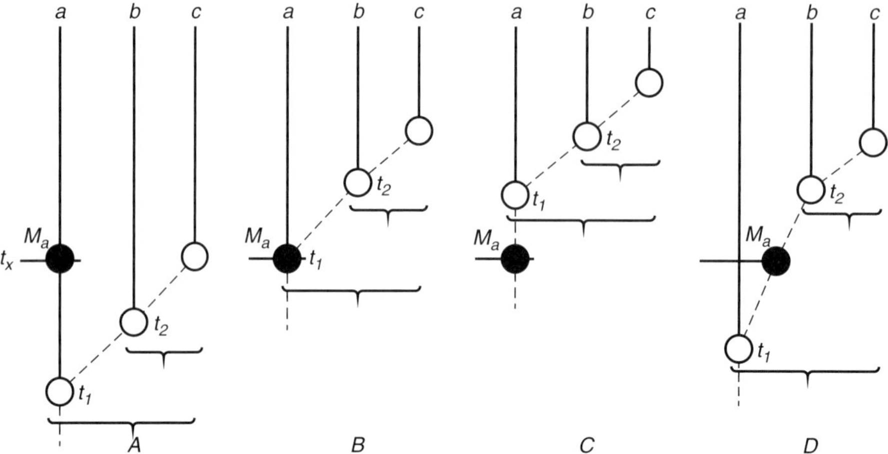

Fig 7.11 Possible phylogenetic positions of a species (black circle) that only has plesiomorphous characters when compared to the subunits of a monophyletic group. After Hennig (1954: Fig 269, for details, see text).

1954: 380). Again, Hennig explained where a fossil may belong that has only plesiomorphous characters as compared to other members of the group to which it belongs (Fig 7.11):

1. It may belong to a recent subgroup which is also plesiomorphous in all of its characters; but alternatively
2. it belongs to the stem species of all the subgroups; or
3. it belongs to an ancestral stage of the stem species.

Furthermore, it can belong to a more apomorphous subgroup of the group as long as this subgroup had not yet developed derived characters recognizable in fossil specimens.

In 1965, Hennig accepted stem groups as preliminary formal elements of phylogenetic systematics (Hennig 1965b: 19). Those fossil species belong to the stem group of a monophyletic unit that originated during the time period between the origin of this group and the splitting of the last stem species of all its extant species. It does not matter whether these are true ancestors or extinct side branches in the tree.

Hennig (1969) later refined his concept of stem groups, explaining that the full complement of constitutive characters of the recent members of a monophyletic group was present in the latest stem species of this group (Hennig 1969: 29, 1981: 24). However, there may have been a long period during which these autapomorphies evolved. Hennig demonstrated this point using examples from insects. The Trichoptera and Lepidoptera form a monophyletic group of higher rank, the Amphiesmenoptera. Hence, there are three stem species which need consideration: the latest stem species of the recent Trichoptera, the latest stem species of the recent Lepidoptera and the latest stem species of the Amphiesmenoptera. In all three groups, a certain period of time will have passed during which their autapomorphies evolved. This led Hennig to the question of how to delimit groups of organisms when fossils are included (Hennig 1969: 29, 1981: 24–26). He discussed two acceptable possibilities:

1. A group could be defined to include the species that have descended from the latest common stem species of this group;
2. A group could be defined in relation to its sister group. Then, it would not only include the latest stem species of all recent members of the group and these recent species, but also all of its ancestors which descended from the stem species that is also the stem species of its sister group. This has the practical advantage that one can assign to a group all those fossils that have at least one constitutive character of the group.

Obviously, a decision as to whether option (1) or (2) is to be preferred will have consequences on determining the age of a group. Hennig chose the second method for assigning fossils to recent groups (Hennig 1969: 24; Hennig 1981: 29).

As a result, Hennig distinguished between three kinds of groups (Figs 7.12 and 7.13):

1. A monophyletic *group which comprises all its recent members (e.g. *Trichoptera in Fig 7.13, marked by an asterisk; this is the crown group of Jefferies 1979, the "groupe apical" of Tassy 1999: 18);
2. A recent monophyletic group which includes all fossils that are more closely related to the recent species of this group than they are to the recent species of its sister group (e.g. Trichoptera, no asterisk; this is the pan-group [Pan-Monophylum] of Lauterbach 1989);
3. Fossils of a certain group that cannot be shown to belong to a *group can be referred to as members of the stem group of the *group (Hennig 1969: 35, 1981: 30).

If study of the fossils [of a stem group] shows that some are much more closely related to the recent species of the whole group than others, the desire to reflect this in the hierarchy of the classification is irresistible (Hennig 1969: 35, Hennig 1981: 31). This may lead to a gradual dissolution of the stem group.

However, because of difficulties in practice, Hennig found that it inevitably led to discriminate between the two types of stem groups: the "echte" and the "unechte Stammgruppe" (Fig 7.13); valid and invalid stem group. The French translations are more meaningful: "groupe-souche véritable" and "groupe-souche factice" (Tassy 1999: 9). Hennig explained that it may be impossible to tell whether fossils belong to the stem group of a particular *group or if they are early members of the stem groups of the subgroups of the *group instead (Hennig 1969: 36–37, Hennig 1981: 33). He

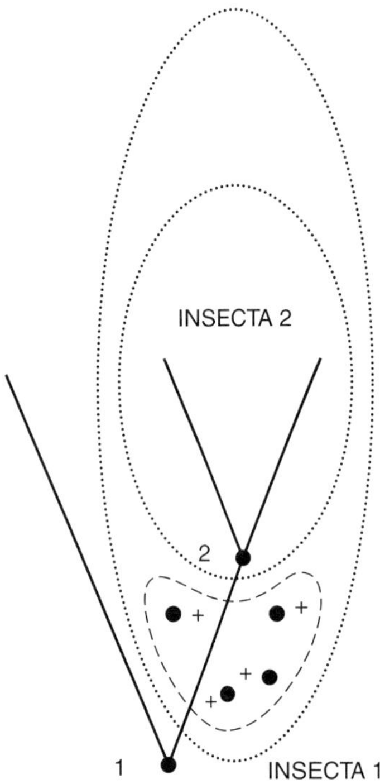

Fig 7.12 Different subunits within Insecta. The *Insecta comprises all its recent members (= Insecta 2 in this figure), the Insecta (no asterisk) includes both, its recent members and its entire stem group (outer dotted line, Insecta 1). Stem group of the insects enclosed by broken line. After Hennig (1983: Fig 3).

called such a mixture of fossils an invalid stem group: "Wherever possible, we must separate these invalid stem groups into their component parts, the valid stem groups" (Hennig 1981: 33, Hennig 1969: 37). The content of a valid stem group was described above: it includes the species which constitute the direct line of ancestors of a particular *group (and only of this group) and also the "side branches" of this line.

The treatment of stem groups led Hennig to discuss ground plans. To recall: in the history of any group there are two important points with respect to time: (1) the point when the group was separated from its sister group; and (2) the point when the last stem species of the recent species of that group ceased to exist. The apomorphic characters of such a group are those features that evolved successively between points (1) and (2). Hence, there are two sorts of groundplan characters (Hennig 1983: 11): those present in the oldest stem species of the group (i.e. at point in time 1) and those of the crown group (*group sensu Hennig). However, it is possible that not a single autapomorphy was present at point 1 in the history of a group. Hennig called the apomorphies of a group that were present at point 2 "derived groundplan characters" (Hennig 1983: 11). Hennig did not address stem groups in his 1966 book.

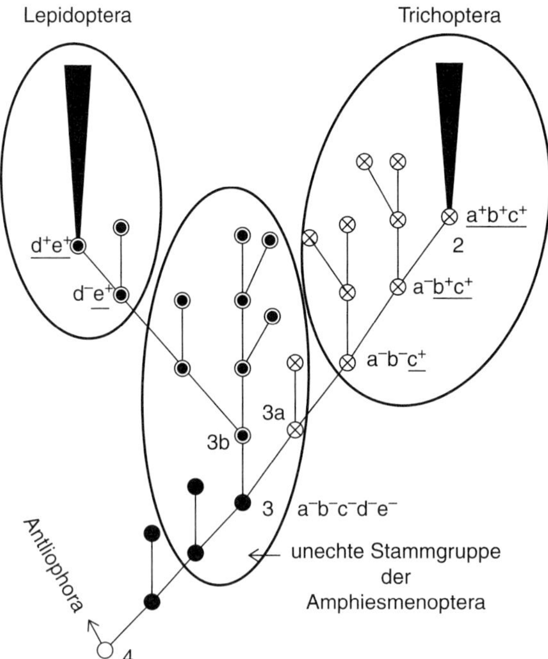

Fig 7.13 Valid and invalid stem groups ("echte und unechte Stammgruppe") using the example of Lepidoptera and Trichoptera. Circles with a black center are members of the valid stem group for Lepidoptera. White circles with a central cross are members of the valid stem group for Trichoptera. Black circles are the stem group representatives of Lepidoptera + Trichoptera (= Amphiesmenoptera). Central ellipse encloses an invalid stem group for Amphiesmenoptera. After Hennig (1969: 37, Fig 8, Hennig 1981: 34, Fig 8).

7.11 Varia

Correctability. Hennig repeatedly pointed out that it is the investigators' duty to present the evidence for their phylogenetic hypotheses because only then can views be corrected or substituted on a well-founded basis and alternatives can only then be discussed. (In the words of Popper, much cited in the phylogeny discussions during Hennig's final years, phylogenetic hypotheses must be falsifiable.) Hennig himself already had to do so in his early phylogenetic entomology papers:

> In my work on larval forms (III, 1952) I have combined *Asiliformia* with *Tabaniformia* in a superordinate group "Tabanomorpha". Today I am ... of the opinion that the *Asiliformia* are more closely related with the 'Muscomorpha', and that the similarities shared with the *Tabaniformia*...must be interpreted as symplesiomorphies. (Hennig 1954: 336, translated)

Wiley explained the logic of such a procedure:

> [A]ny phylogenetic hypothesis is based on hypotheses of synapomorphy, and the synapomorphy logically becomes a proper subset of the phylogenetic hypothesis … Hypotheses of synapomorphy which refute a phylogeny also refute all hypotheses of synapomorphy which form proper subsets of the rejected phylogeny. (Wiley 1975: 238–239, 243)

Griffiths remarked that one "cannot falsify … an hypothesis absolutely. You may think you have falsified it and then subsequently you find that you have not falsified it" (Griffiths 1974: 398; because of new support for a hypothesis that you considered as falsified). Hennig (1972, 1976) himself gave several examples of such cases.

Paraphyletic vs. polyphyletic. Hennig explained that:

> The common feature of polyphyletic and paraphyletic groups is that, unlike mono-phyletic groups, they do not include *all* the descendants of a single stem spe-cies: some of the descendants are excluded from these groups. In fact, there is no sharp distinction between paraphyletic and moderately polyphyletic groups. (Hennig 1981: 6, from the original German in Hennig 1969: 19)

Therefore, Hennig concluded that:

> A distinction between the terms paraphyletic and polyphyletic is possible only on the methodological level … [If] a certain monophyletic group had previously been subdivided into two or more non-monophyletic subgroups … one of two possible errors had been committed (Fig 1, [reproduced here as Fig 7.14]): a subgroup may have been formed on the basis of convergence (such groups have customarily been termed polyphyletic); or a subgroup may have been formed on the basis of symple-siomorphic agreement. (Hennig 1974: 284, Hennig 1975: 247–248)

In the latter case paraphyletic groups were formed:

> Both types of groups […] are similar, for the members of each group-type lack a stem species common only to themselves. As the figure shows for groups of either type, there need be no difference between them in the structure of their genea-logical relationships. The terminological distinction between paraphyletic and poly-phyletic groups is valid, therefore, only when attention is drawn to the particular kind of mistake made in the process of character analysis that led to the formation of the groups. From this standpoint, the terms paraphyletic and polyphyletic are not used for indicating differences in the genealogical relationships between taxa. (Hennig 1975: 248)

7.12 Misunderstandings and unjustified criticisms

What disturbed Hennig to a significant degree throughout his career was that the crucial difference between similarity and phylogenetic relationship (and the incompatibility often encountered between the two) was not understood. An article

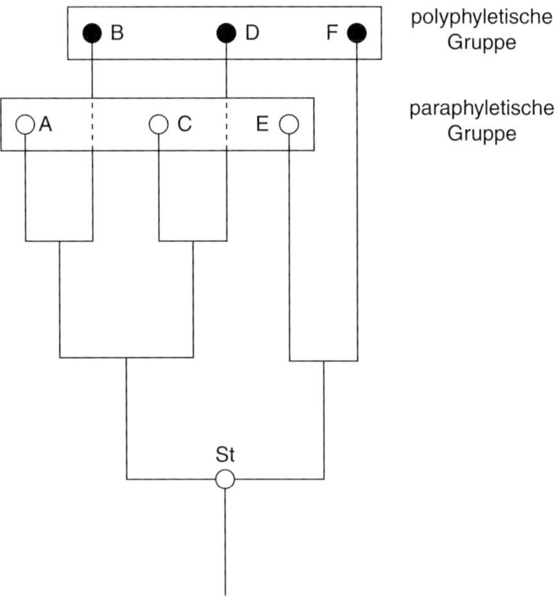

Fig 7.14 Polyphyletic and paraphyletic groups; see text for explanation. After Hennig (1974: 284, Fig 1; Hennig 1975: 248, Fig 1).

published in 1971 reflected Hennig's disappointment and frustration. The main principles of the theory of phylogenetic systematics were still not understood (and, so it appears, this is still the case; Wiley 2009). Hennig (1971b: 12–14) presented an instructive example using a pair of sister groups of flies, the Sciadoceridae (2 species) and Phoridae (2500 species). One species of the Sciadoceridae occurs in New Zealand and Australia, the other in South America, the Phoridae are cosmopolitan. Two fossils described from the lower Cretaceous Canadian amber were shown by their respective authors to be phylogenetically related to the Phoridae in a stem tree (Fig 7.15). Hennig noted that this was obviously correct; the fossils belong to the history of the Phoridae and not the Sciadoceridae. Hennig then pointed out a crucial mistake. In the text the fossils were said to be in the Sciadoceridae because they resembled the Recent sciadocerids more than the Recent phorids. Therefore, the authors concluded that in the Cretaceous, the Sciadoceridae were distributed in the northern hemisphere. However, there is no evidence for this. As Hennig wrote, this was a classical example of the logical mistake that occurs whenever one combines typological systematics with a biogeographical conclusion that superficially appears as if it was based on a phylogenetic system. Hennig wrote that he had tried to convince the authors that the fossils are phylogenetically related to the Phoridae rather than the Sciadoceridae by using their own stem tree. However, their response was: "But they are still Sciadoceridae." Hennig added that even additional

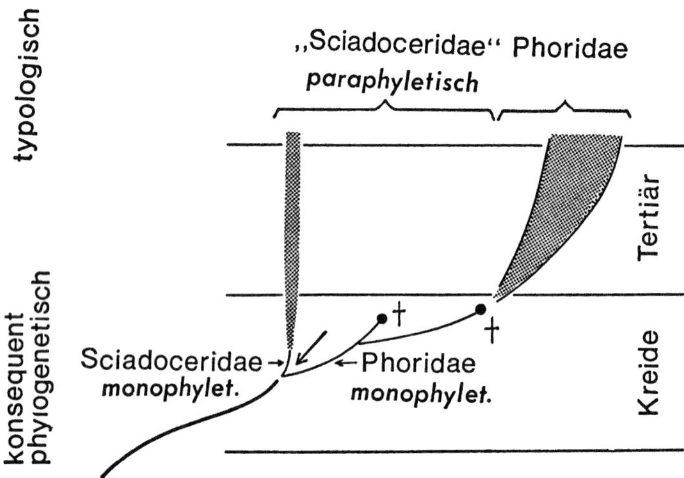

Fig 7.15 "The Cretaceous Phoridae are still Sciadoceridae"; Sciadoceridae as a paraphyletic group. After Hennig (1971b: 13).

explanations had not resulted in any further understanding. He concluded: if it is correct that only the theory of phylogenetic systematics allows unambiguous historical questions to be asked and to allow clear answers, then consideration of phylogenetic systematics should be part of all biological training (Hennig 1971b: 15, modified from the German).

However, up until 1970 only a few systematists were following Hennig's ideas. Nevertheless, his ideas were to cause a scientific revolution in biology. Systematics eventually became synonymous with the reconstruction of phylogeny producing a clear methodology and a true science. "Morphological similarity" was divided into three categories: symplesiomorphy, synapomorphy and convergence (Hennig 1965b: 24: "Die Methode der phylogenetischen Systematik beruht auf der Auflösung des Begriffes der morphologischen Ähnlichkeit in die 3 Kategorien Symplesiomorphie, Synapomorphie und Konvergenz"). As the former two are specified homologies (with the exception of symplesiomorphic "negative" characters; Hennig 1980: 18–19, 1984: 38), it also implies that Hennig had divided homology into two categories: symplesiomorphy and synapomorphy. (In a posthumously published work, Hennig would go so far as to question the value of the concept of "homology," Hennig 1980: 18–19, see above.) Reconstruction of the degrees of phylogenetic relationships via the search for the sister group was based upon discovering synapomorphies. Taxonomy and systematics had merged because any taxonomic statement (e.g. species A belongs to genus X) is, at the same time, a systematic statement. "Classifications" were substituted by the systematization of organisms as the units of a system are natural, individual-like entities with proper names, while the units of a classification are classes. Natural units originate in a phylogenetic splitting event and they cease to exist when its last representative dies. Speciation was recognized as being the occurrence of reproductive isolation which is a splitting event and species are thus delimited by nature and not by convention.

Many biologists did not agree with all of this and defended traditional points of view. Innumerable papers reviewing, commenting, refining, criticizing or transforming Hennig's ideas appeared, often in the form of personal statements. As a result, Hennig (1971b) concluded that current discussions on systematics show that clear issues are in danger to be talked to death.

Before I discuss influential criticisms published immediately prior to Hennig's death, I will comment on a fundamental mistake, first discussed by Illies (1967: 521–522) and Günther (1971: 40). They wrote that one misunderstanding was that the task to systematize organisms is the same as ordering any number of any objects and that the theoretical basis for ordering them can be found in mathematics. Both of them stressed that this is certainly not a valid biological foundation, as it ignores (in contrast to phylogenetic systematics) evolutionary history and thus phylogenetic relationship as the natural order. Hennig (1971b: 9–10) added that the degree of relationship in numerical taxonomy is phenetic or a typological relationship (although the results may agree with those of phylogenetic systematics in certain cases) (see also Hennig 1966b: 79–80).

A basic misunderstandings was that "relationship among organisms, in cladistics, is based upon joint possession of derived features (apomorphous or advanced characters)" (quoted from Bock 1974: 376) or that "a cladistic classification is a hierarchy of ever more embracing synapomorphies" ("Eine kladistische Klassifikation ist eine Hierarchie von immer weiter umfassenden Synapomorphien," Mayr 1990: 268). However, this is not the case. Phylogenetic relationships are the result of (and hence based upon) the unique history of life and the degree of relationship depends on the sequence of phylogenetic splitting events. Phylogenetic relationships are not based upon joint possession of derived features. It is true that in order to detect phylogenetic relationships apomorphic characters as indicators of relationships are needed – but that is all. Systematic relationships cannot be altered, as they are the result of natural processes, while any systematic statement is a refutable (correctable) hypothesis, as one can misinterpret the states of characters chosen for phylogeny reconstruction. Beyond that, phylogenetic relationships may evolve even without the evolution of apomorphic characters in one of two diverging evolutionary lineages, as Hennig demonstrated repeatedly by referring to the deviation rule. This especially makes clear that phylogenetic relationships and characters are not necessarily interconnected. Time is the criterion for relative closeness of the relationship, not characters (Hennig 1948:8).

Mayr contra Hennig: Hennig rarely replied to criticism of his work – with one major exception. When Ernst Mayr published an attack (in fact, it was a renewed attack, see Mayr 1969: 70–71) on phylogenetic systematics (Mayr 1974), Hennig could not ignore it because, as he replied, Mayr had misunderstood a great deal and would cause considerable confusion due to his great reputation (Hennig 1974). Hennig's rebuttal was published in German and considered to be of such importance that it reappeared some months later in *Systematic Zoology* in an English translation (Hennig 1975). It was especially clear from the German version that Hennig was very upset. As Peters (1995: 7) noted, according to his experience, the difficulties of understanding were not Hennig's fault but were anchored in the minds of those who had problems substituting "similarity comparisons" with "analyses of relationships" and "classification" with "systematization."

Many years later, Mayr, this time with Walter Bock as co-author, repeated much of what he said in 1974 (Mayr and Bock 2002). Although that paper was published long after Hennig's death, I refer to it because for the last time it presented a lengthy defense of classification (as opposed to systematization) and because it contained many important misleading statements, as Mayr had done earlier (Mayr 1974).

It is worth mentioning here that Hennig (1980: 23) expressed his dislike of the term "cladistics" as used by Mayr. Very rarely did Hennig allow his feelings to be expressed, but here he wrote: "strict phylogenetic systematics [is] also pejoratively called cladistic by its opponents" (Hennig 1980: 23).

Mayr's misunderstandings were typical for the time, and, unfortunately, some of his misunderstandings have survived.

Mayr stated that phylogenetic systematics (his preferred term was cladistics) was intended to produce "the best classifications" (Mayr 1974: 96, also Bock 1974: 379, Mayr, 1990). However, alternatives are not "best" or "not so good" as such a wording is only possible if one classifies with a purpose (Hennig 1974: 281, referred to Günther, wrote "das Tiersystem der Kochbücher ist das beste für die Zwecke eines Kochbuches"; Hennig 1975: 245, "The system of cook books is the best for the purpose of a cook book"). Phylogenetic systematics, however, reconstructs the phylogenetic relationships and records the results in a written system. As it is a hypothesis, this system reflects the state of our knowledge of relationships at that time. It may be wrong or right, it may be well founded or not, but it cannot per se be "good" or "better." Once the system has been correctly reconstructed (which only is possible with phylogenetic systematics), the question of whether it is good or not is beyond the issue. Mayr wrote that cladistics consists of two "quite different sets of operations": "the reconstruction of the branching pattern of phylogeny" and the construction of a *cladistic classification*" (Mayr 1974: 97, italics in original). Mayr believed that a cladistic analysis "does not automatically provide a classification" (Mayr 1974: 123). This was the view of adherents of evolutionary classifications. In phylogenetic systematics, however, the "cladistic classification" (the textual hierarchical phylogenetic system) is the inevitable outcome of the first set of operations.

Not a different "step," cladistic systematization cannot be a failure ("Fehlschlag" in Mayr 1990: 266) when the reconstruction of the branching pattern is at the same time "an extremely valuable contribution to the method of systematics" which Mayr did admit to.

One of the most disturbing paragraphs in Mayr's paper is the following:

> In order for their method of classification to work, cladists have to make a number of arbitrary decisions, involving a redefinition of well-known terms, a re-interpretation of adaptive evolution, and the proposal of a new species definition. When these arbitrary decisions are rejected, very little support for cladistic classification remains. (Mayr 1974: 100)

If one rejects essential definitions, which are a presupposition for any particular theory (and on which they are based for sound practical work), then, of course, not much of a theory is left. For example, Mayr (1974: 100–101) criticized Hennig for having redefined the word "phylogeny" (see also Mayr 1969: 70: "Users of the recent literature are warned to look out for the misleading use of the term phylogeny by the cladists"). Hennig replied that phylogenesis has never before been unambiguously defined:

> According to evolutionary theory, phyla originate by successive cleavage events within organismic communities of reproduction. Phylogenetic systematics defines phylogenesis in this unambiguous sense. (Hennig 1975: 246)

Another example is the term relationship:

> [R]elationship for the cladist is genealogical kinship. But cladistic kinship alone, for an evolutionist, is a completely one-sided way of documenting relationship, because it ignores the fate of phyletic lines subsequent to splitting. (Mayr 1974: 102)

This is no argument at all. In phylogenetic systematics one intends to reconstruct the phylogenetic system (= phylogenetic relationships). The "fate of phyletic lines subsequent to splitting" can be visualized by pointing to the characters that have evolved after a splitting event. These characters are, in the terminology of phylogenetic systematics, autapomorphies. Autapomorphies are important if one compares sister groups to demonstrate evolutionary divergence. It is, of course, fascinating to know, for example, that birds have accumulated many more autapomorphies than crocodiles and, in Hennig's opinion, "information about the possible amount of adaptiogenetic divergence of sister groups is here even better expressed than in the usual classifications" (Hennig 1975: 249).

Hennig made it clear that Mayr was inconsistent with his own views on classification:

> Mayr would have reason to reject groups such as Chordata, which include divergent adaptiogeneses (e.g. Tunicata and Aves), and Mammalia, which also include divergent adaptiogeneses (e.g., Monotremata and Proboscidea). (Hennig 1975: 249)

Mayr's discussion of the term 'monophyletic' may serve as a third example. He stated that:

> [E]ver since Haeckel [the term monophyletic] has been applied to groups which satisfied two conditions: 1. the component species … are believed to be each other's nearest relatives, and 2. they are all inferred to have descended from the same common ancestor … [So] birds are monophyletic, crocodilians are monophyletic, and reptiles are monophyletic. (Mayr 1974: 104)

Under this definition the three groups are indeed monophyletic because the term "relative" sensu Mayr is not phylogenetically defined and because this definition of monophyly is nonsensical: any two taxa have descended from the same ancestor (Hennig 1975: 247), among them the Reptilia (with the exclusion of birds). Nevertheless, Mayr went on by saying: "Hennig has created enormous confusion by adding to the traditional definition of a monophyletic taxon the following qualification: '… and which includes all species descended from this stem species'" (Mayr 1974: 104). However, a monophyletic group sensu Hennig is not limited by convention. It exists as an individual in nature (Hennig 1950a, 1953). Five years before, Mayr himself had written that "the usual phrasing of the principle of monophyly ('a taxon is monophyletic if its members are descendants of a common ancestor') is too vague" (Mayr 1969: 75).

The importance of Hennig's definition of monophyly is evident from four lines, as cited by Hennig in the following:

> Von der Phylogenese einer Tiergruppe zu sprechen, hat nur Sinn, wenn es sich um eine monophyletische Gruppe handelt. Die Stammesgeschichte einer solchen Gruppe beginnt mit der Entstehung ihrer Stammart, und sie endet mit dem Aussterben des letzten körperlichen Nachkommens dieser Stammart.
>
> [To talk about the phylogeny of an animal group makes sense only if it is a monophyletic group. The phylogeny of such a group begins with the origin of its stem species, and it ends with the extinction of the last physical descendant of this stem species.] (Hennig 1965a: 13)

The convoluted history of the term monophyly in systematics was reviewed by Farris (1990).

Dichotomous splitting. Mayr assumed that in phylogenetic systematics only dichotomous splitting is accepted. However, polytomous splitting was never denied by phylogenetic systematists, but in the practical work on the phylogenetic system it may be impossible to reconstruct the sequence of a series of splitting events. But, as Hennig emphasized, as long as it appears to an investigator that he is dealing with a tri- or polytomy, he can never be certain if there is, in fact, a series of dichotomies hidden in the respective branching area of the cladogram. Thus, any polytomy is a challenge for further research. In Hennig's words:

> [T]he principle that every monophyletic group has only one sister group, although not strictly verifiable empirically, has a high heuristic value: it challenges the

investigator to study carefully every case where no dichotomy has yet been demonstrated. (Hennig 1975: 255–256)

Phylogenetic classification? There is a fundamental difference between a classification and a system:

[T]he construction of a cladogram in accordance with the principles of phylogenetic systematics results in a system rather different in principle from various kinds of possible classifications. Although my original perception of this distinction was somewhat unclear, I have nevertheless avoided speaking of phylogenetic 'classification', preferring instead phylogenetic "system" – but I have sometimes used "classification" under the influence of English usage. (Hennig 1975: 246)

Griffiths (1974) made the issue clear: in nature there is a system of organisms that has been shaped by evolutionary processes over the course of time. If one reconstructs exactly this system, one is not classifying (ordering objects according to certain characters and thus creating classes), but systematizing.

However, Mayr explained:

The evolutionary taxonomist believes that an approach which superimposes a carefully weighted phenetic analysis on a preceding cladistic analysis is better able to establish degree of relationship than either a purely cladistic or an unweighted phenetic approach. And a classification based on such a multiple-based determination of relationship will be more reliable and more predictive than one based on one-sided criteria. (Mayr 1974: 103)

If so, one would not have had so many different "evolutionary" classifications (Simpson 1975: 6: "different classifications can be consistent with the same phylogeny"). Mayr's opinion allowed arbitrary classifications to persist and for the invention of ever more paraphyletic groups. Therefore, it is not surprising to read that later Mayr wrote: "admittedly, it is somehow subjective to determine where to draw the line between animals that are 'still reptile' and those that are 'already mammal'" in an evolutionary classification (Mayr 1990: 275, "wo man die Grenzlinie zwischen 'noch Reptil' und 'schon Säugetier' zieht").

Subjectivity in evolutionary classifications was never a problem for evolutionary systematists: "[T]his subjectivity is a consequence of basing classification on the simultaneous evaluation of two semi-independent variables rather than just one of the two" (Bock 1974: 378).

Bock added that "No reason exists why biological classification must be based on only one variable and why it cannot be based on two (or more) variables be they semi-independent or completely independent" (Bock 1974: 378). However, there *is* a reason: with two variables (similarity and phylogenetic relationship, in this case) one will never know on which basis a certain statement/hypothesis/etc. is made. A classification based on several ordering principles loses its information content (Hennig 1974, Hennig 1966b: 77; see also Wiley 1975:241 and many others). Günther (1971: 33) had stressed a few years earlier that it is logically impossible to group according to more than one criterion. The logic behind this is Aristotle's posit of the uniformity of the *principium* (or *fundamentum*) *divisionis*.

Mayr and Bock pointed out that, in their view, for biological classifications "entities are grouped together as members of a taxon because they share a suite of homologous features" (which are plesiomorphies *and* synapomorphies) (Mayr and Bock 2002: 178). They also wrote that the correct classification is achieved by overall similarity ("classing together the things that possess in common the greatest number of attributes"), providing that overall similarity is not deceiving as, for example, convergence (Mayr and Bock 2002: 178). Note that in their view the fact of deception does not apply to symplesiomorphies. The idea behind this was simple: "Greater phenotypical similarity implies greater genetical similarity and hence closer relationship" (Bock 1974: 377). However, it is a fallacy because here, once again, the term relationship cannot mean phylogenetic relationship as paraphyletic groups often are a result (Mayr 1974: 119: "the retention of a large number of ancestral characters is just as important an indicator of 'relationship' (traditionally defined) as the joint acquisition of a few 'derived' characters"; see also Mayr and Bock 2002: 189). Accepting such phenetically defined assemblages means to "emphasize the evolution that has not occurred" (Rosen 1974: 447, see also Nelson 1989: 66). Paraphyletic taxa are not individuals, while monophyletic groups are (Hennig 1950a: 115, 120, 210–211, 1953: 3).

Mayr also mistreated categorical ranks (Mayr 1974: 105, 113–115; see also Mayr 1990: 271 and other examples):

> It is implicit in his [the cladist's] principles that he is forced to make the prediction that sister lines derived from a stem species will have sufficiently similar evolutionary fates so that the resulting sister groups can be ranked at the same categorical level (= are coordinate). The case of birds and crocodilians is a particularly convincing illustration of the thousands of occasions where this prediction does not come true.

However, Hennig underlined again and again (and formulated the derivation rule to illustrate this) that sister groups do not have the same evolutionary fates, and that this was of no consequence for phylogenetic systematization. Until it was realized that categorical ranks are nonsensical in phylogenetic systematics, sister groups were assigned the same rank because they had the same hierarchical level, having originated through the same phylogenetic splitting event. Mayr and Bock admitted that because "rank in a Darwinian classification simply indicates level of similarity" (Mayr and Bock 2002: 189) and while there is no objective measure of similarity, ranking would be arbitrary (see also Simpson 1961: 17–18, Mayr et al. 1953: 46).

Phylogeny, evolution and classification. Different concepts of taxa are inherent in different approaches to ordering organisms by either the process of evolutionary classification or of phylogenetic systematics. Bock explained:

> Categories, such as species, genus and family, are words and hence are defined … Taxa are groups of organisms and hence are real objects in nature which are recognized, delimited, described and named. Taxa are never defined. Names are never defined. And it is not correct to speak of one's concept of a taxon such as one's concept of the Dinosauria. These taxa exist in nature. (Bock 1974: 384)

This is certainly correct if one accepts that taxa are the products of natural processes, i.e. that they originate through the splitting of one species into daughter species and cease to exist with the extinction of the last descendant of their stem species. Such a taxon is, in short, a stem species and all its descendants (a monophylum sensu Hennig, or a natural taxon sensu Wiley 1979) but would not include paraphyletic groups, as accepted by Bock. The latter are mental constructs.

The same applies to Mayr and Bock who seemingly agreed with Hennig (e.g. Hennig 1947: 279, 1950a: 111–130, 1966b: 81) that

> [A] higher taxon has certain ontological characters (e.g. restriction in time and space, etc.), which are associated by the philosophers with the designation "individual", but it lacks the internal cohesion of an individual. [Higher taxa] are definitely not classes in the Platonic sense, but biopopulations. (Mayr and Bock 2002: 174)

However, if a higher taxon is not strictly monophyletic (in Hennig's sense) then it is not an individual because its restriction in time is not determined by history, but depends on arbitrary decisions as to which groups shall constitute such a taxon.

Predictability. Repeatedly, the idea that a system (or classification) of organisms should allow for maximum prediction was emphasized:

> A classification, which utilizes all potentially available information, is … more predictive than a classification which arbitrarily restricts itself entirely to the information provided by the branching pattern. (Mayr 1974: 122)

> One widely accepted criterion [for judging which system of classification is the best] is that of greatest predictability of unknown characters in known organisms or in newly discovered species. (Bock 1974: 379)

Bock added that "maximum prediction of unknown features represents the general goal of phenetic approaches and of many evolutionary taxonomic approaches to classification" (Bock 1974: 379). I have never understood what this means. If one classifies organisms according to their similarities (and thus creates classes), one finds in these classes what was put into them. However, this has nothing to do with predictability; it is merely a circular argument. If there are individual (or species-specific, etc.) deviations from the character sets used for such a classification, then one cannot make predictions if anything from a certain class is taken. In phylogenetic systematics, it is not a prediction that one taxon of a pair of sister groups is different from the other, as all the structures, behavioral patterns, and so on, is subjected to evolutionary change. In fact, one cannot predict which characters there are in one group if one only knows the other group (see also Farris 1974: 397). However, in phylogenetic systematics it is possible that if a pair of sister groups is known, one can attempt to reconstruct to a certain degree, and thus "predict," the character set of the immediate common ancestor. If, however, "predictability" refers to Popper's criterion of a testable theory, then as Rieppel notes:

> In order to be testable, a theory has to predict empirical facts (experiments or observations). Otherwise, no deductions of testable predictions would be possible. Phylogeny, the evolution of organisms, is not reproducible, however, nor can phylogenetic development be predicted. (Rieppel 1980: 82)

Hennig's species concept. Hennig's view on species was outlined above. In his reply to Mayr (1974) Hennig wrote: "Hitherto I have assumed that the biological species concept, used by me since 1950, does not essentially differ from that of Mayr" (Hennig 1974, 1975: 255). However, in contrast to Mayr, Hennig included in his version of the species concept that species are systems that have a temporal dimension with clear-cut limits. As late as 2000, Mayr (2000a: 27) could not tell where the temporal limits under his conception of the biological species concept are (see also Meier and Willmann 2000: 37). (However, he did stress that species *taxa* have an extension in time: often, the biological species concept –which reflects the non-dimensional situation – and the delimitation of species taxa are confused, Mayr wrote, "but species taxa have, of course, an extension in time.") Nevertheless, Mayr's concept of species implied a measure of arbitrariness in practical taxonomic work as it allowed for gene flow between species ("harmonious gene pools require a protective device ... the so-called isolating mechanisms are this device. Each species is normally reproductively isolated[1] from other species coexisting at the same locality"; footnote 1: "This protection is much more rigidly enforced among animals than among plants"; Mayr 1958: 15) Hennig's species concept, as a strictly non-arbitrary concept, was termed the "consequent biological species concept" (Willmann 1985).

Hennig would have agreed with Bock that operational definitions are poorly suited to historical biology:

> The approach to definition that I favor may be called "theoretical definition" in which the formal definition of the word is the concept and the word, so defined, stands for the concept. Hence, the word 'species' is defined, for example, using the biological species concept ... the species definition can be associated with a definite biological phenomenon, namely lack of gene exchange ... these [theoretical] definitions, in contrast to operational definitions, do not necessarily tell us how to recognize objects in nature to which the word can be applied ... Words are defined,

> but then working methods must be developed by which objects in nature are recognized and the defined words are applied to them. (Bock 1974: 384)

Mayr wrote that under "the Biological Species Concept, species are no longer considered to be classes (natural kinds) that can be defined, but rather concrete particulars … that can be described and delimited but not defined" (Mayr 2000a: 18). Hennig went even further than Mayr by stating that species (as well as monophyletic groups) are individuals (e.g. Hennig 1950a: 115). Therefore, one should add that species cannot be delimited, but rather that it is the task of the taxonomist to detect their limits. In contrast to his earlier emphasis on gene flow, Mayr pointed out that "the essence of the biological species concept is discontinuity due to reproductive isolation" (Mayr (2000b: 94; see also Mayr 1969: 26: "A species is a protected gene pool" and "species" is a relational term: "A is a species in relation to B and C because it is reproductively isolated from them").

Classification, Darwin, Hennig and systematization. Mayr and Bock explained that the term "ordering system" "denotes the general concept that includes classification as one of its subdivisions" (Mayr and Bock 2002: 172). Once again, it is clear that these authors were not interested in an order that reflects the structure of the natural system (which is a product of phylogenesis) and that the systematization of organisms was not their aim. It is no surprise, therefore, they did not even mention this term. This in turn implies that they were not interested in natural taxa, but in taxa that are, in part, mental constructs. In order to defend the *classification* of organisms, Mayr and Bock added that "a cladistic ordering of branches (or parts of branches) of phylogenetic trees" is not a classification (as it happens, contra Mayr 1974) because "these branches are often quite heterogeneous and do not satisfy the definition of classes" (Mayr and Bock 2002: 182).

Mayr and Bock also stated that Darwin classified, while Hennig did not. They suggested (and intended their readers to believe) that Hennig was not part of the Darwinian tradition and that therefore his views must be ill-founded. To underline this, they added that:

> Taxa are traditionally used in a Darwinian classification for a class of similar species but are misleadingly used by most cladists for a branch (clade) or part of a branch of a cladogram. The entities ordered in a cladification are not classes, but clades or cladons. The term taxon applies to taxonomic classes … By not being delimited on the basis of similarity, clades do not qualify as classes … Hennig's system of ordering fails to meet the qualification of a classification. (Mayr and Bock 2002: 182–183; see 189)

That ordering systems, such as phylogenetic trees and dendrograms, are not classifications Mayr and Bock (2002) correctly stated but they used a somewhat strange foundation to illustrate their point. In contrast to their assumption, these ordering systems are not classifications because they are just visualizations of a system. And with respect to Darwin, it has been repeatedly pointed out that his remarks on the philosophy of classification are ambiguous and, according to many authors (e.g. Nelson 1974, Ghiselin 1985, Padian 1999), Darwin's view correspond more closely to Hennig's views than with Mayr's – and that "Mayr's 'synthetic or evolutionary method of classification' is not of Darwinian but rather of Mayrian manufacture" (Nelson 1974: 452). If, however, phylogenetic systematics is not in agreement with Darwin, then Hennig is indeed responsible as he expanded Darwin's view on classification by the recognition of paraphyletic groups and their exclusion from the system.

7.13 Looking backwards in search of the future

There is only one phylogeny and only one natural phylogenetic system underlying the diversity of life. Hennig described the decisive principles of phylogenetic systematics in just a few sentences: the goal of phylogenetic systematics is to present the phylogenetic relationships between species, so that only monophyletic groups are allowed in the system. How to reach that goal is via monophyletic groups that can only be validly recognized by the possession of derived (apomorphous) characters:

> Als Ziel der phylogenetischen Systematik kann es bezeichnet werden, die phylogenetischen Beziehungen der Arten so darzustellen, daß im System nur monophyletische Gruppen geduldet werden. Über den Weg, der zu diesem Ziele führt … gilt (in etwas vereinfachter Formulierung) die grundlegende Feststellung, daß eine monophyletische Gruppe als solche nur am Besitz abgeleiteter (apomorpher) Merkmale zu erkennen ist.
>
> [The goal of phylogenetic systematics, as it may be called, is that only monophyletic groups are tolerated in the system of species phylogenetic relationships. Towards this goal … (in a somewhat simplified formulation) the basic findings are that a monophyletic group is recognized as such only by its derived ('apomorphous') features.] (Hennig 1958: 506)

Only monophyletic groups should be named in order to avoid any confusion (Hennig 1984: 32). It is the task of phylogenetic systematics to reconstruct phylogenetic relationships in order to provide biologists with the information required for further investigations, such as biogeography, host–parasite histories, adaptational studies of structures, comparative biochemical investigations, etc., and finally for understanding the course of evolution. A hypothesis about phylogenetic relationships is not "a phylogeny," but a more or less well-founded scientific statement; that is, a presentation of an attempted reconstruction of the phylogeny of a section of organic diversity. Hennig would have appreciated the sentence that "Nothing in biology makes sense except in the light of evolution, and nothing in evolution makes sense except in the light of phylogeny" (Sytsma and Pires 2001: 726, Mindell et al. 2004: 108) because it implies a high degree of responsibility for phylogenetists.

Every newly published phylogenetic hypothesis requires discussion of previously published assumptions because not every reader is in the position of evaluating the differences between previous hypotheses.

Hennig (1955:29) wished that every systematist would clearly indicate the foundations of his phylogenetic conclusions, i.e. which characters were relied upon and how they were interpreted. This would allow the next worker to check and recheck those characters, add new information and discuss points of agreement and disagreement, and, if necessary, to propose a more well-founded hypothesis of phylogenetic relationships based on new evidence with characters and character analyses being the nucleus in the search for the sister group. Illies (1967: 528) had highlighted the transparency and honesty (!, den "Vorteil der Transparenz und absoluten Ehrlichkeit") of phylogenetic systematics. It was well understood that phylogenetic hypotheses must be testable, i.e. the data must be falsifiable or open to corroboration. If, however, characters used for phylogeny reconstruction (and I mean characters from any source) are not presented or explained, then it is not a true (or at least not a completely presented) scientific hypothesis. A first step in this fatal direction is to value characters per se not as highly as they deserve or, conversely, to neglect their importance. Furthermore, if several trees based on different analytical procedures are presented in the same paper without a qualified comparative evaluation by its authors, then phylogeny reconstruction appears to have reached a stage that undermines both its falsifiability and the ability to corroborate the scheme (see also Futuyma 2004: 35, who says much the same).

Karl Popper said that scientific progress is to a large degree accomplished by the falsification of old views. However, if one presents a new phylogenetic hypothesis of a particular taxon without simultaneously rejecting previous hypotheses on well-founded grounds, such progress cannot be achieved.

Acknowledgements

I am highly indebted to David Williams for inviting me to write this chapter and for his careful revision and the number of valuable comments and suggestions. I would furthermore like to express my sincere gratitude towards Shauna Grassmann and Sophia Willmann who helped editing the text.

References

Ax, P. (1984). *Das Phylogenetische System*. Stuttgart, New York: G. Fischer Verlag.

Ax, P. (1987). *The Phylogenetic System. The Systematization of Organisms on the Basis of Their Phylogenesis*. Chichester, NY: J. Wiley and Sons.

Bock, W.J. (1968). Phylogenetics systematics, cladistics and evolution. *Evolution*, 22, 646–648.

Bock, W.J. (1974). Philosophical foundations of classical evolutionary classification. *Systematic Zoology*, 22, 375–392.

Brauer, F.M. (1885). Systematisch-zoologische Studien. *Sitzungsberichte der Akademie der Wissenschaften mathematisch-naturwissenschaftliche Klasse*, 91 (1), 237–413.

Brundin, L. (1966). Transarctic relationships and their significance, as evidenced by chironomid midges. With a monograph of the subfamilies Podonominae and Aphroteniinae and the Austral Heptagyiae. *Kungliga Svenska Vetenskapsakademiens Handlingar*, Fjärde series, 11, 1–472.

Brundin, L. (1968). Application of phylogenetic principles in systematics and evolutionary theory. In *Current Problems of Lower Vertebrate Phylogeny*, ed. T. Ørvig. Stockholm: Almqvist and Wiksell, pp. 473–495.

Brundin, L. (1972). Evolution, causal biology, and classification. *Zoologica Scripta*, 1, 107–120.

Cracraft, J. (1974). Phylogenetic models and classification. *Systematic Zoology*, 23, 71–90.

Donoghue, M., and Kadereit, J. (1992). Walter Zimmermann and the growth of phylogenetic theory. *Systematic Biology*, 41, 74–85.

Dupuis, C. (1978). La "Systématique Phylogénétique" de W. Hennig (Historique, Discussion, Choix de Références). *Cahiers des Naturalistes, Bulletin des Naturalistes Parisiens*, n. s. 34, 1–71.

Farris, J.S. (1974). Discussion of symposium papers. Comment on Dr. Bock's paper. *Systematic Zoology*, 22, 375–392.

Farris, J.S.(1976). Phylogenetic classification of fossils with recent species. *Systematic Zoology*, 25, 271–282.

Farris, J.S. (1990). Haeckel, history, and Hull. *Systematic Zoology*, 39, 81–88.

Futuyma, D.J. (2004). The fruit of the tree of life. Insights into evolution and ecology. In *Assembling the Tree of Life*, ed. J. Cracraft and M. Donoghue. Oxford: Oxford University Press, pp. 25–39.

Ghiselin, M.T. (1985). Mayr versus Darwin on paraphyletic taxa. *Systematic Zoology*, 34, 460–462.

Griffiths, C.D. (1972). *Studies on the Phylogenetic Classification of Diptera Cyclorrhapha, With Special Reference to the Structure of the Male Postabdomen*. The Hague: Junk (Series entomologica 8).

Griffiths, C.D. (1974). On the foundations of biological systematics. *Acta Biotheoretica*, 23, 85–131.

Griffiths, C.D. (1976). The future of Linnaean nomenclature. *Systematic Zoology*, 25, 168–173.

Günther, K. (1956). Systematik und Stammesgeschichte der Tiere. *Fortschritte der Zoologie*, Neue Folge 10, 36–55.

Günther, K. (1962). Systematik und Stammesgeschichte der Tiere 1954–1959. *Fortschritte der Zoologie*, Neue Folge 14, 268–547.

Günther, K. (1971). Natürliches oder phylogenetisches System? *Erlanger Forschungen*, B 4, 29–41.

Hamilton, A. (2014). Historical and conceptual perspectives on modern systematics. Groups, ranks, and the phylogenetic turn. In *The Evolution of Phylogenetic Systematics*, ed. A. Hamilton. Berkeley, CA: University of California Press, pp. 89–116.

Hennig, W. (1936). Über einige Gesetzmäßigkeiten der geographischen Variation in der Reptiliengattung *Draco* L.: "parallele" und "konvergente"

Rassenbildung. *Biologisches Zentralblatt*, 56, 549–559.

Hennig, W. (1943). Ein Beitrag zum Problem der "Beziehungen zwischen Larven- und Imaginalsystematik". *Arbeiten über Morphologische und Taxonomische Entomologie, Berlin-Dahlem*, 10, 138–144.

Hennig, W. (1947). Probleme der biologischen Systematik. *Forschungen und Fortschritte*, 21/23, 276–279.

Hennig, W. (1948). *Die Larvenformen der Dipteren*. 1. Teil. Berlin: Akademie-Verlag.

Hennig, W. (1949). Zur Klärung einiger Begriffe der phylogenetischen Systematik. *Forschungen und Fortschritte*, 25, 136–138.

Hennig, W. (1950a). *Grundzüge einer Theorie der phylogenetischen Systematik*. Berlin: Deutscher Zentralverlag.

Hennig, W. (1950b). *Die Larvenformen der Dipteren*. 2. Teil. Berlin: Akademie-Verlag.

Hennig, W. (1952a). *Die Larvenformen der Dipteren*. 3. Teil. Berlin: Akademie-Verlag.

Hennig, W. (1952b). Autorreferat: Grundzüge einer Theorie der phylogenetischen Systematik. *Beiträge zur Entomologie*, 2, 329–331.

Hennig, W. (1953). Kritische Bemerkungen zum phylogenetischen System der Insekten. *Beiträge zur Entomologie*, 3, Sonderheft, 1–63.

Hennig, W. (1954). Flügelgeäder und System der Dipteren unter Berücksichtigung der aus dem Mesozoikum beschriebenen Fossilien. *Beiträge zur Entomologie*, 4, 245–388.

Hennig, W. (1955). Meinungsverschiedenheiten über das System der niederen Insekten. *Zoologischer Anzeiger*, 155, 21–30.

Hennig, W. (1957). Systematik und Phylogenese. In *Bericht über die Hundertjahrfeier der Deutschen Entomologischen Gesellschaft Berlin*, ed. H.-J. Hannemann. Berlin: Akademie-Verlag, pp. 50–71.

Hennig, W. (1958). Die Familien der Diptera Schizophora und ihre phylogenetischen Verwandtschaftsbeziehungen. *Beiträge zur Entomologie*, 8, 505–688.

Hennig, W. (1960). Die Dipteren-Fauna von Neuseeland als systematisches und tiergeographisches Problem. *Beiträge zur Entomologie*, 10, 221–327.

Hennig, W. (1962). Veränderungen am phylogenetischen System der Insekten seit 1953. *Bericht über die 9. Wanderversammlung Deutscher Entomologen*, 45, 29–42.

Hennig, W. (1964a). Diskussionsbemerkung. In Ax, P.: Der Begriff Polyphylie ist aus der Terminologie der natürlichen, phylogenetischen Systematik zu eliminieren. *Zoologischer Anzeiger*, 173, 63.

Hennig, W. (1964b). Wirbellose II. Gliedertiere. *Taschenbuch der Zoologie*, Bd. 3, 2. Aufl., Leipzig: Edition.

Hennig, W. (1965a). Wirbellose I. Ausgenommen Gliedertiere. *Taschenbuch der Zoologie*, Bd. 2, 2. Aufl., Leipzig: Edition.

Hennig, W. (1965b). Phylogenetic systematics. *Annual Review of Entomology*, 10, 97–116.

Hennig, W. (1965c). Die Acalyptratae des Baltischen Bernsteins und ihre Bedeutung für die Erforschung der phylogenetischen Entwicklung dieser Dipteren-Gruppe. *Stuttgarter Beiträge zur Naturkunde*, 145, 1–215.

Hennig, W. (1965d). Vorarbeiten zu einem phylogenetischen System der Muscidae

(Diptera: Cyclorrhapha). *Stuttgarter Beiträge zur Naturkunde*, 141, 1–100.

Hennig, W. (1966a). Conopidae im Baltischen Bernstein (Diptera; Cyclorrhapha). *Stuttgarter Beiträge zur Naturkunde*, 154, 1–24.

Hennig, W. (1966b). *Phylogenetic Systematics*. Urbana, IL: University of Illinois Press.

Hennig, W. (1966c). The Diptera fauna of New Zealand as a problem in systematics and biogeography. *Pacific Insects Monograph*, 9, 1–81.

Hennig, W. (1967). Die sogenannten "niederen Brachycera" im Baltischen Bernstein (Diptera: Fam. Xylophagidae, Xylomyidae, Rhagionidae, Tabanidae). *Stuttgarter Beiträge zur Naturkunde*, 174, 1–51.

Hennig, W. (1968). Kritische Bemerkungen über den Bau der Flügelwurzel bei den Dipteren und die Frage nach der Monophylie der Nematocera. *Stuttgarter Beiträge zur Naturkunde*, 193, 1–23.

Hennig, W. (1969). *Die Stammesgeschichte der Insekten*. Frankfurt am Main: Waldemar Kramer and Co.

Hennig, W. (1970). Insektenfossilien aus der unteren Kreide. II. Empididae (Diptera, Brachycera). *Stuttgarter Beiträge zur Naturkunde*, 214, 1–12.

Hennig, W. (1971a). Neue Untersuchungen über die Familien der Diptera Schizophora. *Stuttgarter Beiträge zur Naturkunde*, 226, 1–76.

Hennig, W. (1971b). Zur Situation der biologischen Systematik. *Erlanger Forschungen*, B 4, 7–15.

Hennig, W. (1972a). Eine neue Art der Rhagionidengattung *Litoleptis* aus Chile, mit Bemerkungen über Fühlerbildung und Verwandtschaftsbeziehungen einiger Brachycerenfamilien (Diptera: Brachycera). *Stuttgarter Beiträge zur Naturkunde*, 242, 1–18.

Hennig, W. (1972b). Insektenfossilien aus der unteren Kreide. IV. Psychodidae (Phlebotominae), mit einer kritischen Übersicht über das phylogenetische System der Familie und die bisher beschriebenen Fossilien (Diptera). *Stuttgarter Beiträge zur Naturkunde*, 241, 1–69.

Hennig, W. (1973). Diptera (Zweiflügler). *Handbuch der Zoologie* IV/2 (2. Teil) 31, 1–200. Berlin: Walter de Gruyter.

Hennig, W. (1974). Kritische Bemerkungen zur Frage "Cladistic analysis or cladistic classification?". *Zeitschrift für Zoologische Systematik und Evolutionsforschung*, 12, 279–294.

Hennig, W. (1975). "Cladistic analysis or cladistic classification?": A reply to Ernst Mayr. *Systematic Zoology*, 24, 244–256.

Hennig, W. (1976). Das Hypopygium von *Lonchoptera lutea* Panzer und die phylogenetischen Verwandtschaftsbeziehungen der Cyclorrhapha (Diptera). *Stuttgarter Beiträge zur Naturkunde*, 283, 1–63.

Hennig, W. (1980). Wirbellose I. 4th edition. *Taschenbuch der Speziellen Zoologie*. Frankfurt: Verlag Harri Deutsch.

Hennig, W. (1981). *Insect Phylogeny*. New York: John Wiley and Sons.

Hennig, W. (1982). *Phylogenetische Systematik*. Hamburg: Verlag Paul Parey.[2]

Hennig, W. (1983). Stammesgeschichte der Chordaten. *Fortschritte in der Zoologischen Systematik und Evolutionsforschung*. Beihefte zur

---

[2] This is the German text that was finalized in 1961, translated into English and then published as *Phylogenetic Systematics*, Hennig (1966b)

Zeitschrift für Zoologische Systematik und Evolutionsforschung. Hamburg, Berlin: Verlag Paul Parey.

Hennig, W. (1984). *Aufgaben und Probleme stammesgeschichtlicher Forschung.* Hamburg, Berlin: Paul Parey Verlag.

Hennig, W. (1986). Wirbellose II. *Taschenbuch der Speziellen Zoologie.* Frankfurt: Verlag Harri Deutsch.

Hennig, W. (1982). Vorwort des Herausgebers. In *Phylogenetische Systematik*, ed. W. Hennig. Berlin, Hamburg: Verlag Paul Parey, pp. 5–6.

Hertwig, R. (1914) *Abstammungslehre, Systematik, Paläontologie, Biogeographie.* Leipzig: B.G. Teubner.

Huxley, J.S. (1958). Evolutionary processes and taxonomy with special reference to grades. In *Systematics of today: proceedings of a symposium … in commemoration of the 250th anniversary of the birth of Carolus Linnaeus*, ed. O. Hedberg. Uppsala: Lundequistska Bokhandeln (Acta Universitatis Upsaliensis, 6), pp. 21–39.

Illies, J. (1967). Die Gattung *Megaleuctra* (Plecopt., Ins.). Beitrag zur konsequent-phylogenetischen Behandlung eines incertae-sedis Problems. *Zeitschrift für Morphologie und Ökologie der Tiere*, 60, 124–134.

Jefferies, R.P.S. (1979). The origin of chordates: a methodological essay. In *The Origin of Major Invertebrate Groups*, ed. M.R. House. London, New York: Academic Press, pp. 443–447.

Kiriakoff, S.G. (1955). Le Système phylogénétique: principes et méthodes. *Bulletin et Annales de la Société Royale Belge d'Entomologie*, 91, 147–158.

Kiriakoff, S.G. (1959). Phylogenetic systematic versus typology. *Systematic Zoology*, 8, 117–118.

Königsmann, E. (1975). Termini der phylogenetischen Systematik. *Biologische Rundschau*, 13, 99–115.

Lauterbach, K.-E. (1989). Das Pan-Monophylum: ein Hilfsmittel für die Praxis der Phylogenetischen Systematik. *Zoologischer Anzeiger*, 223, 139–156.

Lauterbach, K.-E. (1992). Phylogenetische Systematik. Bedeutung, Leistung, Anspruch. *Bericht des Naturwissenschaftlichen Vereins für Bielefeld und Umgegend*, 33, 209–240.

Löther, R. (1972). *Die Beherrschung der Mannigfaltigkeit. Die philosophischen Grundlagen der Taxonomie.* Jena: G. Fischer.

Mayr, E. (1958). The evolutionary significance of the systematic categories. In *Systematics of Today: Proceedings of a Symposium … In Commemoration of the 250th Anniversary of the Birth of Carolus Linnaeus*, ed. O. Hedberg. Uppsala, Sweden: Lundequistska Bokhandeln (Acta Universitatis Upsaliensis, 6), pp. 13–20.

Mayr, E. (1969). *Principles of Systematic Zoology.* New York: McGraw-Hill Book Company.

Mayr, E. (1974). Cladistic analysis or cladistic classification? *Zeitschrift für Zoologische Systematik und Evolutionsforschung*, 12, 94–128.

Mayr, E. (1990). Die drei Schulen der Systematik. *Verhandlungen der Deutschen Zoologischen Gesellschaft*, 83, 263–276.

Mayr, E. (2000a). The biological species concept. In *Species Concepts and Phylogenetic Theory. A Debate*, ed. Q.D. Wheeler and R. Meier. New York: Columbia University Press, pp. 17–29.

Mayr, E. (2000b). A critique from the biological species concept perspective: What is a species, and what is not? In *Species Concepts and Phylogenetic Theory. A Debate*, ed. Q.D. Wheeler and R. Meier. New York: Columbia University Press, pp. 93–100.

Mayr, E. and Bock, W.J. (2002). Classifications and other ordering systems. *Journal of Zoological Systematics and Evolutionary Research*, 40, 169–194.

Mayr, E., Linsley, E.G., and Usinger, R.L. (1953). *Methods and principles of systematic zoology*. New York, Toronto, London: McGraw-Hill Book Company, Inc.

Meier, R., and Willmann, R. (2000). The Hennigian species concept. In *Species Concepts and Phylogenetic Theory. A Debate*, ed. Q.D. Wheeler and R. Meier. New York: Columbia University Press, pp. 30–43.

Mindell, D.P., Rest, J.S. and Villarreal, L.P. (2004). Viruses and the tree of life. In *Assembling the Tree of Life*, ed. J. Cracraft and M.J. Donoghue. Oxford: Oxford University Press, pp. 107–118.

Naef, A. (1917). *Die individuelle Entwicklung organischer Formen als Urkunde ihrer Stammesgeschichte*. Jena: G. Fischer Verlag.

Naef, A. (1919). *Idealistische Morphologie und Phylogenetik*. Jena: G. Fischer Verlag.

Nelson, G. (1974). Darwin-Hennig classification: a reply to Ernst Mayr. *Systematic Zoology*, 23, 452–458.

Nelson, G. (1989). Species and taxa: systematics and evolution. In *Speciation and its Consequences*, ed. D. Otte and J.A. Endler. Sunderland, MA: Sinauer Assoc. Inc., pp. 60–81.

Padian, K. (1999). Charles Darwin's views of classification in theory and practice. *Systematic Biology*, 48, 352–364.

Paramonow, S.J., (1935). Die Methoden der modernen Zoosystematik (Zoogeographie). *Gegenwärtige Systematik, ihre Methoden und Aufgaben*, 13, 3–23.

Peters, G. (1995). Über Willi Hennig als Forscherpersönlichkeit. *Sitzungsberichte der Gesellschaft Naturforschender Freunde zu Berlin*, (N.F.) 34, 1–10.

Remane, A. (1952). *Die Grundlagen des natürlichen Systems, der vergleichenden Anatomie und der Phylogenetik. Theoretische Morphologie und Systematik I*. Leipzig: Geest u. Portig.

Richter, S. and Meier, R. (1994). The development of phylogenetic concepts in Hennig's early theoretical publications (1947–1966). *Systematic Biology*, 43, 212–221.

Rieppel, O. (1979). Why to be a cladist. *Zeitschrift für Zoologische Systematik und Evolutionsforschung*, 18, 81–90.

Rieppel, O. (2006). The metaphysics of Hennig's phylogenetic systematics: substance, events and laws of nature. *Systematics and Biodiversity*, 5, 345–360.

Rieppel, O. (2014). The early cladogenesis of cladistics. In *The Evolution of Phylogenetic Systematics*, ed. A. Hamilton. Berkeley, CA: University of California Press, pp. 117–137.

Rosen, D.E. (1974). Cladism or gradism? A reply to Ernst Mayr. *Systematic Zoology*, 23, 446–451.

Schmitt, M. (2003). Willi Hennig and the rise of cladistics. In *The New Panorama of Animal Evolution*, ed. A. Legakis, S. Sfenthourakis, R. Polymeni and M. Thessalou-Legaki, Proceedings of the 18th International Congress of Zoology. Sofia: Pensoft, pp. 369–379.

Schmitt, M. (2013). *From Taxonomy to Phylogenetics: Life and work of Willi Hennig*. Leiden: Brill.

Simpson, G.G. (1951). The species concept. *Evolution*, 5, 285–298.

Simpson, G.G. (1961). *Principles of Animal Taxonomy*. New York: Columbia University Press.

Simpson, G.G. (1975). Recent advances in methods of phylogenetic inference. In *Phylogeny of the Primates*, ed. W.P. Luckett and F.S. Szalay. London, New York: Plenum Press, pp. 3–19.

Stammer, H.J. (1960). Neue Wege in der Insektensystematik. XI. *Internationaler Kongreß für Entomologie. Verhandlungen*, Band I, 1–7.

Sturtevant, A.H. (1942). The classification of the genus *Drosophila*, with descriptions of nine new species. *University of Texas Publications*, 4213, 1–55.

Sytsma, K.J. and Pires, J.C. (2001). Plant systematics in the next 50 years – re-mapping the new frontier. *Taxon*, 50, 713–732.

Tassy, P. (1999). Willi Hennig et l'objet paléontologique. *Geodiversitas*, 21, 5–23.

Thienemann, A. and F. Krüger. (1937). "*Orthocladius*" abiskoensis Edwards und rubicundus (Mg.), zwei "Puppen-Spezies" der Chironomiden (Chironomiden aus Lappland II). *Zoologischer Anzeiger*, 117, 257–267.

Thompson, D'Arcy. (1917). *On Growth and Form*. Cambridge: Cambridge University Press.

Throckmorton, L.H. (1968). Concordance and discordance of taxonomic characters in *Drosophila* classifications. *Systematic Zoology*, 17, 355–387.

Van Valen, L. (1978). Why not to be a cladist. *Evolutionary Theory*, 3, 285–299.

Wiley, E.O. (1975). Karl R. Popper, systematics, and classification: A reply to Walter Bock and other evolutionary taxonomists. *Systematic Zoology*, 24, 233–243.

Wiley, E.O. (1979). An annotated Linnean hierarchy, with comments on natural taxa and competing systems. *Systematic Zoology*, 28, 308–344.

Wiley, E.O. (1981). *Phylogenetics. The Theory and Practice of Phylogenetic Systematics*. New York, Chichester, Brisbane, Toronto: John Wiley and Sons.

Wiley, E.O. (2009). Patrocladistics, nothing new. *Taxon*, 58, 2–6.

Willmann, R. (1985). *Die Art in Raum und Zeit*. Hamburg, Berlin: Paul Parey Verlag.

Willmann, R. (2003). From Haeckel to Hennig: the early development of phylogenetics in German-speaking Europe. *Cladistics*, 19, 449–479.

Woodger, J.H. (1952). From biology to mathematics. *The British Journal for the Philosophy of Science*, 3 (9), 1–21.

Zimmermann, W. (1931). Arbeitsweise der botanischen Phylogenetik und anderer Gruppierungswissenschaften. In *Handbuch der biologischen Arbeitsmethoden, Abteilung IX: Methoden zur Erforschung der Leistungen des tierischen Organismus, Teil 3: Methoden der Vererbungsforschung, Heft 6 (Lieferung 356)*, ed. E. Abderhalden. Berlin: Urban and Schwarzenberg, pp. 941–1053.

Zimmermann, W. (1934). Genetische Untersuchungen an *Pulsatilla* I-III. *Flora*, 129, 158–234.

Zimmermann, W. (1937). Arbeitsweise der botanischen Phylogenetik und anderer Gruppierungswissenschaften. In *Handbuch der biologischen Arbeitsmethoden*, ed. E. Abderhalden, Abt. IX, Teil 3/II. Berlin, Wien, pp. 941–1053.

Zimmermann, W. (1943). Die Methoden der Phylogenetik. In *Die Evolution der Organismen*, ed. G. Heberer. Jena: Gustav Fischer Verlag, pp. 20–56.

Zimmermann, W. (1959). Methoden der Phylogenetik. In *Die Evolution der Organismen*, ed. G. Heberer. Stuttgart: Gustav Fischer Verlag, pp. 25–102.

What we all learned from Hennig

GARETH NELSON

The phylogenetic systematics of Willi Hennig (1913–76) is significant for what it means by relationship. Hennig viewed phylogenetic relationship as a property of species and species groups. Here it is viewed more fundamentally as a property of organisms and their parts. However viewed, it contrasts with earlier prevailing theory about genealogy and phylogeny, particularly exemplified by George Simpson's publications of 1945 and 1961. That theory is based on similarity and its gaps, reflecting what is here termed Simpson's enigma. For Hennig, and what we learned from him, the theory of similarity and gaps is "entirely different" from genealogical and phylogenetic relationship.

Begin with Colin Patterson (1933–98):

> What we all learned from Hennig back in those early days boiled down to just one thing, what relationship means. No one had put it plainly before. Once you agreed what relationship meant, how to recognise it became obvious – synapomorphy – and then it was also obvious what was wrong with systematics as we'd been practising it in the 50s and early 60s, when everyone was preoccupied with polyphyly. Our mistake was thinking in terms of origins rather than relationships. (Patterson 2011: 124)

Patterson wrote this in 1995, near the end of his life, for an Annual Address to the Systematics Association (the SA's 29th); hence it is his mature view of this period of history.

In his address, given 6 December 1995, he does not elaborate on what relationship means in theory, but he does show how it applies to his research, beginning

The Future of Phylogenetic Systematics: The Legacy of Willi Hennig, eds. D. Williams, M. Schmitt and Q. Wheeler. Published by Cambridge University Press.

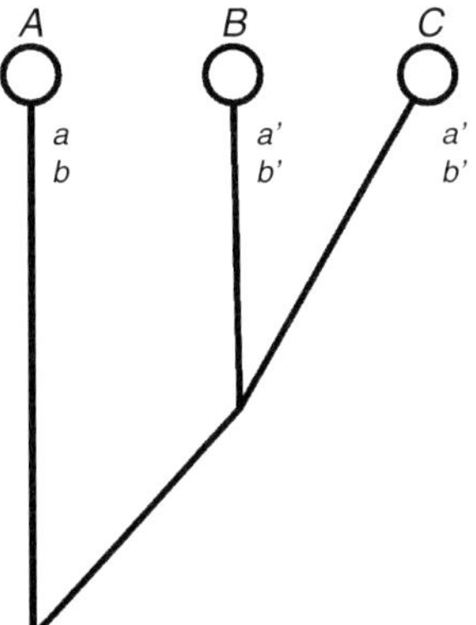

Fig 8.1 Species *B* and *C* are more closely related to each other than to species *A*. Redrawn after Hennig (1966: Fig 25): *A*(*BC*) as evidenced by synapomorphies *a'* and *b'*: *a*(*a'a'*), *b*(*b'b'*).

with his PhD study on chalk (Cretaceous) fishes (1960–61, University of London, thesis published 1964). His long-time colleague in palaeontology at the British Museum (Natural History), known since 1992 as the Natural History Museum, Peter Forey later wrote in an obituary:

> Colin realised quickly that his search for ancestors in chalk fishes had been doomed to failure because of theoretical and methodological difficulties of recognising ancestors. Colin, like several others in the 70's, recast the questions we asked of the fossil record. Instead of asking questions about ancestry and descent – which are questions about process, Colin searched for cladistic sister groups, which can be discovered by examining structures in specimens. (Forey 1998: 5)

Hennig's definition of relationship:

> A particular taxon B is more closely related to another taxon C than to a third taxon A if, and only if, it has at least one stem species in common with C that is not also a stem species of A. When so defined the species and monophyletic groups … are ranked according to the degree of their kinship. (Hennig 1966: 74)

Hennig illustrated his concept of relationship (1966: Fig 25, similar to an earlier one in 1950: Fig 23, the later redrawn here as Fig 8.1). There are three contemporaneous species ABC, interrelated through a stem species that is hypothetical and nameless: A(BC). In a simpler explanation, he used a different figure with a named stem species (1966: Fig 4, no earlier version in 1950), redrawn here as Fig 8.2. In Fig 8.2 a different species A is the ancestor of species B and C. In Fig 8.2 no two species relate more closely. The stem species A is merely a member of a group with species B and C: ABC.

His summary diagram of relationships (1966: Fig 6, no earlier version in 1950), redrawn here as Fig 8.3, shows hologenetic relationships divided into phylogenetic

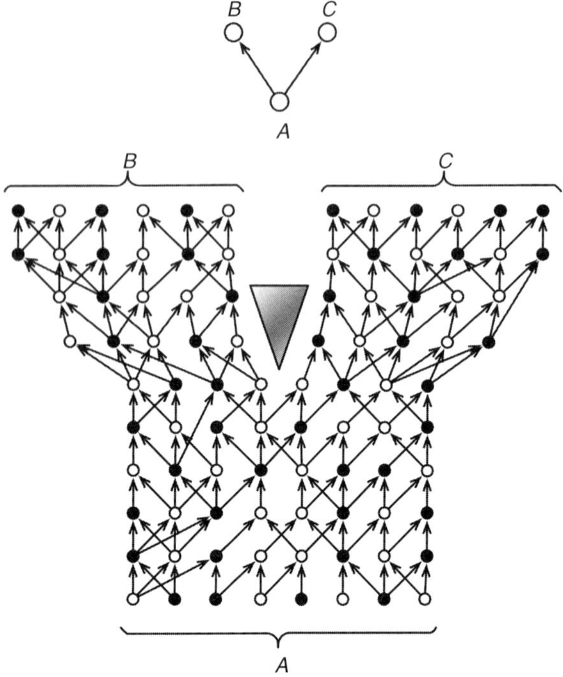

Fig 8.2 "Species cleavage" (redrawn after Hennig 1966: Fig 4). No two species relate more closely than to the third: *ABC*.

(between species) and autogenetic (between organisms), and the latter divided into tokogenetic (between parents and offspring) and ontogenetic (between successive life stages, Hennig's semaphoronts). Shown also is a small diagram illustrating "phylogenetic relationships," with one species as the ancestor of two other species, similar to Fig 8.2.

These two figures of three species, with one the ancestor of the other two, do not exemplify Hennig's definition of relationship because they are too simple, and the timeline is too short. For Fig 8.2, for example, imagine an earlier branching species X, such that X(ABC) is true, with ABC having a hypothetical stem species that is not also a stem species for X; or a later branching of species B, leading to species B1 and B2, such that AC(BB1B2) is true, with B1 and B2 having B as a stem species that is not also a stem species for AC. These comments are not meant as a criticism; the figures are fine, so far as they go.

At times Hennig clearly distinguishes between characters and phylogenetic relationship, which would not apply to characters:

> [W]e can speak of the evolution of an individual character ... but not of its phylogenesis, since a character has no actual individual nature but appears only as a peculiarity of an individual. (1966:198)

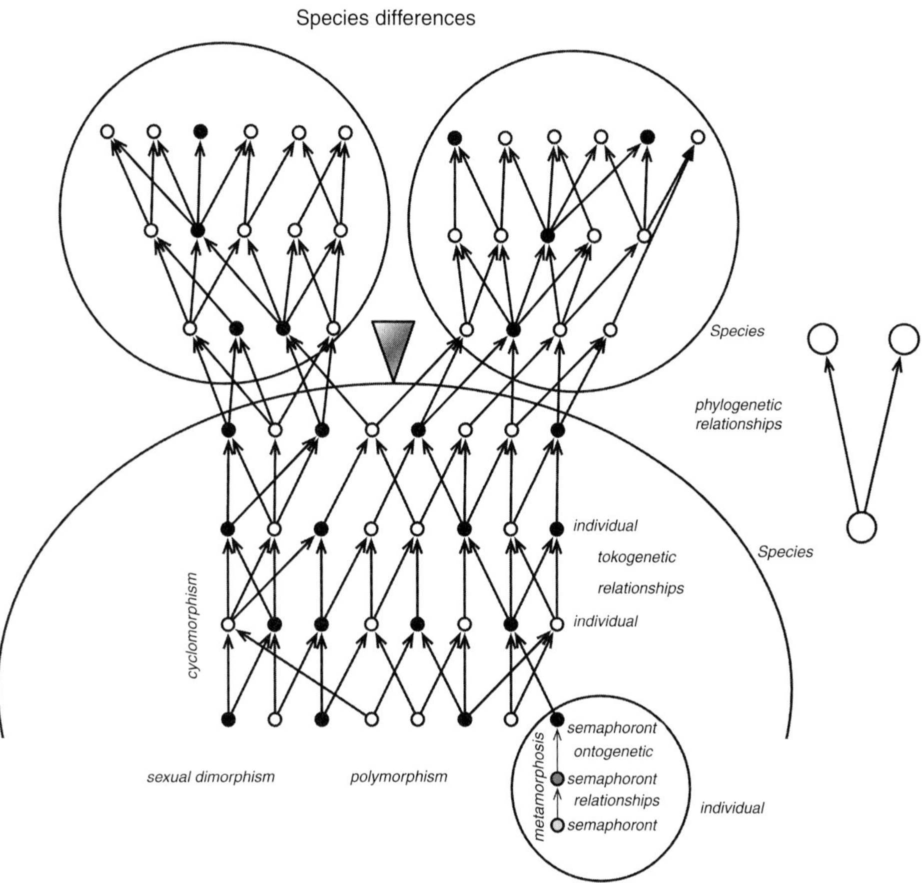

Fig 8.3 "Total structure of hologenetic relationships." Redrawn after Hennig (1966: Fig 6). As in Fig 8.2, no two species are more closely related than they are to the third.

> Phylogenetic kinship exists only between species and species groups. (1966: 122)

At other times Hennig implies that organisms and characters, as parts of organisms, have phylogenetic relationships among themselves:

> By 'phylogenetic system' we mean a system that expresses the phylogenetic relationships of organisms. (1966: 10)

> "phylogenetischen Verwandtschaftsbeziehungen" der Organismen (1982: 17)

> homologous characters (or better, parts of organisms) (1966: 94)

> für Merkmale (oder besser für Teile von Organismen) (1982: 97)

> the phylogenetic relationships of the three gene arrangements (1966: 115)

> Die stammesgeschichtliche Verwandtschaft der drei oben erwähnten Genanordnungen. (1982: 115)

Perhaps the sense of these diverse statements is that, for Hennig, organisms and their parts passively reflect "phylogenetic relationships" of their species and other taxa, to which they and their species belong.

A difficulty with the view that phylogenetic relationships exist only between species, and not between organisms, or parts of organisms ("characters"), is what a species concept exactly means (de Pinna 2014); another is the multiplicity of species concepts. Wilkins lists "26 Species Definitions in the Modern Literature":

> 1 Agamospecies, 2 Autapomorphic species, 3 Biospecies, 4 Cladospecies, 5 Cohesion species, 6 Compilospecies, 7 Composite species, 8 Ecospecies, 9 Evolutionary species, 10 Evolutionary significant unit, 11 Genealogical concordance species, 12 Genic species, 13 Genetic species, 14 Genotypic cluster, 15 Hennigian species, 16 Internodal species, 17 Least Inclusive Taxonomic Unit (LITUs), 18 Monophyletic species, 19 Morphospecies, 20 Non-dimensional species, 21 Nothospecies, 22 Phenospecies, 23 Recognition species, 24 Reproductive competition species, 25 Successional species, 26 Taxonomic species. (Wilkins 2009: x–xi)

I refer again to Colin Patterson, this time on the subject of homology:

> Last year I was asked to write a paper on homology for a Systematics Association symposium on "Problems of Phylogeny Reconstruction" … I accepted because I had always felt that the concept of homology raised vague problems that I had tried to kick back under the rug, but I wondered how on earth there could be anything to say about homology, a relation that had been under discussion for well over 150 years. (quoted in Williams and Ebach 2014:168)

This he wrote for a lecture ("Homology and Phylogeny") at Harvard University in November 1981 (unpublished). His thoughts were later published in a Systematics Association volume, resulting from a Systematic Association symposium (April 1980). In his abstract one reads:

> [H]omology is the relation which characterizes monophyletic groups. … Corollaries of this definition are that synapomorphy and homology are the same; that every worthwhile proposal of homology is a hypothesis of a monophyletic (natural) group. (Patterson 1982: 21)

I take this to mean that a homology statement, which concerns the parts of organisms, is of the form, the simplest for example, A(BC). And that the statement may be seen as a phylogenetic relationship between the parts of the organisms that have them: parts of organisms B and C relate more closely than to parts of organism A; and, therefore, organisms B and C evidently relate more closely than to organism A. By "evidently," I mean that the relationship of the parts is evidence of the relationship of the organisms that have them – and, by extension, of the relationships of the species and higher taxa to which belong the organisms with their parts. From this standpoint, opposite to that of Hennig, the "phylogenetic relationships" of species and other taxa appear as passive reflections of the relationships of their included organisms and, ultimately, of the parts of the organisms.

Reflection of relationship – from organism to species or from species to organism – is its direction meaningful except as metaphysics? Relevant, perhaps, is what functions as the source of evidence. Does knowledge of synapomorphy arise from relationships of parts, or from relationships of species? If the latter, how then can it be evidence of the same?

That homology is a relationship between parts of organisms is inherent in Richard Owen's (1804–1892) treatment of the subject. That homology means phylogenetic relationship has a more troubled history, most recently in terms of "homology as synapomorphy" – an idea the priority for which Edward Wiley lodged a proprietary claim (Wiley and Lieberman 2011:118): "Phylogeneticists, however, treated homology as synapomorphy (Wiley 1975, 1976, Bonde 1977, Patterson 1982, de Pinna, 1991; and others)."

Regardless of proprietorship or priority, Forey notes accurately I think for Patterson:

> And in 1982 he published one of his most significant papers bringing together the disparate ideas of homology to a single explanation that homologies are theories to be tested; and he set up three tests we might apply. That paper too opened new doors into the theoretical rooms of systematics and modern discussions of homology are in Patterson's language. (Forey 1998: 5),

If "'phylogenetic' relationship has the character of genealogical relationships between organisms and groups of organisms" (Hennig 1966:10), the question arises: how are these relationships acquired? Here it is helpful, perhaps, to consider a newborn human, who comes into the world with parents, grandparents, great-grandparents, etc. back forever through the mists of time; also with aunts, uncles, cousins, and so on. The newborn acquires these relatives – these relationships – simply by coming into being. They are inherited. Perforce, most of them would be unknown to anyone living and would ever remain so; even absent from human knowledge, they would still exist as discoverable possibilities.

If genealogical relationships are inherited, what about phylogenetic relationships? Recent years have seen renewed research into the phylogenetic relationships of mammals (McKenna and Bell 1997), summarized as:

```
*Mammalia
   Prototheria: monotremes
   *Theria
      Marsupialia: marsupials
      *Placentalia
      Xenarthra: edentates
      *Epitheria
         Anagalida: elephant shrews, rabbits rodents
```

<pre>
Ferae: pangolins, carnivores
Insectivora: moles, shrews
Ungulata: ungulates
*Archonta
   Chiroptera: bats
   Scandentia: tree shrews
   *Primates
      Dermoptera: gliding lemurs
      *Euprimates
         Strepsirhini: lemurs
         *Haplorhini
            Tarsiiformes: tarsiers
            *Anthropoidea
            Callitrichoidea: marmosets, new world monkeys
            Cercopithecidae: old world monkeys
            *Hominidae
               Hylobatinae: gibbons
               *Homininae
                  Pongini: orangutan
                  *Hominini
                     Gorillina: gorillas
                     *Hominina
                        Pan: chimpanzees
                        *Homo
</pre>

Again it is perhaps helpful to consider the newborn as a *Homo*, a hominin, a hominid, an anthropoid – a member of all the groups marked with an asterisk (*). Again, the newborn acquires these relatives – these phylogenetic relationships – simply by coming into being. They too would be inherited. This is not to say that all apparent relationships are necessarily true; a human after all might relate more closely to the orangutan than to African apes.

If so, it is the organism with its parts that is the basis for phylogenetic relationship. Nothing else inherits, except an organism. Species do not inherit, neither do taxa in general, except metaphorically (Gilmour, 1961: 36).

Patterson mentions that, with respect to Hennig's notion of relationship, "No one had put it plainly before" (see above). The shrouded Zeitgeist of that earlier era is well exemplified by George Simpson (1902–84), and particularly in his classification of mammals:

> When the ancestral group is known, how is it to be classified? Can it be more nearly related to one than to the other of its descendent lines? In a sequence, is a group more nearly related to its ancestors, its descendants, or its contemporaries of like origin; in the human family analogy, is a man more nearly related to his father, son, or brother? (Simpson 1945: 17)

Simpson's Enigma:

Is a man more closely related to his father, son, or brother?

Comment:
Even when the genealogy is assumed to be known, the relationships
of the organisms, or taxa, are ambivalent!

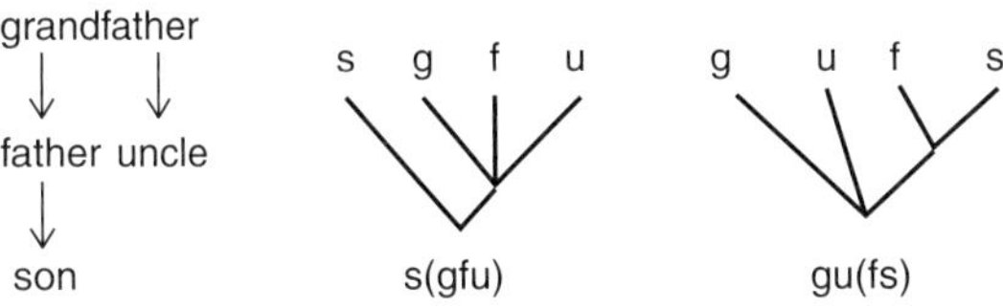

Fig 8.4 Cladistic representation of Simpson's enigma with its only cladistic
resolution: gu(fs). Symbols: g, grandfather; f, father; u, uncle; s, son. Cladistically spurious
alternatives include s(gfu); see text for others, su(gf), sf(gu), sg(fu), g(ufs), etc.

Simpson offered no solution to his enigma, because he apparently believed it has
none. He elaborated it some 15 years later:

> Is a man more closely related to his father, son, or brother? The actual genes
> involved may be quite different, but the degree of genetical relationship to father
> and to son is invariably the same (0.5 in terms of proportion of shared chromo-
> somes). Genetical relationship to a brother is variable – from 1.0 to 0.0 in terms of
> chromosomes, although the probability of those extremes is exceedingly low – but
> the mean value is the same as for father or son. Unfortunately, relationships among
> taxa do not have such fixed a priori expectations, and they cannot be precisely
> measured. The same two kinds of relationships nevertheless exist: among succes-
> sive taxa in an ancestral–descendant lineage, and among contemporaneous taxa
> of more or less distant common origin. In accordance with the usual coordinates of
> tree representation, the former relationships are called *vertical* and the latter *hori-
> zontal*. One kind of relationship is obviously just as objective as the other. (Simpson
> 1961:129)

Simpson never analysed the details of his enigma, which are shown in Fig 8.4,
with its only cladistic resolution. There is a son (s), his father (f), uncle (u), and
grandfather (g). For Simpson, equally valid representations of these relationships
would be a grouping of the elders s(gfu) on the one hand, and of the son and father
on the other gu(fs). From a cladistic viewpoint, however, only the latter contains
information of genealogy. Other variations, equally valid from Simpson's point of
view but also cladistically spurious, would be a grouping of grandfather and father
su(gf), grandfather and uncle sf(gu), father and uncle sg(fu), and uncle, father, and
son g(ufs). So also would be gf(su), fu(sg), f(sug), and u(sgf).

Simpson further developed his enigma in various theoretical diagrams. In one
(1961:190, Fig 19B, redrawn here as Fig 8.5), there are seven unnamed species, here

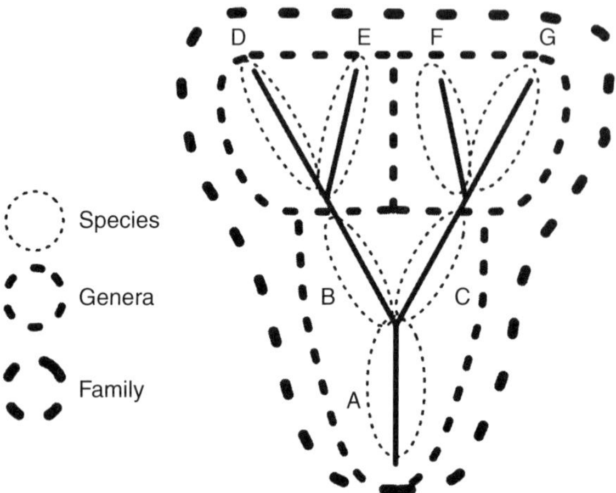

Fig 8.5 "Phylogenetic tree … correctly divided into taxa." Redrawn after Simpson (1961: Fig 19B, p. 190), with species names added (A, B, etc.).

designated A–G. A is the ancestor of B and C, B of DE, and C of FG, with the cladistic representation A(BDE)(CFG). Simpson's preferred, and cladistically spurious, representation is (ABC)(DE)(FG).

In another, and more complicated enigma (1961:197, Fig 22A, redrawn here as Fig 8.6), there are 13 unnamed species, A–M. A is the ancestor of BC, D of E, E of FG, G of HI, H of JK, I of LM, with the cladistic representation of (ABC)(D(EF(G(HJK) (ILM)))). Simpson's preferred, and cladistically spurious, representation is (ABCDEFG)(HIJKLM). In his Fig 22B, he considers a hypothetical example of a dozen species with its accurate cladistic representation, which he terms "showing complication and imbalance." Cladistic representation, however, can be simplified; for Fig 8.6, for example, to (ABC)(DEFGHIJKLM).

The rationale underlying Simpson's grappling with relationship is exposed in another figure (1961:193, Fig 20, redrawn here as Fig 8.7) showing ten unnamed species, represented by dots in a linear array, with variously sized gaps between the dots. The species (dots) are grouped in five genera: a–e. There is otherwise no assumed phylogeny; hence no cladistic representation of it. Simpson, nevertheless offers three possible phylogenies, based on supposed similarities, reflecting the varied gaps in the row of dots: 7B, a(e(d(bc))), is the "most probable phylogeny," 7C, (a(bc))(de), is "a less probable phylogeny," and 7D, (ab)(c(de)), is a "still less probable phylogeny."

"What we all learned from Hennig" is the vacuous nature of such reckoning, which first fails to distinguish between plesiomorphy and apomorphy, and then relies on size of apparent gaps as evidence. Hennig comments (1966: 10): "it is enough to say that the 'phylogenetic' relationship has the character of genealogical relationships [which] are something entirely different from 'similarity.'"

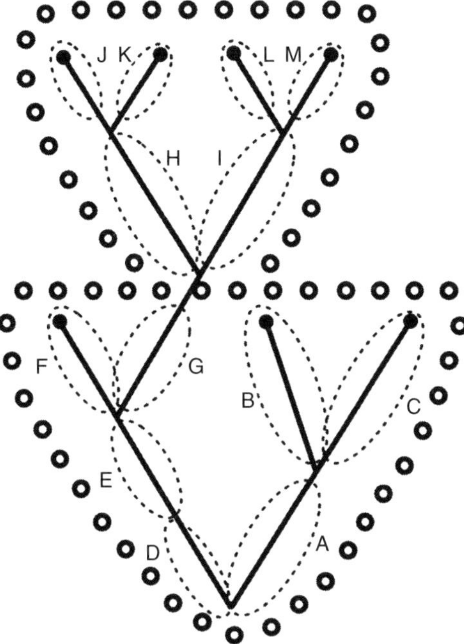

Fig 8.6 "A simple phylogeny divided into taxa." Redrawn after Simpson (1961: Fig 22A, p. 197), with species names added (A, B, etc.).

"What we all learned from Hennig" was real enough in the mid-1970s for a grateful Linnean Society of London (1974) and American Museum of Natural History (1975) to present Hennig with gold medals. Such was the Society's highest award. The Museum's was one of a number authorized by the Trustees for the Museum's centenary, celebrated 9 April 1969. Medals were then given to three Apollo 9 astronauts and five scientific luminaries, four described as "members of the curatorial staff" of the Museum (in its annual report for 1968–69): Theodosius Dobzhansky (1900–75) of Columbia University, Libbie Hyman (1888–1969), Ernst Mayr (1904–2005), Margaret Mead (1901–78), and George Gaylord Simpson (1902–84). Medals were later given to three other Museum staff: 1970 Edwin Colbert (1905–2001), 1971 Horace Stunkard (1889–1989), and 1979 Norman Newell (1909–2005).

Donn Rosen (1929–86), then Chairman of the Department of Ichthyology (1965–75), knew that yet unawarded medals survived within the office of Thomas Nicholson (1922–91), then director of the Museum (1969–89). Through Rosen's effort the Trustees were persuaded to free up a medal for Hennig; he was the first recipient without a close relationship with the museum, although in 1973 he was elected a Corresponding Member (a five-year honorary term). In 1975, a year before Hennig died, Nicholson travelled to Stuttgart to make the presentation with cooperation of the U.S. Department of State (Nicholson 1975).

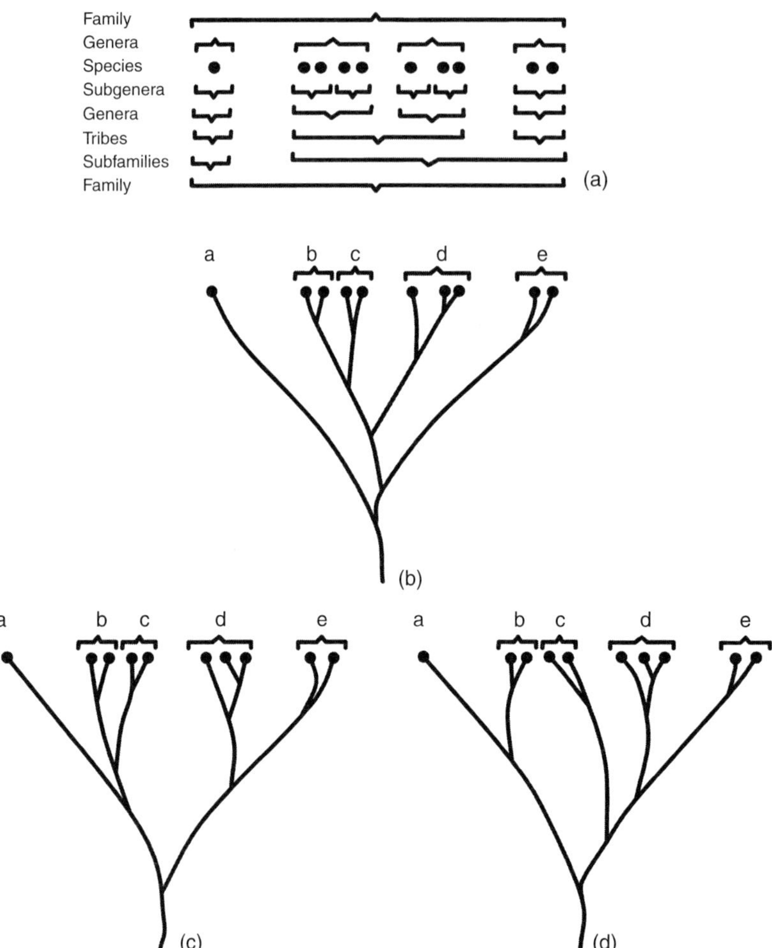

Fig 8.7 (a) "Ten contemporaneous species (black dots), with gaps of various sizes, corresponding with degrees of dissimilarity between them. (b–d) Some possible phylogenies for the species shown in (a). (b) Most probable phylogeny. (c) A less probable phylogeny. (d) Still less probable phylogeny." Redrawn after Simpson (1961: Fig 20, p. 193).

These two awards, 40 years ago, mark the end of the long discussion about "what we all learned from Hennig." One may doubt that we have since learned anything more from him.

Acknowledgements

For the redrawn figures, I am indebted to J.S. Wilkins, and for comment on the manuscript to L.C.S. Assis, A. Brower, M.C. Ebach, A. Gill, P.Y. Ladiges, R.D. Mooi,

L.R. Parenti, M.C.C. de Pinna, N.I. Platnick, O.C. Rieppel, M. Schmitt, R.T. Schuh, P. Tassy, J.W. Wägele, W.C. Wheeler, J.S. Wilkins, and D.M. Williams.

References

Bonde, N. (1977). Cladistic classification as applied to vertebrates. In *Major Patterns in Vertebrate Evolution*, ed. M.K. Hecht, P.C. Goody and B.M. Hecht. New York: Plenum, NATO Advanced Study Institutes Series, Vol. 14, pp. 741–804.

De Pinna, M.C.C. (1991). Concepts and tests of homology in the cladistic paradigm. *Cladistics*, 7, 367–394.

De Pinna, M. (2014). Species tot sunt diversae quot diversas formas ab initio creavit – A dialogue on species. *Arquivos de Zoologia*, 45 (esp.), 25–32.

Forey, P. (1998). Colin Patterson [obituary]. *The Systematics Association Newsletter*, April 1998, no. 9, 5 April, pp. 5–6.

Gilmour, J.S.L. (1961). Taxonomy. In *Contemporary Botanical Thought*, ed. A.M. MacLeod and L.S. Cobley. Chicago, IL: Quadrangle Books, pp. 27–45.

Hennig, W. (1950). *Grundzüge einer Theorie der phylogenetischen Systematik.* Berlin: Deutcher Zentralverlag; reprinted 1980, Koenigstein: Otto Koeltz.

Hennig, W. (1966). *Phylogenetic Systematics.* Urbana, IL: University of Illinois Press.

Hennig, W. (1982). *Phylogenetische Systematik.* Berlin: Verlag Paul Parey.

McKenna, M.C. and Bell, S.K. (1997). *Classification of Mammals above the Species Level.* New York: Columbia University Press.

Nicholson, T.D. (1975). Report of the director. *106th Annual Report, American Museum of Natural History, July 1974 through June 1975*, 6–7.

Patterson, C. (1960–61). A review of Mesozoic acanthopterygian fishes, with special reference to English chalk. (PhD thesis, University of London) University College and Guy's Hospital Medical School [University of London Library catalogue].

Patterson, C. (1964). A review of Mesozoic acanthopterygian fishes, with special reference to those of the English chalk. *Philosophical Transactions of the Royal Society of London, series B, Biological Sciences*, 247 (739), 213–482.

Patterson, C. (1981). Homology and phylogeny. Unpublished notes for a lecture at Harvard University.

Patterson, C. (1982). Morphological characters and homology. In *Problems of phylogenetic reconstruction*, Systematics Association Special Volume No. 21, ed. K.A. Joysey and A.E. Friday. London: Academic Press, pp. 21–74.

Patterson, C. (2011). Adventures in the fish trade [29th Annual Address 1995, Systematics Association]. In *Morphological and Molecular Approaches to the Phylogeny of Fishes: Integration or Conflict*, ed. M.R. de Carvalho and M.T. Craig. *Zootaxa*, 2946, 118–136 (written in 1995, published posthumously).

Simpson, G.G. (1945). The principles of classification and a classification of mammals. *Bulletin of the American Museum of Natural History*, 85, i–xvi, 1–350.

Simpson, G.G. (1961). *Principles of Animal Taxonomy*. New York: Columbia University Press.

Wiley, E.O. (1975). Karl R. Popper, systematics, and classification: a reply to Walter Bock and other evolutionary taxonomists. *Systematic Zoology*, 24, 233–243.

Wiley, E.O. (1976). The phylogeny and biogeography of fossil and recent gars (Actinopterygii: Lepisosteidae). *University of Kansas, Museum of Natural History, Miscellaneous Publication*, 64, 1–111.

Wiley, E.O. and B.S. Lieberman. (2011). *Phylogenetics: The Theory of Phylogenetic Systematics*. Hoboken, NJ: John Wiley & Sons.

Wilkins, J.S. (2009). *Defining Species: A Sourcebook from Antiquity to Today*. American University Studies, series V, Philosophy, vol. 23. New York: Peter Lang.

Williams, D.M. and Ebach, M.E. (2014). Patterson's curse, molecular homology, and the data matrix. In *The Evolution of Phylogenetic Systematics*, ed. A. Hamilton. Berkeley, CA: University of California Press, pp. 151–187.

9

Semaphoronts: the elements of biological systematics

Leandro C.S. Assis

9.1 Introduction

In order to provide a new philosophical and theoretical foundation for systematics, as well as improving its scientific status, Willi Hennig coined several new terms (e.g. synapomorphy, symplesiomorphy, and semaphoront) and reformulated some old ones (e.g. monophyly, phylogeny, and relationship) (Rieppel 2003, 2016). As a consequence, when discussing the impact of Hennig's (1965, 1966) thoughts in biology in general and in systematics in particular, the first point that comes to mind is his reformulation of the concept of monophyly (Schuh and Brower 2009, Rieppel 2010, Vanderlaan et al. 2013, and references therein). Accordingly, within the Hennigian system, supraspecific categories are an expression of phylogenetic relationships, and this implies the replacement of para- and polyphyletic taxa to monophyletic taxa (Hennig 1965, 1966).

Even though the theoretical and practical search for monophyletic groups constitutes one of the main goals of systematic and evolutionary studies, the use of morphological characters of adult organisms, along with having the genome sequenced of all living species, is not enough to accomplish the Hennigian revolution. A reason for this is that we know very little about the semaphoronts, "the elements of biological systematics" (Hennig 1966: 6). Hence, in addition to the topics addressed in the present book, I highlight in this chapter that to understand Willi Hennig's legacy

The Future of Phylogenetic Systematics: The Legacy of Willi Hennig, eds. D. Williams, M. Schmitt and Q. Wheeler. Published by Cambridge University Press.

and the future of phylogenetic systematics it is necessary to understand the role of the semaphoront in support of the inference of species and monophyletic groups. In this respect some fundamental questions come to mind:

1. What is the role of the semaphoronts in phylogenetic systematics?
2. Do semaphoronts constitute a bridge linking phylogenetics and other approaches?
3. Why have semaphoronts been broadly omitted from phylogenetic analyses?
4. What are the theoretical and practical differences between semaphoront-based phylogenetics and DNA-based phylogenomics?
5. Why are semaphoronts so essential for the future of phylogenetic systematics?

   Answers to these questions can be found by focusing on:

1. The semaphoront concept;
2. Hennig's foundations of phylogenetic systematics based on hologenetic (i.e. ontogenetic, tokogenetic, and phylogenetic) relationships;
3. The integration among semaphoronts, character, and homology;
4. The place of semaphoronts within the evolutionary studies, with emphasis on the link between phylogenetics and developmental evolution; and
5. The perception, according to the role of the semaphoront in the Hennigian system, that phylogenetics and phylogenomics are grounded in different theoretical and practical principles.

9.2 Semaphoronts

As Rieppel has aptly remarked in this book:

> To disambiguate phylogenetic systematics, Hennig wanted to replace natural language with a "precision language" (Hennig 1949: 138), and rendered the character bearer as *Merkmalsträger-Semaphoront* (Hennig 1950: 9; "character-bearing semaphoront"; Hennig, 1966: 6).

Hennig presented his definition of semaphoront as follows:

> [W]e should not regard the organism or the individual as the ultimate element of the biological system. Rather it should be the organism or the individual at a particular point of time, or even better, during a certain, theoretically infinitely small, period of its life. We will call this element of all biological systematics, for the sake of brevity, the *character-bearing semaphoront*. Definition of a semaphoront as the individual during a certain, however brief, period of time has the advantage that it may be thought of more simply as acting and showing evidence of life processes. No generally applicable statement can be made about how long a semaphoront exists as a constant systematically useful entity. It depends on the rate which its different characters change. (Hennig 1966: 6, italics on original)

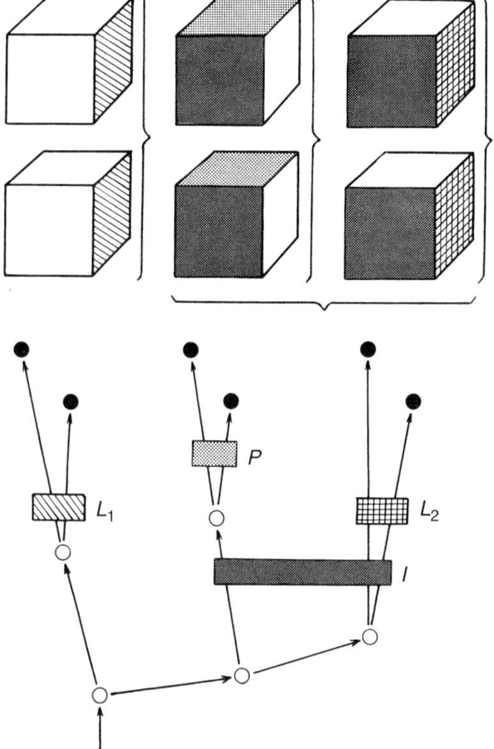

Fig 9.1 Reproduction of Fig 12 (from Hennig 1957), showing monophyletic groups supported by larval, pupal, and imaginal stages.

In his book *Phylogenetic Systematics*, Hennig provided some examples of taxa within Diptera that are empirically circumscribed on the basis of differences and similarities among correspondent semaphoronts of distinct organisms (Hennig 1966: 33–35). Hennig called "metamorphisms" the "differences in form between ontogenetically related semaphoronts" (Hennig 1966: 33). Once he investigated an insect group in which the organisms undergo complex metamorphisms, the use of the semaphoront concept was very appropriate (Wheeler 1990, Rieppel 2003). He illustrated the importance of semaphoront in support of phylogenetic relationships in Fig 35 of Hennig (1966), showing monophyletic groups supported by larval, pupal, and imaginal stages (Hennig 1966: 126) (Fig 9.1).

In relation to the introduction of this new concept in systematics, Rieppel (2003) raised a fundamental question, revealing Hennig's more essential and rebellious motivation:

> [I]t begs the question as to why Hennig (1950: 6) bothered to introduce a confusing new expression, the 'character-bearing semaphoront' (Hennig, 1966: 6). The

> answer is that he wanted more. Hennig wanted to base the logical primacy of the phylogenetic system on a 'concise foundation' of 'unequivocally defined, central concepts' (Hennig, 1974: 282), and the semaphoront was one of those. (Rieppel 2003: 169)

In line with this, Rieppel commented on how Hennig was influenced by the logical and causal lawfulness construction of the world, as expressed in the positivist philosophy of both Rudolf Carnap and Theodor Ziehen:

> Ziehen's reduction process can thus be seen to lie at the root of Hennig's (1950: 6) concept of the semaphoront, revealing the causal lawfulness that governs the relations between organisms. [...] according to Ziehen, the sameness of an individual in time and space must be based on a lawful continuity between its consecutive manifestations [...] From this then must follow that 'each single form amenable to description represents an arbitrary clip of the whole that is determined by the point in time chosen' (Hennig 1950: 8; 1966: 6). (Rieppel 2003: 169, Rieppel 2015)

9.3 Hologenetic relationships

In order to understand the role of the semaphoront in Hennig's plan of argumentation it is necessary to review how he treated biological relationships, from the level of the organism to the level of the species and monophyletic groups. Hennig (1966: 31, figure 6) viewed three kinds of relationships: (1) ontogenetic, at the level of an organism or individual; (2) tokogenetic, at the level of species; and (3) phylogenetic, at the level of a monophyletic group.

Ontogenetic relationships

> We have recognized the semaphoronts as the elements of biological systematics. Some of these semaphoronts are linked together into semaphoront groups (the individuals in everyday language) by genetic relationships, which we called ontogenetic relationships. (Hennig 1966: 29)

Tokogenetic relationships

> There are also genetic relationships between individuals, which we call "tokogenetic relationships," that arise through the phenomenon called "reproduction." [...] Groups of individuals that are interconnected by tokogenetic relationships are called species. (Hennig 1966: 30)

Phylogenetic relationships

> New species arise when gaps develop in the fabric of the tokogenetic relationships. The genetic relationships that interconnect species we call phylogenetic relationships. (Hennig 1966: 30)

Hologenetic relationships

> In spite of these differences in their structural pictures, the phylogenetic, tokogenetic, and ontogenetic relationships are only portions of a continuous fabric of relationships that interconnect all semaphoronts and groups of semaphoronts. With Zimmermann we call the totality of these the "hologenetic relationships". (Hennig 1966: 30)

Although it is not easy to delimit the semaphorontic series of an organism, as well as comparing it with the series of other organisms, ontogeny is a potentially observable process. Conversely, tokogeny and phylogeny are unobservable processes. Their products, i.e. species and monophyletic groups, are inferred through different data sets, methods, theories, concepts, and personal desires (Hennig 1966, Mayden 1997, Cracraft 2000, Wheeler and Meier 2000).

With respect to the nature of the semaphoronts, Hennig remarked:

> The morphological characters of its spatial, three-dimensional body are not the only properties of a semaphoront. Rather these properties encompass the totality of its physiological, morphological, and psychological (ethological) characters. We will call the totality of all these characters simply the total form (or the holomorphy) of the semaphoront, which thus is to be regarded as a multidimensional construct. (Hennig 1966: 7)

Despite recognising the role of all these biological systems, Hennig aimed to posit the phylogenetic system as the general, reference system for biology, as well as regarding the semaphoront as the common element of all systems:

> In the semaphoront we have found the element that must be identical in all conceivable systems of biology insofar as they are related to the living natural objects and not to particular properties of life or to specific processes of life. It is therefore possible to investigate the relations between the different systems, which in themselves are completely and basically equally justified and equally necessary. This is done most usefully by choosing one system as the general reference system with which all systems the other systems are compared. Creating such a general reference system, and investigating the relations that extend from it to all other possible and necessary systems in biology, is the task of systematics. (Hennig 1966: 7)

As Hennig himself said, all these systems are related to each other on the basis of the semaphoront, the basic element of biological systematics. In line with this, the understanding of the semaphoronts depends on understanding their ecological, ethological, genetic, morphological, and physiological properties. This constitutes a laborious task for comparative biology students.

9.4 Semaphoronts, character, and homology

Character and homology represent two key concepts in comparative biology (Rieppel 1988, Hall 1994, Wagner 2001). Although there exists a continuous debate about their meaning in phylogenetic systematics (e.g. Hennig 1966, Patterson 1982, Rieppel 1988, de Pinna 1991, Hall 1994, Brower and Schawaroch 1996, Hawkins 2000, Wagner 2001, Müller 2003, Williams 2004, Love 2007, Sereno 2007, Shubin et al. 2009, Scotland 2010, 2011, Wiley and Lieberman 2011, Brower and de Pinna 2012, Nixon and Carpenter 2012, Williams and Ebach 2012, Assis 2013, 2014a, Brower 2015), the way Hennig combined them with the notion of the semaphoront has rarely been commented on (Nelson 1985, Wheeler 1990, Wolfe and Hegna 2014). In Hennig's words:

> We will call those peculiarities that distinguish a semaphoront (or a group of sema-phoronts) from other semaphoronts 'characters,' keeping in mind that this desig-nation will never include merely morphological characters in the narrow sense, but always means the multidimensional totality. (Hennig 1966: 7)

And with respect to characters and homologies, he wrote:

> Different characters that are to be regarded as transformation stages of the same original character are generally called homologous. (Hennig 1966: 93)

In line with this, there is a relation among semaphoronts, character, and homol-ogy. Semaphoronts of two or more organisms are differentiated by their homolo-gous characters. This relation brings us to an understanding of homology in which the ontogenetic information has a fundamental role in the taxonomic circumscrip-tion of species and monophyletic groups (Hennig 1966, Nelson 1978, 1985, Wheeler 1990, Wolfe and Hegna 2014) supported by tokogenetic homologies and synapo-morphies, respectively (Assis 2014a). Indeed, once Hennig emphasized the primacy of the semaphoront over the organism,

> [W]e should not regard the organism or the individual as the ultimate element of the biological system. Rather it should be the organism or the individual at a par-ticular point of time, or even better, during a certain, theoretically infinitely small, period of its life,

a semaphorontic view of homology implies that:

> we should not regard the [*part*] or the [*organ*] as the ultimate element of [*hom-ology*]. Rather it should be the [*part*] or the [*organ*] at a particular point of time, or even better, during a certain, theoretically infinitely small, period of its [*ontogeny*]. (modified from Hennig 1966: 6)

Hennig also considered the "criterion of ontogenetic precedence" relevant information for "character phylogeny," making reference to both Haeckel and Naef:

> This assumes that the transformation of a character during ontogeny 'recapitu-lates' the phylogenetic transformation of this character, so that the direction of the transformation from plesiomorphous to apomorphous condition during phylogeny

can be determined from the sequence of ontogenetic stages. This criterion belongs in the sphere of ideas of the biogenetic law. In the same sense, Naef speaks of a primacy of ontogenetic precedence, and formulates his "law of terminal deviation" accordingly: "In the course of phylogenetic change, the stages of a morphogenesis are the more conservative the earlier they occur in the ontogenetic series, and the more advanced the latter they occur." Zimmermann (1953) calls Haeckel the true founder of the law of recapitulation. (Hennig 1966: 95)

[T]he "criterion of ontogenetic character precedence" remains an important aid in phylogenetic systematics, provided it is not uncritically evaluated more highly than other aids that, under certain circumstances, may lead to different conclusions. Complete rejection of the law of recapitulation (de Beer 1959) is certainly unjustified. (Hennig 1966: 96)

Arthur (2011) stressed that Ernst Haeckel's 'law' and its association with the 'laws' of Karl Ernst von Baer have frequently been misunderstood, and that Gavin de Beer was one of the main biologists responsible the misunderstanding. In order to clarify this issue Arthur wrote:

Von Baerian divergence […] is a pattern of early similarity giving way to latter differences. Von Baer gave four laws to describe this pattern; but it can adequately be described by the first of these: "The general features of a large group of animals appear earlier in the embryo than the special features." (Arthur 2011: 16)

According to Haeckel:

[T]hat Ontogeny is a recapitulation of Phylogeny; or, somewhat more explicitly: that the series of forms through which the individual organism passes during its progress from egg cell to its fully developed state, is a brief, compressed reproduction of long series of forms through which the animal ancestors of that organism (or the ancestral forms of its species) have passed from the earliest periods of so-called organic creation down to the present time. (Haeckel 1896: 7, cited in Arthur 2011: 16)

Based on Haeckel's explanation, Arthur raised two important questions – and attempted to answer them:

First, did Haeckel mean that the ontogenies of descendant species went through stages similar to the adult of forms of their ancestors? Second, and related to that, is Haeckel's law compatible or incompatible with von Baer's law?

[…]

"Regarding the first question, here is the answer provided by de Beer [1940], who interprets Haeckel's law as follow […]: 'The adult *stages* of the ancestors are repeated during the development of the descendants.' De Beer's next section is entitled 'The rejection of the theory of recapitulation.' But should we accept that a thorough student of embryology, such as Haeckel, really believed this? Probably not. (Arthur 2011: 16, italics in original)

Again, Arthur returns to Haeckel:

> The fact that an examination of the human embryo in the third or fourth week of its evolution [= development] shows it to be altogether different from the fully developed Man, and that it exactly corresponds to the *undeveloped embryo-form* presented by the Ape, the Dog, the Rabbit, and other Mammals, at the same stage of their Ontogeny. (Haeckel 1896: 18 as cited in Arthur 2011: 17, italics in original)

Concluding the point, Arthur (2011: 17) stated that both Haeckel and von Baer had the same idea, not like the "caricature recapitulation," but that of embryonic divergence, and that the use of the word "recapitulation" by Haeckel potentially created a historical misunderstanding.

Despite Hennig (1966: 95–96) endorsing the use of the "criterion of ontogenetic precedence" in character phylogeny, he did not discuss it exhaustively. A decade later, Nelson (1978) revived the criterion by comparing the impact of both the ontogenetic and paleontological arguments when applied to character phylogeny. In so doing, he restated the biogenetic law as a way to infer character polarity and, consequently, phylogenetic relationships:

> The biogenetic law may, therefore, be restated as follow: *given an ontogenetic character transformation, from a character observed to be more general to a character observed to be less general, the more general character is primitive* [plesiomorphic] *and the less advanced* [apomorphic]. (Nelson 1978: 327, italics on the original)

Thus, "[p]hylogenetic reconstruction in its entirety appears to be an extrapolation of the orderliness of development" (Nelson 1978: 324). Subsequently, Nelson (1985) presented a paper in response to criticisms of the ontogenetic criterion provided by Brooks and Wiley (1985) and Kluge (1985). With respect to the relation between ontogeny and systematics Nelson revisited an important issue, confessing that he "was never fully satisfied" with his previous efforts to integrate ontogeny and systematics. Nelson (1985: 42) cited Danser as a way to "fill what had before been empty in my understanding":

> Many false conceptions have arisen in systematics because living beings were treated as objects which the scientist proposed to classify. Once for all attention must be drawn to the fact that the systematist never classifies objects but life cycles. When he speaks of a plant or an animal he always means by it the whole life cycle of that plant or animal. The conception *Ranunculus acer* includes not only the body of the plant such as we find it at a given moment dead or alive, but also the complete history of its development and all the stadia of its life, therefore, the seed and the seedling as well as the full-grown plant, the development of the one from the other, indeed functions of every organ this including the ecology of the plant and its parts e.g. the entire history of its propagation. The objects which we use as material such as herbarium plants, stuck insects, skins of mammals, and fossils are nothing but an often inevitable substitute for the living beings represented by them. The missing links of the life cycle in these objects must be completed as far

as possible by the knowledge of the systematist if his systematics is to preserve its scientific character. Even the structure of the full-grown stage, on which, in many groups is almost exclusively based, must, however, great at times its importance, be only looked upon as the representation of the life cycle. (Danser 1950: 118)

For Nelson:

[W]hat was clarified is the relation between taxon (as ontogeny, or life cycle), character (as transformation, or part of life cycle), and ontogeny. It seemed immediately obvious then, and has remained so since, that all characters (synapomorphies) are potentially understandable as ontogenetic transformations. And as a transformation, each character is related through ontogeny to its homologs. Hence its relations ('polarity') are phenomena open to directly empirical investigation. (Nelson 1985: 42)

In addition, Nelson (1985) considered Hennig's (1966) concept of the semaphoront narrower than Danser's (1950) concept of the life cycle, as the former is applied to an individual and the latter to a taxon. It is interesting to note that Danser (1950) was occasionally cited in Hennig's (1966) *Phylogenetic Systematics,* but not with respect to his thoughts on the relationship between taxa and life cycles. Indeed, Willi Hennig's plan of argumentation – based on hologenetic relationships – was strongly influenced by the work of Walter Zimmermann, who created the "hologenetic spiral" illustrating the close relationship between phylogeny and ontogeny (Donoghue and Kadereit 1992, Claßen-Bockhoff 2001):

[Zimmermann] was also careful to point out that the process of phylogenetic character transformation is continuous and does not allow a clearcut distinction between phylogeny and ontogeny. This argument foreshadowed his 'hologenetic spiral,' mentioned above. He also noted that for purposes of phylogeny reconstruction we are forced to select particular stages of the life cycle for comparison. Here, Zimmermann's outlook is similar to that of Hennig (1966: 6), who used the word 'semaphoront' for the individual at a particular point in its development. (Donoghue and Kadereit 1992: 77–78).

In his paper entitled "Ontogeny and character phylogeny," Wheeler (1990) addressed the theoretical and practical implications of the concept of the semaphoront in systematics. He aptly revisited the discussion concerning ontogeny and its relation to the concepts of character, semaphoront and character polarity, and elucidated the relationships among species of the genus *Agathidium* (Coleoptera: Leiodidae), using both ontogenetic (Nelson 1978) and outgroup comparisons (Watrous and Wheeler 1981). In contrast to Nelson (1985), Wheeler (1990) argued that Hennig's (1966) notion of the semaphoront and Danser's (1950) notion of life cycles are not different, despite the fact that the former has a character-centered view and the latter a taxon-centered view. He termed "Nelson's rule" the new version of von Baer's law as proposed by Nelson (1978), as well as concluding that this

Finally, it is worth noting that semaphoronts have not been only ignored in molecular analyses but also in many morphological analyses. Again, understanding the ontogenetic stages for some taxa is extremely difficult. In some cases, ontogenetic limits from one stage to another cannot be easily established. In addition, because of the time of work and specimens availability, the number of taxa to be included in ontogeny-based phylogenies tends to be drastically inferior when compared with those based on adults and DNA sequences. Therefore, some of the major challenges for phylogenetic systematists of the twenty-first century is to balance evidence quantity and with quality, as well as the relative speed one gains data, with respect to the multiple hypotheses of relationship supported by thousands of genomic characters and dozens or hundreds of semaphorontic behavioural, ecological, morphological, and physiological characters (cf. Nelson 2004, Wheeler et al. 2013, Assis 2014b).

Acknowledgments

I am grateful to David Williams, Michael Schmitt, and Quentin Wheeler for inviting me to participate in the meeting "Willi Hennig (1913–1976): His Life, Legacy and the Future of Phylogenetic Systematics" and writing a chapter for this book. I also thank David Williams, Olivier Rieppel, and an anonymous referee for constructive comments on it.

The author research is supported by FAPEMIG APQ-01225-13.

References

Abzhanov, A., Kuo, W.P., Hartmann, C., et al. (2006). The calmodulin pathway and evolution of elongated beak morphology in Darwin's finches. *Nature*, 442, 563–567.

Arthur, W. (2011). *Evolution: a Developmental Approach*. Oxford: Wiley-Blackwell.

Arthur, W. and Chipman, A.D. (2005). The centipede *Strigamia maritima*: what it can tell us about the development and evolution of segmentation. *BioEssays*, 27, 653–660.

Assis, L.C.S. (2013). Are homology and synapomorphy the same or different? *Cladistics*, 29, 7–9.

Assis, L.C.S. (2014a). Species and tokogenetic homologies: 10 sutras. *Cladistics*, 30, 10.

Assis, L.C.S. (2014b). Testing evolutionary hypotheses: from Willi Hennig to Angiosperm Phylogeny Group. *Cladistics*, 30, 240–242.

Assis, L.C.S. and Santos, L.M. (2014). Phylogenetics is not phylogenomics. *Cladistics*, 30, 8–9.

Barrett, C.F., Davis, J.I., Leebens-Mack, J., Conran, J.G. and Stevenson, D.W. (2013). Plastid genomes and deep relationships among commelinid monocot angiosperms. *Cladistics*, 29, 65–87.

Brooks, D.R. and Wiley, E.O. (1985). Theories and methods in different approaches to phylogenetic systematics. *Cladistics*, 1, 1–11.

Brower, A.V.Z. (2015). Transformational and taxic homology revisited. *Cladistics*, 31, 197–201. DOI: 10.1111/cla.12076

Brower, A.V.Z. and Schawaroch, V. (1996). Three steps of homology assessment. *Cladistics*, 12, 265–272.

Brower, A.V.Z. and de Pinna, M.C.C. (2012). Homology and errors. *Cladistics*, 28, 529–538.

Claβen-Bockhoff, R. (2001). Plant morphology: the historic concepts of Wihelm Troll, Walter Zimmermann, and Agnes Aber. *Annals of Botany*, 88, 1153–1172.

Cracraft, J. (2000). Species concepts in theoretical and applied biology: a systematic debate with consequences. In *Species Concepts and Phylogenetic Theory: A Debate*, ed. Q.D. Wheeler and R. Meier. New York: Columbia University Press, pp. 3–14.

Cracraft, J. (2005). Phylogeny and evo-devo: characters, homology, and the historical analysis of the evolution of development. *Zoology*, 108, 345–356.

Danser, B.H. (1950). A theory of systematics. *Bibliotheca Biotheoretica*, 4, 117–180.

de Beer, G.R. (1940). *Embryos and Ancestors*. Oxford: Clarendon Press.

de Pinna M.C.C. (1991). Concepts and tests of homology in the cladistic paradigm. *Cladistics*, 7, 367–394.

Dikow, R.D. and Smith, W.L. (2013). Complete genome sequences provide a case study for the evaluation of gene-tree thinking. *Cladistics*, 29, 672–682.

Donoghue, M.J. and Kadereit, J.W. (1992). Walter Zimmermann and the growth of phylogenetic theory. *Systematic Biology*, 41, 74–85.

Gilbert, S.F. (2014). *Developmental Biology*. Sunderland, MA: Sinauer Associates Inc.

Gilbert, S.F. and Epel, D. (2009). *Ecological Developmental Biology: Integrating Epigenetics, Medicine, and Evolution*. Sunderland, MA: Sinauer Associates Inc.

Grandcolas, P., Deleporte, P., Desutter-Grandcolas, L. and Daugeron, C. (2001). Phylogenetics and ecology: as many as characters possible should be included in the cladistics analysis. *Cladistics*, 17, 104–110.

Haeckel, E. (1896). *The Evolution of Man: A Popular Exposition of the Principal Points of Human Ontogeny and Evolution*. New York: Appleton.

Hall, B.K. (ed.) (1994). *Homology: the Hierarchical Basis of Comparative Biology*. San Diego: Academic Press.

Hawkins, J.A. (2000). A survey of primary homology assessment: different botanists perceive and define characters in different ways. In *Homology and Systematics. Coding Characters for Phylogenetic Analysis*, ed. R. Scotland and R.T. Pennington. London: Taylor & Francis, pp. 22–53.

Hennig, W. (1949). Zur Klärung einiger Begriffe der phylogenetischen Systematik. *Forschungen und Fortschritte*, 25, 136–138.

Hennig, W. (1950). *Grundzüge einer Theorie der phylogenetischen Systematik*. Berlin: Deutscher Zentralverlag.

Hennig, W. (1957). Systematik und Phylogenese. In *Bericht über die Hundertjahrfeier der Deutschen Entomologischen Gesellschaft Berlin*, ed. H.-J. Hannemann. Berlin: Akademie-Verlag, pp. 50–71.

Hennig, W. (1965). Phylogenetic systematics. *Annual Review of Entomology*, 10, 97–116.

Hennig, W. (1966). *Phylogenetic Systematics*. Urbana, IL: University of Illinois Press.

Hennig, W. (1974). Kritische Bemerkungen zur Frage "Cladistic analysis or cladistic classification?" *Zeitschrift für Zoologische Systematik und Evolutionsforschung*, 12, 279–294.

Klingenberg, C.P. (2008). Morphological integration and developmental modularity. *Annual Review of Ecology, Evolution, and Systematics*, 39, 115–132.

Kluge, A.G. (1985). Ontogeny and phylogenetic systematics. *Cladistics*, 1, 13–27.

Love, A. (2007). Functional homology and homology of function: biological concepts and philosophical consequences. *Biology and Philosophy*, 22, 691–708.

Mayden, R.L. (1997). A hierarchy of species concepts: the denouement in the saga of the species problem. In *Species: The Units of Biodiversity*, ed. M.A. Claridge, H.A. Dawah and M.R. Wilson. London: Chapman and Hall, pp. 381–424.

Miranda, L.S., Collins, A.G. and Marques, A.C. (2010). Molecules clarify a cnidarian life cycle – the "hydrozoan" *Microhydrula limopsicola* is an early life stage of the "staurozoan" *Haliclystus antarcticus. PloS One*, 5, 1–9.

Müller, G.B. (2003). Homology: the evolution of morphological organization. In *Origination of Organismal Form: Beyond the Gene in Developmental and Evolutionary Biology*, ed. G.B. Müller and S.A. Newman. Cambridge, MA: The MIT Press, pp. 51–69.

Nelson, G. (1978). Ontogeny, phylogeny, paleontology, and the biogenetic law. *Systematic Zoology*, 27, 324–345.

Nelson, G. (1985). Outgroups and ontogeny. *Cladistics*, 1, 29–45.

Nelson, G. (2004). Cladistics: its arrested development. In *Milestones in Systematics*, ed. D.M. Williams and P.L. Forey. Boca Raton, FL: CRC Press, pp. 127–147.

Nixon, K.C. and Carpenter, J.M. (2012). On homology. *Cladistics*, 28, 160–169.

Patterson, C. (1982). Morphological characters and homology. In *Problems of Phylogenetic Reconstruction*, ed. K.A. Joysey and A.E. Friday. London: Academic Press, pp. 21–74.

Rieppel, O.C. (1988). *Fundamentals of Comparative Biology*. Basel: Birkhäuser Verlag.

Rieppel, O. (2003). Semaphoronts, cladograms and the roots of total evidence. *Biological Journal of the Linnean Society*, 80, 167–186.

Rieppel, O. (2005). Modules, kinds, and homology. *Journal of Experimental Zoology (Molecular and Developmental and Evolution)*, 304B, 18–27.

Rieppel, O. (2010). Species monophyly. *Journal of Zoological Systematic and Evolution*, 48, 1–8.

Reippel, O. (2016). Willi Hennig as philosopher. In: *The Future of Phylogenetic Systematics: The Legacy of Willi Hennig*, eds. D. Williams, M. Schmitt and Q. Wheeler. Cambridge: Cambridge University Press, pp. 357–277.

Schuh, R.T. and Brower, A.V.Z. (2009). *Biological Systematics: Principles and Applications*. New York: Cornell University Press.

Scotland, R.W. (2010). Deep homology: a view from systematics. *BioEssays*, 32, 438–449.

Scotland, R.W. (2011). What is parallelism? *Evolution and Development*, 13, 214–227.

Sereno, P.C. (2007). The logical basis for morphological characters in phylogenetics. *Cladistics*, 23, 565–587.

Shubin, N., Tabin, C. and Carroll, S. (2009). Deep homology and the origins of evolutionary novelty. *Nature*, 457, 418–423.

Theissen, G. and Melzer, R. (2007). Molecular mechanisms underlying origin and diversification of angiosperm flower. *Annals of Botany*, 100, 603–619.

Vanderlaan, T.A., Ebach, M.C., Williams, D.M. and Wilkins, J.S. (2013). Defining and redefining monophyly: Haeckel, Hennig, Ashlock, Nelson and the proliferation of definitions. *Australian Systematic Botany*, 26, 347–355.

Wagner, G.P. (2001). Characters, units, and natural kinds: an introduction. In *The Character Concept in Evolutionary Biology*, ed. G.P. Wagner. San Diego: Academic Press, pp. 1–10.

Watrous, L.E. and Wheeler, Q.D. (1981). The out-group comparison method of character analysis. *Systematic Zoology*, 30, 1–11.

Webster, M. and Zelditch, M.L. (2005). Evolutionary modifications of ontogeny: heterochrony and beyond. *Paleobiology*, 31, 354–372.

West-Eberhard, M.J. (2003). *Developmental Plasticity and Evolution*. Oxford: Oxford University Press.

Wheeler, Q.D. (1990). Ontogeny and character phylogeny. *Cladistics*, 6, 225–268.

Wheeler, Q.D. and Meier, R. (eds.) (2000). *Species Concepts and Phylogenetic Theory: A Debate*. New York: Columbia University Press.

Wheeler, Q., Assis, L. and Rieppel, O. (2013). Heed the father of cladistics. *Nature*, 496, 295–296.

Wiley, E.O. and Lieberman, B.S. (2011). *Phylogenetics. The Theory and Practice of Phylogenetic Systematics* 2nd edition,. Hoboken, NJ: J. Wiley and Sons, Inc.

Williams, D.M. (2004). Homologues and homology, phenetics and cladistics: 150 years of progress. In *Milestones in Systematics*, ed. D.M. Williams and P.L. Forey. Boca Raton, FL: CRC Press, pp.191–224.

Williams, D.M. and Ebach, M.C. (2012). Confusing homologs and homologies: a reply to "On homology". *Cladistics*, 28, 223–224.

Wolfe, J.M. and Hegna, T.A. (2014). Testing the phylogenetic position of Cambrian pancrustacean larval fossils by coding ontogenetic stages. *Cladistics*, 30, 366–390.

Zimmermann, W. (1953). *Evolution : Die Geschichte ihrer Probleme und Erkenntnisse*. Freiburg: Karl Alber.

10

Why should cladograms be dichotomous?

RENÉ ZARAGÜETA BAGILS and SOPHIE PÉCAUD

10.1 Introduction

A cladogram is usually considered as resolved when all its branching points are bifurcations (see e.g. Hennig 1966: 211). The question we address in this chapter is: upon which foundations does this "principle of dichotomy" rest? Here, we face an alternative: either the principle of dichotomy is *empirical* (i.e. it emerges from a set of empirical investigations, for instance concerning the processes of speciation), or it is *methodological* (i.e. it emerges from general considerations about how scientific inquiry should be pursued). But each of these answers contains several possibilities: upon what *kind* of empirical foundation could the principle of dichotomy be founded? Or upon what *kind* of methodological foundation?

The debate as it has developed in the history of cladistics has trapped itself into a restrictive dilemma: either the principle of dichotomy is based upon Ernst Mayr's allopatric speciation model (the sole empirical foundation taken into consideration), or it is reducible to a very general requisite of scientific methodology, which conforms to Karl Popper's opinion according to which a theory or a set of hypotheses (such as a cladogram) is all the more falsifiable when it carries a greater amount of empirical information (the sole methodological foundation taken into consideration). Those positions are exemplified, for instance, by Lars Brundin (1966) for the former, and Norman Platnick (1979) for the latter.

Our aim is to escape from this dilemma. Founding the principle of dichotomy on an empirical basis does not necessarily mean it should be grounded on Mayr's

The Future of Phylogenetic Systematics: The Legacy of Willi Hennig, eds. D. Williams, M. Schmitt and Q. Wheeler. Published by Cambridge University Press.

allopatric speciation model; neither does founding it on a methodological basis mean it should necessarily be grounded on Popper's normative philosophy of science. In the first and second part of this chapter, we will examine the two traditional ways cladists have used to justify the principle of dichotomy: the empirical way, with Hennig (1966) and, above all, Brundin (1966), and the methodological/Popperian way, with Platnick (1979). An additional inquiry into Nelson and Platnick (1981) will pave the way for yet another, original, solution to the problem of the foundation of the principle of dichotomy, which we will outline in the third part of this chapter.

10.2 The temptation of an empirical foundation

In his article about "Hennig's dichotomization of nature" (Rieppel 2010), Olivier Rieppel listed the supporters of Willi Hennig who had opted, after him and because of him, for an empirical justification of the principle of dichotomy – a justification grounded in hypotheses about speciation processes. He mentioned in particular the Swedish entomologist Lars Brundin (1907–93), an early promoter of Hennig's phylogenetic systematics (Dupuis 1978: 4–5), best known for his important work in biogeography (Brundin 1966). Let us examine Brundin's position first, for it is much less ambiguous than that of Hennig himself, and exhibits in a much clearer way the theoretical option that has been attacked by neo-Darwinian systematists as "unrealistic" (e.g. Darlington 1970 and Mayr 1974).

Brundin's clearly empirical foundation

Brundin's reasoning can be reduced to three claims. Turning his back on an old-fashioned conception of systematics – an "art" rather than a "science" (he cites Simpson 1961: 107) –, he first asserted that phylogenetic systematics must be entirely subordinated to evolutionary biology, and especially to population genetics, one of the most lively domains of biological research of his time. In his view, every principle, every rule, and every concept of phylogenetic systematics ought to be "logical consequences" of population genetics' results:

> The definitions and concepts [of phylogenetic systematics] are logical consequences of the results of the population genetics; and the method of argument follows strict rules, thus giving reasonable guarantee for a phylogenetic system which is on the whole free of inconclusive judgment and able to function as a reliable general reference system. (Brundin 1966: 22).

The idea of the phylogenetic system becoming a "general reference system" of biology thanks to its being in accordance with the latest discoveries of contemporary biological research exists in Hennig's works (in this respect, Brundin is indeed a disciple of the German entomologist); but despite Brundin's rejection of Simpson's conception of systematics, his bold idea of an entire subordination of systematics to evolutionary biology is rather neo-Darwinian than Hennigian – see, for instance, the relatively late character of the main neo-Darwinian works on systematics (e.g. Huxley 1940, Mayr 1942, Simpson 1961 and Mayr 1969).

Brundin's second claim was that the allopatric speciation model proposed by Ernst Mayr is the "essence" of speciation:

> The essence of speciation is the production of two well-integrated gene pools from a single parental one (Mayr 1963). The process presupposes the development of (geographical) barriers and a complete break in the gene flow between earlier contiguous populations. (Brundin 1966: 14)

That claim illustrated what Hennig himself wrote the very same year in *Phylogenetic Systematics*: for the biology of the 1960s, the generality of the allopatric speciation had become "dogma" (Hennig 1966: 229), i.e. literally, an article of faith no longer submitted to critical analysis.

Brundin finally concluded – his third claim – that the *generality* (if not the *universality*) of allopatric speciation implied the *necessity* (he uses the term) of a hierarchical and dichotomous representation of the phylogenetic system:

> The gaps in the genetic connections now existing between different species would be bridged if we could go back in time. For example, if we follow the flow of genetic connection existing between the individuals of the species *B* back in time, and do the same with respect to the species *C*, we will eventually arrive at the species *A*, which can be designated as the common ancestral species of *B* and *C*. Genetic flow in time, connecting different species in the way just mentioned, do form the phylogenetic connections between the actual species [...] Consequently we arrive by necessity at the following picture [Fig 10.1] of the structure of the phylogenetic connections which, according to the evolutionary theory, must be generally valid. (Brundin 1966: 14)

Hennig's hesitation

Was Brundin correct when he attributed to Hennig the foundation of the principle of dichotomy upon Mayr's allopatric speciation model? The answer is complex. While it is true that Hennig did not reduce the principle of dichotomy to a mere methodological principle, he was much more cautious than Brundin when it came to linking it to specific hypotheses about speciation processes – and he became all the more cautious over time. Below we examine what he wrote.

Rieppel wrote, "In ... 1950, Hennig went to great length to provide a biological basis to the principle of dichotomy rooted in theories of speciation, which in hindsight proved flawed, not the least for their orthogenetic implications" (Rieppel, this volume). In 1966, Hennig seemed to take a step back. Rieppel wrote: "This may well be the reason why in his *Phylogenetic Systematics*, Hennig (1966: 210) reduced the principle of dichotomy [...] to a merely methodological one" (Rieppel, this volume).

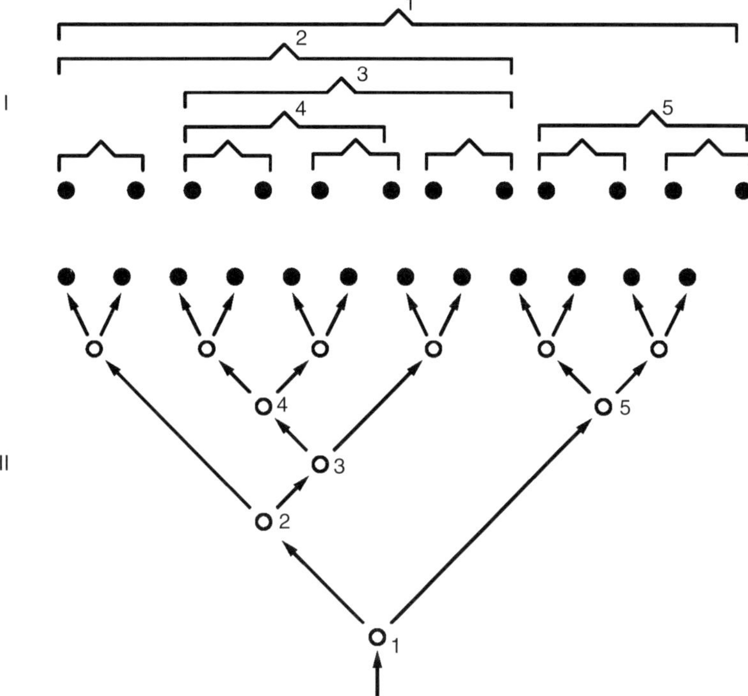

Fig 10.1 Different presentations (I, II) of the phylogenetic connections between the species of a monophyletic group. From Hennig (1957: Fig 4), redrawn and modified from Brundin (1966: 15).

The situation is in fact more complex when properly scrutinized. In 1966, Hennig developed an argument in three parts (Hennig 1966: 209–213).

The first part was the affirmation of the principle of dichotomy as methodological. Hennig began by expressing his astonishment that

> The view is often advanced that phylogenetic systematics presupposes a dichotomous structure of the phylogenetic tree. Because dichotomy is not the rule, it is said that a system that gives the impression of a continuously dichotomous differentiation cannot be regarded as a true presentation of the actual kinship relations. (Hennig 1966: 209)

On the contrary, he emphasized that

> If phylogenetic systematics starts out from a dichotomous differentiation of the phylogenetic tree, this is primarily no more than a methodological principle. (Hennig 1966: 210)

Dichotomy, Hennig wrote, is a logical consequence of the method of phylogenetic systematics itself. Consider three taxa, A, B, and C. The aim of phylogenetic analysis is to decide which of them – (A,B), (B,C), or (A,C) – are most closely related. Each time a character state displaying an evolutionary novelty (an apomorphy) indicates that a number of groups (e.g. species) form a monophyletic taxon, the method demands that the analysis be pursued: new synapomorphies must be identified, allowing us to form less inclusive monophyletic taxa, until each group has finally found its ultimate sister group. The result, Hennig wrote, is "a dichotomous system of which no one would argue that it does not represent correctly the relationships of the [groups]" (Hennig 1966: 211).

The second part of Hennig's argument was a reflection on unresolved cases. It is a fact, he wrote, that one sometimes struggles in vain to resolve phylogenetic relationships between taxa – the "phylogenetic tree" (i.e. cladogram) expressing their relationships remains "polytomous" rather than becoming dichotomous. But are those cladograms *irreducibly polytomous*? What concerns us here is what this question *presupposes*. In Hennig's mind, a cladogram can be irreducibly polytomous in two cases: either because (1) it endeavors to bring to light relationships which are *truly polytomous*, in other words those which result from radiation (i.e. multiple speciation) processes; or because (2) we lack data (in this case, synapomorphies) to reconstruct the sequence of dichotomous speciations which actually took place (see the example in Hennig 1966: 212). In the first case, the cladogram includes what we now usually refer to as "hard" polytomies; it is polytomous *de jure*. In the second case, it only includes "soft" polytomies: it is polytomous *de facto*, which means that one would hope that the method be refined in such a way that would finally be able to reconstruct the sequences of dichotomous speciations that are currently unresolved.

The existence of (2) – cases in which the polytomous structure of the cladogram results from a lack of data rather than from an actual radiation – led Hennig to be extremely cautious as to the very *existence* of events of radiation (an empirical claim), and in any case to the *relevance* of the concept of radiation in phylogenetic analysis (a methodological claim). As for the latter, he wrote that "the impossibility of determining with certainty the sequence of dichotomous cleavages in a group never means that all the species arose simultaneously (by radiation) from one stem species" (Hennig 1966: 211). As for the former, he wrote: "A priori it is very improbable that a stem species actually disintegrates into several daughter species at once" (Hennig 1966: 211); and, more implicitly: "If the distinction between 'radiation' and 'dichotomy' is to make any sense at all it can only be that we speak of radiation when the sequence of cleavages within a particular time span cannot be determined" (Hennig 1966: 213) – thus admitting that dichotomous speciation must be considered the rule, and radiation the exception, if such an evolutionary event exists at all.

What does (2) tell us about Hennig's position concerning the principle of dichotomy? Even though Hennig claimed that the principle of dichotomy is "primarily no more than a methodological principle" (Hennig 1966: 210), he nonetheless linked it immediately to empirical hypotheses (or even strong beliefs) about speciation processes: a dichotomous cladogram represents dichotomous speciations; a (truly) polytomous cladogram represents radiations. (2) – cases in which the polytomous structure of the cladogram results from our lack of data – then tempers (1) – cases in which the polytomous structure of the cladogram results from an actual radiation. It is true that "phylogenetic systematics [does not] presuppos[e] a dichotomous structure of the phylogenetic tree" (Hennig 1966: 209); it could represent events of radiation (by the means of polytomous cladograms), if we had confirmation that such evolutionary events happen. However, it is the dichotomous or polytomous character of speciation that actually and ultimately justifies the dichotomous or polytomous structure of the cladogram.

In the third part of his argument, Hennig exchanged the point of view of the theoretician with that of the practitioner. The entomologist in him wondered why the phylogenetic relationships of older groups are so frequently patently dichotomous, while those of more recent groups more often seem polytomous. The difficulty is so overwhelming that he even asked:

> May not the picture of a radiation [...] be the correct one for these older groups too, and the picture of a clear dichotomy actually be only an artefact of our method of recognizing kinship relations? (Hennig 1966: 213)

He answered:

> In older groups a clear and appropriate picture of dichotomous differentiation [...] may result from the random extinction of species or their descendants that actually arose approximately at the same period. (Hennig 1966: 213)

That third part of Hennig's argument is quite puzzling: Hennig seems to take another step back by conceding that radiation actually exists, and might be fairly common. The key to a proper understanding of that passage is the term "approximately": here, Hennig widened the notion of simultaneity ("approximate" rather than strict) in order to give an operational meaning to that of radiation (the "approximately" simultaneous arising of several species). (The indeterminacy of the notion of "simultaneity" is in fact the main obstacle for an operational definition of the notion of radiation, as remarked by Hennig himself, followed by Platnick 1979: 539.) It can be assumed that his aim here was to show consideration for what field systematists – like himself – notice when trying to reconstruct the phylogenetic relationships of recent taxa: apparently generalized events of radiation. Hennig's very pragmatic claims in this passage cannot therefore be considered as important as the other ones.

Was Mayr (1974) right, then, when he attacked Hennig on the empirical front? (According to him, Hennig's "principle of dichotomy" should be rejected because "all the indications are that a simple dichotomy into two daughter species is not the rule" [Mayr 1974: 110][1]). In any case, Hennig laid himself open to such an attack. It is true that he was more cautious than Brundin when it came to justifying the principle of dichotomy with empirical assumptions about speciation processes. But in *Phylogenetic Systematics*, he had it both ways: the methodological way (cladograms should be dichotomous because we should never jump to the conclusion of the existence of an actual radiation), and the empirical one (cladograms should be dichotomous because dichotomous speciation is the rule). Hennig's response to Mayr – the principle of dichotomy should be considered "heuristic" (Hennig 1975: 256) – therefore expressed either dishonesty, or a late and never explicitly elucidated rejection of the empirical justification of the principle of dichotomy which was present in 1966 as well as in 1950, as we have seen.

10.3 The principle of dichotomy as a requirement of scientific methodology

Platnick's Popperian attempt

Mayr was not the only one to criticize the principle of dichotomy. Another and earlier attack was that of American entomologist Philip Darlington (1904–83) (Darlington 1970). In 1971, a response to Darlington did not come from Hennig himself, but from a young supporter of phylogenetic systematics, Gareth Nelson. Darlington (1970) rejected several principles of Hennig's method. Nelson, however, only responded to one of them, that which concerned the principle of dichotomy, for it is the only one he thought was "really [...] essential to the philosophy of Hennig and Brundin" (Nelson 1971: 374).

In 1971, Nelson only kept Hennig's methodological strategy of defense, and did not even mention Hennig's empirical hesitations: inasmuch as we are never sure if the polytomy of a given cladogram is due to our lack of data, or to an actual event of radiation, we should never jump to the conclusion of the existence of an actual radiation. In Nelson's terms:

> An argument for multiple speciation can rest only upon negative evidence, and is equivalent to the acceptance of a null hypothesis, i.e. an hypothesis of no significant difference between all possible theories embodying dichotomous speciation. The use of dichotomous speciation as a methodological principle is required before an hypothesis of multiple speciation is even tentatively acceptable. (Nelson 1971: 373–374)

---

[1] Let us note, in passing, the irony of the reproach, for it is precisely the reference to Mayr as the promoter of the generality of allopatric speciation which had led Hennig, and above all Brundin, to found their principle of dichotomy upon the hypothesis of the generality of such a process...

In 1979, Norman Platnick followed Nelson in his general project – justifying the principle of dichotomy without any reference to empirical assumptions:

> If classifications (that is, our knowledge of patterns) are ever to provide an adequate test of theories of evolutionary process, their construction must be independent of any particular theory of process. The question with regards to the methodological preference for dichotomous hypotheses, therefore, is whether the preference can be justified by arguments that do not depend on any particular view of the mechanism of speciation. (Platnick 1979: 539)

Platnick's originality was that he tried to give a more precise meaning to – if not definitely prove – Hennig's intuition according to which a dichotomous cladogram is *more heuristic* than a polytomous one. To do so, he chose to *compare the empirical content* of both types of cladograms by "looking at the information content of [the] cladistic hypotheses" (Platnick 1979: 539) they are composed of. His underlying assumption was that the larger the empirical content of an hypothesis (e.g. a cladogram) is, the more heuristic it becomes, since, as Popper had shown, statements which have a larger empirical content say more about the world, and are easier to test (see e.g. Popper 2002: 128); Platnick (1979: 540) cited Popper's disciple Imre Lakatos (1970), who referred to this as "Popper's supreme heuristic rule: 'devise conjectures which have more empirical content that their predecessors.'"

Platnick presented two cladograms, one dichotomous, the other trichotomous (Fig 10.2). The trichotomous cladogram (Fig 10.2a) is composed of four hypotheses, numbered from 1 to 4, which, according to Popper, can also be expressed as "proscriptions" or "prohibitions" (Popper 2002: 48). Hypothesis 1 says that there is a certain number of character states shared exclusively (i.e. synapomorphies) by scorpions, whipspiders, and spiders; correlatively, it forbids that we find character states shared exclusively by some members of the group "scorpions, whipspiders, and spiders" and any other organism outside this group. For instance, "it prohibits the existence of any synapomorphies shared uniquely by spiders plus ticks, or spiders plus elephants" (Platnick 1979: 540). Hypothesis 2 says that there are a certain number of character states shared exclusively (i.e. synapomorphies) by scorpions; correlatively, it forbids that we find character states shared exclusively by some scorpions and any other organism outside the taxon of scorpions. It is the same with hypotheses 3 and 4, which concern, respectively, whipspiders and spiders.

Let us consider now the dichotomous cladogram (Fig 10.2b):

> It's obvious that all four of the hypotheses expressed by the first cladogram are also expressed in the second, but that there is an additional one, no. 5, predicting that we will find some synapomorphies only true for whipspiders and spiders, and that we will not find any synapomorphies unique to any others organisms plus only some members of the group Labellata (whipspiders plus spiders). In short, the dichotomous cladogram contains [...] five hypotheses instead of four [...] and this additional information content is in itself justification enough for a preference

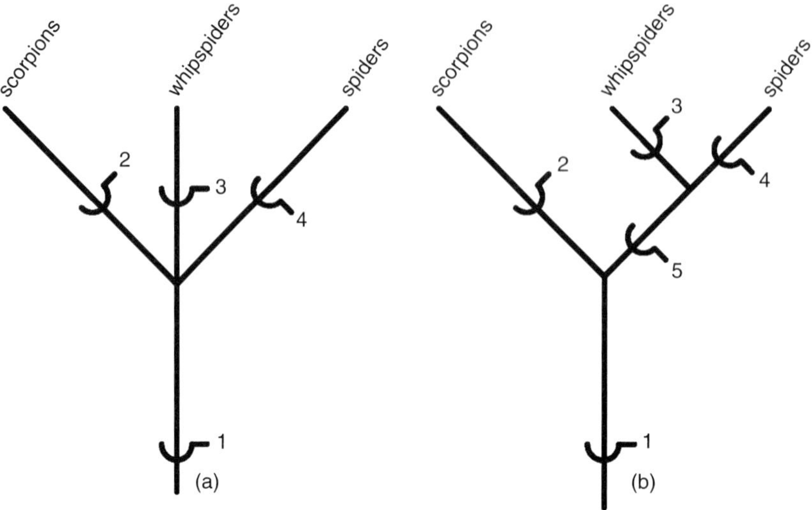

Fig 10.2 Information content of a trichotomous (a) and dichotomous (b) cladogram (redrawn and modified from Platnick 1979: 540).

> for dichotomous hypotheses, without recourse to any knowledge claims about the mechanisms of speciation. (Platnick 1979: 540)

Hennig's intuition therefore received a precise meaning: dichotomous cladograms are more heuristic because they subordinate a larger number of hypotheses; in that sense, they permit us to formulate a larger number of predictions, and prohibit a larger number of facts. The Popperian framework that Platnick adopted was crucial to his argument: the reason why dichotomous cladograms are preferable to trichotomous ones is that "the amount of empirical information conveyed by a theory, or its *empirical content*, increases with its degree of falsifiability" (Popper 2002: 96).

Platnick's argument seems rather convincing, but it is very dangerous in the polemical context in which the discussions about the principle of dichotomy usually take place. Hennig's neo-Darwinian challengers could indeed turn such a Popperian argument to their advantage (as Bock 1973, for instance, has done in the past): it could probably be said that the hypotheses which compose neo-Darwinian phylogenetic trees (hypotheses of ancestor–descendant relationships plus hypotheses about the degree of divergence of taxa) carry even more empirical content than those which are subordinated by dichotomous cladograms (mere "cladistic" hypotheses). Does it mean that they are more heuristic? We doubt that Platnick would give a positive answer to that question. What is interesting to us, though, is that Platnick's attempt was not a mere renunciation (to Hennig's empirical justification of the principle of dichotomy), but also a re-interpretation and a continuation (of Hennig's methodological justification). The few paragraphs he dedicated to the problem at stake must hence be complemented with Nelson and Platnick's subsequent reflection on the principle of dichotomy, featured in *Systematics and Biogeography* (Nelson and Platnick 1981: 257–265).

Nelson and Platnick: the cladogram as an analytic tool

In the early 1980s, Nelson and Platnick (1980, 1981) offered a reflection on the numerous cases in which a cladogram can be polytomous. Their starting point was the same distinction as Hennig's: either polytomy

> may reflect nothing more than ignorance of certain character distributions [or it is] caused by cases of "simultaneous" multiple speciation, instances of hybridization, or groups wherein one species is ancestral to two or more others (as in speciation by the sequential isolation of two or more peripheral population without change in the central population of a "mother" species, or cases in which studied fossil species are actually the ancestors of other studied species). (Nelson and Platnick 1981: 257)

Let us note that, just like Hennig, Nelson and Platnick made room for the existence of a "true" (or "hard") polytomy, i.e. a polytomy which does not result from a lack of data, but which is justified by the very nature of the relationships it endeavours to represent. They nonetheless introduced a subtle yet fundamental change in perspective: polytomy is not only likely to represent occurrences of radiation (as Hennig and Brundin believed), but also occurrences of *hybridization* and even *ancestor–descendant relationships*. How can we understand such a shift, and why is it important?

In Hennig's works, cladograms or, rather, "scheme[s] of argumentation of phylogenetic systematics" (Hennig 1966: 90) were not yet very well distinguished from phylogenetic trees in the neo-Darwinian sense (i.e. trees representing ancestor–descendant relationships plus degrees of divergence). Branches still represented the transformation of species from one into another (or at least the transformation of characters from one state into another); and nodes still represented events of speciation. As a consequence, in Hennig's mind, a true polytomy could only be the representation of a radiation; a hybridization would be represented by a reticulate pattern, and an ancestor–descendant relationship would require the ancestor to be placed at a node of the tree. Nelson and Platnick, on the contrary, worked with cladograms that had been entirely released from their neo-Darwinian heritage: a cladogram was nothing more than a tool permitting us to synthesize and put in order data about taxa, and to formulate hypotheses about their phylogenetic relationships. In that sense, nodes represented mere *distribution of character states*, which means that if they indeed pointed *to evolutionary events*, the nature of those events could not be "read" directly from the cladogram, and still needed to be hypothesized. To clarify: a true polytomy did not necessarily point to a radiation. Other events, such as hybridizations, also produced such a pattern when their result (the current organisms and their character states) was submitted to a cladistic analysis; it was also the case when an actual ancestor was analysed along with its descendants (see the explanation in Nelson and Platnick 1981: 259 *et seq.*).

The main point that differentiates Nelson and Platnick from Hennig is that the former rejected the problem of the interpretation of polytomous cladograms outside the realm of cladistics. Cladistics is exclusively interested in the building of cladograms on the basis of the distribution of character states. Some of those distributions result in polytomous cladograms. In such cases, it is always wiser to start by applying Hennig's rule and try to resolve the cladogram before assuming the occurrence of a "true" polytomy. When such a solution is not possible, one can try to interpret it in terms of evolutionary events (and especially speciation events): is it the result of radiation, hybridization? Is there a "true" ancestor among the taxa under analysis? To those questions, though, cladistics cannot answer; another discipline must take over, which Nelson and Platnick suggested that we call "arboristics":

> If cladistics is that part of systematics concerned with cladograms, then perhaps it is time to speak of "arboristics" as that part concerned with trees and, specifically, modes of speciation in general, as well as particular histories of speciation. One might conceive of an "arboristic analysis" which attempts to determine what tree is the cause of a particular instance of trichotomy or multiple branching, and operates by investigating the particular character distributions found in a given instance and their relative compatibility with various evolutionary scenarios. (Nelson and Platnick 1981: 265).

What can we retain from this brief account of Nelson and Platnick's solution? Their main contribution consisted in the release of the cladogram from its neo-Darwinian heritage, and the recognition of its proper nature and usefulness: that of a tool allowing us to test cladistic hypotheses – and, even more precisely, the hypotheses of synapomorphy that they are composed of (see the third part of this chapter). They led us further than Hennig. Although their starting point was the same – polytomy is either due to a lack of data, or due to an evolutionary event producing an actual polytomous pattern – they showed that radiation is not the sole evolutionary event likely to produce such a pattern: hybridizations, or the presence of actual ancestors among the taxa under analysis, can also produce, when analysed, a polytomous cladogram. That conclusion is only possible when the cladogram has been recognized for what it is: an analytic tool; it is therefore impossible from Hennig's perspective.

Nelson and Platnick's argument nonetheless still lack something: what about *dichotomy*? Its case is only *implicitly* evoked: by showing that polytomy is not necessarily implied by a radiation, Nelson and Platnick let us suppose that, symmetrically, dichotomy is not necessarily implied by a dichotomous speciation (e.g. Mayr's allopatric speciation). But then, what does dichotomy emerge from? Why should Hennig's rule be considered valid (or at least heuristic)? In *Systematics and Biogeography*, Nelson and Platnick did not give any other answers to that question than the ones they gave in their respective articles from the 1970s: Nelson's one (Nelson 1971), which only repeated Hennig's rule, and Platnick's Popperian one (Platnick 1979), which was not entirely convincing, and which is in any case, as we have seen, a bit dangerous from a strategic point of view.

What we want to do now is to fill in the gap left by Nelson and Platnick. We keep and make our own argument about polytomies: that they do not necessarily result from radiations, just as dichotomies do not necessarily result from dichotomous speciations. But we want to go further as far as the *positive justification* of dichotomy is concerned. In this respect, we are not entirely satisfied with Hennig's rule; neither are we with Platnick's Popperian attempt. Hence we now want to return to Hennig's first intuition, but in order to develop it in a way that significantly differs from his: the principle of dichotomy is a requirement of the cladistic method/theory itself.

10.4 A third way

Theoretical abnormality

Any biologist who sees a dichotomously rooted tree may immediately identify it as a phylogenetic tree. However, s/he will be unable to explain the method used to obtain it, to know how to interpret its results or which are the theoretical assumptions incorporated into that particular enquiry. The inability to determine the theory and method of a scientific result is a flaw, probably unique to phylogenetics, the origin of which lies, we speculate, in Hennig's initial ambiguity.

We focus here on the justification of dichotomy. By justification we mean the arguments that establish its legitimacy: what knowledge and beliefs do we possess on dichotomy and why do we have them? Are these reasons valid?

Three theories may produce rooted, dichotomous trees accounting for the evolutionary relationships of taxa:

1. Phenetics. Under this theory (when interpreted as evolutionarily grounded), overall resemblance, as a relevant estimation of phylogenetic relationships, is used to group taxa together.
2. Cladistics. Phylogenetic relationships are argued through the maximisation of the congruence of different hypotheses of homology.
3. Probabilities. Phylogenetic relationships are given by the most probable character state distribution, given an evolutionary model.

Each of those theories is interpreted through different methods that translate abstract concepts into particular operations. Some phenetic methods (e.g. unweighted pair group method with arithmetic mean [UPGMA], Sneath and Sokal 1973; neighbor-joining, Saitou and Nei 1987) can justify dichotomy as a methodological constraint. Nevertheless, phenetics cannot assign character states to nodes and does not account for evolutionary relationships, so we will not further discuss it.

We will show that at least one cladistic interpretation, i.e. three-item analysis, may also justify dichotomous trees. We think, however, that no valid justification has ever been provided for parsimony or any probabilistic method in phylogeny. Moreover, we think that there is no such valid justification.

Each method may be implemented through a number of computer programs. Thus, a scientific result performed by a computer program depends on the accuracy of the implementation of a method which, in turn, depends on the accuracy of the interpretation of a theory. The relevance of the result depends ultimately on the connection of the theory and the evolutionary history of the sample.

Systematists should have been astonished to discover that the three available theories that produce phylogenetic trees, with all the available methods that derive from them, often with different implementations, lead to results having the same structure. The result is supposed to correspond to the evolutionary history of the taxa represented in the phylogenetic tree. However, as the three theories make different assumptions about the history that the tree is supposed to represent, it is not clear in which sense these theories are in correspondence with the real world.

Moreover, the exact opposite is verified. Molecular phylogenetics has traditionally been presented through the results produced by diverse methods derived from different theories. Recently (Rindal and Brower 2011), a thorough analysis of papers using parsimony (a cladistic method) and maximum likelihood (ML) (a probabilistic method) showed that more than 99% of the analysed papers had obtained not only the same structure but also the same empirical relationships or, at least, they were not strongly contradictory. The authors defended the idea that the use of different methods is superfluous. By doing this, they considered modern parsimony cladistic analysis and probabilities using a ML approach as theory-independent.

We disagree with that interpretation. If the authors reanalyse the original hypothesis with phenetic methods like UPGMA and find non-contradictory results, should they interpret those as wrong?

On the reverse, if the results found with UPGMA were the same, should this be an argument for the relevance of UPGMA or of the empirical results? Should the number of different methods giving non-contradictory results be a criterion of optimality? Then, would a generally rejected theory implemented through an irrelevant method give an argument of choice?

We answer negatively to all of those questions. We interpret that result as surprising and quite unique in science. Different theories producing not only the same structure, i.e. a rooted, dichotomous tree, but also non-contradictory empirical results, should be surprising and should be explained. Moreover, there is no known justification for the connection between the diverse theoretical foundations of those methods and reality.

Nomological incommensurability

In the second half of the last century, Thomas Kuhn introduced the idea of nomological incommensurability (Kuhn 1962). Kuhn analysed the history of the dynamics of scientific theories, especially the physical and chemical ones (Kuhn 1962). He discovered that at least two theories dealing with the same empirical data sets often coexisted for long periods of time. The notion of incommensurability is linked to the idea of under-determination of theories by empirical data, i.e. the fact that at least two theories may explain as well the same dataset, for any given set of empirical data (e.g. Quine 1951). Coexisting theories show a relationship of nomological incommensurability, literally the existence of laws with no common measurement. In other words, coexisting theories appeal to different theoretical entities in order to explain the same facts. We think it is also the case in phylogenetics. Hennigian theory is based on hypotheses of homology, whereas ML or Bayesian methods are based on the notions of probabilities and evolutionary models. In cladistic analysis, the idea of probability or model (in the sense used in probabilistic phylogenetics) is not defined, not only because of its lack of meaning but also (and above all) because of its absence of relevance. In the same way, "homology" is a hollow term in the underlying theory of probabilistic methods. It may be argued that alignment of sequences imply hypotheses of homology. However, alignment is also needed in phenetic analysis, obviously with no homology statements intended. Both theories are thus incommensurable because they produce their objectivity through non-translatable terms. This means that those theories cannot refute each other on this ground.

However, that idea also challenges the assumption of realism in phylogenetic research. The fact that methods implementing probabilistic, cladistics, and phenetic theories of phylogenetic reconstruction often produce the same results based on incommensurable theoretical terms seems to defy rationality. Is phylogenetics a theory-empty endeavor, simply translating purely observational facts into dichotomous rooted trees? Or does the emptiness of theoretical justification of some results give us an opportunity to find a criterion for choosing one of the theories?

Regardless, there is still no justification for dichotomy in similarity. Dichotomy is imposed onto these methods with no explicit theoretical assumption of its meaning.

Epistemology

We may still wonder about the source of knowledge in all of those different theoretical approaches. The epistemic access we have to evolutionary relationships in

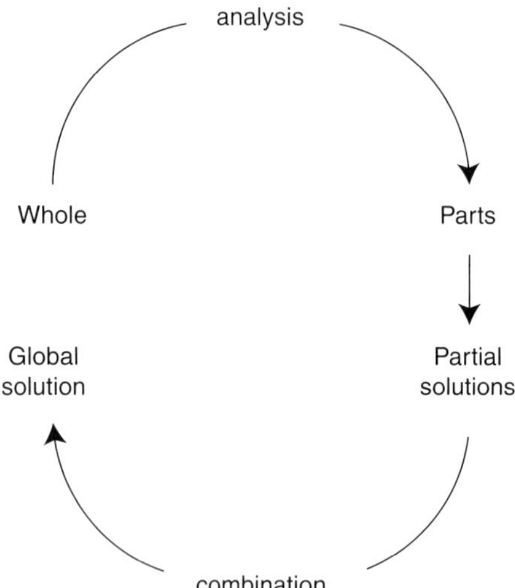

Fig 10.3 Drawing representing the Cartesian analytical method. A problem concerning a whole is usually solved by decomposing it into its parts. Partial solutions are found and combined into a general one. What is known through the general solution is, thus, only the combination of what is known about the parts. In other terms, the only epistemic access we have to the global solution is through the combination of partial solutions.

phylogenetics and biogeography and its consequences has surprisingly been one of the most neglected subjects in the literature of systematics. One may argue that the subject is at the centre of many relevant theoretical discussions (i.e. Hennig 1950, Farris 1983, Nelson and Platnick 1981). However, as we have seen previously, those discussions fail to address in a satisfactory manner the problem of the dichotomous nature of phylogenetic trees: is it based on the evolutionary process? Does it imply that the course of evolution is dichotomous?

Phylogenetics always proceeds by analysis. Analysis is perhaps the most common methodological procedure used in science. It has been used ever since the beginning of the European scientific endeavour. However, it was with the works of René Descartes (1637) that the method was first theorised. Cartesian analysis consists of the decomposition of a problem into its parts (Fig 10.3); then, the formulation of partial solutions; and finally, the combination of these partial solutions into a global or total solution, the synthesis.

The main consequence of Cartesian analysis is that all we know about the global solution is no more than the combination of the partial solutions. In phylogenetic analysis the general problem concerns finding, or arguing, the phylogenetic relationships of terminal taxa. This is done by decomposing taxa into hypotheses of homology or characters. Thus, ultimately, anything we know about taxa relationships comes from characters.

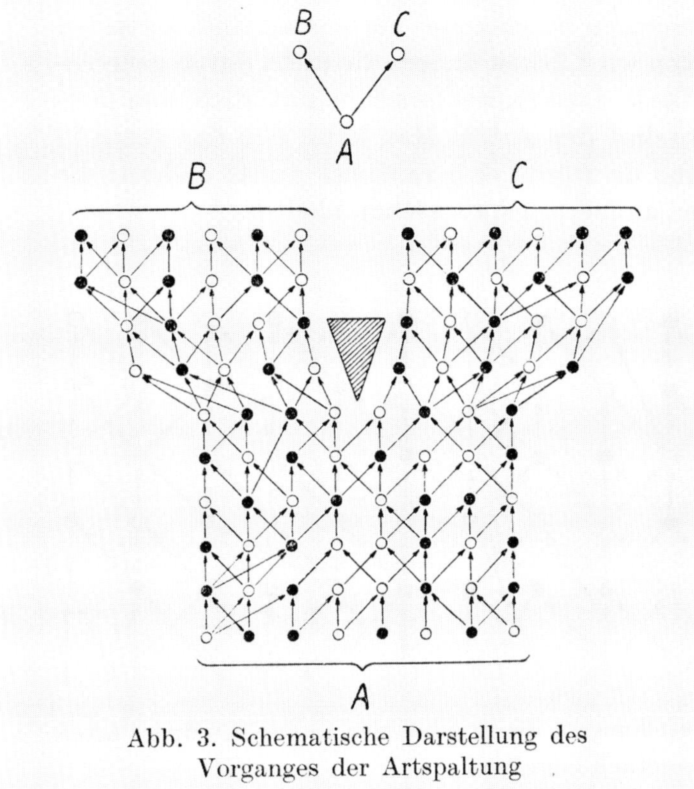

Abb. 3. Schematische Darstellung des
Vorganges der Artspaltung

Fig 10.4 Reproduced from Hennig (1957: 58, Fig 3).

In Willi Hennig's *Phylogenetic Systematics*, two figures show the two possible epistemic justifications of dichotomy in phylogenetic trees (Hennig 1966: 19, Fig 4; 31,
Fig 6, reproduced as Figs 10.4 and 10.5). In the first figure, Hennig justified dichotomy as a result of a dichotomous process (Hennig 1966: 19, Fig 4; Hennig's Fig 6 is
discussed below). As seen in the first part of this chapter, he introduced this justification as inherited from current evolutionary or phylogenetic theory: an extrapolation
of population genetics may explain the phylogenetic relationships of taxa. Hennig's
figures are found at the beginning of his book. They are thus placed as background
knowledge that Hennig adapted to his new theory. However, some aspects of what
Hennig considered as inherited from current evolutionary theory and his phylogenetic systematics seem to be in strong contradiction. Moreover, Hennig's figs 4 and
6 may have severe problems:

1. Are physical barriers intelligent? The adoption Mayr's standard allopatric speciation model led Hennig to insist on the importance of establishing physical barriers as a cause of interruption of the genetic flux between parts of a pre-existing

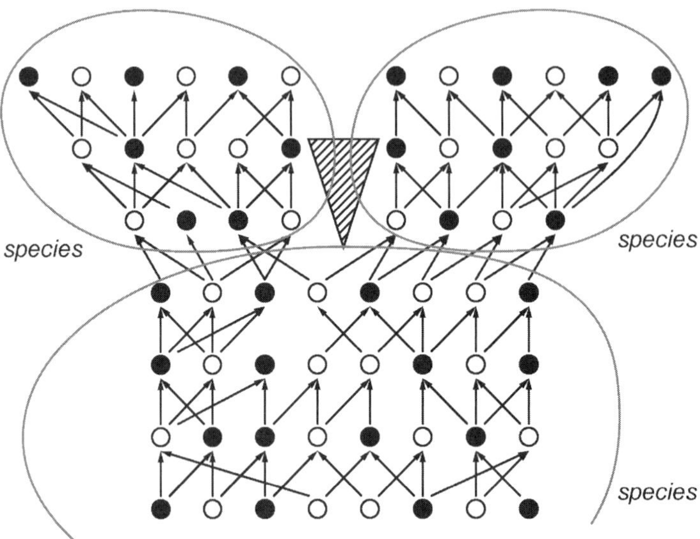

Fig 10.5 Redrawn and modified from Hennig (1982: Fig 6).

population. As a consequence, those barriers cause the split of the original population into two new populations. Nevertheless, we see no reason that physical barriers should cut populations into two, and only two, subpopulations. We see no reason either to suppose that this happens often, or in general, or with a greater or lesser probability.

A collateral comment on the general acceptance of this model as the standard model of speciation is that it is highly congruent with the fundamentals of vicariance biogeography. Indeed, it is not plausible to suppose that each physical barrier will affect one, and only one, population independently from the rest of the biota. However, dichotomous processes are assumed in general by phylogeographers, who usually appeal to dispersal as the standard process.

2. The process does not seem to be Darwinian. The most basic principles of Darwinian evolutionary theory state that natural selection is the cause of taxic differentiation. Natural variation produces differential average reproductive success and lifespan of the individuals of a population, leading to the emergence of particular features. Eventually, the accumulation of these fixed evolutionary novelties results in the differentiation of new taxa. However, in the process drawn in Hennig's figures, natural selection is, at best, an effect of taxic differentiation, its cause being the establishment of a barrier. The figures are compatible with a process where only genetic drive or neutral changes are responsible for taxic differentiation.

3. Definition of monophyly does not stand. Two standard definitions of monophyly exist. According to the first, a monophyletic taxon is composed of all the descendants of an ancestral species (e.g. Hennig 1968: 98, Darlu and Tassy 1993). In the second, a monophyletic taxon is an ancestor and all of its descendants (e.g. Hennig 1968: 95–96, Kitching et al. 1998, Judd et al. 2001, Lecointre and Le Guyader 2001). In either case, the definition of taxon B of Hennig's fig 4 (Hennig 1966: 31) includes taxa A and C, and its definition overlaps with that of taxon C.

4. Apomorphy and plesiomorphy cannot be discerned. Hennig's phylogenetic systematics was grounded on the distinction between apomorphic and plesiomorphic character states. In Hennig's figs 4 (Hennig 1966: 19) and 6 (Hennig 1966: 31), there cannot be a synapomorphy of the newly differentiated taxa, labelled "B" and "C" in fig 4. Any character state that characterizes those taxa was necessarily present in the ancestral "A" toxon. Thus, any synapomorphy is, in fact, a symplesiomorphy and the theoretical distinction made by Hennig is at odds with this supposed empirical evidence: Hennig's theoretical foundations and the empirical background represented by figs 4 and 6 cannot be simultaneously true.

5. Characters are absent. All Hennig's theory was based on characters. They are the only epistemic access we have to the results of taxic evolution. However, in Hennig's representation of the alleged real process, no characters are involved at any level. Note that from a methodological point of view, Hennig advocated for the use of Cartesian analysis, i.e. decomposing taxa relationships into character–state relationships. In addition, his introduction of taxic evolution derived from standard population genetics of his time resulted in a logical contradiction. Whereas characters are assumed to transform one into another, taxa are supposed to differentiate from more general to more particular.

Hennig drew a very different figure later in the same book (Hennig 1966: 71, Fig 18, reproduced as Fig 10.6), showing a totally different representation. Indeed, in that figure, taxa are hierarchically structured.[2] The figure shows the mathematical equivalence that exists between hierarchical classifications (which are, as set of clusters, usually represented by "set diagrams") and hierarchical n-trees (usually represented by "arrow diagrams"). In graph theory, a hierarchical n-tree is a finite fully ramified rooted tree with a single root and n leaves (what is usually referred to as a "rooted tree"). Hierarchical n-trees are therefore a particular case of finite directed trees, the latter being themselves a particular case of finite simple directed trees, i.e. a particular case of finite directed rooted 1-graphs. In other words, branches (i.e. the edges of G) are above all defined as links connecting nodes

---

[2] We refer here to hierarchical classifications contrary to Hennig, who referred to another, though not disconnected, concept of hierarchies taken from both Woodger (1952) and Gregg (1954) (Hennig 1965: 98, Hennig 1966: 16–20). A hierarchical classification can be defined as a classificatory structure H of a set A, i.e. a covering set of non-empty subsets of A, the latter being the clusters of H, such that (1) there is a cluster in H that contains all the elements of A, (2) for any element x of A, $\{x\}$ is in H, and (3) the intersection between any two clusters of H is either empty or one of the two (see Cao et al. 2007 and Lebbe 1991).

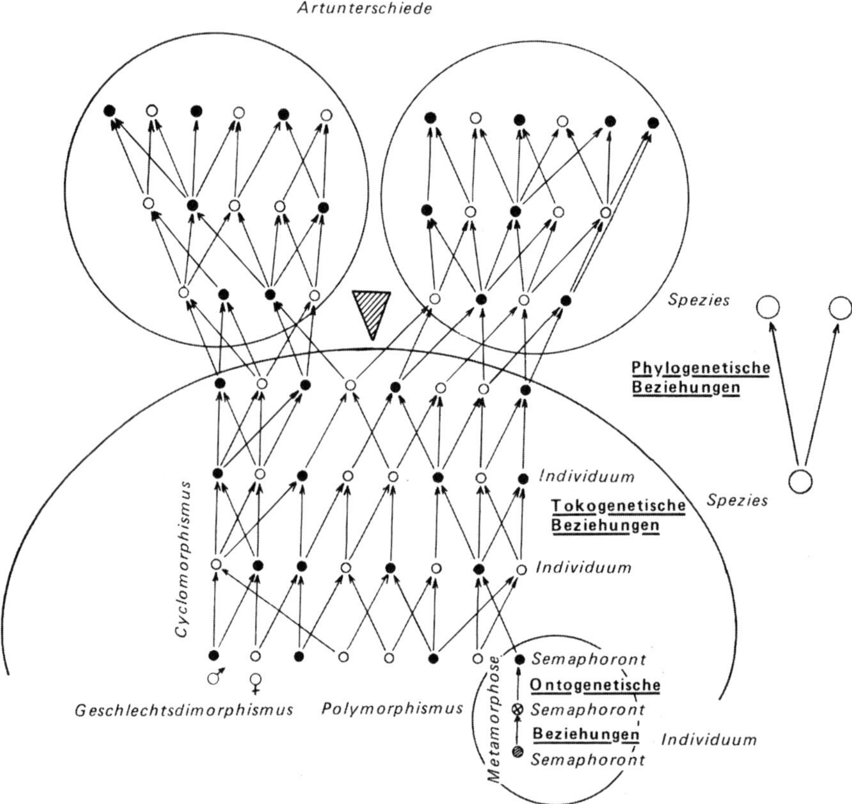

Abb. 6. Schema des Gesamtgefüges der hologenetischen Beziehungen und der seinen einzelnen Abschnitten zugeordneten Gestaltunterschiede

Fig 10.6 Reproduced from Hennig (1957: 59, Fig 4).

(sometimes with an associated length). If Hennig's theory deals with hierarchical trees, then only nodes convey any evolutionary sense. Branches do not convey the sense Mayr or Hennig supposed. If Hennig's theory concerns the kind of networks represented in his figs 4 and 6, then only "branches" convey any sense, the nodes only being speciation events with no empirical content. That solution has been generally favoured, despite the contradictions it causes concerning what is understood by monophyly, synapomorphy, natural selection as the cause of taxic evolution, etc. As a consequence, it is assumed that the justification of dichotomy exists because speciation is dichotomous.

Hennig's fig 18 (here Fig 10.6), however, is compatible with any standard evolutionary mechanism, as genetic drive, natural selection or a mix of both; allopatric or sympatric diversification or a mix of both; with the standard definition of monophyly; with a distinction between apomorphy and plesiomorphy, and with a causal role of natural selection.

However, it is still incompatible with Hennig's view of characters as transformation series. Transformation series can only be represented as a hierarchy if there is no diversification, i.e. if the modification can be represented by a pectinate tree of character states (see Zaragüeta Bagils et al. 2004 for a relevant analogy). As shown by Zaragüeta Bagils et al. (2004), a pectinate tree may be converted to an ordered partition (the relevant classificatory structure to represent a transformation series). However, the kind of differentiation represented by a balanced tree may not be interpreted as a transformation series. In any transformation, the transformed thing is not the untransformed object. In differentiation, the transformed object is part of the untransformed thing. A transformation process is better represented by a partition, for example a scale, whereas a differentiation cannot be represented by an ordered partition but may be represented as a hierarchy. In other words, a transformation series of collective entities (e.g. fins–limbs–wings) with paraphyletic ancestral parts (fins and limbs in the example) can only be represented by a partition.

Homologues, homology, characters, taxa, and cladograms

Any phylogenetic study begins with a selection of terminal taxa, or of specimens representing terminal taxa. In any case, these taxa are considered objects, i.e. they can be empirically measured and the results are supposed to be observable data, despite their intrinsic variability. Because the problem of finding how these taxa are related has always been too complicated to be done directly, we decompose them into sets of parts. Systematists hypothesize that the parts, however different, are the same, i.e. may be named by the same noun.

This property of sameness (see e.g. Williams 2004 and references within) corresponds to Richard Owen's definition of homologue: the same part in several organisms despite variation in form and function. In other terms, a homologue is something that is several things, those things being the same but different in any way that they might be described. In consequence, it is certainly a complex concept that systematists discovered long ago, at least since Pierre Belon's *Histoire de la Nature des Oyseaux* (1555).

Being the same and being different at the same time is not a question that can be affirmatively or negatively answered (Fig 10.7). It is obvious to modern systematists that among the parts of organisms, some of them can be considered "the same thing" although they remain different (as singular parts). Nevertheless, the question might be answered by considering relationships among homologues as a relative, at least ternary, relationship. The question: "Are these things the same or different?" has to be changed to: "Among these named things, which are more or less the same and which are just different?" The question might be recursively asked until reduced

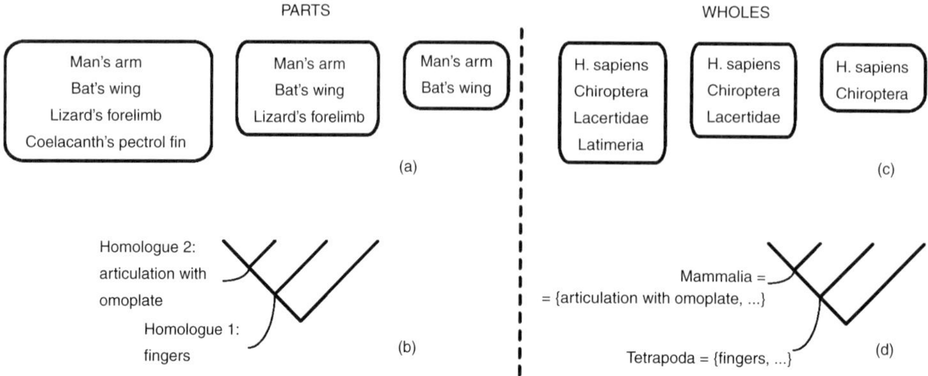

Fig 10.8 The Cartesian analytical structure of cladistic analysis. (a) Terminal taxa are decomposed into homologues and those are grouped in sets. (b) The classificatory structure that links them is called a hypothesis of homology. Different hypotheses of homology are combined through hybrid classificatory structures called characters, where terminal homologues are replaced with terminal taxa (wholes) and internal nodes are homologues, called character states. The "whole" part of the analytical method, obtained by combining characters, consists of: (c) Classes of homologues, the taxa. Taxa intension is thus composed by homologues or character states; (d) Their hierarchical relationships, i.e. the cladogram. When totally resolved, the dichotomous structure of the cladogram is justified by the dichotomous structure of hypotheses of homology (or the elementary statements they are composed of) and not by the existence of a dichotomous process or by a measurement of information.

relationships among the parts differs from the structure of the relationships among the wholes: states of characters may transform but taxa do not. Following Hennig's theory, it may be said that scales transformed into feathers, i.e. feathers are not scales; sternums transformed into the carinas, which are not sternums, snouts transformed into beaks and forelimbs into wings. However, it would be incorrect, in the same theory, to say that dinosaurs transformed into birds. Birds are said to "be" dinosaurs, i.e. birds differentiated among dinosaurs. The epistemic justification of that hypothesis is based on hypotheses of differentiations of homologues: feathers are differentiated scales, carinae are differentiated sternums, beaks are snouts and wings are forelimbs. Thus, birds are differentiated dinosaurs. Fig 10.8 shows the analytical structure of cladistic analysis based on its epistemology: everything that is known about taxa derives from characters, thus from hypotheses of homology. Homologues are the parts of taxa, homology the parts of cladograms. Thus, dichotomy is justified by the epistemic structure of phylogenetic systematics.

Hennig obviated the epistemic aspect of phylogenetic research. In his theory, the epistemic access is strictly limited to the observation of actual parts and their grouping into sets, i.e. homologues. His idea of transformation of character states was taken from external theories and introduced without critical examination. Hennig seems to have developed his theory in two independent parts: phylogenetic systematics as a new theory with new theoretical terms, correctly defined and with their relevant justification; and a priori link to the evolutionary theory of his time as justification of the same theory. We speculate, as a simple conjecture, that this step was a second stage of the development of his ideas. However, because it was introduced at the beginning of *Phylogenetic Systematics* (Hennig 1966), it probably hindered him from discovering the inconsistent justification of dichotomy introduced.

Consequences

Amazingly, almost all authors have agreed with Hennig's incorrect justification, i.e. that the justification of dichotomous trees lies in a dichotomous speciation process (e.g. Samadi and Barberousse 2005, 2006), or have failed to provide an alternative justification of the source of dichotomy.

This assumption, empirically empty and theoretically meaningless, has been the base of some criticism of the phylogenetic enterprise as a whole. We will make only some short comments on a few selected mainstream researches that are based on the false assumption that the justification of dichotomy in phylogenetic trees lies in a dichotomous speciation process.

Networks: Recently, a new domain of phylogenetic research has arisen from the criticism of trees as a relevant representation of phylogenetic relationships. That criticism is based on the discovery of lateral gene transfer and hybridization as a relatively common process. Some authors have seen in this discovery a reason to draw lines across branches of phylogenetic trees (e.g. Doolitle and Bapteste 2006). However, the assumption of network-like phylogenetic relationships is based on the assumption that phylogenetic trees are extrapolations of the tokogenetic process represented by Hennig in his figs 4 (Hennig 1966: 19) and 6 (Hennig 1966: 31). Following Nelson and Platnick (see the second part of this chapter), we have shown that this is not the case. Networks defenders criticize tree builders because the latter are supposed to assume that evolution is dichotomous because speciation must be dichotomous. Phylogenetic trees do not represent dichotomous speciation processes and, thus, current phylogenetic theory, at least cladistic theory, is not affected by the existence of hybridizations or lateral gene transfers, which should just be among the possible sources of homoplasy.

Temporal calibration: The second problem we would like to mention concerns the temporal calibration of nodes using DNA sequences (see e.g. Benton et al. 2009). This procedure is based on the assumption, also found in Hennig's phylogenetic systematics, that both sister groups have the same age of origin. That supposition is based, in turn, on the idea that the justification of dichotomy depends on the fact that nodes are dichotomous speciation events. However, as nodes do not indicate dichotomous speciation events, equal age of sister groups is not justified. Other problems, such as taxic temporal paralogy (Zaragüeta Bagils et al. 2004), may also have an impact on the relevance of current node dating methods.

10.5 Conclusion

In conclusion, we have shown that the justification of dichotomy lies in the formal structure of hypotheses of homology proposed through Cartesian analysis. The only epistemic access we have to relationships among taxa is homology. Homology, i.e. rooted trees indicating relative identity relationships among homologues, and characters may be decomposed to the simplest hypothesis involving three items. Three-item statements are intrinsically dichotomous.

Dichotomy is thus justified by a theory implying the existence of relative identity and by a method of arguing that identity using Cartesian analysis and therefore allowing epistemic access to these relationships represented by a hierarchy.

Hennigian phylogenetic systematics has all the requirements to produce the justification of dichotomy in the resulting cladograms. However, Hennig (1975) was not capable of producing such a justification when that aspect of cladistic analysis was criticized by Mayr (1974). We have speculated about the possibility of Hennig being incapable of retrieving his own reasoning because of his deep belief that the justification of cladograms should be searched for in population genetics theory.

If taxa originated through the accumulation of different alleles, a mechanism of interruption of the genetic flux among the different parts of the original population became fundamental. Hennig took Mayr's allopatric speciation process as the justification of tree-like relationships among taxa. However, those processes did not imply dichotomy, as Mayr (1974) noted. Hennig's figs 4 (Hennig 1966: 19) and 6 (Hennig 1966: 31), however, made the assumption that evolution as a tree was not a metaphor or a highly abstract theoretical construction, as it indeed was in his own theory, but simply an observational fact. In assuming that dichotomy was derived from the process and not from the analytical abstraction of taxa, he lost the justification.

The same assumption seems to have been made by most, if not all, systematists, given the absence of discussion of this aspect in the literature. However, it seems clear that if the only knowledge we may have of evolutionary relationships among taxa comes from homologues and their relationships, i.e. homology and characters, dichotomy has to be found in characters.

We have shown that it may be the case, without needing to produce any other ad hoc hypothesis. The taxic approach to phylogenetic systematics (e.g. Williams 2004, and references therein), even if it is not strictly Hennigian, can explain dichotomy.

Either way, that approach may eliminate the logical contradictions existing in the original development of cladistic theory.

We have thus shown that the cladistic theory may justify dichotomy on theoretical grounds. We have shown that other phylogenetic theories are, at least, incommensurable with cladistics. Probability approaches such as ML and Bayesian inferences do not have the theoretical entities into which homology is translated. As such, they cannot justify, from our point of view, dichotomy. Those theories seem to derive dichotomy from the process, nodes often being considered as speciation (dichotomous) events. All process-based approaches, like phylogeography, evolutionary relationships using networks, molecular dating of nodes, etc., that are founded on the supposition that dichotomous trees imply dichotomous speciation events, may have serious issues with this assumption.

Acknowledgements

We thank Stéphane Prin for comments and improvements, mainly on the second part of the text. We are grateful to Rhys Chatham and Eglantine Lemense for English proofreading and style improving. We thank David Williams for the invitation to one of us (RZB) to participate in the Linnean/Systematics Association meeting on Willi Hennig in October 2013.

References

Belon, P. (1555). *L'Histoire de la nature des oyseaux, avec leurs descriptions et naïfs portraicts retirez du naturel, escrite sept livres, etc.* Paris: Gilles Corrozet.

Benton, M., Donoghue, P.C.J. and Asher, R.J. (2009). Calibrating and constraining molecular clocks. In *Timetree of Life*, ed. S.B. Hedges and S. Kumar. Oxford: Oxford University Press, pp. 35–86.

Brundin, L. (1966). Transantarctic relationships and their significance, as evidenced by chironomid midges. *Kungliga Svenska Vetenskapsakademiens Handlingar*, Fjärde series, 11, 1–472.

Bock, W.J. (1973). Philosophical foundations of classical evolutionary classification. *Systematic Zoology*, 22, 375–392.

Cao, N., Zaragüeta Bagils, R. and Vignes-Lebbe, R. (2007). A hierarchical representation of the hypothesis of homology. *Geodiversitas*, 29, 5–15.

Darlington, P.J., Jr. (1970). A practical criticism of Hennig-Brundin "Phylogenic [sic] systematics" and Antarctic biogeography. *Systematic Zoology*, 19, 1–18.

Darlu, P. and Tassy, P. (1993). *La reconstruction phylogénétique. Concepts et méthodes.* Paris: Masson.

Darwin, C. (1859). *On the Origin of Species by Means of Natural Selection, or the Preservation of Favoured Races in the Struggle for Life.* London: John Murray.

Descartes, R. (1637). *Discours de la méthode pour bien conduire sa raison et chercher la vérité dans les sciences, plus la Dioptrique, les Météores et la Géométrie qui sont des essais de cette méthode.* Leyde: Ian Maire.

Dupuis, C. (1978). Permanence et actualité de la systématique: la "systématique phylogénétique" de Willi Hennig. *Cahiers des Naturalistes*, 34, 1–69.

Doolitle, W.F. and Bapteste, E. (2006). Pattern pluralism and the Tree of Life hypothesis. *Proceedings of the National Academy of Sciences of the United States of America*, 104, 2043–2049.

Farris, J.S. (1983). The logical basis of phylogenetic analysis. In *Advances in cladistics, II*, ed. N.I. Platnick and V.A. Funk. New York: Columbia University Press, pp. 7–36.

Gregg, J.R. (1954). *The Language of Taxonomy: An Application of Symbolic Logic to the Study of Classificatory Systems.* New York: Columbia University Press.

Hennig, W. (1950). *Grundzüge einer Theorie der phylogenetischen Systematik.* Berlin: Deutscher Zentralverlag.

Hennig, W. (1957). Systematik und Phylogenese. In *Bericht über die Hundertjahrfeier der Deutschen Entomologischen Gesellschaft Berlin*, ed. H.-J. Hannemann. Berlin: Akademie-Verlag, pp. 50–71.

Hennig, W. (1965). Phylogenetic systematics. *Annual Review of Entomology*, 10, 97–116.

Hennig, W. (1966). *Phylogenetic Systematics.* Urbana: University of Illinois Press.

Hennig, W. (1968). *Elementos de una sistemática filogenética.* Buenos Aires: Editorial Universitaria de Buenos Aires.

Hennig, W. (1975). Cladistic analysis or cladistic classification?: a reply to Ernst Mayr. *Systematic Zoology*, 24, 244–256.

Hennig, W. (1982). *Phylogenetische Systematik.* Hamburg : Verlag Paul Parey.

Huxley, J.S. (ed.) (1940). *The New Systematics.* Oxford : Clarendon Press.

Judd, W.S., Campbell, C.S., Bouharmont, J., et al. (2001). *Botanique systématique: Une perspective phylogénétique.* Louvain-la-Neuve, Paris: De Boeck Université.

Kitching, I.J., Forey, P.L., Humphries, C.J. and Williams, D.M. (1998). *Cladistics: The Theory and Practice of Parsimony Analysis*, 2nd edition. Oxford: Oxford University Press.

Kuhn, T. (1962). *The Structure of Scientific Revolutions.* Chicago, IL: University of Chicago Press.

Lakatos, I. (1970). Methodology of scientific research programs. In *Criticism and the Growth of Scientific Knowledge*, ed. I. Lakatos and A. Musgrave. Cambridge: Cambridge University Press, pp. 91–196.

Lebbe, J. (1991). *Représentation des concepts en biologie et médecine. Introduction à l'analyse des connaissances et à l'identification assistée par ordinateur.* Ph.D. Thesis in Life Sciences. Paris: Université Paris 6 Pierre-et-Marie-Curie.

Lecointre, G. and Le Guyader, H. (2001). *Classification phylogénétique du vivant.* Paris: Belin.

Mayr, E. (1942). *Systematics and the Origin of Species.* New York: Columbia University Press.

Mayr, E. (1963). *Animal Species and Evolution.* Cambridge, MA: Harvard University Press.

Mayr, E. (1969). *Principles of Systematic Zoology.* New York: McGraw-Hill.

Mayr, E. (1974). Cladistic analysis or cladistic classification? *Zeitschrift für Zoologische Systematik und Evolutionsforschung*, 12, 94–128.

Nelson, G.J. (1971). "Cladism" as a philosophy of classification. *Systematic Zoology*, 20, 373–376.

Nelson, G.J. (1994). Homology and systematics. In *The Hierarchical Basis of Comparative Biology*, ed. B.K. Hall. San Diego, CA: Academic Press, pp. 101–149.

Nelson, G.J. and Platnick, N.I. (1980). Multiple branching in cladograms: two interpretations. *Systematic Zoology*, 29, 86–91.

Nelson, G.J. and Platnick, N.I. (1981). *Systematics and Biogeography: Cladistics and Vicariance*. New York: Columbia University Press.

Platnick, N.I. (1979). Philosophy and the transformation of cladistics. *Systematic Zoology*, 28, 537–554.

Popper, K.R. (2002 [1935]). *The Logic of Scientific Discovery*. London: Routledge Classics.

Quine, W. v. O. (1951). Two dogmas of empiricism. *The Philosophical Review*, 60, 20–43.

Rieppel, O. (2010). Willi Hennig's dichotomization of nature. *Cladistics*, 26, 1–10.

Rieppel, O. 2016. Willi Hennig as philosopher. In *The Future of Phylogenetic Systematics – The Legacy of Willi Hennig*, ed. D.M. Williams, M. Schmitt and Q.D. Wheeler. Cambridge: Cambridge University Press, pp. 357–377.

Rindal, E. and Brower, A.V.Z. (2011). Do model-based phylogenetic analyses perform better than parsimony? A test with empirical data. *Cladistics*, 27, 331–334.

Saitou, N. and Nei, M. (1987). The neighbor-joining method: a new method for reconstructing phylogenetic trees. *Molecular Biology and Evolution*, 4, 406–425.

Samadi, S. and Barberousse, A. (2005). L'arbre, le réseau et les espèces. Une définition du concept d'espèce ancrée dans la théorie de l'évolution. *Biosystema*, 24, 53–62.

Samadi, S. and Barberousse, A. (2006). The tree, the network, and the species. *Biological Journal of the Linnean Society*, 89, 509–521.

Simpson, G.G. (1961). *Principles of Animal Taxonomy*. New York: Columbia University Press.

Sneath, P.H.A. and Sokal, R.R. (1973). *Numerical Taxonomy*. San Francisco, CA: Freeman.

Williams, D.M. (2004). Homologues and homology, phenetics and cladistics: 150 years of progress. In *Milestones in Systematics*, ed. D.M. Williams and P.L. Forey, Boca Raton: CRC Press, pp. 191–224.

Woodger, J.H. (1952). From biology to mathematics. *The British Journal for the Philosophy of Science*, 3, 1–21.

Zaragüeta Bagils, R., Lelièvre, H. and Tassy, P. (2004). Temporal paralogy, cladograms, and the quality of the fossil record. *Geodiversitas*, 26, 381–389.

11

Hennig's auxiliary principle and reciprocal illumination revisited

RANDALL D. MOOI and ANTHONY C. GILL

11.1 Introduction

The availability of Hennig's views to an English-speaking audience was instrumental in the revolution in paleontology and systematics, and his influence continues through the present day (Rosen et al. 1979, Carvalho and Craig 2011). As might be expected over the 50 years since their introduction, many of Hennig's specific methods regarding character analysis and biogeography have been significantly altered – "Phylogeneticists have also moved beyond solely employing Hennig's argumentation schemes" (Wiley and Lieberman 2011: xiii) – or even dismissed over the years (e.g. the chorological method, Hennig 1966: 133). However, Hennig's founding principles have been almost wholly maintained, to the extent that they are frequently quoted verbatim by modern workers in support of their approach and in introductory texts on phylogenetics. Of course, as for all founding works, the interpretation of Hennig's principles and their application can vary from worker to worker for the very reason that methods and context evolve over time. And many systematists rely on the interpretation of these principles by others and follow 'best practices' as deemed by a smaller group of perceived authorities; we have done so ourselves (Mooi 1995, Mooi and Gill 2004).

Two major concepts introduced into systematics by Hennig (1966) remain prominent in the primary and secondary literature: his auxiliary principle (Hennig

The Future of Phylogenetic Systematics: The Legacy of Willi Hennig, eds. D. Williams, M. Schmitt and Q. Wheeler. Published by Cambridge University Press.

1966: 121) and method of reciprocal illumination (Hennig 1966: 21). These founding principles are the basis, in part, for identifying and differentiating the three kinds of similarity among characters: synapomorphy, symplesiomorphy, and convergence. The auxiliary principle was introduced to counter the assumption that the prevalence of parallelism or convergence made phylogenetic systematics untenable. Reciprocal illumination provided a mechanism for continually exploring data and conclusions while having each inform the other without dismissal as circularity.

Despite the general acceptance of these two concepts, their application has not been consistent in theory or practice. In some instances, practices have changed so dramatically since Hennig's time that it is unclear why the principles are seen as necessary justification for a given approach. Here we explore Hennig's auxiliary principle and its implications for how incongruence among characters is interpreted, with particular emphasis on reversals. This necessarily leads to an examination of the application of reciprocal illumination in modern methods and how that reflects on our approaches to characters and the value of empirical evidence.

11.2 Hennig's auxiliary principle and reciprocal illumination: how they work together

Hennig's auxiliary principle states that:

> the presence of apomorphous characters in different species "is always reason for suspecting kinship [i.e., that the species belong to a monophyletic group], and that their origin by convergence should not be assumed a priori" (Hennig 1953). (Hennig 1966: 121)

Hennig (1953: 18, translation) felt that without it, "phylogenetic systematics would not have a leg to stand on." This view has generally been accepted in systematics (e.g. Wiley 1981, Carvalho and Craig 2011, Wiley and Lieberman 2011), although frequently with slight modifications that appear to have had an impact on how systematics is approached and practiced. These will be discussed below.

The original formulation of the auxiliary principle was in the context of reciprocal illumination (Hennig 1966: 21) where, if such apomorphous characters were later shown to be incongruent with other such characters, there is a method of checking, correcting, and rechecking whereby character conflict is resolved, or at least some accounting of the conflict is made "by trying to bring the relationships indicated by the several series of characters into congruence" (Hennig 1966: 122). The implication is that the initial hypothesis of homology for each character (or character state tree) is compared to those of others, and where they differ the characters would be re-examined to tease out mistaken observations or interpretations. The interpretation of each character is discussed in light of the indicated relationship and the relationships implied by other characters. Character discussion and interpretation was a significant part of Hennig's empirical work (e.g. Hennig 1953).

11.3 Changing principles (and their implications)

With the incorporation of numerical approaches in systematics, Hennig's auxiliary principle was reformulated for computational purposes (Farris et al. 1970: 174): "In the absence of evidence to the contrary, any state corresponding to a step shared by a group, G, of OTUs is taken to have arisen just once in G." This is a subtle change wherein the value of the character as evidence has been shifted to reside in the tree as a step rather than as a shared derived state as evidence for a group. This means that sharing "apomorphous characters in different species" (Hennig 1966: 121) no longer necessarily indicates monophyly of all taxa exhibiting that state, but rather a shared state can be used as a step for as many independent 'local' tree phenomena as required by parsimony (or other clustering method). The character/step remains consistent with the reformulated auxiliary principle through reference to a particular group (G) as long as it is internally consistent (within G), regardless of the broader distribution of the state. Hennig specifically referred to the presence of *apomorphous* characters (or states, Hennig interchangeably used "character" for "character state," see Grant and Kluge 2004) as a reason to hypothesize monophyly of the taxa possessing them. Hennig's version means that every character has merit as an initial hypothesis of relationship among the taxa in question. With the change introduced by Farris et al. (1970), the auxiliary principle for a particular character has meaning only in relation to the character state distribution as determined through parsimony; the information content of a character changes from being *within* the character as a separate entity to being referable to information of *all* characters within a topology – information content ('evidence') becomes reliant on congruence with other characters.

In recent times, Hennig's principle has been altered more explicitly, with De Laet (2005: 86) renaming his version as the Hennig–Farris auxiliary principle, which reads: "homology should be presumed in absence of evidence to the contrary." Wiley and Lieberman (2011: 118) provide a similar definition. De Laet (2005: 86) specifies that "homology refers to similarities among organisms that have arisen historically through inheritance from a common ancestor, irrespective of these similarities being apomorphic or plesiomorphic." Hennig's requirement of apomorphy in his original formulation of the principle has thus been eliminated. This removes any a priori requirement for a character to support a hypothesis of relationship except in the very broadest of terms, and might even be considered self-affirming; states recognized as a character are states of that character [i.e. states within a column in a matrix can float at random on the tree – "freely reversible" (Wiley and Lieberman

2011: 168)]. The trivial conclusion of this application is that "characters considered homologous [...] are homologous" Nelson (1994: 127).

How would the states in such a character ever be shown in conflict with a tree? Or in conflict with other such characters? Characters have no initial integrity as hypotheses, but are arranged most parsimoniously on a network and patterns emerge from the re-arrangements of the parsimony algorithm. With such an approach, there is no testing of initial hypotheses as there is none available for this purpose. Incongruence, the signal that a second look might be prudent, vanishes (Hennig's "checking," 1966: 122), and the extra steps become additional 'evidence' for the most parsimonious tree. Of course, extra steps could be used as a cue for Hennig's rechecking, but accepted practice, and the disappearance of character discussion in most instances, has eliminated this critical step of reciprocal illumination.

Grant and Kluge (2004, 2009) and Kluge (2005) embrace the character-as-step interpretation of Farris et al. (1970) and expand this to an ideographic approach to phylogenetic inference, where a character is a series of singular historical events. This seems unproblematic, but the emphasis is now on discovered transformation events via parsimony as the evidence for relationship. These events, however, are invisible and should be deduced from empirical data (Nelson 2011). Rieppel (2007: 357) suggested "Hennig recognized the shortcomings" of such a strictly ideographic concept of characters; we see it removing the auxiliary principle from the equation entirely. It has led Grant and Kluge (2004: 27) to argue (our additions in square brackets and emphasis in italics): "Synapomorphy, in contrast [to homology], refers to the shared occurrence of a derived (apomorphic) character, *whether or not that shared occurrence resulted from the same transformation event (homology) or different transformation events* [homoplasy!]." For Grant and Kluge, synapomorphy = potential homology OR potential homoplasy. Hennig's own definition of synapomorphy contradicts this interpretation (our emphasis) (Hennig 1966: 89): "We will call [...] the presence of apomorphous characters synapomorphy, always with the assumption that the compared characters *belong to one and the same transformation series.*" Grant and Kluge's change to the definition of synapomorphy might surprise many systematists, but, in fact, it has been widely accepted *de facto* as an operational condition for numerical cladistics, with its reliance on parsimony to drive objectives and results. This removal of Hennig's auxiliary principle from the phylogenetic program and the redefinition of synapomorphy permits homoplasy to be interpreted as synapomorphy and permits the topology to determine homology a posteriori.

The implications of these changes to, or even removal of, the auxiliary principle are found, for example, in Wiley et al. (2011a: 9) where:

> [I]f the distribution [of a character state] on a tree does indicate homoplasy, then it
> is possible that two or more subclades have the synapomorphic property [...] What

is homoplastic at one level (the entire transformation series) may be locally synapomorphic (i.e. homoplasy at one level may be homology at another...).

Indeed, this is the usual interpretation; characters that are not congruent are simply redefined. Such redefinition gives the (false) impression of meeting Hennig's goal (1966: 122) of "trying to bring the relationships indicated by the several series of characters into congruence," but given the discovery of incongruence, our work as systematists has really just begun. The incongruence has actually shown that the "synapomorphic property" of the character as originally recognized has been misidentified (if the topology is accepted); its original hypothesis of synapomorphy as determined via Hennig's auxiliary principle has been falsified. That it is "locally synapomorphic" is an ad hoc assumption resulting from the clustering procedure and is not an hypothesis based on any empirical observation. One cannot (or should not) re-apply Hennig's auxiliary principle to this "local synapomorphy" because the initial observation/hypothesis of the character has not changed. In Hennig's terminology, reciprocal illumination has *checked* the character and found it wanting (falsified); it now requires *correcting* and *rechecking* (Hennig 1966: 122). "Correcting" is recognition of a mistake, and one that cannot be fixed by merely changing character state distribution (optimizing) or re-scoring based on evidence from other characters. Employing optimized or re-scored characters is no different than using mapped characters as synapomorphies where they "cannot lend support to the inferences of monophyly, because such 'synapomorphies' are empirically empty: they can never be shown to be wrong" (Assis and Rieppel 2011: 97). Any apparent incongruence and inherent error in a hypothesis of homology should be addressed by re-examination and re-evaluation of the character itself through as many similarity tests as possible, and then rechecked via congruence. Continued incongruence, despite re-examination and testing, might force the investigator to settle for ad hoc assumptions of convergence or, if the character is convincing enough, to question the topology upon which the incongruence is based. That is the nature of *reciprocal* illumination; testing goes both ways – tree to characters and characters to tree.

Modern approaches do not address the implications of incongruence of characters and, instead, emphasize the incongruence of topologies, as if the topologies are evidence rather than the characters upon which they are based. Topologies are merely summaries of evidence, not evidence in themselves. Without applying Hennig's auxiliary principle and reciprocal illumination, we are left arguing about topologies and the methods by which they were produced rather than focusing on the evidence they summarize. Without that focus, the "solutions" (topologies) are produced in intellectual and factual isolation from each other; reciprocal illumination can play no role in resolving incongruence. We are left with two (or more) different topologies where the arguments for or against rest on number of taxa, number of characters, optimization methods, and support values rather than how hypotheses of character homology, the evidence, can inform one another through reciprocal illumination. A recent example involves the flatfishes, Pleuronectiformes, where the monophyly of these fishes based on their shared and spectacular asymmetry (Chapleau 1993, Friedman 2008, 2012) has been alternately rejected or corroborated by various molecular studies (summarized in Betancur-R. and Ortí 2014). In contributions appearing almost simultaneously, Betancur-R. et al. (2013) used molecular data to support flatfish monophyly, whereas Campbell et al. (2013) argued non-monophyly. However, neither of these papers nor their subsequent rebuttals/replies (Betancur-R. and Ortí 2014, Campbell et al. 2014a, 2014b) examined how particular characters (base pairs) were distributed on different trees and altered topologies; instead, they emphasized differences in analyses, support values, etc. But what was happening at the character level? Did the character 'evidence' that supported one topology change to 'non-evidence' in the other, or was this "previous evidence" ignored and did 'new evidence' take its place? Mooi and Gill (2010) provide additional examples where a focus on topology over evidence removes any role for reciprocal illumination. Without examining specific character distributions and how they influence one topology and change in another, topological comparisons have little value. Without the auxiliary principle and without reciprocal illumination, what are we left with that resembles Hennigian phylogenetics?

Computational necessity and the construction of unrooted trees has led to a migration away from the emphasis that Hennig placed on *apomorphous* characters (our emphasis):

> But it often happens that only one character can certainly or with reasonable probability be interpreted as *apomorphous*, while other characters are either obviously plesiomorphous or in the present state of our knowledge not certainly identifiable in the group in question. (Hennig 1966: 121)

Hennig's auxiliary principle is relied upon to permit the one character to provide evidence of relationship. If the character is found to be globally incongruent, the auxiliary principle cannot reasonably be re-applied locally to legitimize the use of homoplasy as synapomorphy. This can only be done through re-examination of the character and an actual re-interpretation that does not rely on its congruence with other characters. Otherwise the tree will never be tested – the concept of reciprocal illumination is lost. The tree shines new light on the character, but the reverse can never happen. The Farris et al. (1970) reformulation was a step towards the general opinion that character evidence resides in the tree through parsimony rather than in the characters (observations) themselves. "The test of congruence is a necessary but not also a sufficient condition for the discovery of synapomorphy" (Assis and Rieppel 2011: 97, also Rieppel 2006 and Cao et al. 2007).

11.4 Hennig's auxiliary principle and parsimony

Hennig's auxiliary principle provides integrity and explanatory power to individual characters as evidence for relationship. And this approach is consistent with cladistic methods because Hennig formulated his principle based on *apomorphic* characters. It is derived character states that are indicators of relationship among taxa.

Wiley and Lieberman (2011: 118) assert that the auxiliary principle is "just a restatement of [...] the parsimony principle." This seems a misinterpretation. Parsimony does not require the partitioning of character states into plesiomorphic or apomorphic; it minimizes steps. This is why an unrooted tree can be constructed. Taxa would merely be placed into piles sharing character states with no a priori expectation of apomorphy; Wägele (2005) has called such an approach phenetic cladistics. But Hennig specified that the auxiliary principle applies to *apomorphous* characters – these are the ones that "arouse the suspicion of affinity" (Hennig 1953: 18, translation). For an individual character, trees are already minimal and rely on the empirical evidence of similarity (based on several criteria as outlined or modified by several authors: Remane 1952, Patterson 1982, Wägele 2005) to assign homologues, and distribution to hypothesize apomorphy (and thereby monophyly and homology) – a character can hardly be incongruent with itself (unparsimonious). If such an apomorphic character is found to be incongruent with others so hypothesized, it requires re-examination – this is reciprocal illumination. Retaining its altered interpretation as discovered on an unrooted or even rooted tree offers no illumination other than the fact that its original observation does not agree. In no way should it be construed to offer evidence of local synapomorphy through the auxiliary principle; otherwise the auxiliary principle is being used as argument for convergence, the exact explanation for which it was formulated to counter!

Hennig (1953: 18, translation) suggested that, "The burden of proof must rather be placed upon the assertion that certain synapomorphies [...] can be based on convergence." Despite the incongruence, that burden remains. The original character observation needs to be re-examined *in the specimens* and an alternate explanation for the character distribution found, or not. If the latter, the ad hoc assumption for the distribution can be maintained as convergence, although if the feature is of enough significance, this might be somewhat unsatisfying. For example, recent topologies suggesting (weakly) that asymmetry has arisen twice among a paraphyletic Pleuronectiformes (flatfishes) beg for developmental and anatomical evidence to support this contention (Campbell et al. 2013, 2014b).

Asymmetry in pleuronectiforms is quite strong evidence for monophyly (one of Hennig's characters "certainly interpretable as apomorphous," Hennig 1966: 121[1]), so is a case where a 'single' character is seen as reciprocally illuminating for the

---

[1] Note the original 1966 translation, "The more certainly characters interpretable as apomorphous..." was reinterpreted by Farris et al. (1970:174) as, "The more characters certainly interpretable as

parsimonious tree – a situation that is very rarely identified. How often are parsimonious trees accepted when incongruent characters are not as striking as asymmetry, but might have an equally justifiable claim to challenge the topology? And, for that matter, without examining character evidence the number of characters actually providing support for the initial topology remains unknown. Because current practice applies parsimony as final arbiter as luminary to incongruent features, the reciprocity of illumination is ignored; trees (parsimony) change characters, but characters rarely (if ever) change trees. It should be remembered that all trees can really do is "point up incongruences and allow us to predict that mistakes have been made, and that more intensive study will reveal them" (Nelson and Platnick 1981: 305). More data and more taxa will not make these incongruences disappear; they *might* make them seem less significant if new data make a particular tree more robust, but reciprocal illumination aims to "bring the relationship indicated by the several series of characters into congruence" (Hennig 1966: 122), i.e. explaining the incongruences or errors in empirical observation where we have identified homologues that were not or apomorphies that were not (Hennig's checking and correcting).

11.5 Congruence and testing

It has been argued, one would like to think accurately, that, "[a]ll methods depend on the skill and experience of the investigator to form transformation series (data columns in a matrix) with robust hypotheses of initial homology statements using solid biological criteria" (Wiley et al. 2011a: 10). In contrast, it has also been suggested (Wiley and Lieberman 2011: 188–9) that "specialists in a group will avoid certain kinds of characters because they have the a priori notion that these characters are subject to homoplasy, but in principle, this should not be necessary because such characters should sort themselves out via the test of congruence when parsimony is applied." So is a character matrix carefully constructed with robust initial homology statements, or is it potentially full of mistakes, or at least unknowns, that will "sort themselves out"? Certainly, relying on characters to sort themselves out has not applied an auxiliary principle and does not employ Hennigian methods (1966: 121) where "characters certainly interpretable as apomorphous" address the question and those that are "obviously plesiomorphous or in the present state of our knowledge not certainly identifiable" hold no sway. Regardless, as noted by Wägele (2005: 224), the criterion of congruence cannot test "whether a scientist has worked

---

apomorphous…" This change was incorporated into the 1979 reissue under the advisement of G. Nelson and retained in the 1999 reissue (G. Nelson, pers. comm. 2014).

hard," whereas explicit hypotheses of character homology via the auxiliary principle can provide such a measure.

Of course, the "homology statements" of Wiley et al. (2011a) only place character states into columns of data and do not commit these to particular apomorphic or plesiomorphic designations; for them (and most workers) these are determined via parsimony (and outgroup choice) so that all characters, robust or not, are sorted out. Some of the issues surrounding this approach were discussed previously. Other authors have also recognized that membership of character states in a transformation series is not tested beyond initial similarity criteria (e.g. Rieppel 1988: 68). Congruence cannot evaluate the assignment of states to a character, it can only count the number of times a hypothesized transformation occurs on a preferred topology (Britz and Johnson 2011: 66); as Richter (2005: 115) noted: "the 'column' is never tested."

What is the role of congruence/parsimony in homology estimation? For de Pinna (1991: 372) there is a duality of generation of hypotheses of homology through various similarity criteria for individual characters followed by their "legitimation" via congruence. This approach grants congruence the final verdict on homology, whereas reciprocal illumination would seem to be more egalitarian and would permit each to influence the other. Rieppel and Kearney (2002: 60) asserted that "character hypotheses must be testable in their own right, and [...] that such testability can only be achieved by due consideration of structural complexity in character analysis." Grant and Kluge (2004: 28) suggested that similarity testing and congruence are decoupled and complementary. In some instances, phenotypic data might exhibit sufficient structural and/or developmental complexity "to defensibly choose among competing hypotheses of homology." Such might be the case for asymmetry in pleuronectiforms, for example. Hennig (1966: 116) also argued that complexity "suggests a way of dealing with the difficulties that arise in systematics." However, congruence (of course) remains as an additional test. It is noteworthy, perhaps, that the complementarity described by Grant and Kluge (2004) invalidates the claim that hypotheses of homology not tested via congruence are not viable (contra Craig 2011: 34 (his emphasis): "...they [Gill and Mooi 2010] provided no *test* of their hypotheses of homology. There is no phylogenetic tree..."). The misconception that a tree is required to test homology is widespread (e.g. Chakrabarty (2010: 514): "No phylogenetic analysis is associated with the hypotheses of relationships, so again there are no tests of their characters.").

In other instances, parts of organisms might be so similar as to be indistinguishable (e.g. nucleotides in DNA sequences) or treated in such a way to make them seem indistinguishable (e.g. number of spines in dorsal fins of fishes), no matter their actual historical identities (homology) – "separate tests of homology are inert in such cases" (Grant and Kluge 2004: 28). Here, congruence is the only option for choosing among competing hypotheses. This has implications for the severity of testing; characters for which both similarity and congruence testing can be applied might seem to carry more weight. Others have come to a similar conclusion (e.g. Neff 1989, Bryant 1989, Britz and Johnson 2011, Wägele 2005). Richter (2005: 105) concurred, "Complex characters possess a higher empirical content than less complex characters because they are more severely testable."

Wiley et al. (2011a: 9–10) describe congruence testing as: "simply another manifestation of Hennig's concept of reciprocal illumination, pitting one hypothesis of synapomorphy against another in the arena of what is now called 'optimization.'" This seems reasonable in the case of complex characters where Hennig's original formulation of the auxiliary principle holds. However, the modified principle as almost universally employed, has replaced Hennig's requirement of apomorphy with one of homology, defined by users as apomorphic *or* plesiomorphic similarities among organisms (see above). A priori hypotheses of apomorphy are no longer in the arena, having been replaced by columns of presumed homologous character states that are sorted by parsimony, not tested by it. Because characters have been given no specific history, only identified as homologous in the broadest sense, they have been robbed of sufficient explanatory power to reciprocally illuminate, i.e. to challenge the topology. Because apomorphy and plesiomorphy designation is deemed plastic within a column of data ("freely reversible characters," Wiley and Lieberman 2011: 168), they can only become informative through parsimony.

This approach to characters, the alteration of the auxiliary principle, and the acceptance of congruence testing through parsimony as the final arbiter has eliminated an important element of reciprocal illumination; the tree does, indeed, influence our interpretation of characters, but individual conflicting characters no longer influence the tree. Conflict has all but disappeared as we now need only create columns of data that are parsimony informative (i.e. variable) and apply congruence testing. Homoplasy is re-interpreted as 'local synapomorphy' and reversal seen to contribute gainfully to tree resolution (Källersjö et al. 1999, Kluge 2005: 27 – "an incongruent character state can in fact increase phylogenetic structure; for example a reversed state can be diagnostic of a monophyletic group" – of course, being diagnostic is quite different from being evidence of relationship, see Cao et al. 2007 or any dichotomous key). Wiley et al.'s (2011a: 10) "robust hypotheses of initial homology statements using solid biological criteria" have been overturned and the "skill and experience of the investigator to form transformation series (data columns in a matrix)" has been tossed aside rather cheaply.

There has been some discussion as to how powerful a test of homology congruence actually provides. Neff (1986: 116) suggested that "cladistic analysis itself [tree building] is relatively trivial: it is only summarizing the information already entirely contained within the characters." Bryant (1989) agreed that cladistic analysis cannot "decide" relationships because it is an inductive procedure summarizing data and acts only to identify incongruence. The critical step is "resolution of incongruence through additional character analysis" (Bryant 1989: 225). However, both Bryant (1989) and Neff (1986: 115) lamented that "the greatest deficit in phylogenetic analysis today is the widespread failure to treat characters themselves as hypotheses subject to test and possible refutation prior to their use [...] in a cladistic analysis." 'Today' can be defined as a 30 year period (and counting).

Why should not characters that have been most rigorously tested through character analysis (via similarity criteria) be considered more informative than those that have not been, or even cannot be? Neff (1986) suggested that this might form the basis for methods of character weighting and Bryant noted,

> Characters with higher information content, be it structural, positional or ontogenetic, can be more adequately tested [...] more detailed character analysis and weighting based on testability should reduce the amount of incongruence [...] and increase the reliability of resulting phylogenetic patterns. (Bryant 1989: 223–224)

This seems in keeping with Hennig (1966: 121), who emphasized characters "certainly interpretable as apomorphous" over those for which apomorphy is "in the present state of our knowledge not certainly identifiable." Discovering apomorphy through application of the auxiliary principle and reciprocal illumination is the goal of phylogenetics; discovering apomorphy uncovers relationships.

11.6 Hierarchy in characters

That characters are hierarchical and change over time is generally recognized, but is a difficult concept to express because we recognize and record characters and their states empirically as discrete and static observations (Grant and Kluge 2004, Rieppel 2006, Assis and Rieppel 2011). A common example of a transformation series is pectoral fins > arms > bird wings. This might be coded for six taxa as 001122 in a column of a matrix and represented as 00(11(22)) in a topology. The tree makes clear what we do not see in the initial transformation series: that the synapomorphies for the tree are arms + bird wings (1 + 2) at the first step and bird wings (2) at the second, the synapomorphies becoming less inclusive and hierarchical. There has been some debate as to whether or not the zeros (00) hold meaning for systematics (Mooi and Gill 2010 and Mooi et al. 2011, contra Wiley et al. 2011a, b), although many workers have commented on this previously (e.g. Nelson 1994). The confusion appears to stem from a frequently stated axiom that is meant to clarify the hierarchical nature of characters, but appears to have had the opposite result (de Pinna 1991: 372): "every symplesiomorphy is a synapomorphy at a higher level." Wiley and Lieberman (2011: 14) provide a more recent iteration: "All symplesiomorphies at one restricted level of the entire tree of life are synapomorphies at one or more higher levels where they diagnose monophyletic groups..." The statement is false. One need only look at the tree to see that the zeros as representing the character

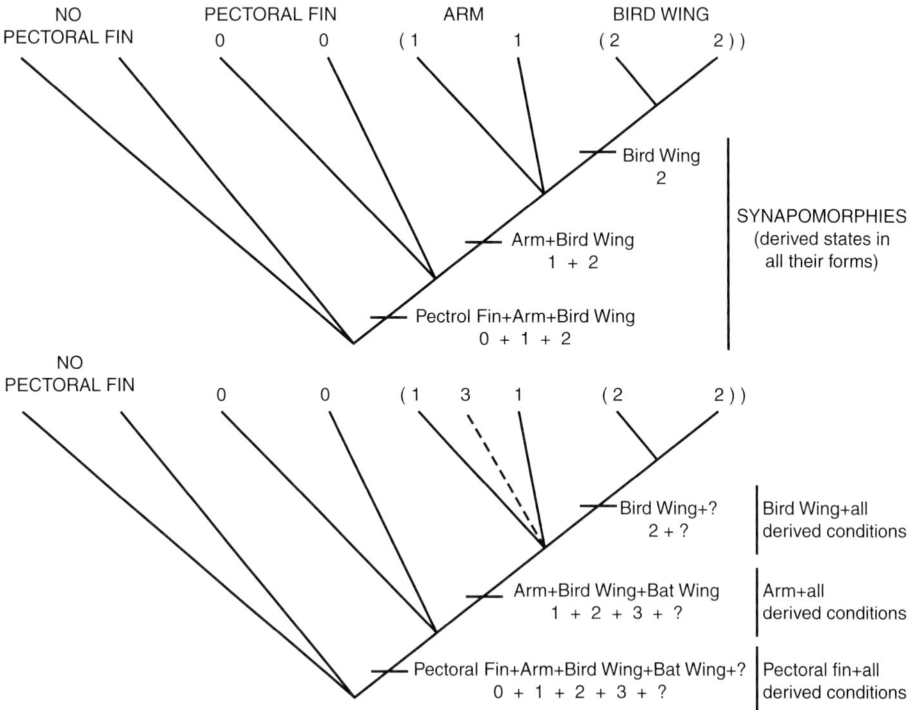

Fig 11.1 Cladograms depicting character states as terminals with their interpretation as synapomorphies marked at the bars. Bottom cladogram shows the insertion of a third state (bat wing) and an unknown modified bird wing (?) arising from either discovery of new taxa or future evolution. Note that the zero (0) state never attains synapomorphy status and that the addition of new states (via new taxa) does not change the relative relationships.

state 'pectoral fins' can never perform the function of synapomorphy (Fig 11.1). If they provide no evidence for relationship (monophyly) for the subset of taxa bearing them, how could they be a synapomorphy for these taxa along with those that do not share that state!

As we move down the tree, the character state 'pectoral fin' (0) becomes *a part* of a synapomorphy at a higher level along with its associated homologous conditions, 'arms' and 'bird wings.' The synapomorphy at a higher level is pectoral fins + arms + bird wings (0 + 1 + 2), not merely pectoral fins (0). This follows what we know, that a synapomorphy is a derived character in all of its states: "It makes no difference whether the synapomorphy consists in the fact that an apomorphous character (a') is present identically in all species or whether it is present in different derived conditions (a' and a")" (Hennig 1966: 90). Despite the introduced confusion, de Pinna (1991: 371) recognizes this, too, suggesting there is a "clear understanding of symplesiomorphy as a subset of synapomorphy." So we should restate and clarify, then, that all symplesiomorphies at one restricted level of the entire tree of life form part of a synapomorphy, along with all its derived conditions, at a higher level where they diagnose more inclusive monophyletic groups. As a corollary, all synapomorphies at a higher level of the tree of life are, in part, symplesiomorphies at a lower level (for example, arms+bird wings as synapomorphy at a higher level is arms as plesiomorphy at a lower level) (Fig 11.1). Recognition of this takes into account the hierarchical and dynamic nature of evolution and admits that synapomorphies (or autapomorphies) that we identify in species today form the basis for synapomorphy (and plesiomorphy) as these species change over time, by including any future evolution and apomorphy. In essence, the synapomorphy arms + bird wings (1 + 2) includes additional apomorphies that have not yet been recognized (e.g. bat wings, whale flippers) or that will evolve.

For complex characters with robust homology and apomorphy hypotheses, adding taxa, even those with their own apomorphies, should not alter relative relationships. Mooi and Gill (2010) were brought to task for suggesting this and considered ignorant of the scientific method (Craig 2011: 34). Hennig, however, is not so ignorant:

> Theoretically it is possible to ask and answer the question, which species is more closely related to the other than to the third, with three arbitrarily chosen species. We can also say in this case that the two species that are more closely related to each other are in a sister group relationship, and that both are in a sister group relationship to the third species. One can add a third or fourth species and investigate how they can be placed into the system of the three initial species. (Hennig 1984: 54, translation)

It is easy to see that introducing a third state, bat wings, to a carefully tested character can be accommodated without problem and foster some confidence in the hypothesis (Fig 11.1). It is when the addition of taxa (and character states) does suggest topological changes that things need to be examined carefully; change can be good, but not accepted without weighing evidence.

How did systematists so readily embrace the platitude that plesiomorphy (0) becomes synapomorphy (1) without considering that the higher level synapomorphy must include all of its "different derived conditions" as so clearly recognized by Hennig? The strong and continuing influence of ancestor–descendant thinking makes the idea of a linear pectoral fins > arms > bird wings relationship, rather than a hierarchical rendering, easy to accept. As we move down the tree, the plesiomorphic condition (pectoral fins, 0) is moved into an "ancestral" position, although we know that ancestors (and their characters) are inaccessible (Nelson 1994, 2004). But we are building a topology of relationships for organisms as we see them today, not at some point in the past. That pectoral fins were potentially an apomorphy to themselves in the past is equally inaccessible and irrelevant; pectoral fins evolved and no longer characterize a monophyletic group unless all subsequently derived states of pectoral fins are included. What is moved down the tree to become synapomorphy is not only the plesiomorphy, but all of the different conditions of that character exhibited by the monophyletic group (Fig 11.1). As noted by Nelson,

> Cladists routinely reject statements such as birds evolved from reptiles, and accept statements such as bird wings evolved from arms, when the two statements, to the extent that they mean anything at all, mean exactly the same thing: such as, the relationship among birds (and some of their parts) is one and the same node of life's hierarchy. (Nelson 1994: 127)

Further, it seems clear from the confusion inherent between the hierarchy of plesiomorphy and apomorphy and 'ancestral characters' that we should heed Nelson's solution:

> A cladogram differs from other types of phylogenetic trees in placing all organisms, both fossil and present, in terminal positions, implying that ancestral taxa are artifacts. Cladistics may possibly be improved if parts of organisms were treated in the same fashion in character (state) trees, with the implication that ancestral characters, too, are artifacts. (Nelson 1994: 137)

This implication is further explored below.

Matrices of character states do a poor job of depicting this dynamic relationship and hierarchy (Williams and Ebach 2006, Cao et al. 2007). In a data matrix, a character is a column of presumed homologous states (in the sense of Wiley et al. 2011a, where homology = apomorphy + plesiomorphy). In the simple case of a binary character with state 0 in two ingroup taxa and state 1 in two ingroup taxa with state 0 shared by outgroups, the cladistic relationship of the ingroup is unequivocally 00(11) employing Hennig's auxiliary principle. However, with changes to the principle introduced by Farris et al. (1970) and De Laet (2005), the character in a matrix is initially interpreted as (00)(11), permitting congruence via parsimony to determine which is apomorphy. In the extreme view of Grant and Kluge (2004), where the auxiliary principle has been effectively eliminated, the character can be (0)(0)(1)(1), where relations among even similar states need not have any historical implications ("synapomorphy" can result from the same transformation event or different transformation events). Indeed, incongruence might suggest this to be so. Without applying the auxiliary principle and the full cycle of reciprocal illumination, this approach to characters and their states permits congruence (via parsimony) to determine a hypothesis of homology when the empirical evidence suggests otherwise.

Multistate characters provide a more complicated situation. Assume a character in an ingroup with state 0 in two taxa, state 1 in two taxa, and state 2 as a modified form of state 1 in two taxa, with state 0 shared by the outgroups. The cladistic relationship is unequivocally 00(11(22)) using Hennig's original auxiliary principle. Again, the modified auxiliary principle provides (00)(11)(22) as the interpretation. The latter treats character states as only homologous, but with no particular history, so options are 00(11(22)), 00(22(11)), 11(00(22)), 11(22(00)), 22(00(11)), 22(11(00)), or no interrelated hierarchy at all, 00(11), 11(00), 00(22), 22(00), etc. For example, numbers of dorsal-fin spines are assumed homologous at the level of "having dorsal spines" and are aligned in a dorsal-fin column; the particular numbers of spines are separate states allowed to float on a tree within this homologous space. In essence, each spine number is treated as a unique and unrelated character. Such a character compromises Hennig's use of "characters certainly interpretable as apomorphous" and avoidance of characters where apomorphy is "in the present state of our knowledge not certainly identifiable" (Hennig 1966: 121). What is preferred for a character is a hierarchical arrangement, or, if this is not possible, at least a clear statement of apomorphy, 00(11,22) rather than (00)(11)(22). A maximally testable multistate character would be maximally hierarchical, 00(11(22)). Meristic characters (and nucleotides) do not permit such a historical treatment because there is no inherent hierarchy, only replacement. In theory, meristic characters could be redefined through more detailed examination of homology through developmental and other studies, although this is rarely undertaken.

Carine and Scotland (1999) identified complement and paired homologues to examine relationships. Complement homologues (following Patterson 1982) are designated as absence/presence characters and deemed unproblematic because the 'absence' cannot provide evidence, whereas paired homologues exhibit conditions where each homologue can provide evidence of groups. There are several problems with how Carine and Scotland examine homologues and homology, one being that the terminology and issues are tied to the methods employed rather than to biological concepts. Their approach has ignored Hennig's auxiliary principle. Absence and presence is not a matter of zeros and ones in a matrix, it is a matter of absence or presence of hypothesized apomorphy in the character. One element of the pair is necessarily apomorphic relative to the other. If, on the other hand, each element of the paired homologue is considered potentially equally informative and can identify monophyletic groups at the same hierarchical level, they have separate histories and are not homologous. Tables 1 and 2 in Carine and Scotland (1999) do not report the outgroup state; without reference to an outgroup the paired homologue is, in essence, a two-taxon statement and uninterpretable. Hennig's auxiliary principle cannot be applied to such paired homologues and we are left with characters whose relative apomorphy cannot be assessed/hypothesized.

The paired homologues of Carine and Scotland are characters for which no hierarchy is discoverable (or at least they have not provided character details that suggest such a hierarchy might exist). Carine and Scotland treat each different state as an independent character, which might work computationally, but makes little sense biologically or historically; character independence is only one of several possibilities. This approach is minimally hierarchic and introduces maximum "evidence" of monophyly; each state is interpreted as unique and with no historical relationship with

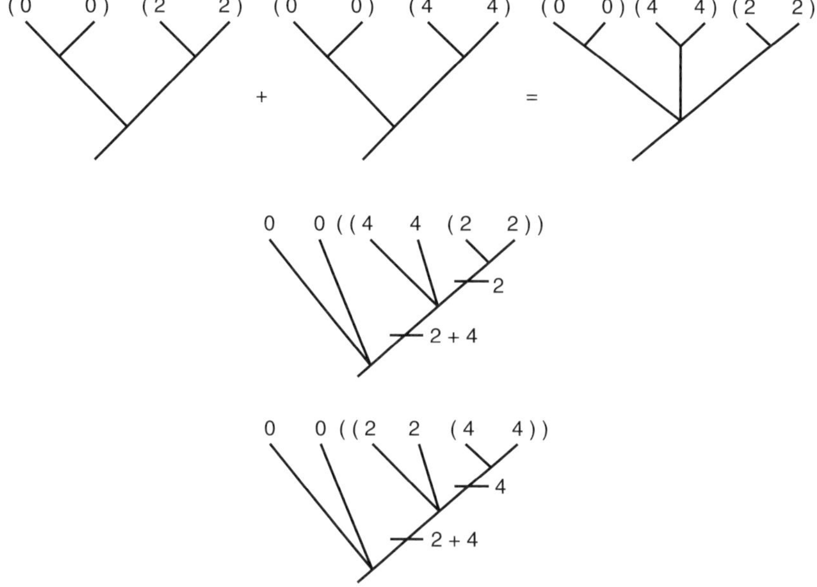

Fig 11.2. Character state trees for six taxa for the hypothetical character 'stamen number' as employed by Carine and Scotland (1999). Coding is: 0 for an unknown number of stamens in the outgroup taxa (not noted by Carine and Scotland), 2 and 4 stamens in four ingroup taxa. Upper series show character trees with paired coding and all states considered equally informative for two stamens, four stamens and a combined tree for what are really two characters. This treatment only identifies taxa as having a particular state. Two bottom trees show character trees for two possible hierarchical hypotheses of apomorphy if one were possible based on similarity tests. The first is a 0 > 4 > 2 and the second a 0 > 2 > 4 hierarchy of apomorphy. Note that the outgroup condition is shown as plesiomorphous relative to 2 and 4, but this would have had to have been based on some similarity tests and cannot be assumed. Unless either of the two bottom scenarios is viable, the character is one to be avoided in systematics.

other states, with all the inherent problems that were discussed previously for meristic characters. For example, flower color is coded as (not red)(red) [(00)(red red)] or (not blue)(blue) [(00)(blue blue)]; they have effectively made the colors separate homologues. Similarly, stamen number is coded (not two)(two) [(00)(22)]and (not four) (four) [(00)(44)]. With such coding, other relationships are precluded, for example, (two(four)) [(22(44))] or (four(two)) [(44(22))] or ((four)(two)) [((22)(44))] (Fig 11.2). As Carine and Scotland (1999: 121) make clear, the various designations as paired and complement and various coding schemes are devised to permit "all data to be considered potentially informative." Mishler (2005: 57) has suggested, "more-complicated models for tree building can then be seen for what they are: attempts to compensate for marginal data." Hennig (1966: 121) also suggested avoiding marginal data, particularly those for which apomorphy or plesiomorphy are "in the present state of our knowledge not certainly identifiable in the group in question."

11.7 Reversals

Reversal is a simple concept: in any particular transformation series, an apomorphic state returns to its plesiomorphic state, or there is a new origin of a previous state. How an apparent reversal is interpreted is not so simple.

With the application of Hennig's auxiliary principle, it would seem uncontroversial that reversals, being a form of convergence, are identified secondarily through congruence testing. This would suggest that 'reversal' is a description or, at best, an ad hoc assumption regarding a character distribution determined by other characters. In and of itself, it would seem that a reversal is empirically empty; the character's original hypothesis of apomorphy has been rejected and it is mapped on a tree built using other evidence (Assis and Rieppel 2011). As has been noted in the previous sections, however, modifications of the auxiliary principle and a willingness (or lack of perseverance) to employ only a partial application of reciprocal illumination has left congruence through parsimony (or other clustering procedures) as the final arbiter of character distribution and interpretation (see also Bryant 1989). Use of characters that cannot be subjected to similarity tests (via complexity) and are only subject to the congruence test has influenced (and encouraged?) the contention that characters as interpreted on topologies provide evidence unto themselves. Grant and Kluge (2004, 2009) have brought this approach to its ultimate conclusion by suggesting that any change indicated on a parsimony tree is, by definition, a synapomorphy, and no auxiliary principle need be applied. This is the essential approach of the majority of (numerical) systematists, despite rhetoric to the contrary.

Hence, the suggestion that reversals are merely secondary reinterpretations via parsimony, and are ad hoc assumptions providing no evidence for phylogeny (at least without more study) has raised hackles. A general introduction to systematics, presumably aimed at students, warns that methods that imply "rejection of reversals as synapomorphies," employ a priori character ordering, and rely on irreversibility are "outside the evolutionary paradigm" and not phylogenetic (Wiley and Lieberman 2011: 202, contra Wiley et al. 2011a, b).

So what did Hennig say? This is not an appeal to authority, but a place to begin examining this issue. The contention that Hennig used reversals as synapomorphies has been expressed by many, particularly Farris (e.g. 1997, 2011). The oft-quoted passage by Hennig (1966: 95), as edited by Farris, reads:

> For example, the absence of the wings in fleas is undoubtedly an apomorphous character in comparison with the presence of wings in other holometabolic insects.

> On the other hand, the possession of wings is an apomorphous character in comparison to their absence in the so-called "Apterygota"… The absence of wings in the Anoplura and Mallophaga is a synapomorphous character, whereas in the Collembola, Protura, etc. it is a symplesiomorphous character. (Farris 2011: 211)

This does seem convincing, except that the term used by Hennig is "absence" and not "reversal," and the entire paragraph provides additional context and reads thusly:

> Finally, the concepts of symplesiomorphy and synapomorphy go somewhat beyond the range of what are ordinarily called "homologous characters." We started from the idea that a, a', a" are different characters in a transformation series. We can speak without reservation of homologous characters if a, a', a" are transformation stages of the organ. But the transformation a-a'-a" may also consist in the complete reduction of the organ. For example, the absence of the wings in fleas is undoubtedly an apomorphous character in comparison with the presence of wings in other holometabolic insects. On the other hand, the possession of wings is an apomorphous character in comparison to their absence in the so-called "Apterygota." In general we speak only of the homology of organs, but a "character" may also be the absence of an organ. This discrepancy between the concepts "organ" and "character" explains the tortured impression produced by many phylogenetic discussions that try to make do with concepts such as "special homology," "limited homology," and so on (instead of "synapomorphy"). It is completely unequivocal to say that the absence of wings in the Anoplura and Mallophaga is a synapomorphous character, whereas in the Collembola, Protura, etc. it is a symplesiomorphous character. This cannot be expressed in an equally unequivocal way by saying that the absence of wings is a "special homology" in the Anoplura and Mallophaga, but not in the Collembola, Protura, etc. (Hennig 1966: 94–95)

Hennig wrote this section to demonstrate the utility of the terms synapomorphy and symplesiomorphy without reference to a discovery method. However, it is clear that by referring to the absence of wings in fleas as part of a monophyletic higher taxon that this is a secondary absence or loss, and not a reversal to a previous state (Carine and Scotland 1999). The absence is also referring specifically to a transformation series resulting in the "complete reduction of the organ," and as such is a loss of wings, not an absence. Absence and loss might look the same, at least superficially, but absence means a feature was never there whereas loss means just that – a feature once there has disappeared. Nelson and Platnick came to this conclusion early on:

> [T]here is a difference between the absence of a character and the loss of a character […] whereas springtails and silverfish merely lack wings, fleas have lost them. In other words, some characters which might appear to be simple absences are the result of character transformations. (Nelson and Platnick 1981: 30)

Grant and Kluge (2009) interpret Hennig's view of the loss of wings in fleas as a new state in a unique transformation (a-a'-a"; absence of wings – wings – loss of wings) whereas Farris interprets the absence as a new origin of a previous state (a-a'-a; absence of wings – wings – absence of wings).

Hennig himself was a little less clear on how he came to see the lack of wings in fleas. We agree with Grant and Kluge (2009) that Hennig's (1966) reference to a new state in a transformation and reduction of an organ certainly implies the loss scenario, but Hennig's metaphysics and methods are somewhat muddled (Rieppel 2007). However, looking at Hennig's systematic work, there are clues that he did see the lack of wings in fleas as a loss (a new condition) and not a reversal (return to previous state). There is little doubt that Hennig based his concept of a monophyletic Pterygota (winged insects) on more than the presence of wings (Hennig 1953: 20, translation): "The Pterygota form a true descent-community [Abstammungsgemeinschaft] (a monophyletic group). Numerous synapomorphies of the component groups, combined under this designation, prove that." Further, he rejected attempts by others to suggest that wingless cockroaches are "primarily apterous" (Hennig 1953: 21), i.e. that they share the plesiomorphic condition, and he certainly included fleas and other wingless pterygotes as derived members of that taxon. It appears that Hennig saw in fleas a loss of wings and not a return to the plesiomorphic condition. This is clarified later:

> The term homology can be applied without question really only to structures that are present (Positivmerkmale). A "character" is by definition also the absence of a structure. Then, however, it is of great importance for phylogenetic systematics if this "negative character" is the result of a reduction of a certain structure or its primary absence. The simple homology term takes no notice of this important question. (Hennig 1984: 38, translation)

Reversal to the plesiomorphic condition seems not an option.

Hennig does deal with the possibility of reversals:

> The significance of synapomorphies [...] is not fully understood without consideration of the so-called irreversibility rule. If the evolution of characters and structure-plans could transpire in the reverse, then one could take into consideration that apparent symplesiomorphies, in the same way as this is possible in apparent synapomorphies, had come into being through convergence. This, certainly, could be assumed in some cases. The extent of the validity of the irreversibility rule has been much discussed (Hennig, 1950; Remane, 1952; in entomology under the name Meyrick's Law by Tillyard, 1919, see also Sachtleben, 1951), with the result that it certainly cannot be regarded as unlimited. *But the impact upon taxonomic work is not extremely great for, in general, it can be assumed that the reverse development does not go so far that no synapomorphies of any kind remain among actually related groups.* (Hennig 1953: 17–18, translation, our emphasis)

Hennig did not consider reversals to be synapomorphies, but instead, because these characters were indistinguishable from plesiomorphies, they were neutral and provided no information about relationships. Information about relationships comes instead from characters that could be considered synapomorphic a priori. Synapomorphy would override any signal from symplesiomorphies. Hennig (1966: 116) made clear that issues of reversibility are dealt with by relying on complexity, the auxiliary principle, and congruence – reversal undoubtedly occurs (at least as reduction or loss), but can be overcome.

Regardless, whether actual reversal or loss, detection is through congruence testing, i.e. either will show up as incongruence or apparent error. If truly a reversal where the character states are indistinguishable (a-a'-a), there is little that re-examination and reciprocal illumination can do to make this incongruence available as evidence, as opposed to ad hoc assumption. As noted above, and recognized by Hennig, reversals are symplesiomorphies (shared zeros in a typical matrix) and cannot provide evidence of relationship. Loss (a-a'-a"), although superficially similar to absence, might be detectable through more rigorous similarity tests (development, atavism, vestigial morphology) and provide direct evidence for relationship (e.g. 'limbless' snakes). However, the majority of characters employed in phylogenetics today (nucleotides, most 'paired homologues') do not have such similarity tests available to them to discover that they are synapomorphies providing evidence, rather than symplesiomorphies secondarily mapped on a tree constructed from other evidence.

What is done with 'reversal' in research? Is it used as synapomorphy? Molecular studies do not, for the most part, provide character analysis, so it is difficult to survey the literature for how frequently apparent reversal is treated as synapomorphy. It is very likely to be frequent given the prevailing methods. Reversal is sometimes identified in morphological studies. Britz et al. (2014), in a detailed study of developmental truncation in a miniaturized fish, referred to primary absence (= absence) and secondary absence (= loss), although referred far more frequently to reduction and loss, directly appealing to Hennig for its interpretation as synapomorphy. They provided only one reference to reversal and that appeared in a quote from another source. Britz et al. (2014) clearly saw loss as a derived feature rather than a return to a plesiomorphic condition; they did not equate 'reversal' with "synapomorphy" and went to extraordinary and impressive lengths to demonstrate that apparent absences of characters in their study organisms are synapomorphies (losses), not plesiomorphy.

In an effort to diagnose monophyletic teleost groups (most 'fishes'), Wiley and Johnson (2010) listed potential synapomorphies for almost 120 major groups. Among the hundreds of character descriptions, reference to reversal in some form appears 25 times. For eight of these, characters are described as "unique and unreversed," implying a level of quality or explanatory power where there is no subsequent reversal. The 17 times that reversal is mentioned, it is as an exception to the usual condition (i.e. as conflict), seemingly more as a cautionary note and something that might deserve re-examination. Reversals are not specifically identified as potential synapomorphy, although that might be the implication. However, in at least some instances, loss or reduction rather than strict reversal might be a more promising avenue of study. Even Wiley seems reluctant to use reversals as evidence, or at best it is evidence with an asterisk.

So it seems Hennig did not use reversals as synapomorphies. He also insisted that a priori character ordering was a fundamental principle of phylogenetics: "I have therefore called it an 'auxiliary principle' that the presence of apomorphous characters in different species 'is always reason for suspecting kinship (i.e., that the species belong to a monophyletic group) ...'" (1966: 121) He also relied on irreversibility of complex organs, where "Dollo's law applies without restriction" (1966: 116) to deal with convergence and reversal. If Hennig should be banished from the "phylogenetic tent" as pitched by Wiley and Lieberman (2011: 201–2), we prefer to camp with Hennig.

Nelson and Platnick (1981: 28–30) explain clearly why reversals are uninformative: the absence of a character is uninformative. A reversal is the absence of an apomorphy. How can the absence of an apomorphy be used to form a group? At face value, the absence of legs is not useful evidence to create a taxon (like snakes) because many organisms don't have legs. However, through reciprocal illumination we might be able to determine that it is not a case of absence of legs, but that there has been a loss of legs as an apomorphic condition. This might seem semantic, but only if characters are considered phenetically and not hierarchically.

11.8 Conclusion

Although Hennig's auxiliary principle is generally described as a fundamental concept in systematics, it has seen modifications that have significantly changed its meaning and influence. Its original emphasis on apomorphous characters has been altered to refer to homologous characters, where such characters are defined as being both apomorphous and plesiomorphous. This has removed the primary role of Hennig's auxiliary principle to form hypotheses of apomorphy, homology, and relationships, and shifted these functions to congruence testing. Congruence is an important test, but is not the only test. A priori hypotheses of apomorphy have lost any role in the process of reciprocal illumination. Hence, reciprocal illumination, another of Hennig's primary contributions, has become unidirectional – congruence testing influences hypotheses of character evolution, but characters themselves are passive in what should be a dynamic interplay of similarity testing and congruence testing. Some workers have shifted the focus further from character to event, theoretically sound perhaps, but empirically unattainable (i.e. events are not visible). Such a view makes any and every change as mapped on a minimum-length tree informative, giving the topology ultimate determination of homology and apomorphy. This weakened role has been driven by the preferred methods (parsimony) and type of data (physically indistinguishable regardless of historical identities, e.g. nucleotides; or characters coded to appear indistinguishable, e.g. meristics) employed by most practitioners of systematics, rather than by any theoretical underpinning. Congruence testing has become primary, in large part due to application of character types inaccessible to similarity testing; states replace other states rather than forming hierarchical relations. The result is that a priori hypotheses of apomorphy have lost any role in systematics, and character history is determined through congruence via parsimony or other clustering procedures. Hawkins et al. summarized:

> To allow primary homology [apomorphy] decisions to emerge as part of these results [parsimony] is to retreat from the vital task of primary homology assessment, and tends towards operationalism. Any retreat from the fundamentally important task of homology assessment is a retreat from theory. (Hawkins et al. 1997: 282)

Changes introduced to Hennig's original principles were introduced to permit and then justify an approach where apomorphy no longer needs to be identified and characters that exhibit inaccessible variation can still be applied. We remain unconvinced that Hennigian argumentation is mirrored by parsimony. Kluge agreed:

> The most significant difference between the two approaches concerns the fact that Phylogenetic Systematics [Hennigian argumentation] estimates only the cladistic parameter and in doing so it uses only derived states. Quantitative Phyletics [as applied in parsimony algorithms] estimates both cladistics and patristics and it uses all states, derived and primitive, in those estimations [...] to ignore similarity based on shared primitive states can lead to significantly different phylogenetic hypotheses. (Kluge 1976: 43)

Further, parsimony does not require apomorphy identification; Hennig does. Parsimony requires characters as matrix columns of states, ordered or not; Hennig requires hierarchical states with plesiomorphy and apomorphy identified. Parsimony (or any other clustering procedure) does not require reciprocal illumination, and modern approaches ignore it; Hennig requires checking, correcting, and rechecking and permits characters to test topologies as much as topologies test characters.

Ultimately, systematists might not differ all that much. We can agree with Wiley and Lieberman (2011: 189) who said: "Hennig (1966) stressed that critical phylogenetic inquiry is not simply a process of building a tree; it is a process of reciprocal illumination where the investigator is constantly questioning both the data and the results." Similarly, Grant and Kluge (2004: 29): "Like any scientific hypothesis, proposed character-states and the transformation series through which they are related are open to refutation and refinement through cycles of empirical testing and re-testing." And Kluge:

> [I]ncongruence [...] suggests the need for further study, a posteriori, and further testing of the incongruences may lead to a re-interpretation of the data, such as a redefinition of the characters and character states, and ultimately to a more severely tested and better-supported hypothesis [...] one must be careful to maintain testability at all levels of analysis and re-analysis, because it is easy for reciprocal clarification to be born out of utilitarianism. (Kluge 2005: 39)

Words are good, but practice is better – do these cycles actually occur and is utilitarianism avoided? Finding modern examples is a challenge. An exception is Britz et al. (2014), a study that exhibits careful examination of characters as individually meaningful hypotheses of relationship, testing through congruence and re-evaluation through similarity testing and tested by congruence yet again. General application of their approach and techniques would benefit systematics (see also Cruickshank 2011, Wägele 2005).

There will always be some debate about how to deal with character conflict and whether or not it holds information in and of itself. There should, however, be little debate regarding the use of falsified hypotheses as evidence. Of course, the argument is that the falsified hypothesis has been 're-interpreted' through congruence/optimization to become evidence. Perhaps the argument is not concerning the falsified hypothesis, but that having falsified the hypothesis it has been demonstrated that the observations were wrong and that observations shown to be incorrect cannot be used as evidence for anything. They require the correcting and retesting of Hennigian argumentation. Incorporating identified errors without further justification than the congruence test is methodologically unsound, and empirically empty (Cao et al. 2007, Assis and Rieppel 2011). As Platnick (1979: 544) noted: "whereas phenetic methods are willing to accept incongruence between characters as a feature of the real world, cladistic methods regard the discovery of apparent incongruence as an indication that the taxonomist has made a mistake." The majority of workers today seem quite willing to accept incongruence and redefine it post hoc as evidence rather than identify an error.

Perhaps all this can be summarized as an appeal to look carefully at characters. Patterson intoned:

> But what matters in systematics, or matters most, is looking at and comparing specimens, as carefully and in as much detail as you can, searching for synapomorphies. If you neglect that, your primary duty, and concentrate on what is secondary, manipulating the matrix and drawing conclusions from it, you can get in a horrible mess (Patterson 2011: 129)

Mishler (2005: 57) noted that, "Debates over more-complicated models for tree building can then be seen for what they are: attempts to compensate for marginal data." Patterson (1994: 185) had said much the same thing many years before, that models of character evolution are attempts to "wring truth from recalcitrant data." We see marginal/recalcitrant data to include those for which a priori hypotheses of apomorphy have not been formed; we also see data that are subject to similarity and congruence tests ("certainly interpretable as apomorphous" and maximally hierarchical) as being more informative to those reliant on congruence tests alone. We disagree with Mishler (2005: 69) that "everything interesting has already been encoded in the matrix"; matrices do not represent hierarchy well and frequently misrepresent complexity (Cao et al. 2007).

What are the goals of cladistics? Is it simply about discovering the shortest tree for a particular set of data, the primary focus of numerical systematists? Or is it about character discovery? Nelson and Platnick provide a useful perspective:

> To the extent that conflicting positive occurrences can be studied and re-interpreted, conflicting occurrences disappear – if not in fact, at least in one's best judgement. If all conflict is resolved, such that all positive occurrences are combinable in a single cladogram, the choice of the most efficient summary is unproblematical: it is that cladogram that includes all positive occurrences as single lines. Yet as long as there is conflict among positive occurrences, there is a problem that may be investigated: namely, of the conflicting occurrences, which are real and which not? This residual problem cannot be solved, except perfunctorily, through the use of a clustering procedure. Its solution is possible only through the study of organisms and new knowledge of, or new insight into, their real characteristics. (Nelson and Platnick 1981: 199)

Hennig summarized his phylogenetic systematics in this way:

> Thus the question of whether kinship relations based on a single character or a single presumed transformation series of characters correspond to the actual phylogenetic relationships of the species is tested by means of other series of characters: *by trying to bring the relationships indicated by the several series of characters into congruence.* (Hennig 1966: 122; our emphasis)

Clustering procedures, no matter how sophisticated, cannot meet this goal – "they attempt to cope with the incongruence among characters rather than attempting to reduce or eliminate it" (Bryant 1989: 223). The aim of phylogenetics should always be reduction of incongruence through additional character analysis by application of similarity criteria, Hennig's auxiliary principle, and reciprocal illumination. This appears to have been Hennig's goal as well (1966: 122): "In the final analysis this is again the method of 'checking, correcting, and rechecking.'" Cladistics has strayed from its founding principles as it has focused on finding exact solutions to marginal data or those inaccessible to similarity testing, and as it has attempted to compensate for marginal quality by increasing quantity. Re-dedication to the original formulation of the auxiliary principle and application of reciprocal illumination as originally conceived, i.e. as a focus on apomorphous character analysis and re-analysis (via re-examination of characters in specimens), would go a long way to avoiding Patterson's "horrible mess" and remaining true to Hennig's legacy.

Acknowledgements

We thank David Williams for inviting us to contribute to this volume – even though he knows he will regret it. Thanks to G. Nelson for comments on translations in reissues of Hennig (1966). Thanks are also due to R. Britz for many suggested improvements to the manuscript and for his provision of translations of portions of Hennig (1984). The approaches and opinions here are, of course, our own, but have grown and evolved from many discussions and debates with colleagues throughout our careers, both those who agree and disagree – thanks (and apologies) to all.

References

Assis, L.C.S. and Rieppel, O. (2011). Are monophyly and synapomorphy the same or different? Revisiting the role of morphology in phylogenetics. *Cladistics*, 27, 94–102.

Betancur-R., R., Li, C., Munroe, T.A., Ballesteros, J.A. and Ortí, G. (2013). Addressing gene tree discordance and non-stationarity to resolve a multi-locus phylogeny of the flatfishes (Teleostei: Pleuronectiformes). *Systematic Biology*, 62, 763–785.

Betancur-R., R. and Ortí, G. (2014). Molecular evidence for the monophyly of flatfishes (Carangimorphariae: Pleuronectiformes). *Molecular Phylogenetics and Evolution*, 73, 18–22.

Britz, R. and Johnson, G.D. (2011). Comments on the establishment of the one to one relationship between characters as a prerequisite for homology assessment in phylogenetic studies. In *Morphological and Molecular Approaches to the Phylogeny of Fishes: Integration or Conflict?*, ed. M.R. De Carvalho and M.T. Craig. *Zootaxa*, 2946, 63–70.

Britz, R., Conway, K.W. and Rüber, L. (2014). Miniatures, morphology and molecules: *Paedocypris* and its phylogenetic position (Teleostei, Cypriniformes). *Zoological Journal of the Linnean Society*, 172, 556–615.

Bryant, H.N. (1989). An evaluation of cladistic and character analyses as hypothetico-deductive procedures, and the consequences for character weighting. *Systematic Zoology*, 38, 214–227.

Campbell, M.A., Chen, W-J. and López, J.A. (2013). Are flatfishes (Pleuronectiformes) monophyletic? *Molecular Phylogenetics and Evolution*, 69, 664–673.

Campbell, M.A., Chen, W-J. and López, J.A. (2014a). Molecular data do not provide unambiguous support for the monophyly of flatfishes (Pleuronectiformes): A reply to Betancur-R and Ortí. *Molecular Phylogenetics and Evolution*, 75, 149–153.

Campbell, M.A., López, J.A., Satoh, T.P., Chen, W-J. and Miya, M. (2014b). Mitochondrial genomic investigation of flatfish monophyly. *Gene*, 551, 176–182.

Cao, N., Zaraguëta Bagilis, R. and Vignes-Lebbe, R. (2007). Hierarchical representation of hypotheses of homology. *Geodiversitas*, 29, 5–15.

Carine, M.A. and Scotland, R.W. (1999). Taxic and transformational homology: different ways of seeing. *Cladistics*, 15, 121–129.

Carvalho, M.R. de and Craig, M.T. (2011). Morphological and molecular approaches to the phylogeny of fishes: integration or conflict? *Zootaxa*, 2946, 1–2.

Chakrabarty, P. (2010). The transitioning state of systematic ichthyology. *Copeia*, 2010, 513–514.

Chapleau, F. (1993). Pleuronectiform relationships: a cladistic reassessment. Bulletin of Marine Science, 52, 516–540.

Craig, M.T. (2011). The ghost of crises past: a reply to Mooi and Gill. In *Morphological and Molecular Approaches to the Phylogeny of Fishes: Integration or Conflict?*, ed. M.R. De Carvalho and M.T. Craig. *Zootaxa*, 2946, 33–35.

Cruickshank, R.H. (2011). Exploring character conflict in molecular data. In *Morphological and Molecular Approaches to the Phylogeny of Fishes: Integration or Conflict?*, ed. M.R. De Carvalho and M.T. Craig. *Zootaxa*, 2946, 45–51.

De Laet, J.E. (2005). Parsimony and the problem of inapplicables in sequence data. In *Parsimony, Phylogeny, and Genomics*, ed. V.A. Albert. Oxford: Oxford University Press, pp. 81–116.

de Pinna, M.G.G. (1991). Concepts and tests of homology in the cladistics paradigm. *Cladistics*, 7, 367–394.

Farris, J.S. (1997). Cycles. *Cladistics*, 13, 131–144.

Farris, J.S. (2011). Systemic foundering. *Cladistics*, 27, 207–221.

Farris, J.S., Kluge, A.G. and Eckardt, M.J. (1970). A numerical approach to phylogenetic systematics. *Systematic Zoology*, 19, 172–191.

Friedman, M. (2008). The evolutionary origin of flatfish asymmetry. *Nature*, 454, 209–212.

Friedman, M. (2012). Osteology of †*Heteronectes chaneti* (Acanthomorpha, Pleuronectiformes), an Eocene stem flatfish, with a discussion of flatfish sister-group relationships. *Journal of Vertebrate Paleontology*, 32, 735–756.

Grant, T. and Kluge, A.G. (2004). Transformation series as an ideographic character concept. *Cladistics*, 20, 23–31.

Grant, T. and Kluge, A.G. (2009). Parsimony, explanatory power, and dynamic homology testing. *Systematics and Biodiversity*, 7, 357–363.

Gill, A.C. and Mooi, R.D. (2010). Character evidence for the monophyly of the Microdesminae, with comments on relationships to *Schindleria* (Teleostei: Gobioidei: Gobiidae). *Zootaxa*, 2442, 51–59.

Hawkins, J.A., Hughes, C.E. and Scotland, R.W. (1997). Primary homology assessment, characters and character states. *Cladistics*, 13, 275–283.

Hennig, W. (1953). Kritische Bemerkungen zum phylogenetischen System der Insekten. *Beiträge zur Entomologie*, 3, 1–85. [English translation available at: www.canacoll.org/Diptera/pdfs/ Hennig_1953_Critical_Remarks_in_ Relation_to_the_Phylogenetic_System_ of_insects.pdf]

Hennig, W. (1966). *Phylogenetic Systematics*. Urbana, IL: University of Illinois Press.

Hennig, W. (1984). *Aufgaben und Probleme stammesgeschichtlicher Forschung*. Berlin: Paul Parey.

Källersjö, M., Albert, V.A. and Farris, J.S. (1999). Homoplasy increases phylogenetic structure. *Cladistics*, 15, 91–93.

Kluge, A.G. (1976). Phylogenetic relationships in the lizard family Pygopodidae: An evaluation of theory, methods and data. *Miscellaneous Publications Museum of Zoology, University of Michigan*, 152, 1–72.

Kluge, A.G. (2005). What is the rationale for 'Ockham's razor' (a.k.a. parsimony) in phylogenetic inference? In *Parsimony, Phylogeny, and Genomics*, ed. V.A. Albert. Oxford: Oxford University Press, pp. 15–42.

Mishler, B.D. (2005). The logic of the data matrix in phylogenetic analysis. In *Parsimony, Phylogeny, and Genomics*, ed. V.A. Albert. Oxford: Oxford University Press, pp. 57–70.

Mooi, R.D. (1995). Revision, phylogeny, and discussion of biology and biogeography of the fish genus *Plesiops* (Perciformes: Plesiopidae). *Royal Ontario Museum Life Science Contributions*, 159, 1–107.

Mooi, R.D. and Gill, A.C. (2004). Notograptidae, sister to *Acanthoplesiops* Regan (Teleostei: Plesiopidae: Acanthoclininae), with comments on biogeography, diet and morphological convergence with Congrogadinae (Teleostei: Pseudochromidae). *Zoological Journal of the Linnean Society*, 141, 179–205.

Mooi, R.D. and Gill, A.C. (2010). Phylogenies without synapomorphies: a crisis in fish systematics: time to show some character. *Zootaxa*, 2450, 26–40.

Mooi, R.D., Williams, D.M. and Gill, A.C. (2011). Numerical cladistics, an unintentional refuge for phenetics: a reply to Wiley et al. In *Morphological and Molecular Approaches to the Phylogeny of Fishes: Integration or Conflict?*, ed. M.R. De Carvalho and M.T. Craig. *Zootaxa*, 2946, 17–28.

Neff, N.A. (1986). A rational basis for a priori character weighting. *Systematic Zoology*, 35, 110–123.

Nelson, G.J. (1994). Homology and systematics. In *Homology: The Hierarchical Basis of Comparative Biology*, ed. B.K. Hall. San Diego, CA: Academic Press, San Diego, pp. 101–149.

Nelson, G.J. (2004). Cladistics: its arrested development. In *Milestones in Systematics. The Systematics Association Special Volume Series 67*, ed. D.M. Williams and P.L. Forey. Washington DC: CRC Press, pp. 127–147.

Nelson, G.J. (2011). Resemblance as evidence of ancestry. In *Morphological and Molecular Approaches to the Phylogeny of Fishes: Integration or Conflict?*, ed. M.R. De Carvalho and M.T. Craig. *Zootaxa*, 2946, 137–141.

Nelson, G.J. and Platnick, N.I. (1981). *Systematics and Biogeography: Cladistics and Vicariance*. New York: Columbia University Press.

Patterson, C. (1982). Morphological characters and homology. In *Problems of Phylogenetic Reconstruction*, ed. K.A. Joysey and A.E. Friday. London: Academic Press, pp. 21–74.

Patterson, C. (1994). Null or minimal models. In *Models in Phylogeny Reconstruction*, ed. R. Scotland, D.J. Siebert and D.M. Williams. Oxford: Oxford University Press, pp. 173–192.

Patterson, C. (2011). Adventures in the fish trade. In *Morphological and Molecular Approaches to the Phylogeny of Fishes: Integration or Conflict?*, ed. M.R. De Carvalho and M.T. Craig. *Zootaxa*, 2946, 116–134.

Platnick, N.I. (1979). Philosophy and the transformation of cladistics. *Systematic Zoology*, 28, 537–546.

Remane, A. (1952). *Die Grundlagen des Natürlichen Systems, der Vergleichenden Anatomie und der Phylogenetik*. Leipzig: Geest and Portig.

Richter, S. (2005). Homologies in phylogenetic analyses: concept and tests. *Theory in Biosciences*, 124, 105–120.

Rieppel, O. (1988). *Fundamentals of Comparative Biology*. Basel: Birkhäuser.

Rieppel, O. (2006). Willi Hennig on transformation series: metaphysics and epistemology. *Taxon*, 55, 377–385.

Rieppel, O. (2007). The metaphysics of Hennig's phylogenetic systematics: substance, events and laws of nature. *Systematics and Biodiversity*, 5, 345–360.

Rieppel, O. and Kearney, M. (2002). Similarity. *Biological Journal of the Linnean Society*, 75, 59–82.

Rosen, D.E., Nelson, G. and Patterson, C. (1979). Foreword. In *Phylogenetic Systematics*, ed. W. Hennig. Urbana: University of Illinois Press, pp. vii–xiii [1999 reissue].

Wägele, J.-W. (2005). *Foundations of Phylogenetic Systematics*. Munich: Verlag Dr. Friedrich Pfeil.

Wiley, E.O. (1981). *Phylogenetics. The Theory and Practice of Phylogenetic Systematics*. New York: Wiley-Interscience.

Wiley, E.O. and Johnson, G.D. (2010). A teleost classification based on monophyletic groups. In *Origin and Phylogenetic Interrelationships of Teleosts*, ed. J.S. Nelson, H.-P. Schultze and M.V.H. Wilson. Munich: Verlag Dr. Friedrich Pfeil, pp. 123–182.

Wiley, E.O. and Lieberman, B.S. (2011). *Phylogenetics. Theory and Practice of Phylogenetic Systematics*, 2nd edition. Hoboken, NJ: Wiley-Blackwell, John Wiley and Sons.

Wiley, E.O., Craig, M.T., Davis, M.P. Holcroft, N.I., Mayden, R.L. and Smith, W.L. (2011a). Will the real phylogeneticists please stand up? In *Morphological and Molecular Approaches to the Phylogeny of Fishes: Integration or Conflict?*, ed. M.R. De Carvalho and M.T. Craig. *Zootaxa*, 2946, 7–16.

Wiley, E.O., Craig, M.T., Davis, M.P. Holcroft, N.I., Mayden, R.L. and Smith, W.L. (2011b). A response to Mooi, Williams and Gill. In *Morphological and Molecular Approaches to the Phylogeny of Fishes: Integration or Conflict?*, ed. M.R. De Carvalho and M.T. Craig. *Zootaxa*, 2946, 36–40.

Williams, D. and Ebach, M.C. (2006). The data matrix. *Geodiversitas*, 28, 409–420.

12

Dispersalism and neodispersalism

Malte C. Ebach and David M. Williams

12.1 Introduction

Willi Hennig's *Phylogenetic Systematics* (Hennig 1950, 1966) and the molecular clock divergence dating have had a greater impact on dispersalism than the combined theories of continental drift, land bridges and, later, plate tectonics. The result, *neodispersalism*, is dependent on Hennig's *progression rule* and the determination of molecular divergence dates, which have made it seemingly immune to conflicting geological, oceanographic and geographical evidence. Proponents of neodispersalism have re-written the history of dispersal, introducing the notion that vicariance, rather than dispersal, has been the dominant biogeographical doctrine (herein *doctrinaire vicariance*) since the establishment of plate tectonic theory in the 1970s. More precisely, the claim is that vicariance 'changed many biologists views of the history of life, and the way they approached their science' (de Queiroz 2014: 14). Neodispersalists, like de Queiroz, also claim that a conflict between proponents of vicariance and dispersal has been evident since the early 1980s, with vicariance biogeographers seemingly rejecting the reality of long-distance dispersal.

The received history above is one we will challenge. The debates between present-day dispersalists and vicariance biogeographers mirror those from the 1950s. The issue at stake here is not the conflict between vicariance versus dispersal, but something deeper and more significant – a conflict between those that believe biology is an all-inclusive field, unaffected by Earth processes, and those who understand life and Earth to have evolved together (*sensu* Croizat 1964). This division alone explains the existence of these many debates, from the 1940s to the 1980s through to today. In this chapter, we present a history of these historical

The Future of Phylogenetic Systematics: The Legacy of Willi Hennig, eds. D. Williams, M. Schmitt and Q. Wheeler. Published by Cambridge University Press.

biogeographical debates and show how, contrary to received history, *dispersalism* rather than vicariance remains the dominant biogeographical theory.

Before we delve into this literature, it is worth exploring the nub of the matter, namely dispersal, why biogeographers use it and what justifications they use and have used. We will show that the justifications for dispersal today are the same used in the 'pre-tectonic' times of the 1940s, even though they were effectively debunked in the 1970s. With all this in mind, why would biogeographers continue defending this history? The answer lies in how biogeographers *reason* when evidence is simply missing.

12.2 Dispersal and dispersion

Every weekday, when Londoners buy their morning coffee on their way to work, little thought goes toward how the bean of a tropical plant made it to the cold temperate regions of Europe. The coffee plant, *Coffea arabica*, cannot grow in UK fields and so is usually cultivated in warmer climates, such as Arabia, Central America and Brazil. The bean must have been transported to the UK somehow. We can assume it was transported to the UK, but we may never find out how, unless we locate the supplier, the wholesaler and the cargo manifest, possibly from a shipping company. Without any documented evidence, we will never know how the coffee bean got to the UK. We just know it did *somehow*. This is the first problem with dispersal – it is something that is reasoned in the absence of evidence, in this case, the provenance of the coffee bean. But what of the provenance of other organisms, which occur in far-flung places, like iguanas that are found on either side of the Pacific? Without evidence for provenance we can only speculate through reasoning. But reasoning alone is not evidence. What if we discover that coffee can be grown commercially in the warmer regions of Devon? Suddenly we have two conflicting ways to explain how coffee came to be in our morning drink.

What we do know is that coffee beans, like the people drinking coffee, move around. What we do not have is evidence for one provenance over another. Udvardy (1969) called this *dispersion*, which is similar to Croizat's usage of distribution (see Platnick 1976). It stands to reason that organisms move within their areas of distribution (i.e. dispersion), but the direction of the movement and where they have originated (i.e. dispersal), is open only to debate and is rarely supported by evidence. While there are many definitions of dispersal, most of which will be discussed below, herein we use the following definition throughout the chapter:

> Dispersal is a biogeographical narrative[1] to explain species histories over space and time.

---

[1] We, however, use the Oxford English Dictionary to define the term 'narrative' as 'an account of a series of events [...] given in order and with the establishing of connections between them; a narration, a story an account' (narrative, n. OED Online 2016).

Dispersion, however, is defined throughout 'as a property of individuals, the process by which an organism is able to spread from its place of origin to another locality' (Platnick 1976: 294). In short, both dispersion and dispersal are ways to explain organismal distribution. While dispersion is a descriptive term for the observed movement of individuals within their areas of distribution, dispersal is a narrative associated with the provenance (i.e. origin and geographical history) of any particular taxon. We refer to the latter as *doctrinaire dispersalism*. However, it is dispersal, rather than dispersion, that has been the default explanation for biogeographical distributions since the 1940s. Why then do some biogeographers argue otherwise?

In this chapter we challenge the fallacy of *doctrinaire vicariance*, namely that 'vicariance biogeography' or vicariance explanations, have dominated biogeographical thought and practice during 1970s and 1980s (*sensu* Waters et al. 2013, de Queiroz 2005, 2014).

We claim the *opposite* to be true. Dispersal was, and still is, the default explanatory device for disjunct distributions. In order to demonstrate the inherent negationist claims of neodispersalists, namely that *doctrinaire vicariance* was the default rather than the exception, we outline the history of dispersalism before and after the impact of Hennig's *Phylogenetic Systematics*. We focus on several conferences held between 1949 and 1988. A summary of the discussions held in these conferences will show how drift, land bridges and vicariance were disqualified as plausible narratives early on in favour of dispersal, and how the acceptance of plate tectonics had little if no effect on dispersal hypotheses.

12.3 Dispersal as the default explanation in biogeography (1949–1988)

> Biogeographers a-talking: 'and of ever more colloquia and symposia'. (Gray and Boucot 1976b: ix)

In the introduction to the 1976 *Proceedings of the 37th Annual Biology Colloquium*, palaeontologists Jane Gray and Art Boucot had this to say about plate tectonics:

> Between 1947 and 1976, less than two generations, the world has witnessed a scientific revolution that rightly or wrongly, has had far-reaching consequences for many disciplines as the revolution precipitated by Darwin and Wallace in the mid-1800's. (Gray and Boucot 1976a: v)

Ironically, Gray and Boucot used the same dispersal models on newly reconstructed maps of the Palaeozoic based on palaeomagnetic data, as if plate tectonics had no 'far-reaching consequences' whatsoever. By 1976, pre-tectonic biogeographical dispersalism still dominated. In order to understand why dispersal has remained as the biogeographical paradigm during twentieth-century biogeography, it is important to show how its practitioners portrayed it at the time. We concentrate our discussion on several biogeography symposia and meetings where the big issues of the day were debated.[2] Moreover, we will show the justifications for dispersal in the early to mid-twentieth century were completely unaffected by plate tectonics. We will also show how these justifications for dispersal have changed little, with the exception of Hennig's progression rule, an implementation that was introduced in the 1960s, most notably by Brundin (1966), and used as the default way to read direction in phylogenetic trees.

The role of the South Atlantic Basin in biogeography and evolution, New York City, 1949[3]

> Ernst Mayr [...] points out that the symposium was organized, not to defend or disprove any particular hypothesis, but for the presentation and discussion of current evidence in one restricted but critical field of a much broader subject [...] Dr. Mayr is satisfied... that there is no need to postulate former land connexions between South America and Africa to explain the distribution of mammals and or birds, the available facts being 'diametrically opposed to the possibility of such a connection, (Holmes 1953: 699–670).

On 28 and 29 December 1949, the Society for the Study of Evolution (SSE) held its Fourth Annual Meeting in New York. The topic of the meeting, 'The role of the South Atlantic Basin in biogeography and evolution, with special reference to the Mesozoic', chaired by Ernst Mayr, was ideally organized by the SSE 'because it has in its ranks representatives of all the sciences interested' in the 'outstanding unsolved problem' of 'whether the southern continents have received all their faunistic and floristic elements from the north or have been at times in direct contact with each other; furthermore, in case such a contact existed, whether it was effected by land

---

[2] Not all symposia and meetings are included in this account, as they did not discuss the role of dispersal or because they ignored the conflict between dispersal, land bridges and drift altogether. These meetings include the Eighth Annual Biology Colloquium (Theme: Biogeography), 19 April 1947 at Oregon State College, Corvallis, published as 'Biogeography' (Antev 1947); Tethys, an ancestral Mediterranean (The evolution of a biotic region), held the jointly by the Systematics Association and the Palaeontological Association at the University of Leicester, 21–23 September 1966, published as a Systematics Association publication (Adams and Ager 1967); and the meeting of the American Institute of Biological Sciences and the Pacific Division of the American Association for the Advancement of Science (26–27 August 1957) and the Indianapolis meeting of the American Association for the Advancement of Science (28 December 1957). The proceedings of both were published as *Zoogeography* (AAAS 1958); and 'Biogeographie et Evolution en Amerique Tropicale', published by the Laboratoire de Zoologie de l'ecole normale supérieure (Descimon 1976). Other ecological biogeographical symposia and meetings (e.g. on island biogeography) have not been included.

[3] The meeting proceedings were published as an issue of the *Bulletin of the American Museum of Natural History*, entitled *The Problem of Land Connections across the South Atlantic, with Special Reference to the Mesozoic* (Mayr 1952a).

bridges or by continuity, subsequently broken by continental drift. The problem is an old one, but so many new facts have come to light that a fresh attack is overdue' (Mayr 1952a: 85). Although Mayr does not specify what exactly is being attacked, we may guess from Holmes' note above, that it is coming to the problem from a fixist perspective.

The 1949 meeting (and the 1952 publication of the proceedings) is a mix of geological and biogeographical papers that attempt to justify that biotic evolution in the southern continents during the Mesozoic was a result of dispersal, land bridges (or a mix of the two) or, continental drift (Fig 12.1). 'In order to reach a more concrete foundation for the discussion, the very broad subject was restricted in space and time to the possible connections between South America and Africa and to the Mesozoic Era, since for the Tertiary it seems sufficiently certain that no transatlantic connections did exist' (Schindewolf 1954: 157), hence the meeting's 'special reference to the Mesozoic'. Attending the meeting was geologist Ken Caster who upon seeing the final program was dismayed at the 'manifestly stacked symposium', which was 'composed mainly of those paleontologists and biologists as speakers who were outstanding opponents of continental drift, or specialists in groups of organisms, especially vertebrates and higher plants, whose record would be irrelevant or indicate disjunct evolution' (Caster 1981 in Frankel 2012). These 'outstanding opponents' included Ernst Mayr, Philip J. Darlington, George G. Simpson, Alfred Romer and Daniel I. Axelrod. Of these, Mayr's and Simpson's contributions are perhaps the most revealing as each symbolizes the important role dispersal played in the continental drift debate. For Simpson, dispersal acted as an independent biological explanation, one that was free from (and immune to) geological evidence and theory. As such, its role was to counter any geological/geographical explanations for biotic distribution.

Mayr's 'methodological principle' or, the immunity of biological conclusions to geological evidence

Returning to Holmes' review of the 1952 proceedings, we find that, 'Dr. Mayr himself recalls the methodological principle that biological conclusions must be based on biological evidence, and geological conclusions on geological evidence' (Holmes 1953: 669).

What Holmes is referring to is Mayr's zoogeographical conclusions:

> First of all, it is now universally realized that zoogeographical conclusions must be based on biological (including paleontological) evidence, and geological conclusions on geological evidence in order to avoid circular reasoning [...] Subsequently, the independent conclusions of both fields must be compared to see whether or not they are in agreement. If several solutions are equally probable according to the evidence of one field, the one should be favored that is most probable according to the evidence of the other field. (Mayr 1952b: 255)

THE PROBLEM
OF
LAND CONNECTIONS
ACROSS THE SOUTH ATLANTIC,
WITH SPECIAL REFERENCE
TO THE
MESOZOIC

ERNST MAYR (EDITOR), MAURICE EWING, WALTER H.
BUCHER, KENNETH E. CASTER, CARL O. DUNBAR,
MARSHALL KAY, GEORGE GAYLORD SIMPSON,
DANIEL I. AXELROD, THEODOR JUST,
WENDELL H. CAMP, P. J. DARLINGTON, JR.,
ALFRED E. EMERSON, BOBB SCHAEFFER,
DAVID H. DUNKLE, EDWIN H. COLBERT,
AND ALFRED S. ROMER

BULLETIN
OF THE
AMERICAN MUSEUM OF NATURAL HISTORY
VOLUME 99 : ARTICLE 3   NEW YORK : 1952

Fig 12.1 Title page of the publication *The Problem of Land Connections Across the South Atlantic, With Special Reference to the Mesozoic* published in the *Bulletin of the American Museum of Natural History* for the meeting 'The role of the South Atlantic Basin in biogeography and evolution' (Mayr 1952a).

Mayr provides an example:

> The zoogeographer may be able to present weighty evidence in favor of a former (more or less direct) connection between two areas, but he cannot decide between the various possible geological means (land bridge or pre-drift contact) by which such a contact could have been achieved. (Mayr 1952b: 255).

Mayr also extends this to the geologist who, 'must realise that there are many potential pitfalls in the interpretation of the biological evidence' (Mayr 1952b: 255). What Mayr is telling us is not 'to underestimate the powers of dispersal'. But why?

Essentially Mayr sees biology and geology as separate, not open to reciprocal illumination. For example, the distribution of the fossil plant *Glossopteris*, which is found in most southern landmasses, needs to be assessed using biological mechanisms, processes and hypotheses. That is, the distribution of *Glossopteris* cannot be explained by a geological process in as much as it cannot be used as evidence for a geological process (such as continental drift). Mayr's 'methodological principle' (as termed by Holmes) is authoritarian with little bearing on scientific evidence.[4] Restricting the method to 'biological (including paleontological) evidence' is keeping it immune from conflicting geological evidence. It is dispersal, not other forms of evidence, that takes preference within this highly constrained 'methodology'.

By freeing dispersal from the threat of geological evidence in drawing biological conclusions, Mayr creates a professional demarcation. Geologists have no say in biological conclusions (e.g. land bridges, drift) and vice versa. For example,

> The theory of continental drift is a brilliant and very persuasive attempt to explain many puzzling situations in the history of Earth. However, [...] the facts of submarine geology cannot be harmonized with drift. There is no geological evidence for large-scale horizontal movements. On the other hand, there is a rapid accumulation of evidence for recent large-scale vertical crustal movements in the Caribbean, in the central Pacific, and in the East Indies. It is hoped that these recent findings will not precipitate a new boom of land-bridge constructions. (Mayr 1952b: 255–256).

The absence of a mechanism for drift severely hampered the pro-drift debate. Most of the convincing evidence was found in fossil distributions and stratigraphy (i.e. palaeontology). Since Mayr disregards any 'biological (including paleontological) evidence' in drawing geological conclusions – drift is dismissed. Accordingly Mayr's 'method' is unable to draw biological conclusions from land bridges, even if they support dispersal. Perhaps this explains Mayr's dismissal of 'a new boom of land-bridge constructions'. Given this, how do biogeographers, following Mayr's 'method', use evidence to draw their hypotheses?

Here we can turn to Simpson's 'all-or-none' proposition:

> Premise: the dispersal of a given group of organisms either (a) requires a land connection, or (b) does not. Obligatory alternatives: (a) If dispersal does not require

---

[4] Historian Henry R. Frankel also notes that Mayr 'gave no indication that he thought a landbridge, let alone continental drift, was needed to explain the affinities between African and South American Triassic reptiles. He cautioned that former hypothetical land connections should not be accepted unless consistent with all evidence, as if nature and capable scientists were in need of restraint. He apparently felt, as the authority, that he had to say something to keep things on an even keel, to prevent things from getting out of his control' (Frankel 2012: 119).

> land, distribution of the group has no bearing on past land connections; (b) if dispersal does require land, disjunctive areas occupied by the group have been connected by continuous land. (Simpson 1952: 163)

Simpson considered the logic behind the 'all-or-none' propositions simple. 'It makes every man his own palaeogeographer. To reconstruct a land connection one needs only a map, a ruler' and distributional information. The result leads to 'conclusions that are extremely complicated and contradictory, to the point of utter chaos' (Simpson 1952: 163). Simpson like Mayr makes an authoritative claim:

> There is no group of organisms that cannot be dispersed across water. Less striking but equally true is the factual contradiction of the other branch of the dichotomy: there is no group of organisms the dispersal of which is not influenced in some way and to some degree by the presence of intervening water [...] The predictive or inferential situation is not one of absolute alternatives but one of degree. It is a matter of probability. (Simpson 1952: 163)

The claim that dispersal can happen *no matter how improbable* may seem immune to evidence to the contrary. Simpson is making an authoritative claim, one that is political and not scientific – dispersal occurs because 'I say so', regardless how improbable and regardless of any contradictory geological evidence.

Rules of dispersal: Simpson's 'shades of gray': It would be unfair to say that *doctrinaire dispersalism* lacks a scientific methodology. Apart from Holmes' 'methodological principle' of Mayr, dispersalism contains several assumptions. For instance, there needs to be a point of origin, a physiological adaptation (e.g. wings on seeds, the ability to survive on a raft) and a dispersal route (e.g. across an ocean or mountain range). Evidence is also required in the form of a disjunct distribution. The evidence together with these assumptions form the dispersal hypothesis, but Simpson quite rightly warns us:

> Biogeographic inferences will lead to false and contradictory conclusions unless some account is taken of probability in each case. The picture cannot be painted in black and white when nature has neither, but only *infinite gradations of darker and lighter gray.* (Simpson 1952: 164, our emphasis)

Simpson's 'shades of gray', namely a probability between $p = 0.001$ and $p = 0.999$, means that a high probability dispersal is most likely made by organisms over favourable environments, whereas low probability dispersals are made by, say, large land mammals across vast ocean distances. It is important to remember that Simpson includes all known distributions in his gradations of 'darker and lighter gray'. That is to say, the more improbable the dispersal – such as an elephant dispersing across 20 000 km of ocean – then the fewer number of long-distance dispersals. Perhaps this is why we find very few elephants in Australia. If, on the other hand, there are a greater number of related organisms on either side of a barrier, then the probability of dispersal, in Simpsons view, is more likely. Ironically, the evidence used by some biogeographers, like Alexander du Toit (1878–1948), for a land connection between two continents is used by Simpson to indicate that there was a higher likelihood of dispersal:

> It suggests that dispersal crossed a major barrier rather than followed a usual migration route. In the case of land organisms, it suggests overseas dispersal as against existence of a land bridge. All this emphasises the need to evaluate the total situation. Taken by itself, presence of a group of land organisms on two continents inevitably suggests a land connection, and such cases have been cited with this conclusion in innumerable biogeographic studies. Within the total picture, the particular case may join with others as an item of the evidence for a land bridge or it may actually become evidence against that inference. (Simpson 1952: 175–176).

The mind boggles.

The 1949 meeting was about arbitrating biogeography. By asserting that only biological evidence can be used for biological conclusions, all disjunct distributions are dismissed as evidence for continental drift, and geological hypotheses, such as land bridges. Rather dispersal, taken as a given probability (no matter how *improbable*), could logically be used to explain *any* distribution. By the time Willi Hennig's work appeared in English, dispersal had become indoctrinated in the minds of biogeographers worldwide. Perhaps this is why Mayr's 'Conclusion' chapter to the meeting proceedings (1952c) reads like the attack on drift is done-and-dusted, as if the storm of geological theory had finally abated once and for all.[5]

Pacific Basin Biogeography, Honolulu, Hawai'i, 1961[6]

> [A] naturalist getting hold of Gressitt's Symposium [...] is struck by the striking dualism of doctrine permeating its pages. At one end of the field stand contributors for whom *dispersal is by chance*, and, at the other end, authors for whom *dispersal is by land bridges*. (Croizat 1994: 158, original emphasis)

Mayr's attitude prevails through the Pacific Basin Biogeography Symposium in which dispersal was a major theme. Land bridges, connecting islands to continents, were by this time all but rejected by most Pacific biogeographers as 'recent progress in knowledge of the historical geology of the mid-Pacific tends to argue against continuous land connections to presently existing islands' (Gressitt 1963: 2). But then again continental drifters never claimed land connects between Pacific islands, particularly places like Hawai'i. Rather, the debate between land connections was concentrated in the southern hemisphere, particular with reference to the Gondwanan

---

[5] Gareth Nelson (pers. comm. 2014) recalls Bobb Schaeffer mentioning that Mayr had considered the 1949 meeting to have finally 'killed off' the continental drift theory (see also Frankel 2012).

[6] The Tenth Pacific Science Congress was held at the Bishop Museum, Honolulu, Hawai'i and organised by J. Linsley Gressitt. The Proceedings were published as *Pacific Basin Biogeography* in 1963 (Gressitt 1963).

continents: 'In the Antarctic subsymposium some diverse view are presented on the possible influence of the Antarctic on spread of life, on southern continent relationships, and particularly on biotic affinities between New Zealand and southern South America' (Gressitt 1963: 3).

The 'Antarctic Relationships' subsymposium included some influential people: Charles Fleming, Lucy Cranwell (pre-vicariant), Lars Brundin and J. Linsley Gressitt. It also included a geologist, Raymond J. Adie, who, having just organised the Matthew Pontaine Maury Memorial Symposium on Antarctic Research (Adie 1962), had several new perspectives on Antarctic connections to South America (via the Scotia Arc) and a possible, although tenuous, connection to New Zealand. Adie (1963) however remained hesitant:

> From the brief geological evidence already given it seems that the problems concerning the distribution of faunas and floras of the southern Hemisphere fall into two clear-cut time categories: pre-Cretaceous and post-Cretaceous, each of which should perhaps be attacked by different methods. For the purposes of their investigation of pre-Cretaceous faunal and floral distributions, it seems more apt for biologists to use a more closely knit land mass as a reconstruction of the southern continents, whereas the present-day land distribution is essential for the post-Cretaceous faunal and floral distribution patterns. (Adie 1963: 460–461).

The Cretaceous marks a distinct boundary in the palynological record of New Zealand's palaeoflora, namely the fossil records of many, if not all, extant taxa do not extend into the pre-Cretaceous. With post-Cretaceous isolation (and the rejection of land bridges), most if not all taxa seemingly arrived in New Zealand via dispersal, even if drift had occurred. Fleming, however, takes this one step further:

> What kept out the land dinosaurs,[7] the early mammals, and the snakes from New Zealand? This question, rather than any geological or geophysical difficulties, inclines the writer to give wavering support to the view that the dispersal of Paleoaustral [pre-Cretaceous] organisms, like that of the Neoaustral element [post-Cretaceous], was across sea. (Fleming 1963: 382)[8]

[7] The first New Zealand dinosaurs were found in the 1980s (Molnar 1981; see Agnolin et al. 2010) and recently, the first fossil mammals (Worthy et al. 2006). It appears that dispersal seems a more likely form of distribution when evidence is missing.

[8] In his review of the Symposium, Philip J. Darlington Jr. compares the biogeography of New Zealand with that of the West Indies: 'A long generation ago most biogeographers interested in the West Indies thought that some plants and animals on the islands must have dispersed across land. Now, most of those working on West Indian biogeography think that somehow the entire biota has been derived across water. I think that the debate will probably go this way in the Pacific, and in the Antarctic too, in the end. But I cannot be sure …' (Darlington 1964: 708). Darlington was correct. The debate certainly did go that way regardless of the significant discoveries and theories of the South West Pacific made by geoscientists thanks to Mayr's 'methodological principle' (see discussion on the drowning of New Zealand below).

Remarkably, by 1963, tectonic events had no impact on how southern hemisphere biogeographers viewed their field, having changed little since Mayr (1952a). Perhaps this is what led McVean to remark, in his review of *Pacific Basin Biogeography*, that:

> If there is one outstanding lesson to be learned from such a collection of papers it is this – that the problems of historical biogeography must be regarded as a whole and all relevant evidence taken into account. Hypotheses founded on evidence from one branch of biology alone, or even on all biological evidence, are of limited value however neatly they account for the biological facts if they are at variance with geophysical findings. (McVean 1967: 236–237)

The same attitude prevailed a decade later at the next symposium on historical biogeography.

Biogéographie et liaisons intercontinentales au cours du Mésozoique, Monaco 1972[9]

> Indeed, the stated purpose of the symposium was to answer the question 'Are enough geological and geographical facts available to make it possible to decide whether or not the concept of continental drift can be used by the student of animal and plant distribution as a working hypothesis with reasonable confidence?' (p. 4). Not much space is devoted to that question. But most authors embrace continental drift and try to relate biogeographical information to it, and in that sense answer the question affirmatively. (Nelson 1976: 496)

As the title suggests, the symposium concerned itself mostly with the Mesozoic break up of Pangea and the distribution of organisms during the late Cretaceous and early Palaeogene. In contrast to the Mesozoic symposium held in 1949, 'continental drift is now an accepted fact' (Cracraft 1975a: 31). But like the 1949 symposium, all biogeographical contributions follow a similar line:

> [T]he construction of hypotheses about the location of ancestral species. Such a procedure would necessarily follow a phyletic analysis and would then permit statements to be made about centers of origin and pathways of dispersal. (Cracraft 1975a: 36).

For others plate tectonics has changed little:

> Providing the concepts of continental rifting and actual drift are not confused, new data from sea-floor spreading and application of tectonic plate theory to the area do not very appreciably alter the previously published picture of the New Zealand Mesozoic largely developed by Fleming. (Gaskin 1975: 94)

Dispersal was in 1949 as it is in 1975, the primary distribution mechanism. Vicariance, or relict taxa, were not important in explaining present-day distributions

---

[9] The symposium was held at the XVIIe Congrès International de zoologie in Monaco on 25–30 September 1972 and published three years later (Anonymous 1975).

of taxa. Like fixed continents, moving plates were just another geological feature that taxa dispersed across.

Biogeography: the 21st Systematics Symposium, Missouri Botanical Garden, 1974[10]

> I dare-say that it is not the 'opportunities of migration,' but rather the evidence of large-scale vicariance, that makes this symposium volume one of the more interesting that have so far appeared'. (Nelson 1976: 502)

What may have irked Nelson (1976) in his review of the symposium was how little impact plate tectonics and vicariance had on dispersalists. Nelson had cited American entomologist George F. Edmunds' (1920–2006) concern:

> The number of biogeographers who confidently drew dispersal routes on fixed continent maps ten or more years ago and now just as confidently draw dispersals of the same organisms on continental drift maps must cause us to seriously question the procedures of biogeographers. (Edmunds' 1975: 251)

Indeed. Clearly there was a mechanism that had changed the dynamic of ocean basins, islands and continents. Strangely those whom one would assume to adopt drift wholeheartedly were loath to acknowledge its role:

> Few concepts have enjoyed more of a band-wagon among scientists with biogeographic interests than continental drift, which is now being mindlessly touted as the 'philosophers stone', the solution to all problems of distribution past and present. (Gray and Boucot 1975a)

Why were even palaeontologists unable to accept that drift played an enormous role in explaining fossil distributions?

The quote above is from the 1976 *Proceedings of the 37th Annual Biology Colloquium*, which as its title was 'Historical biogeography: plate tectonics and the changing environment'. This large volume is testament to Edmunds criticism above: 'dispersals of the same organisms on continental drift maps'. Palaeontologists, who were possibly the most ardent of the fixists, reluctantly accepted the concept of a dynamic Earth, but like many of their neontologist counterparts, were not giving up dispersalist hypotheses. The reason is that dispersalist hypotheses are great ways to explain distribution when evidence is absent. Even the new plate tectonic theory presented more questions than answers. What exactly were the mechanisms responsible for uplift, orogeny and basins? By removing an old theory, it also removed the mechanisms usually used to explain many geological and geographical features. In fact, only until the 2000s have geophysicists been able to determine that tectonics can explain extraordinarily fast processes like mountain building in

---

[10] The symposium was held on 18–19 October 1974 and published in the *Annals of the Missouri Botanical Garden* (62: 225–385, 1975).

areas that are nowhere near plate margins (e.g. Sandiford 2007; Quigley et al. 2009). In the mid-1970s much less was understood. Dispersal still had its original role of trying to explain with recourse to evidence.

The 1974 Missouri Botanical Garden conference, however, did split with the long-held belief that biological evidence be used explicitly for biological theories. The seemingly dogmatic insistence of Mayr and Simpson to isolate biological evidence from geological theories had less credence in 1975 than it did in 1949. Plate tectonics did, then, have an impact after all, but only a gradual one.

Joel Cracraft's contribution, however, was quite different:

> I believe most biogeographers would subscribe to the belief that the biotic and geologic worlds have evolved together, that major distributional patterns of both plants and animals should be similar to each other and relate to major historical changes in geography and climate in a parallel manner. (Cracraft 1975b: 227).

Would they? Clearly the Croizatian dictum of life and Earth evolving together was not altogether the main drive of most biogeographers at the 1974 symposium. There was a band-wagon of sorts, which Peter Raven himself, had joined (Raven and Axelrod 1974). The result is recent history re-written as evidenced in Raven's summary:

> Earlier, when the geological evidence was less clear, analyses of the ranges of plants and animals contributed useful hypotheses about past geological events. This was, of course, especially true when these connections represented the sort of vacariant [sic, vicariant] relationships discussed so ably by Cracraft (this symposium). (Raven 1975: 383)

Yet Raven clearly saw geological data as providing useful information about dispersal, and not 'the sort of vicariant relationships discussed so ably by Cracraft':

> There have been far greater opportunities for migration in the Southern Hemisphere than most biogeographers accepted as recently as a decade ago [...] Now, however, geology has arrived at a degree of sophistication where biologists must pay more attention to that evidence, to the age of their groups, and begin to integrate them into more realistic patterns. (Raven 1975: 380, 383–384)

For dispersalists, geological evidence was there to gain greater insight into dispersal, not vicariance. Ironically, it was that 'geological evidence' that was discredited by biogeographers because of the seeming evidence-free plausibility of dispersal. Compare Raven's statement with that made by Thorne from his 1963 *Pacific Basin Biogeography* contribution:

> Perhaps the most important, although a much neglected, working principle of biogeography is that biogeographic inferences and conclusions should be based on the known geographic distribution of all biotic groups. (Thorne 1963: 311)[11]

---

[11] By the time of the 1975 Systematics Symposium, Thorne has almost reverted back to a Buffonian approach to geography in sorting out problematic, misnamed or mislabeled taxa: 'a careful study

Geological data is therefore secondary, providing supporting information about the ages of dispersal events. However, the 'geographic distribution of all biotic groups', as Thorne describes it above, is not evidence for 'major geological vicariance events, primarily continental rifting and shifting' (Thorne 1982: 823).

Cracraft's claim that 'most biogeographers would subscribe to the belief that the biotic and geologic worlds have evolved together' became more tenable in 1979 during a Vicariance Symposium held in New York. Rather than having a 'stacked symposium' in favour of dispersal, this time it was stacked in favour of vicariance.

Vicariance Biogeography Symposium, New York City, 1979[12]

> Less credible and realistic are the vicarists, like Platnick, Springer, Schuh, Mankiewicz, Patterson, Parenti, Croizat, and Nelson, who disregard dispersal and seek to explain nearly all disjunctions as the result of major geological vicariance events, primarily continental rifting and shifting. (Thorne 1982: 823).
>
> Further, vicariance biogeographers do not deny the possibility of dispersal over barriers, but they consider it to be a random event. (Ferris 1980: 72)

The Symposium of the Systematics Discussion Group of American Museum of Natural History held at the AMNH in New York, between 2 and 4 May 1979, was a completely different affair to those discussed above. The emerging fields of palaeomagnetics and geophysics had together created plate tectonics, the process that supports the notion of drifting continents. Comparative biology, too, had a new philosophy – cladistics – which challenged the hegemony of fossil evidence and the importance of the modern synthesis, both championed by Simpson and Mayr, for determining the phylogenetic relationships among plant and animals. The 1949 meeting and the 1979 symposium did have some similarities.[13] Each of the published chapters from the 1979 proceedings had one or more responses (Nelson and Rosen 1981; as did the oral presentations, see Ferris 1980). Most importantly, however, was that dispersal, in theory and practice, remained the same despite the passage of 30 years.

of plant geography can be most helpful in placing dubious taxa. Doubtful genera "out of place" geographically should be restudied with geography in mind' (Thorne 1975: 364).

[12] The symposium proceedings were published as the edited book *Vicariance Biogeography: A Critique* (Nelson and Rosen 1981).

[13] Both the 1979 Vicariance Symposium and 1949 SSE meeting were held in New York City and hosted by members of the American Museum of Natural History (AMNH): Norman D. Newell organised the 1949 meeting and Gary Nelson and Donn Rosen the 1979 symposium. However, the irony fades as the 1949 conference was held in the 'main lecture room of Brander Mathews Hall at Columbia [University]' (Anon 1949a: 67; Anon 1949b: 614, 617), whereas the 1979 symposium was held at the AMNH.

The 1979 symposium attended by a mix of some modern synthesists[14] and several younger cladists, was to be an odd affair. In the introduction to the 1981 edited symposium volume, Donn Rosen claimed that:

> 'For many systematists opposition to the narratives and scenarios of Darwinian evolutionism was mandated by the ideas of phylogenetic systematics espoused by Willi Hennig' (Rosen 1981: 2).

The cladists, many staunch followers of Hennig's phylogenetic systematics, were at loggerheads with the traditional Darwinian evolutionists (for the most part these were the modern synthesists). It happened that Hennig had been awarded the AMNHs gold Medal in 1975, at the same time Croizat's panbiogeography started to emerge as a dynamic field of enquiry within the English-speaking biogeography community (Croizat et al. 1974).

> That year, however, the battle lines were instantly reformed when Croizat's panbiogeography (and its consequences for Darwinian dispersalism) was brought into the debate. Following so closely on the heels of the stern challenge by cladists, the advocacy of panbiogeography and its explicit criticism of traditional dispersalism proved to be intolerable for the museum's ardent Darwinians. (Rosen 1981: 3).

This is where everything seems odd. True, the cladists and Darwinians did disagree about the role of fossils and ancestors in systematics, but Hennig, however, was an 'ardent dispersalist'. What was going on?

Adopters of any philosophy do not necessarily adopt all the methods associated with it. Hennig's phylogenetic systematics was adopted but only in part. The progression rule, for instance, which states that 'a cladogram of species-relationships is sometimes sufficient in itself to indicate the center-of-origin and direction of dispersal' (Nelson and Platnick 1981: 514) merely replaced Darlington's 'rule of thumb', namely 'the distribution centre of a race is determined by the habitat of its most primitive species (Matthew 1914: 201, Darlington 1957)[15]. The array of ideas subsumed within Croizat's panbiogeography, however, was also cherry-picked. The idea of taxon tracks (patterns of distribution laid out on a map) and that life and Earth evolving together was adopted, whereas Croizat's orthogenetic theory of evolution was ignored by almost all. By 1979 vicariance biogeographers were not Hennigian – but neither were they Croizatian. They were a mixture of both, something that did not sit well with either Croizat or, for that matter, Lars Brundin, Hennig's advocate

---

[14] Rosen wrote in the Introduction: 'For the readers of this volume who may wonder why the views of certain leading Neo-Darwinians are not represented, it should be recorded that P.J. Darlington Jr., E. Mayr and GG. Simpson were invited to participate as principal speakers but declined' (Rosen 1981: 4). In the end, only J.A. Wolfe, a member of that school, was present.

[15] Ironically, William Diller Matthew thought the opposite. He believed that climate controls distribution and that the most 'progressive species of a race', the one which can tolerate a change in climate, would be the one found at the centre of origin, 'the most primitive and unprogressive species will be those remote from the center' (Matthew 1914: 201).

at the meeting: 'The editors of the proceedings of the Symposium in question [...] relegated my paper to the very end of the gathering, when it could no longer be discussed' (Croizat 1982: 291).

What could not be discussed, according to Croizat (1982), was the relevance of Nelson's approach, where he 'incorporated the Panbiogeography – hopefully indeed – with Hennig's Systematic Phylogeny, renaming the mixture "Vicariance Biogeography"' (Croizat 1982: 294–295). For Croizat, the melding of panbiogeography with Hennig's *Phylogenetic Systematics*

> has been dragged in with Hennigism to the very extent of publicly losing its identity under the improper designation of 'Vicariance Biogeography' [...] It subtly works to confuse Panbiogeography with 'Vicariance Biogeography', which has had catastrophic results on biological thinking by fomenting needless doubts, unjustified questions etc. (Croizat 1982: 296, 299)[16]

For Croizat, panbiogeography was a scientific field in its own right, a synthesis grounded in its own philosophy (i.e. life and Earth evolving together), theory (i.e. orthogenetic evolution) and methodology (i.e. tracks, main massings, etc.). Panbiogeography was an alternate modern synthesis and not merely a theory in search of a method. For the modern synthesists, however, Hennig's phylogenetic systematics was a theory in search of a methodology, or at least an implementation. While it did offer three methods, namely the progression rule, the deviation rule and the chorological method, it had few within the cladistic community who adopted it. For example, Nelson, Rosen and Platnick rejected these methods in favour of their own interpretation of Croizat's work. Cladistics effectively broke into two camps: the traditional Hennigian cladists and the newer 'transformed' cladistics that rejected the underlying philosophy of the modern synthesis.

The progression rule and Brundin's phylogenetic biogeography: Hennig never lived to see the 1979 symposium. Rather, it was Swedish entomologist Lars Brundin, the most vocal supporter of Hennig's methods, who attended the 1979 symposium; he did not approve of vicariance biogeography. Rather, Brundin proposed *phylogenetic biogeography*, a method based on 'Hennig's phylogenetic systematics and its biogeographic implications' (Brundin 1981a: 95). The main method in phylogenetic

---

[16] Regardless, many present-day biogeographers still confuse panbiogeography and vicariance biogeography. For example, Renner, in a review of the cladistic biogeography textbook *Comparative Biogeography* (Parenti and Ebach 2009), suggested that 'if you are looking for a glimpse into the curious world of panbiogeography, this is the best volume to get to obtain that information" (Renner 2010). Of the 312 pages in *Comparative Biogeography*, only 3 pages are dedicated to panbiogeography. In our view, the best and most recent volume to obtain information on panbiogeography is the 240 page *Panbiogeography: tracking the history of life* (Craw et al. 1999). Croizat offered a much fuller critique of Hennig's phylogenetic systematics in Croizat (1978).

biogeography was Hennig's deviation rule, the chorological method and the progression rule, the latter having been misinterpreted by several authors.[17]

The main thrust of the progression rule is that cladograms can be used to indicate the direction of dispersal, if and only if, evidence for dispersal exists. The progression rule, then, is not a way to discover dispersal:

> the progression rule, according to my arguments, can be applied in those cases where there are indications that dispersal has occurred (sympatry, multiple-step morphoclines that can be followed over the map etc.). A more adequate name for my biogeography would be 'Phylogenetic biogeography according to the vicariance/ dispersal model'. (Brundin 1981b: 153)

Brundin's claim is interesting because it requires dispersal to be known, meaning a source other than phylogenetic trees is required (e.g. sympatry, multiple-step morphoclines). Odd, then, that the progression rule is cited as the main cause of dispersal hypotheses by its detractors (e.g. Platnick 1981). The reason is that 'indications that dispersal has occurred' are directly derived from the phylogenetic trees themselves. Nelson, in an attempt to formalise both Hennig's and Brundin's procedure, noted two types of erroneous assumptions, which he termed Type-I and Type-2 errors: 'Resolving dispersal where none occurred might be termed a "Type-I error"; not resolving dispersal when dispersal did occur might be termed a "Type-II error." Type-I errors seem to be by far the more common in biogeography (Nelson 1974: 555, footnote 2).

In other words, Hennig's and Brundin's procedures, namely the progression rule, the deviation rule and the chorological method, together and individually, result in Type-I errors – and Hennig's progression rule defers back to Darlington's rule, that the oldest taxon (or in Hennig's case the most basal taxon) represents the centre of origin. Darlington's method, like the progression rule 'depends on the same assumption as the oldest fossil method – that the fossil record is complete' (Patterson 1981: 488).

For Brundin, however, vicariance biogeographers, like Platnick and Nelson, represented 'an extreme type of vicariance biogeography in which there is no room for the study of dispersal [and] the application of the progression rule [is] rejected' (Brundin 1981b: 152). This stands in stark contrast to Platnick's claim that Brundin's 'pioneering (1966) study of trans-Antarctic relationships' without which 'none of us would today be attending a symposium devoted to the synthesis of cladistics, biogeography and historical geology' (Platnick 1981: 144) – clearly vicariance biogeographers have chosen to adopt only parts, rather than all of the trappings, of phylogenetic biogeography.

---

[17] The progression rule is repeatedly misinterpreted. For example, simply reading directionality based on the age of islands or age of fossils without evidence for dispersal is a common mistake (see Losos and Rickleff 2009: 109).

Vicariance biogeographers had created a whole new field, one that uses the idea of vicariance from panbiogeography and phylogenetic relationships from *Phylogenetic Systematics*. However, vicariance biogeography rejected the a priori assumptions inherent in each. Croizat's panbiogeography assumed orthogenesis as a mechanism of species origin, while Brundin's phylogenetic biogeography required Darlington's rule, neither of which was acceptable to the vicariance biogeographers. For Brundin and Croizat, this seemed odd – why adopt the methods but reject the overall theory? Perhaps Croizat was right, vicariance biogeography was a theory in search of a method. In any case, both Brundin, like Croizat, saw vicariance biogeographers creating a hybrid field, devoid of certain a priori assumptions, which in their view would result in 'catastrophic results on biological thinking by fomenting needless doubts, unjustified questions'. As we shall see below, quite the opposite was true.

Patterson, dispersal and branching diagrams: The primary aim of Patterson's contribution to the 1979 symposium was to establish the role fossils played in historical biogeography, given that, so it seemed, dispersal and vicariance biogeography required different sorts of evidence from fossil remains, crucial with respect to dispersal, negligible with respect to vicariance. Prior to his discussion of fossils, Patterson set out to investigate if there was any 'Analogy between historical biogeography and systematics', his first chapter sub-heading (Patterson 1981: 447). Patterson's previous investigations on the basis of taxon systematics underpinned his understanding of biogeography. He reduced the workings of systematics to six words: 'interpret the distribution of homologies parsimoniously' (Patterson 1981: 448). That is, 'insisting on one property – *homology*, and one criterion – *parsimony*, analytical systematics seems to me to come as close to a science as is likely, and these two factors distinguish it from history' (Patterson 1981: 449[18]). Seeking the analogy in biogeography, Patterson wrote, 'So I believe that we shall not develop an analytical historical biogeography until we can codify biogeographical method in some simple phrase like 'interpret the distribution of homologies parsimoniously" (Patterson 1981: 449). Patterson was, then, probably among the first to attempt an exploration of homology in the context of biogeography ('can "homology" be replaced, for example, by "dispersal routes," "generalized tracks," "monophyletic groups," or "areas of endemism"?', Patterson 1981: 449). Patterson reviews several methods from this perspective, with the 'Analogy between historical biogeography and systematics' being the search for biogeographical homology. Patterson ended up rejecting most methods of biogeography primarily because no such notion of homology was evident to make it applicable to a general approach.

---

[18] It is worth noting here that amongst those who still retain an interest in analytical systematics, homology and parsimony are as contentious a subject now as it was then (Williams and Ebach 2012).

Not surprisingly, he found the method outlined just a few years previous by Platnick and Nelson (1988) to be the most attractive. Here their focus changed not so much to what process – vicariance or dispersal – was prevalent, but to how one might distinguish between the two. That could be accomplished by acquiring sets of congruent taxon cladograms: dispersal would apply only to individual taxa, derived from cladograms or otherwise, whereas vicariance is a general explanation applicable to 'a cladogram of areas, and tested by cladograms of taxa' (Patterson 1981: 465). Here Patterson developed, or applied, the method of general cladograms, as outlined by Platnick and Nelson (1978), to distinguishing or teasing out what kind of evidence applied to what kind of explanation. The 'Analogy between historical biogeography and systematics' meant 'Congruent cladograms of individual taxa occupying those areas correspond to synapomorphies, congruent character distributions in systematics' (Patterson 1981: 465). Here, then, was a way to interpret biogeographical homologies in an analogous fashion to those of characters (or character state trees) in systematics.

Of course, as in systematics, the development of analytical procedures in biogeography opened the floodgates for countless numerical methodologies, many, if not most, blind or near myopic, to the notion of spatial homology (as in much of the post-cladistic systematic method development). But for the theme of this chapter, Patterson's closing remarks retain their initial resonance. While Patterson acknowledged that a worthy aim may be to correlate the general (area) cladogram with historical geology, he offers the view that, following Croizat, 'the biological evidence stands, whether or not it matches fashionable geology. The effort of paleobiogeographers is better spent in analysing the relationships of taxa and biotas [than] in reconciling distribution with today's historical geology' (Patterson 1981: 489). The proposal here, regardless of what 'fashionable geology' may offer, is to let biology speak for itself, and let geology speak for itself.

Symposium moderator Virginia Ferris asked in her summary of the event 'So, what was accomplished by the symposium? Did it prove that biogeographers are divided among those who seek solutions to real-world problems and those who regard biogeography as a science in search of a paradigm [...]?' (Ferris 1980: 75). The same set of questions could be asked today.

Alternative Hypotheses in Biogeography, Seattle, Washington, 1980[19]

> These nine contributions are a diverse lot. I classified them in accordance with Endler's scheme: ecological determinism 6; dispersal 2; vicariance 1. As a first step toward a new synthesis, this symposium leaves the reader tottering on the one

---

[19] "Symposium on Alternative Hypotheses in Biogeography presented at the Annual Meeting of the American Society of Zoologists and the Society of Systematic Zoology, 27–30 December 1980, at Seattle, Washington" (Endler 1982: 349, footnote 1). Contributions to the Symposium were published in *American Zoologist* (1982, 22, 349–471). In the same year the symposium Arctic

> foot of ecological determinism. Endler's perception of three schools that largely ignore one another reduces to one school that ignores, as best it can, the other two. Searching for elements of synthesis I could find none except Endler's reductionism to natural selection as explanation of everything. (Nelson 1982: 342)

Patterson's analogy between systematics and biogeography (more accurately, Parenti's comparison between cladograms and area character state-trees) did not bode well for those working on single taxa. For them, Hennig's progression rule was far more useful as it requires only a single cladogram and some reasoning in regards to direction of dispersal. Edmunds also attended the 1980 symposium and stated that:

> Most organisms analysed exhibit marked asymmetry of successful dispersal from north to south versus south to north. There is not a consistent directional pattern. It is probable that comparative life history patterns, functional roles in the community and general competitive positions need analysis for each group. (Edmunds 1982: 374).

Three things merit comment:

1. Edmunds' earlier 1975 criticism that the 'number of biogeographers who confidently drew dispersal routes on fixed continent maps 10 or more years ago and now just as confidently draw dispersals of the same organisms on continental drift maps must cause us to seriously question the procedures of biogeographers' (Edmunds 1975: 251). But that appears to be exactly what Edmunds does when referring to the bi-directionality of the organisms he studied (mayflies).
2. The lack of a pattern, returning us to Parenti's claim that each cladogram is merely a character-state tree for biogeography that tells us 'something about the genealogy of the areas' (Parenti 1981: 493). A single cladogram may be a pattern for a set of character state trees, but in biogeography it merely tells us about the relationships of a single group of taxa, not a *pattern of relationship* for many taxa. Given that Edmunds' cladogram is not a pattern of relationship, and that dispersal does not form aggregate patterns within multiple cladograms, his attempt to

---

Refugia and the Evolution of Arctic Biota was held as part of the Second International Congress of Systematic and Evolutionary Biology, University of British Columbia, Vancouver, Canada, 17–24 July 1980. Most presentations from each symposium were published as *Evolution Today* (Scudder and Reveal 1981), while several of the conference proceedings were published elsewhere, notably (Crovello 1981). In no case was there any significant discussion about vicariance. However a review of conference proceedings by Patton (1982) mentions a paper by Hoffmann (1981) on mammals, which contains a 'view with relevance to the ongoing discussion of vicariant versus dispersalist schools of biogeography' (Patton 1982: 287). We believe Patterson (1983) erroneously refers to this as Biogeography of Regional Biotas and Communities (Patterson 1983: 1). We are unclear as to why he did so as no such symposium was advertised (see Reveal and Scudder 1979, Anon 1978: 472).

use Hennig's progression rule is merely drawing 'dispersals of the same organisms on continental drift maps' and cladograms.

3. Non-biological evidence, such as geology, tectonics and geography are not considered.

American ornithologist Joel Cracraft, also a contributor to the 1980 symposium, had presented a vicariance biogeographical analysis using cladistics (i.e. cladistic biogeography). This analysis is in stark contrast to the remaining nine papers in that it offered an analytical method. Cracraft's analysis, however, was not free of dispersal: 'a vicariance pattern between the east and south, but other evidence suggests these elements may have attained this disjunction by northward dispersal' (Cracraft 1982: 419). The 'evidence' comes directly from the progression rule, based on the hierarchy of taxa on a cladogram. Even for practitioners who heartily embraced vicariance, dispersal is still seen as a valid inference based on the topology of cladograms. Moreover, geological evidence is perhaps unwittingly dismissed through the emphasis on ecology and climate:

> Vicariance biogeography suffers from not having all of its general predictions tested. Its major prediction, that concordant cladograms should result from concordant vicariance sequences, is rejected. Concordant cladograms can only result from shared geographic patterns of selection on the characters, and thus contain no information about their biogeographic history. (Endler 1982: 450).

One wonders whether Endler is also referring to concordant character trees? After all, area cladograms in the 1980s were based on concordant vicariance sequences in morphology (i.e. homologues). Could they be a result of 'shared geographic patterns of selection' based on ecological factors? If so, homology is seemingly rejected in favour of Lamarckian environmental determinism.

By the early 1980s, dispersal had trumped vicariance. Biogeographers have been content in demanding physical evidence for vicariance while, at the same time, accepting dispersal in the absence of evidence. The logic follows as so: lack of evidence for vicariance = evidence for dispersal. Even after the revolutions of plate tectonics and cladistics, biogeographers have not changed how they employ dispersal in interpreting distributions. Rather they have slowly adopted a way to imprint dispersal onto tectonics and cladistics.

Evolution, Time and Space: The Emergence of the Biosphere, London 1981[20]

> Rather than focus on a particular topic the present book attempts to deal with the entire field of biogeography [...] Several symposium papers were published elsewhere (Noonan 1984: 124).

---

[20] The Proceedings of a Symposium on Biogeography organized with the British Museum (Natural History), London 6–10 April 1981, on the occasion of the centenary of the museum" (Sims et al.

> The only common theme of this volume is that each chapter seems quite different from every other, although most do focus on the analysis of real-world data. No one ideology or methodology predominates; no common problems are identified or discussed. The organizers wanted a symposium that maximized differences– under some circumstances, a highly laudable goal– but unless there is a restricted set of themes or problems on which to focus that diversity, the results will be a symposium either of individual interests or of participants talking past one other. Both outcomes are in evidence within this volume. (Cracraft 1984: 351).

> Be warned, however, that a substantial proportion of the symposium proceedings are scattered in other journals. (S.C.M[21]. 1984: 376).

In 1981, the British Museum (Natural History) (now the Natural History Museum) and the Systematics Association jointly organised a biogeography symposium to celebrate the museum's centenary (Fig 12.2). Published 2 years later, the edited volume appeared to have no focus, as suggested by the quotations from Noonan (1984) and Cracraft (1984) above.

Firstly, several of the more empirically based talks were published elsewhere in the *Biological* and *Zoological Journals of the Linnean Society*.[22] Secondly, no central theme or goal could be identified for the meeting as Patterson emphasised in his opening chapter:

> What should be the aims of a symposium on biogeography in 1981? [...] I ask these questions, and discuss possible answers to them, in order to bring this meeting into focus ... (Patterson 1983: 1).

The editors responded thus:

> Despite the implications in Patterson's paper to the contrary (ch. 1), we *did* go into the preparation of the programme of this symposium, the fourth major one involving biogeographic themes in the last 26 months. (Sims et al. 1983: viii, original emphasis)[23]

The 'last 26 months' would include the meeting from 1979, in which vicariance, cladistics, dispersal, plate tectonics and other alternative hypotheses

---

1983: frontispiece), was undertaken jointly with the Systematics Association, London, and published in their *Systematics Association Special Volume* series as *Evolution, Time and Space: The Emergence of the Biosphere* (Sims et al. 1983). The edited volume excluded five talks that were published elsewhere (Sims et al. 1983: xiii).

[21] S.C.M. is most probably the English palaeontologist Simon Conway Morris.

[22] The excluded papers were considerably larger than those that appeared in the symposium volume. The primary reason for their exclusion was the page limit for the final book agreed upon in the contract between the Systematics Association and the editors. Inclusion of these contributions would have far exceeded that limit (Syst. Assoc. Archives). Nevertheless, the empirical depth of the excluded contributions made these papers seem of a different kind.

[23] To be fair, the primary aim of the symposium was to celebrate the Natural History Museum's centenary with a topical subject, biogeography being chosen to fit that bill.

THE SYSTEMATICS ASSOCIATION & BRITISH MUSEUM (NATURAL HISTORY)

On the occasion of
the centenary of the British Museum (Natural History),
a Joint International Symposium on Biogeography

TIME AND SPACE IN THE EMERGENCE OF .THE BIOSPHERE

will be held at

British Museum (Natural History)

6—10 April 1981

The new palaeogeography, the dogmas of vicariance and cladism,
the theory of faunal and floral regions and the dynamic equilibrium theory
are among the front runners of current biogeographical thinking.
The symposium will provide a forum for discussion on what is perhaps the most
exciting yet least well-defined, derived discipline of contemporary biology.
The whole spectrum of biological science, in both pure and applied aspects,
will be represented by international speakers on botany,
palaeobotany, palaeontology and zoology.

Details from: R. W. Sims,
Department of Zoology, British Museum (Natural History),
Cromwell Road, London SW7 5BD.

Fig 12.2 Original flyer for the *Evolution, Time and Space: The Emergence of the Biosphere* symposium held in London 1981, hosted by the British Museum (Natural History) and the Systematics Association, jointly organised to celebrate the Museum's centenary.

were discussed. Apparently by 1983, these topics had not resolved themselves. Biogeography had clearly retained its multidisciplinary nature and continued along its chaotic merry path:

> It would be unjust in the extreme to give the impression that the meeting developed as a mutual admiration or commiseration society. But it was the clearest indication yet of the sincere, often only groping, movements in the direction of mutual aid and understanding in seeking productive parallels and contributions from *all* available methods and data, from whichever groups of organisms. We have progressed beyond infancy in many aspects simply by the recognition that our views and data are still essentially infantile. (Sims et al. 1983: vii, original emphasis)

If vicariance had been the dogma of the day, then surely there would not have been disarray. But clearly there was no consensus, no single methodology or 'dogma':

> In conclusion, what emerges is that despite a hundred years' research biogeography is still in a formative stage. (S.C.M. 1984: 376).

S.C.M. is perhaps, wittingly or unwittingly, paraphrasing Ball (1976):

> I find it convenient to recognise biogeography as passing through three phases: the descriptive or empirical phase, a narrative phase and an analytical phase, the last two comprising historical biogeography. (Ball 1976: 408)

Where then was historical biogeography? By 1981 vicariance biogeography was a theory in search of a method. Dispersalism, in the guise of the progression rule, was a narrative without theory. Was vicariant theory lurking in the minds of biogeographers while narrative dominated the pages of scientific journals? After all, the vicariance biogeographers engaged many biogeographers in debate about *theory*, not implementation. Perhaps this is why is there an incorrect received history that biogeography (for the most part in the 1970s and 1980s) was dominated by *doctrinaire vicariance.*

We believe that by the 1980s theoretical biogeography was mostly dominated by vicariance biogeographers. In contrast, there was also practical biogeography, the biogeography that biogeographers actually did, but was dominated by dispersalist narratives. In reading Ball (1983), one wonders whether dispersalism ever reached the analytical stage, forever stuck in a descriptive or narrative phase. Moreover, biogeographers focusing on a single taxon may never get a chance to use a method or use geological evidence. For example, vicariance or plate tectonics would mean very little to a biogeographer working on a widespread and highly mobile taxon. Why then would such a biogeographer be interested in debating the theoretical aspects of vicariant patterns? Scientists seem to believe that comparative biology offers general laws that all organisms seemingly obey. The notion that comparative biology has general laws is a fallacy. For every rule there will be an exception, and for many taxa that are widespread, vicariance may not be relevant. As Ball points out, biogeography has no unifying principle 'that can be discovered by the correct logical procedures' (Ball 1983: 422). Given Ball's view, it is not surprising to find that biogeographers simply do biogeography without any theory beyond that of simple range expansion or empirical methodology. Like taxonomy, biogeography is a science of description: biogeographers describe their units of classification (i.e. endemic areas/taxon distribution/s) and provide narratives (i.e. dispersal, extinction, etc.) in the same way taxonomists describe taxa and evolution (i.e. character variation). Descriptions, however, are not discoveries (Parenti and Ebach 2013). Not until the Cladistic Revolution have taxonomists been able to discover that mammals, for example, are monophyletic, that is, a natural group. Biogeographers of the 1970s and 1980s were in the same position. Until a method provided a way to discover geographical congruence, biogeography was firmly stuck in Ball's descriptive and narrative phases.

Dispersal and distribution: An International Symposium, Hamburg 1982[24]

> The prelude to the Hamburg symposium is thus extremely interesting, and the book under consideration is opened with excitement: Will the 'dispersionists' strike back? Will the conflicting ideas be discussed, and is a synthesis (like Mayr's attempts within systematics) possible? The reader will be disappointed on this point. The conflicts are almost completely killed by silence. There are neither references to Nelson, Platnick, Rosen, Brundin nor to the preceding symposia. Croizat is mentioned once, as one of several who believed African *Rhipsalis* to be indigenous. And that is all. (Nordal 1985: 14)

A year after the Evolution, Time and Space Symposium, any discussion of vicariance was seemingly stymied. The Dispersal and Distribution symposium, avoided vicariance altogether:

> The more recent confirmation of the theory of plate tectonics and its application to the problems of plant and animal distribution doubtlessly has led to a better understanding of the spatial evolution of lineages, but at the same time a highly theoretical framework has been developed of which the opposing viewpoints between vicariance versus dispersal biogeography is but one example. (Kubitzki 1983: 5)

For Kubitzki, 'plausible biogeographical explanations are more likely to be obtained by common-sense investigations than by formalised approaches or abstract thinking', in short, plate tectonics and vicariance. The symposium sought to 'avoid such conceptual narrowness' and dedicate itself to 'the problems of dispersal and

---

[24] The symposium was held at the University of Hamburg 10–12 June 1982 and published as *Dispersal and Distribution: An International Symposium* edited by Kubitzki (1983).

distribution, especially the interrelationships between the two' (Kubitzki 1983: 5). The symposium could have been held in 1942. No contribution considered plate tectonics or vicariance, but why should it? After all the symposium was about dispersal which for plants at least 'exists in three different modes, viz. specialised, generalised and chance, which of the latter two can be likened to a constant 'background dispersal noise', that greatly modifies the effects of the first' (Berg 1983: 32). But Berg does raise an important point about dispersal, namely that it represents a narrative hypothesis (*sensu* Ball 1976) 'and is one of probability and not certainty. However, and most important, the answer cannot be used as a proof or a disproof of the basic assumption' (Berg 1983: 31). Other mechanisms were considered. Van Zanten (1983), for example, describes how plate tectonic theory 'has become fashionable to explain many distribution patterns on the basis of this theory, perhaps to the extent that its importance in this respect has been overestimated' (van Zanten 1983: 52). Van Zanten does not provide citations, but rather attempts to disassociate vicariance with species level distributions. All contributions claim that dispersal has had a significant effect on distribution, but also acknowledge that very little is known about the physiological and evolutionary mechanisms responsible for dispersal. As Berg suggests, dispersal 'cannot be used as a proof or a disproof of the basic assumption' of distribution. We wonder why not explore other mechanisms for which there is evidence, like vicariance?

In the 'Summary Lecture', at the end of the symposium volume, Ehrendorfer returns to Berg's 'noise'. In a review of the symposium volume, Krahulec notes: 'the shift in the mind of biogeographers is expressed by the last sentence of F. Ehrendorfer's contribution: "There must not be an explanation for everything." I think this is a rather new view in contemporary biogeography' (Krahulec 1985: 216). Krahulec's 'new view' however, seems to stem from critics of long-distance and chance dispersal. For instance, it appears in Nelson's earlier summary of the Vicariance Biogeography: A Critique Symposium: 'Vicariance biogeography is not an attempt to explain everything, or even the greater part, of the geographical distribution of plants and animals, for its focus is allopatric differentiation as manifested by the phenomenon of endemism' (Nelson 1981: 525). Nelson had expressed this view earlier 'that means of dispersal do not explain everything' (Nelson 1978: 290). The same view appears in Craw (1984): 'Biogeography is just telling a 'story': unique narrative explanations are proposed for the historical development of each group studied. But using this approach one can explain everything and anything, and faced with anomalies like organisms with no obvious means of dispersal living on isolated islands, Darwinian biogeographers attribute their presence to 'chance" (Craw 1984: 10). Brian Rosen also shares the same view: 'Taking together all the mechanisms and all the *ad hoc* adjustments mentioned by Briggs [1984], a novice biogeographer would have to conclude that almost anything explains everything (Rosen 1985: 385).

Biogeography of the Tropical Pacific, Honolulu, Hawaii, 1982[25]

> This work is refreshingly free of recent biogeographical polemics. Most of the contributors accept a dispersalist or modified–dispersalist hypothesis. (Schmid 1985: 748)

The Biogeography of the Tropical Pacific symposium, held 20 years after the *Pacific Basin Biogeography Symposium*, is interesting for two reasons. Firstly, it was held in a world in which plate tectonics is fully accepted as a geological mechanism. Secondly, tectonic theory had no effect on how biogeographers formulated their biogeographical hypotheses. Dispersal also remained without theory and relied solely on narrative. Newly established syntheses, such as island biogeography, 'population dynamics, species replacement, evolution and extinction on islands', had helped increase the 'knowledge of the history of the Pacific and its insular biota' (Gressitt 1984: viii[26]). Here we return to the neodispersalist claim that, vicariance 'changed many biologist's views of the history of life, and the way they approached their science' (de Queiroz 2014: 14), which was barely mentioned at all. Holloway (1984) in his study of Lepidoptera in the Melanesian arcs concludes that biogeographical 'patterns in the Lepidoptera could be explained by the geological hypothesis rather than support it' (Holloway 1984: 159). Vicariance was seen as a polemic that started in the 1970s and ignored by many biogeographers. Schmid's claim of a polemic-free biogeography again emphases the position of dispersal as the default explanatory hypothesis among historical biogeographers and vicariance as the interfering newcomer. The polemics are a sad outcome of the vicariance/dispersal debate, so too is dogmatism, particularly that of the dispersalist camp (Mayr 1952a, Boucot 1983).

Symposium on Biogeography and Plate Tectonics in the SW Pacific, Dunedin 1983[27]

> The energy expended in the vicariance–dispersal controversy should now be applied to a search for methods and evidence by which the relative roles that these processes have played can be assessed (Thornton 1983a: 560)

By 1983, it seems, vicariance was accepted as a valid mechanism to explain biotic disjunction. The summary of the symposium, by entomologist Ian Thornton, is strongly in support of developing methods to measure the likelihood of vicariance and dispersal. Thornton's new attitude reflects that of present-day attitudes within

---

[25] Symposium on Biogeography of the Tropical Pacific, held at the Bernice P. Bishop Museum, Honolulu, Hawaii, 23–26 May 1982. The proceedings were published in 1984 as *Biogeography of the Tropical Pacific: Proceedings of a Symposium* and edited by Frank J. Radovsky, Peter H. Raven, and S.H. Sohmer.

[26] The text was written in 1982 the same year Gressitt died (Radovsky 1983).

[27] Published as 'Proceedings of Symposium on Biogeography and Plate Tectonics in the SW Pacific' (Thornton 1983b) in *GeoJournal* 7(6), 479–567. The symposium was part of the 15th Pacific Science Congress. The symposium was dominated by entomological contributions.

biogeography as seen in many of the modelling programmes available, such as DIVA and LeGrange. Thornton also points out that much theoretical debate had ensued over the past decade, leaving biogeography methodologically vacuous. Dispersal, as Thornton correctly states, was 'frequently misunderstood'. The distinction between different types of range extension would for the better part help biogeographers develop the techniques and methods to use both dispersal and vicariance in biogeographical models and hypotheses. The symposium barely reflects Thornton's attitude, with only J.P. Duffels employing a cladistic biogeographical method to find patterns in cicadas (although redefining biotic dispersal as 'vicariant dispersal'). Other contributions however are indistinguishable from earlier descriptive biogeographical studies that employ no method, such as that of Sinzo Masaki, G. Kuschel, George Gibbs, B.J. Donovan and J.D. Holloway.

With the exception of Thornton's and Duffels's contributions, the symposium resembled that of a typical mid-twentieth-century biogeography conference. It was depauperate in methods, full of narrative and typically in favour of dispersal. Thornton's attitude, modern by today's standards, is exceptional in 1983. Moreover, why would he claim that 'no biogeographer offering dispersal explanations has, to my knowledge, denied the possibility of the fragmentation mode of allopatric speciation [sensu vicariance]' (Thornton 1983a: 559)? Most of the symposium contributions do not consider allopatry or vicariance as possible mechanisms. Thornton's attempt to answer 'Vicariance and dispersal: confrontation or compatibility?' is a warning to dispersalists that engaging in methods, rather than relying solely descriptions and narrative, will bring some form of empirical credibility to biogeography.

Vicariance Biogeography: Theory, Methods and Application New Orleans 1987[28]

The Vicariance Biogeography Symposium was held as part of the Society of Systematic Zoology's programme just after the recent passing of Donn Eric Rosen (1929–86) (Nelson et al. 1987), 'Chief among the architects of vicariance biogeographic theory and methodology [and] President of the Society of Systematic Zoology during the years 1976–1977' (Cracraft and Shipp 1988: 219).

By the late 1980s, several cladists had formed a vicariance biogeography program, one that mostly focused on the methods and theory behind geographical congruence (i.e. aggregate patterns in multiple areagrams), rather than on specific distributional histories of species. For example, in the two issues of *Systematic Zoology* devoted to the Symposium, there were a total of 12 contributed papers, of which only one was theoretical (Sober 1988), six were comparisons, tests or critiques of various methods (Cracraft 1988, Page 1988, Wiley 1988, Craw 1988, Kluge 1988, Platnick

[28] Published in *Systematic Zoology* in 1988–89 as a two-part Special Issue called *Vicariance Biogeography* (Volume 37, Numbers 3–4).

and Nelson 1988) and four were on case studies involving vicariance methods. Of the four case studies, three considered dispersal as a plausible process within the vicariance biogeographical methods (Mayden 1988, Rauchenberger 1988, Liebherr 1988), and one was sceptical that vicariance biogeography worked at all (Noonan 1988). Once again the claim that vicariance was a dominant mechanism shaping the distribution of taxa does not hold up: not even within a vicariance symposium.

National Museum Symposium on the Panbiogeography of New Zealand, Wellington 1988[29]

> Incongruously these ideas took firmest root in New Zealand. An entire school of panbiogeography arose there and set about adapting and extending Croizat's models (and the cladistics of Willi Hennig) to the New Zealand region [...] We now have an emerging and competing third paradigm in which the effect of dispersal is played down further in favour of increased Gondwanic influence together with a new element, the rafting in of biotas on slivers ('terranes') of migrating continental rock. (Caughley 1991: 190)

> [A} biogeographer must be a vicarist in principle and a dispersalist in detail, case by case according to the merits of each case. I was of this opinion in 1964, and am still of it today. (Croizat 1982: 297)

In contrast to many of the latter symposia, the *Symposium on Panbiogeography of New Zealand* was entirely about a single synthesis: panbiogeography. Many authors see panbiogeography as the bastion of vicariance, but surprisingly, vicariance biogeographers were at odds with panbiogeogaphers. The issue was about endemic areas and their development: panbiogeographers concerned more with tracks and distributions did not necessarily share the same aims and goals as the vicariance biogeographers who were interested in the relationships between endemic areas. In other words, panbiogeographers were interested in dispersal as a mechanism of distribution, whereas vicariance biogeographers were interested in vicariance as the mechanism responsible for the formation of endemic areas. Panbiogeographers sought to uncover tracks, main massings and nodes, vicariance biographers sought to find geographical congruence between areagrams. Regardless, both fields were tarred with the same brush: the rejection of dispersal. For example:

> [T]he possibility of long-distance biological dispersal and establishment is typically discounted or discarded at the outset of the panbiogeographic exercise [...] Unfortunately, this a priori rejection of long-distance dispersal as an explanation for multitaxon biogeographic pattern ignores abundant evidence supporting this important process [...] We note that a near-exclusive focus on vicariance and dismissal of evidence supporting alternative explanations are not unique to proponents of

---

[29] Published as *Special Issue on Panbiogeography* (Matthews 1989 [1990]) in the *New Zealand Journal of Zoology*; not to be confused with an earlier *Special Issue on Panbiogeography: Space–Time–Form* published in *Revista di Biologia* (Craw and Sermonti 1988).

> panbiogeography: such views are also shared by several cladistic biogeographers.
> (Waters et al. 2013: 495).

Ironically, neither panbiogeographers nor vicariance biogeographers reject dispersal. Rather the issue is 'that means of dispersal do not explain everything' (Nelson 1978: 290); congruence between many areagrams is evidence for allopatry. For example, in the *Special Issue on Panbiogeography* (Matthews 1989) dispersal is still considered to be a valid biogeographical process:

> Nelson and Platnick have followed Croizat's lead and shifted the focus of biogeographic analysis away from a priori questions of vicariance versus dispersal to methodological questions of area relationship. (Page 1989: 477).

> The significance of dispersal across barriers in relation to vicariism is considered in panbiogeographic studies. (Craw 1989: 541)

> Yet again, once beyond the Livingstone Fault dispersal is widespread. (Heads 1989: 568)

> In the section that follows [...] brief summaries are given of their morphological relationships and dispersal. (Climo 1989: 617)

> Thus these moss distribution patterns are consistent with other plant and animal groups, supporting a general pattern of dispersal for the biota of the Tasman Sea region. (Tangney 1989: 678)

The accusation that the 'near-exclusive focus on vicariance and dismissal of evidence supporting alternative explanations are not unique to proponents of panbiogeography' is clearly untrue. Even at a panbiogeographical symposium dispersal had considerable sway as a biogeographical process.

12.4 Doctrinaire vicariance, doctrinaire dispersal

Throughout the 1980s vicariance was tainted with the view that it was polemical and *doctrinaire*, dividing the biogeographical community into 'dispersal' and 'vicariance biogeographers'. In order to understand and assess the roles of dispersal and vicariance, it is important to look at facts, rather than the polemics, no matter how severe. Boucot's review of the 1979 Vicariance Symposium is a good example of the effect of polemics has over facts:

> The book should be available in university libraries in order that both students and faculty can find out what doctrinaire vicariance is like, and as a companion piece to the doctrinaire dispersalist views of Darlington which are already shelved there. (Boucot 1983: 345)

While cutting, the review had little to offer the reader as to what the 1981 symposium proceedings meant in terms of progress in biogeography. Most new ideas, methods and theories arrive via younger and possibly idealistic practitioners. The more a field is ingrained with a doctrine like dispersal, the harder it is to get alternative ideas heard – hence the polemic. More recently in biogeography there has been the addition of phylogeography, molecular clocks and Bayesian analysis. While the latter two are not new, they are novel approaches within biogeography. These approaches were also accompanied by a strong polemic that attempted to undermine morphological studies as too 'homoplastic' and the comparisons of areagrams to geological reconstructions as 'circular' (Renner 2005). Boucot is correct in claiming a 'doctrinaire vicariance' in Nelson and Rosen (1981) as much of 1970s and 1980s biogeographical theory was dominated by vicariance biogeographers. The same can be said for present-day biogeography: theory is dominated by so-called 'vicariance biogeographers', while practitioners are mostly dispersalists or apply a mixture of both dispersal and vicariance. The division between vicariance and dispersal is mostly one between theory and practice: flaws may be pointed out in the Progression Rule, but basal nodes are still declared as centres of origin (= ancestral areas, Bremer 1992). But at present dispersalists are attempting to re-formulate Hennig's and Brundin's progression rule, but discovering and justifying the age of cladogram nodes by using molecular clock estimates (see Heads 2012 for a summary and Graur and Martin 2004 for a general discussion of clocks: 'Because the appearance of accuracy has an irresistible allure, non-specialists frequently treat these estimates [clock age determinations] as factual'). The dispersalism of the 1970s and 1980s, as shown above, was dominated by Mayr's 'methodological principle', Simpson's 'shades of gray' and the progression rule. Now molecular clocks have been added to the mix and doctrinaire dispersal has been reformulated as neodispersalism – and so, too, has its history. Neodispersalists have committed themselves to re-writing biogeographical history, both in the scientific (Waters et al. 2013) and popular literature (de Queiroz 2014). The neodisperalists claim that biogeography has been and still is dominated by *doctrinaire vicariance*. They also claim that long-distance dispersal is a significant driving force of biogeographical patterns, but lack any evidence that leads to its discovery. Neodispersalist have pinned their hopes on this new reformulation, that further narratives incorporating molecular clock dating, will somehow be the smoking gun to discovering dispersal.

12.5 Neodispersalism

Defining neodispersalism. In 2010 and 2012, New Zealand biogeographer Michael Heads coined the terms *neodispersalism* and *neodispersalist* to refer to a new wave of biogeographers who believe that all distributions are new and only explainable through dispersal. Heads notes:

> In New Zealand, for example, the traditional dispersalists accepted an 'old element' that had evolved there. The neodispersalists argue instead that everything dispersed. (Heads, *pers comm.* 22 March 2014)

Neodispersalism can be defined further as *the default explanatory mechanism for recent organismal distributions as depicted in branching diagrams. The direction and age of dispersal events are based on tree thinking and molecular clocks. Neodispersalism is immune to contradictory geological evidence.*

Molecular clocks. The 'invention' of the molecular clock divergence dating has allowed biogeographers to enjoy an apparently sophisticated dating method that, they believe, can be used as evidence in proposing dispersal explanations (Renner 2005). The clock was first proposed by Zuckerkandl and Pauling (1965) for inferring rates of genetic mutations in mDNA that are calibrated based on the minimum age of a fossil. Change the age of a fossil, change the divergence date; change the data (to nucleic DNA, for example), change the rates. Molecular clocks have been criticised for extrapolating inferences where there is little to no evidence to support many of the divergence dates (Graur and Martin 2004). In fact, variations in clock calibrations have revealed vastly different ages in *Nothofagus* (Sauquet et al. 2012). Moreover, using fossils as clock calibrations provide evidence for a minimum age for divergence. Any interpretation of these clocks serve as narratives, and not evidence, for dispersal. The molecular clock, like Hennig's progression rule, uses a branching diagram as an explanatory tool. Evidence for dispersal has remained the same since the 1940s, however the number of narratives continues to increase. Dispersal remains immune to discovery and is solely the product of narrative. Without a coherent theory, neodispersalists will only repeat the mistakes of Mayr, Simpson and Darlington.

Immunity to evidence. Dispersalism has long been immune to contradictory evidence. Simpson's 'shades of gray' and Mayr's 'methodological principle' have carried *doctrinaire dispersal* through much of the twentieth century. Combined with molecular clock dating and Darlington's rule of thumb/Hennig's progression rule, dispersalism has not changed much in the twenty-first century. For example, dispersalists still claim immunity to geological evidence:

> Fortunately for the growth of the discipline, this outlook is fading fast as molecular phylogeneticists, using so-called relaxed clock methods, are discovering that many plant and animal groups are simply too young for their disjunctions to have been caused by continental drift. (Donoghue 2011: 6342)

> In particular, molecular dating of lineage divergences favors oceanic dispersal over tectonic vicariance as an explanation for disjunct distributions in a wide variety of taxa, from frogs to beetles to baobab trees. (de Queiroz 2005: 68)

> Extant lineages of *Nothofagus*, traditionally considered to be ancient relicts of a prior warm and wet environment, may represent a radiation in response to the same climatic changes that are thought to have allowed radiations in groups thought of as 'recent'. (Cook and Crisp 2005: 2542)

The role of *doctrinaire neodispersal* is to reassert dispersalism as the default hypothesis within twenty-first century biogeography, replacing vicariance as a default explanation. Ignoring vicariant patterns and recent developments in the geosciences (e.g. neotectonics, geochemistry, oceanography), neodispersalism has effectively isolated the field from Earth processes. As Donoghue's (2011) comment above shows, many present-day biologists are not familiar with modern geological or even oceanographic theory. Perhaps this is why they refer to old concepts like 'West Wind drift', the oceanic currents claimed to be responsible for long-distance dispersal (Sanmartín et al. 2007, Waters 2008). Modern oceanographers refer to the Southern Ocean current as the Australian Circumpolar Current, which consists of westward flowing gyres rather than a single current moving east (Rintoul et al. 2001): the gyre between New Zealand and Australia flows west, thereby contradicting the hypothesis that ocean currents are responsible for an eastward dispersal (*contra* Sanmartín and Ronquist 2004, Sanmartín et al. 2007, Waters 2008). While biogeographers like Waters and Craw (2006) insist that geological and organismal evidence be assessed independently, they seemingly are unwilling to understand them.

Neodispersal is a return to the days of Simpson and Mayr where biogeographical 'conclusions must be based on biological (including paleontological) evidence, and geological conclusions on geological evidence in order to avoid circular reasoning' (Mayr 1952b). Compare Mayr's dictum to that of Waters and Craw:

> Vicariant biogeographic inferences should ideally be based on a combination of biological and geological information. When such distinct fields of scientific research intersect, however, there is potential for associated 'multidisciplinary' inferences to be clouded by circular reasoning. (Waters and Craw 2006: 351)

The circular reasoning that both Mayr and Waters and Craw refer to relates directly to the notion of 'independent evidence'. For example, land bridges, drift, vicariance all are explanations that require geological, not biological evidence. The notion that biological evidence, such as geographical congruence of taxic distributions, can support geological processes is deemed 'circular'. For example:

> We suggest that NZ biogeography represents another case in which geological and biological explanations apparently lack independence. Specifically, we argue that the 'Gondwanan' ancestry typically ascribed to components of NZ's biota […] lacks geological support. Instead, the geological 'evidence' has largely been driven by the general assumption that NZ has a Gondwanan biota […], which results in circular logic. (Waters and Craw 2006: 351)

> However, constraining nodes in a phylogenetic tree by geological events risks circularity in biogeographic analyses because it already assumes that those events caused the divergence, rather than testing temporal coincidence. (Renner 2005: 552).

The examples above argue that biogeographical hypotheses derive from geology (e.g. tectonic events). For example, the Gondwanan ancestry of New Zealand biota is merely descriptive of the distributions. If, for example, *Nothofagus* in Australia shares a closer relationship to New Zealand species than it does to South American species, we would call this a Gondwanan distribution or pattern. We may also call it an Austral pattern. If the relationships between *Nothofagus* were described as 'Austral', then Waters and Craw would be criticising an Austral (i.e. southern hemisphere) ancestry, something that could only be disproven if we found an endemic northern hemisphere species of *Nothofagus*.

Waters and Craw's argument, like that of Renner, is a case of special pleading for independent evidence. The geological evidence in Water and Craw's case is merely the fragmentation of an Austral distribution. Clearly this is not geological evidence, but rather a case of 'vicariance = geological process' – as if geology alone is responsible for fragmented distributions. Here we see the apparent rise of *doctrinaire vicariance* as though it lurks in all assumptions about disjunct southern hemisphere distributions. Renner's argument for circularity is similar as though geological events are directly responsible for allopatry, that is, vicariance.[30] The spectre of vicariance that haunts dispersal biogeographers is of their own making. The main assumption in vicariance biogeography is that geographical congruence between many different monophyletic taxa *may* result in allopatry and disjunct distributions of *biota*. In other words, geographical congruence is evidence for vicariance in biota and not dispersal in single taxa. As such the methods employed are designed to find vicariance not dispersal, hence *vicariance biogeography*. *Doctrinaire vicariance* is nothing more than a misunderstanding of vicariance, as though all allopatry and disjunction is by default vicariance.

The war of words between *doctrinaire neodispersal* and *doctrinaire vicariance* represents is at best a distraction. Biogeographers do what they do, and use the methods they use because they believe it suits their needs. Few biogeographers take it upon themselves to do biogeography as a form of retaliation or rebellion against some form of dogma. The Oligocene drowning of New Zealand (i.e. Waters and Craw 2006, Trewick et al. 2007, Landis et al. 2008) may be one of the few attempts at an active rebellion against *doctrinaire vicariance*. New virgin territory cleansed of old Gondwanan vicariant taxa was open to new immigrants from Australia. This new utopia of doctrinaire dispersal was, alas, not to be. In 2013, at the VII Southern

---

[30] Alan de Queiroz (2014) notes that it was Renner 'who has probably done more than anyone else in recent years to convince people of the ubiquity of plant dispersal [...] Her graduate school supervisor, Klaus Kubitzki, had a strong belief in the importance of long–distance dispersal, and from him, we assume, she picked up that conviction and never let it go. During a stint as a postdoctoral researcher at the Smithsonian Institution in the 1980s, she was in contact with the other side, the vicariance biogeographers, but she says she always found their extreme view "dogmatic and a bit silly" ' (de Queiroz 2014: 170).

Connections Conference in Otago New Zealand, the drowning theory was finally debunked (see Mildenhall et al. 2014, *ad nauseam*).

12.6 Conclusions

Hennig's *Phylogenetic Systematics* had a profound influence on the history of dispersal. The progression rule has been the default paradigm in reading trees in biogeography. In addition to the introduction of molecular clocks, these are now the main source of neodispersal narratives in historical biogeography.

Dispersal, however, has not changed its central paradigm since the 1940s, other than the addition of Hennig's progression rule and molecular clocks. Dispersal is still the default mechanism or narrative for organismal distribution, based on Mayr's method of 'independent evidence' and Simpson's inevitable probability or 'shades of gray'. Together these dictums form the basis for neodispersalism, a new seemingly invincible narrative immune to contradictory evidence and the default mechanism for twenty-first century biogeography. Regardless, *doctrinaire vicariance*, is still seen as the unfathomable enemy of neodispersal: caught up in a mythical battle of 'circular reasoning' in the minds of neodispersalists.

Acknowledgements

We thank Michael Heads and Gary Nelson for helpful feedback on previous drafts.

References

Adams, C.G. and Ager, D.V., eds. (1967). *Aspects of Tethyan Biogeography*. London: Systematics Association.

Adie, R.J. (1962). The geology of Antarctica. *American Geophysical Union Monographs*, 7, 26–39.

Adie, R.J. (1963). Geological evidence on possible Antarctic land connections. In *Pacific Basin Biogeography: A Symposium*, ed. J.L. Gressitt. Honolulu: Bishop Museum Press, pp. 455–463.

Agnolin, F.L., Ezcurra, M.D., Pais, D.F. and Salisbury, S.W. (2010). A reappraisal of the Cretaceous non–avian dinosaur faunas from Australia and New Zealand: evidence for their Gondwanan affinities. *Journal of Systematic Palaeontology*, 8, 257–300.

Anonymous (1949a). Program of the New York Meeting with Abstracts of Papers, Thirty–Fourth Annual Meeting of the Ecological Society of America with the American Association for the Advancement of Science, the Botanical Society of America, the American Society of Limnology and Oceanography, the American Society of Zoologists and the American Society of Naturalists, New York City. *Bulletin of the Ecological Society of America*, 30, 45–72.

Anonymous (1949b). Preview of the 116th Meeting, AAAS: New York City, 26–31 December 1949. *Science*, 2867, 605–614+617–624.

Anonymous (ed.) (1975). Biogéographie et Liaisons Intercontinentales au Cours du Mésozoïque. *Mémoires du Muséum National d'Histoire Naturelle*, série A, Zoologie, 88, 237 pp.

Anonymous (1978). Second International Congress of Systematic and Evolutionary Biology (ICSEB–II). *Brittonia*, 30, 472.

Antev, E. (1947). *Biogeography*. Eighth Annual Biology Colloquium. Oregon State Chapter of Phi Kappa Phi. Corvallis, OR: Oregon State College, Corvallis.

Ball, I.R. (1976). Nature and formulation of biogeographical hypotheses. *Systematic Zoology*, 24, 407–430.

Ball, I.R. (1983). Planarians, plurality and biogeographical explanations. In *Evolution, Time and Space: The emergence of the Biosphere*, ed. R.W. Sims, J.H. Price and P.E.S. Whalley. Systematic Association Special Volume 38. London: Academic Press, pp. 409–430.

Berg, R.Y. (1983). Plant distribution as seen from plant dispersal – General principles and the basic modes of plant dispersal. In *Dispersal and Distribution: An international Symposium*, ed. K. Kubitzki, Sonderbände de Naturwissenschaftlichen Vereins in Hamburg. Hamburg : Verlag Paul Parey, pp. 13–36.

Boucot, A.J. (1983). Book reviews. *Palaeogeography, Palaeoclimatology, Palaeoecology*, 41, 345–352.

Bremer, K. (1992). Ancestral areas: a cladistic reinterpretation of the center of origin concept. *Systematic Biology*, 41, 436–445.

Brundin, L. (1966). Transarctic relationships and their significance, as evidenced by chironomid midges. With a monograph of the subfamilies Podonominae and Aphroteniinae and the Austral Heptagyiae. *Kungliga Svenska Vetenskapsakademiens Handlingar*, Fjärde series, 11, 1–472.

Brundin, L. (1981a). Croizat's panbiogeography versus phylogenetic biogeography. In *Vicariance biogeography, a critique*, ed. G. Nelson and D.E. Rosen. New York: Columbia University Press, pp. 94–143.

Brundin, L. (1981b). Response. In *Vicariance biogeography, a critique*, ed. G. Nelson and D.E. Rosen. New York: Columbia University Press, pp. 149–158.

Brady, R.H. (1985). On the independence of systematics. *Cladistics*, 1, 113–126.

Briggs, J.C. (1984). *Centres of Origin in Biogeography*. Biogeography Monographs 1, Leeds: Biogeography Study Group.

Brooks, D.R. (1988). Scaling effects in historical biogeography: a new view of space, time, and form. *Systematic Zoology*, 37, 237–244.

Caughley, G. (1991). [Review] Panbio-geography. *Special Issue of New Zealand Journal of Zoology* 16(4) 1989, 815 pp. *Australian Geographic Studies*, 29, 189–191.

Climo, F. (1989). The panbiogeography of New Zealand as illuminated by the genus *Fectola* Iredale, 1915 and subfamily Rotadiscinae Pilsbry, 1927 (Mollusca: Pulmonata: Punctoidea: Charopidae). *New Zealand Journal of Zoology*, 16, 665–678.

Cracraft, J. (1975a). Mesozoic dispersal of terrestrial faunas around the southern end of the world. *Mémoires du Muséum National d'Histoire Naturelle*, Série A, Zoologie, 88, 29–54.

Cracraft, J. (1975b). Historical biogeography and earth history: Perspectives for a future synthesis. *Annals of the Missouri Botanical Garden*, 62, 227–250.

Cracraft, J. (1982). Geographic differentiation, cladistics, and vicariance biogeography: Reconstructing the tempo and mode of evolution. *American Zoologist*, 22, 411–424.

Cracraft, J. (1984). Book review. *The Quarterly Review of Biology*, 59, 350–351.

Cracraft, J. (1988). Deep-history biogeography: retrieving the historical pattern of evolving continental biotas. *Systematic Zoology*, 37, 221–236.

Cracraft, J. and Shipp, R.L. (1988). Vicariance biogeography: theory, methods, and applications: Introduction to the symposium. *Systematic Zoology*, 37, 219–220.

Craw, R. (1988). Continuing the synthesis between panbiogeography, phylogenetic systematics and geology as illustrated by empirical studies on the biogeography of New Zealand and the Chatham Islands. *Systematic Zoology*, 37, 291–310.

Craw, R. (1989). New Zealand biogeography: a panbiogeographic approach. *New Zealand Journal of Zoology*, 16, 527–547.

Craw, R.C., Grehan, J.R. and Heads, M.J. (1999). *Panbiogeography: Tracking the History of Life*. Oxford: Oxford University Press.

Craw, R.C. and Sermonti, G. (1988). Special issue on panbiogeography. *Revista di Biologia/Biology Forum*, 81 (4).

Cook, L.G., and Crisp, M.D. (2005). Not so ancient: the extant crown group of *Nothofagus* represents a post–Gondwanan radiation. *Proceedings of the Royal Society B: Biological Sciences*, 272, 2535–2544.

Crisp, M.D., Trewick, S.A. and Cook, L.G. (2011). Hypothesis testing in biogeography. *Trends in Ecology and Evolution*, 26, 66–72.

Croizat, L. (1964). *Space, Time, Form: The Biological Synthesis*. Caracas: Published by the author.

Croizat, L. (1978). Hennig (1966) entre Rosa (1891) y Løvtrup (1977): medio siglo de "Sistemática Filogenética." *Bolletin Academia de Ciencias Físicas, Matematicas y Naturales, Caracas*, 38, 59–147.

Croizat, L. (1982). Vicariance/vicariism, panbiogeography, "vicariance biogeography," etc.: a clarification. *Systematic Zoology*, 31, 291–304.

Croizat, L. (1994). "Pacific Basin biogeography": Marginal notes to a symposium. *Kirkia*, 15, 157–246.

Croizat, L., Nelson, G. and D.E. Rosen. (1974). Centers of origin and related concepts. *Systematic Zoology*, 23, 265–287.

Crovello, T.J. (1981). Quantitative biogeography: an overview. *Taxon*, 30, 563–575.

Darlington, P.J., Jr. (1957). *Zoogeography: The Geographical Distribution of Animals*. New York: John Wiley and Sons.

Darlington, P.J., Jr. (1964). Biogeography of half the world. *Science*, 144, 708–709.

Descimon, H. (1976). *Biogeographie et Evolution en Amerique Tropicale*. Paris: Publications du Laboratoire de Zoologie de l'ecole normale supérieure, no. 9.

Donoghue, M.J. (2011). Bipolar biogeography. *Proceedings of the National Academy of Sciences*, 108, 6341–6342.

Edmunds, G.F. (1975). Phylogenetic biogeography of mayflies. *Annals of the Missouri Botanical Garden*, 62, 251–263.

Endler, J.A. (1982). Alternative hypotheses in biogeography: Introduction and synopsis of the symposium. *American Zoologist*, 22, 349–354.

Ehrendorfer, F. (1983). Summary lecture. In *Dispersal and Distribution: An International Symposium*, ed. K. Kubitzki. Sonderbände de Naturwissenschaftlichen Vereins in Hamburg, Hamburg: Verlag Paul Parey, pp. 391–398.

Ferris, V.R. (1980). A science in search of a paradigm? Review of the symposium, "Vicariance Biogeography: A Critique." *Systematic Zoology*, 29, 67–76.

Fleming, C.A. (1963). Paleontology and southern biogeography. In *Pacific Basin Biogeography: A Symposium*, ed. J.L. Gressitt. Honolulu: Bishop Museum Press, pp. 369–386.

Frankel, H. (1984). Biogeography, before and after the rise of seafloor spreading. *Studies in History and Philosophy of Science*, 15, 141–168.

Frankel, H. (2012). *The Continental Drift Controversy*. Cambridge: Cambridge University Press.

Gaskin, D.E. (1975). Reconsideration of New Zealand biogeographical problems with special reference to the Mesozoic. *Mémoires du Muséum National d'Histoire Naturelle*, Série A, Zoologie, 88, 87–97.

Graur, D. and Martin, W. (2004). Reading the entrails of chickens: molecular timescales of evolution and the illusion of precision. *Trends in Genetics*, 20, 80–88.

Gray, J. and Boucot A.J. (1976a). Preface. In *Historical Biogeography, Plate Tectonics, and the Changing Environment*, ed. J. Gray and A.J. Boucot. Corvallis, OR: Oregon State University Press, pp. v–vi.

Gray, J. and Boucot A.J. (1976b). Editor's disclaimer. In *Historical Biogeography, Plate Tectonics, and the Changing Environment*, ed. J. Gray and A.J. Boucot. Corvallis, OR: Oregon State University Press, pp. vii–ix.

Gressitt, J.L. (1963). *Pacific Basin Biogeography: A Symposium*. Honolulu: Bishop Museum Press.

Gressitt, J.L. (1984). Foreword. In *Biogeography of the Tropical Pacific: Proceedings of a Symposium*, ed. F.J. Radovsky, P.H. Raven and S.H. Sohmer. Honolulu: Association of Systematics Collections and the Bernice P. Bishop Museum, pp. viii–viv.

Heads, M. (1989). Integrating earth and life sciences in New Zealand natural history: the parallel arcs model. *New Zealand Journal of Zoology*, 16, 549–545.

Heads, M. (2010). Biogeographical affinities of the New Caledonian biota: a puzzle with 24 pieces. *Journal of Biogeography*, 37, 1179–1201.

Heads, M. (2012). *Molecular Panbiogeography of the Tropics*. Berkeley, CA: University of California Press.

Hennig, W. (1950). *Grundzüge einer Theorie der phylogenetischen Systematik*. Berlin: Deutscher Zentralverlag.

Hennig, W. (1965). Phylogenetic systematics. *Annual Review of Entomology*, 10, 97–116.

Hennig, W. (1966). *Phylogenetic Systematics*. Urbana, IL: University of Illinois Press.

Holloway, J.D. (1984). Lepidoptera and the Melanesian arcs. In *Biogeography of the Tropical Pacific: Proceedings of a Symposium*, ed. F.J. Radovsky, P.H. Raven and S.H. Sohmer. Honolulu: Association of Systematics Collections and the Bernice P. Bishop Museum, pp. 129—169.

Hoffmann, R.S. (1981). Different voles for different holes: Environmental restrictions on refugial survival of mammals. In *Evolution Today: Proceedings of the Second International Congress of Systematic and Evolutionary Biology*, ed. G.G.E. Scudder and J.L. Reveal,. Pittsburg, CA: Hunt Institute for Botanical Documentation, pp. 25–45.

Holmes, A. (1953). The South Atlantic: land bridges or continental drift? *Nature*, 4355, 669–671.

Kearey, P., Klepeis, K.A. and Vine, F. (2009). *Global Tectonics*, 3rd edition. West Sussex: Wiley-Blackwell, John Wiley and Sons.

Kluge, A.G. (1988). Parsimony in vicariance biogeography: a quantitative method and a Greater Antillean example. *Systematic Zoology*, 37, 315–328.

Krahulec, F. (1985). [Review of] *Dispersal and Distribution: An International Symposium* by Klaus Kubitzki. *Folia Geobotanica and Phytotaxonomica* 20 (2), 215–216.

Kubitzki, K. (1983). *Dispersal and Distribution: An International Symposium*. Sonderbände de Naturwissenschaftlichen Vereins in Hamburg. Hamburg: Verlag Paul Parey.

Landis, C.A., Campbell, H.J., Begg, J.G., et al.(2008). The Waipounamu erosion surface: questioning the antiquity of the New Zealand land surface and terrestrial fauna and flora. *Geological Magazine*, 145, 173–197.

Liebherr, J.K. (1988). General patterns in West Indian insects, and graphical biogeographic analysis of some Circum-Caribbean *Platynus* Beetles (Carabidae). *Systematic Zoology*, 37, 385–409.

Lohman, D.J., de Bruyn, M., Page, T., et al. (2011). Biogeography of the Indo–Australian Archipelago. *Annual Review of Ecology, Evolution, and Systematics*, 42, 205–226.

Losos, J.B. and Ricklefs, R.E. (2009). *The theory of island biogeography revisited*. Princeton, NJ: Princeton University Press.

Matthew, W.D. (1914). Climate and evolution. *Annals of the New York Academy of Sciences*, 24, 171–318.

Matthews, C. (1989 [1990]). Panbiogeography special issue. *New Zealand Journal of Zoology*, 16, 1–815.

Mayden, R.L. (1988). Vicariance biogeography, parsimony, and evolution in North American freshwater fishes. *Systematic Zoology*, 37, 329–355.

Mayr, E. (1952a). The problem of land connections across the South Atlantic, with special reference to the Mesozoic. *Bulletin of the American Museum of Natural History*, 99, 83–258.

Mayr, E. (1952b). Introduction. In *The Problem of Land Connections across the South Atlantic, With Special Reference to the Mesozoic*, ed. E. Mayr. *Bulletin of the American Museum of Natural History*, 99, 85.

Mayr, E. (1952c). Conclusion. In *The Problem of Land Connections across the South Atlantic, With Special Reference to the Mesozoic*, ed. E. Mayr. *Bulletin of the American Museum of Natural History*, 99, 255–258.

Mildenhall, D.C., Mortimer, N., Bassett, K.N. and Kennedy, E.M. (2014). Oligocene paleogeography of New Zealand: maximum marine transgression. *New Zealand Journal of Geology and Geophysics*, 57, 107–109.

McVean, D.N. (1967). Pacific Basin Biogeography. A Symposium by J. Linsley Gressit. *The Journal of Pacific History*, 2, 235–237.

Molnar, R.E. (1981). A dinosaur from New Zealand. In *Gondwana*

Five: Proceeding of the Fifth International Gondwanan Symposium, ed. M.M. Cresswell and P. Vella. Rotterdam: Balkema, pp. 91–96.

Nelson, G. (1974). Historical biogeography: An alternative formalization. *Systematic Zoology*, 23, 555–558.

Nelson, G. (1976). Biogeography, the vicariance paradigm, and continental drift. (reviews of): *The Geography of the Flowering Plants* (R. Good); *Ecology and Biogeography in India* (M.S. Mani, ed.); *Angiosperm Biogeography and Past Continental Movements* (P.H. Raven and D.I. Axelrod); *Taxonomy Phytogeography and Evolution* (D.H. Valentine,ed.); *Biogeography and Ecology in New Zealand* (G. Kuschel, ed.); *Biogeographie et Liaisons Intercontinentales au cours du Mesozoique* (Anon., ed.); *Biologie de l'Amerique Australe* (Debouteville and Rapoport, eds.); *Atlas of Palaeobiogeography* (A. Hallam, ed.); *Implications of Continental Drift to the Earth Sciences* (Tarling and Runcorn, eds); *The Age of the Ocean Basins* (W.C. Pitman et al.); *Floristics and Paleofloristics of Asia and Eastern North America* (A. Graham, ed.); *Biogeography: The Twenty-First Systematics Symposium* (G. Davidse, ed.); *Introduction to Zoogeography* (J. Illies); *Principles and Problems of Zoogeography* (P. Banarescu); *Introduction to Biogeography* (B. Seddon); *25 Years of Botany* (M.R. Crosby, ed.). *Systematic Zoology*, 24, 490–504 (erratum in *Systematic Zoology*, 25, 200).

Nelson, G. (1982). Review: alternative hypotheses in biogeography. *Systematic Zoology*, 31, 341–342.

Nelson, G. and N.I. Platnick. (1981). *Systematics and Biogeography: Cladistics and Vicariance*. New York: Columbia University Press.

Nelson, G. and Rosen D.E. (1981). *Vicariance Biogeography, A Critique*. New York: Columbia University Press.

Nelson G., Atz, J.W., Kallman, K.D. and Smith, C.L. (1987). Donn Eric Rosen, 1929–1986, *Copeia*, 1987 (3), 541–547.

Noonan, G.R. (1984). Review of *Evolution, Time and Space: The Emergence of the Biosphere. Systematic Zoology*, 33, 123–126.

Noonan, G.R. (1988). Biogeography of North American and Mexican insects, and a critique of vicariance biogeography. *Systematic Zoology*, 37, 366–384.

Nordal, I. (1985). Book Review. *Nordic Journal of Botany*, 5, 14.

OED Online. December 2014. Oxford University Press. www.oed.com/view/ Entry/125146?rskey=wSsbbK&result=1 (accessed 19 January 2015).

Page, R.D.M. (1988). Quantitative cladistic biogeography: Constructing and comparing area cladograms. *Systematic Zoology*, 37, 254–270.

Page, R.D.M. (1989). New Zealand and the new biogeography. *New Zealand Journal of Zoology*, 16, 471–483.

Parenti, L.R. (1981). Discussion. In *Vicariance Biogeography, A Critique*, ed. G. Nelson and D.E. Rosen. New York: Columbia University Press pp. 490–497.

Parenti, L.R. and Ebach, M.C. (2009). *Comparative Biogeography. Discovering and Classifying Biogeographical Patterns of a Dynamic Earth*. Berkeley, CA: University of California Press.

Parenti, L.R. and Ebach, M.C. (2013). Evidence and hypothesis in biogeography. *Journal of Biogeography*, 40, 813–820.

Patterson, C. (1981). Methods of paleobiogeography. In *Vicariance Biogeography, A Critique*, ed. G. Nelson and D.E. Rosen. New York: Columbia University Press, pp. 446–489.

Patterson, C. (1983). Aims and methods in biogeography. In *Evolution, Time and Space: The Emergence of the Biosphere*, Systematic Association Special Volume 38, ed. R.W. Sims, J.H. Price and P.E.S. Whalley. London: Academic Press, pp. 1–28.

Patton, J.L. (1982). Evolutionary mechanisms. *Science*, 216, 287–288.

Platnick, N.I. (1976). Concepts of dispersal in historical biogeography. *Systematic Zoology*, 25, 294–295.

Platnick, N.I. (1981). Discussion. In *Vicariance Biogeography, A Critique*, ed. G. Nelson and D.E. Rosen. New York: Columbia University Press, pp. 144–150.

Platnick, N.I. and Nelson, G.J. (1978). A method of analysis for historical biogeography. *Systematic Zoology*, 27, 1–16.

Platnick, N.I. and Nelson, G. (1988). Spanning-tree biogeography: shortcut, detour, or dead-end? *Systematic Zoology*, 37, 410–419.

de Queiroz, A. (2005). The resurrection of oceanic dispersal in historical biogeography. *Trends in Ecology and Evolution*, 20, 68–73.

de Queiroz, A. (2014). *The Monkey's Voyage*. New York: Basic Books.

Quigley, M.C., Clark, D. and Sandiford, M. (2010). Tectonic geomorphology of Australia. *Geological Society, London, Special Publications,* 346(1), 243–265.

Radovsky, F.J. (1983). J. Lindsay Gressitt (1914–1982). *International Journal of Entomology*, 25, 1–10.

Rauchenberger, M. (1988). Historical biogeography of Poeciliid fishes in the Caribbean. *Systematic Zoology*, 37, 356–365.

Raven, P.H. (1975). Summary of the Biogeography Symposium. *Annals of the Missouri Botanical Garden*, 62, 380–385.

Raven, P.H., and Axelrod, D.I. (1974). Angiosperm biogeography and past continental movements. *Annals of the Missouri Botanical Garden*, 61, 539–637.

Ree, R.H., Moore, B.R., Webb, C.O. and Donoghue, M.J. (2005). A likelihood framework for inferring the evolution of geographic range on phylogenetic trees. *Evolution*, 59, 2299–2311.

Renner, S.S. (2005). Relaxed molecular clocks for dating historical plant dispersal events. *Trends in Plant Science*, 10, 550–558.

Renner, S.S. (2010). Book Review. *The Quarterly Review of Biology*, 85, 224.

Reveal, J.L. and Scudder, G.G.E. (1979). The second international congress of systematic and evolutionary biology. *Taxon*, 28, 287–288.

Rintoul, S., Hughes, C. and Olbers, D. (2001). The Antarctic circumpolar current system. In *Ocean Circulation and Climate*, ed. G. Siedler, J. Church and J. Gould. New York: Academic Press, pp. 271–302.

Ronquist, F. (1997). Dispersal–vicariance analysis: a new approach to quantification of historical biogeography. *Systematic Biology*, 46, 195–203.

Rosen, B.R. (1985). Fact, hypothesis, or idea? *Journal of Biogeography*, 12, 383–385.

Rosen, D.E. (1981). Introduction. In *Vicariance Biogeography, A Critique*, ed. G. Nelson and D.E. Rosen. New York: Columbia University Press, pp. 1–5.

Sandiford, M. (2007). The tilting continent: a new constraint on the dynamic topographic field from

Australia. *Earth and Planetary Science Letters*, 261, 152–163.

Sanmartín, I. and Ronquist, F. (2004). Southern hemisphere biogeography inferred by event-based models: plant versus animal patterns. *Systematic Biology*, 53, 216–243.

Sanmartín, I., Wanntorp, L. and Winkworth, R.C. (2007). West Wind Drift revisited: testing for directional dispersal in the southern hemisphere using event-based tree fitting. *Journal of Biogeography*, 34, 398–416.

Sauquet, H., Ho, S.Y.W., Gandolfo, M.A., et al. (2012). Testing the impact of calibration on molecular divergence times using a fossil–rich Group: The case of *Nothofagus* (Fagales). *Systematic Biology*, 61, 289–313.

Schindewolf, O.H. (1954). Review: The Problem of Land Connections Across the South Atlantic, With Special Reference to the Mesozoic by Ernst Mayr. *The Quarterly Review of Biology*, 29, 157.

Schmid, R. (1985). Reviews: Biogeography of the Tropical Pacific: Proceedings of a symposium by F.J. Radovsky; P.H. Raven; S.H. Sohmer; Pacific basin biogeography by J. Linsley Gressitt. *Taxon*, 34, 747–749.

S.C.M. (1984). Book review. *Geological Magazine*, 121, 376–377.

Scudder, G.G.E. and Reveal, J.L. (1981). *Evolution Today: Proceedings of the Second International Congress of Systematic and Evolutionary Biology*. Pittsburg, CA: Hunt Institute for Botanical Documentation.

Simpson, G.G. (1952). Probabilities of dispersal in geologic time. In *The Problem of Land Connections Across the South Atlantic, With Special Reference to the Mesozoic*, ed. E. Mayr. *Bulletin of the American Museum of Natural History*, 99, 163–176.

Sims, R.W., Price, J.H. and Whalley, P.E.S. (1983). *Evolution, Time and Space: The Emergence of the Biosphere*. Systematic Association Special Volume 38, London: Academic Press.

Sober, E. (1988). The conceptual relationship of cladistics phylogenetics and vicariance biogeography. *Systematic Zoology*, 37, 245–253.

Stelbrink, B., Albrecht, C., Hall, R. and von Rintelen, T. (2012). The biogeography of Sulawesi revisited: is there evidence for a vicariant origin of taxa on Wallace's "anomalous island"? *Evolution*, 66, 2252–2271.

Stoddart, D.R. (1985). Book review. *Annals of the Association of American Geographers*, 75, 604–605.

Tangney, R.S. (1989). Moss biogeography in the Tasman Sea region. *New Zealand Journal of Zoology*, 16, 665–678.

Thorne, R.F. (1963). Biotic distribution patterns in the tropical Pacific. In *Pacific Basin Biogeography: A Symposium*, ed. J.L. Gressitt. Honolulu: Bishop Museum Press, pp. 311–354.

Thorne, R.F. (1975). Angiosperm phylogeny and geography. *Annals of the Missouri Botanical Garden*, 62, 362–367.

Thorne, R.F. (1982). Reviews: New Taxonomies: Vicariance Biogeography: A Critique by Gareth Nelson; Donn E. Rosen; Systematics and Biogeography: Cladistics and Vicariance by Gareth Nelson; Norman Platnick. *Bioscience*, 32, 822–823.

Thornton, I.W.B. (1983a). Vicariance and dispersal: confrontation or compatibility? *GeoJournal*, 7, 557–564.

Thornton, I.W.B. (1983b). Preface. *GeoJournal*, 7, 479.

Trewick, S.A., Paterson, A.M. and Campbell, H.J. (2007). Hello New Zealand. *Journal of Biogeography*, 34, 1–6.

Udvardy, M.D.F. (1969). *Dynamic Zoogeography, With Special Reference to*

Land Animals. New York: Van Nostrand Reinhold Co.

van Zanten, B.O. (1983). Possibilities of long-range dispersal in bryophytes with special reference to the southern hemisphere. In *Dispersal and Distribution: An international Symposium*, ed. K. Kubitzki. Sonderbände de Naturwissenschaftlichen Vereins in Hamburg. Hamburg: Verlag Paul Parey, pp. 49–64.

Waters, J.M. (2008). Driven by the West Wind Drift? A synthesis of southern temperate marine biogeography, with new directions for dispersalism. *Journal of Biogeography*, 35, 417–427.

Waters J.M., Trewick S.A., Paterson, A.M., et al. (2013). Biogeography off the tracks. *Systematic Biology*, 62, 494–498.

Waters J.M. and Craw, D. (2006). Goodbye Gondwana? New Zealand biogeography, geology, and the problem of circularity. *Systematic Biology*, 55, 351–356.

Wiley, E.O. (1988). Parsimony analysis and vicariance biogeography. *Systematic Zoology*, 37, 271–290.

Williams, D.M. and Ebach, MC. (2012). Confusing homologs as homologies: a reply to "On homology". *Cladistics*, 28, 223–224.

Worthy, T.H., Tennyson, A.J.D., Archer, M., et al. (2006). Miocene mammal reveals a Mesozoic ghost lineage on insular New Zealand, southwest Pacific. *Proceedings of the National Academy of Sciences*, 103(51), 19419–19423.

Zuckerkandl, E. and Pauling, L. (1965). Evolutionary divergence and convergence in proteins. *Evolving Genes and Proteins*, 97, 97–166.

13

Molecular data in systematics: a promise fulfilled, a future beckoning

WARD C. WHEELER and GONZALO GIRIBET

13.1 Introduction

Systematics is on a runway today. Advances in DNA sequencing, non-invasive 3D-imaging techniques, and computation are transforming the field. Where for a generation there was a distinction between molecular and morphological or neontological and paleontological, divisions among sources of data and taxonomic perspective are dissolving. The field is becoming more united in outlook and empirical content and the emphasis of practicing systematists changed from technical achievement in molecular data collection to comparative biological hypothesis testing.

Very soon, given tremendous reductions in sequencing costs (Fig 13.1), it will be a given that for whatever problem a systematist may be working on, there could be complete genomes for all the extant taxa under study. Yet, systematics must go beyond extant taxa only, and total evidence approaches incorporating large amounts of molecular data from extant taxa and morphological data from extant and extinct taxa are the future of systematic analysis (e.g. Sharma and Giribet 2014, Arcila et al. 2015).

Even with these technical advances, several ideas and concepts articulated by Hennig (1950, 1966) underlie, or are a component of, modern analysis. Among

The Future of Phylogenetic Systematics: The Legacy of Willi Hennig, eds. D. Williams, M. Schmitt and Q. Wheeler. Published by Cambridge University Press.
© The Systematics Association 2016.

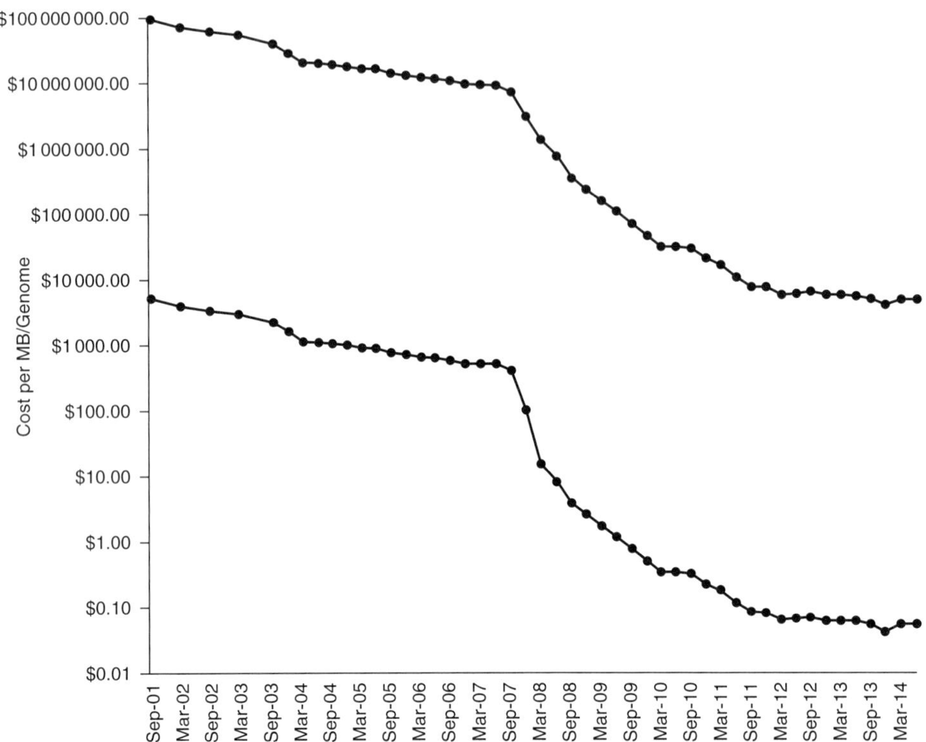

Fig 13.1 Sequencing costs over time. The solid, upper line is per whole genome, the lower, dashed line is per megabase. The initial, log-linear decrease is due to improvements in Sanger-type sequencing. The dramatic drop in cost in 2007 is due to the advent of NextGen sequencing technologies (initially Roche 454 and later Illumina). Data from www.genome.gov/sequencingcosts.

these are monophyletic classification based on recency of common ancestry, the coeval origin of sister taxa, and an ecumenical view of evidence.

13.2 Systematics today

While molecular phylogenetic studies have grown exponentially since the 1980s, most studies published today still rely on PCR-based approaches and Sanger sequencing technology (i.e. a targeted-gene approach). This approach has a number of limitations, including cost and time efficiency, and yields only a handful of loci. Blind approaches to sequencing, including Sanger-based EST (expressed sequence tags) approaches initially (e.g. Dunn et al. 2008), and then so-called 'next-generation sequencing' approaches (McCormack et al. 2013) during the past few years (mainly based on 454 and Illumina technologies), are now yielding enormously larger data sets for the analysis of organismal relationships (e.g. Lee et al. 2011, Misof et al. 2014).

These approaches have allowed for thorough testing of phylogenetic hypotheses by combining deeply sequenced transcriptomes and available genomes (e.g. Kocot et al. 2011, Smith et al. 2011, Cannon et al. 2014, Fernández et al. 2014a, Weigert et al. 2014). In other areas of research, genomes are accumulating at a fast pace for systematic and population genetic studies (e.g. Clark et al. 2007, Gravel et al. 2013, Freedman et al. 2014, Zhan et al. 2014). Currently, transcriptomic analysis is the state of the art for systematic analysis mainly due to its broad genomic sampling and relatively straightforward analytical extension of multilocus approaches. It is made more difficult, however, by its reliance on fragile RNA as opposed to hardier, and more readily available, DNA (discussed below). Even with this technical challenge, transcriptomic systematics has adduced enormous data sets and allowed broad ranging, robust systematic analysis.

A technique that responds to two of the perceived limitations of transcriptomic analysis (RNA and gene orthology) is based on the use of long, common primers much like those used for PCR-based analysis. Referred to as 'anchored phylogenomics' (Lemmon et al. 2012), 'ultra-conserved elements' (McCormack et al. 2013), or 'target enrichment' (Mamanova et al. 2010), the central idea is that there are conserved elements distributed throughout the genome that are flanked by more variable segments (Smith et al. 2014). The conserved areas can be used as starting points for sequencing the variable areas and due to their conservation, locus homology is ensured, this also avoids issues with splice variation. Although these data do allow the use of DNA, there are several shortcomings inherent in the method (Peloso et al. 2016). Firstly, the conserved, primer sequences are determined from taxa where whole genomes or good quality transcriptomes are available. As study taxa increase in distance from these, the conserved segments become less so and are increasingly difficult to use in initiating sequencing. Secondly, since the sequences are in a single direction 'out' from the primers, they can be highly variable in length with high error rates. In general, the data quality does not yet seem to match that of recent transcriptomic data sets (e.g. Peloso et al. 2016).

Until very recently, sequenced whole genomes were only available for a few selected organisms – the so-called model organisms. With the advent of new sequencing technologies, genome availability has been greatly increased. This has helped to push new areas in science, including genomics based on highly degraded DNA (Meyer et al. 2012). These technologies allow, for the first time, the genomic sequencing from small volumes of tissue or from decayed samples, and not just from organisms that can be cultured or for which large amounts of high-quality DNA can be extracted (e.g. Meyer et al. 2012). These developments thus suggest that most museum specimens may be available in the future for genomic work. This sudden, incomparable source of specimens is now, in principle, amenable for genomic research, and could be an important force in the sorely needed revitalization of natural history collections (Kemp 2015).

In the past, use of museum specimens for molecular work (due to the high levels of DNA and RNA degradation in old specimens) was based on PCR-amplification of not more than a handful of genes or genomes from organelles (e.g. Cooper et al. 2001). The advantage of the new techniques is that they can work with short fragments and non-amplified material, thus opening new horizons for highly degraded museum samples. This is important not just for the enhanced quantity of available specimens, but also for their sampling of taxa and localities that may no longer exist.

While the availability of whole genomes for large numbers of non-model organisms is certainly on the horizon, most large systematic studies rely on a crude but efficient proxy for whole genomes – transcriptomes. Transcriptomes are faster and cheaper to generate than genomes, and assembly is in general much easier with existing computational resources. However, there are also drawbacks of using transcriptomes, especially the need for RNA (genomes can be sequenced from DNA samples, which abound in many collections and laboratories around the globe). Perhaps most important, however, is splice variation, hence many transcriptomes have large numbers of isoforms. Distinguishing between the many isoforms and paralogues has become a significant challenge in evolutionary biology (Spitzer et al. 2006), and an area that will see many developments in the near future. Methods for detecting paralogy during orthology assignment have been developed recently (Dunn et al. 2013, Kocot et al. 2013).

Few studies have compared transcriptome completeness across distantly related taxa (e.g. Riesgo et al. 2012, Francis et al. 2013), and with current Illumina technologies using 150 bp pair-end reads, 20 to 30 million reads can provide fairly complete transcriptomes (Francis et al. 2013, Riesgo et al. 2014) at a relatively low cost (less than US$500 per transcriptome). This cost and effort efficiency has pushed the number of taxa in phylogenomic analyses to large numbers in the past five years as sequencing technologies continue to improve (Illumina is now providing 250 bp reads).

Another issue with transcriptomics is data matrix completeness. While the first EST-based phylogenies had large numbers of genes with incomplete data matrices (Dunn et al. 2008, Hejnol et al. 2009), or were restricted to just a few genes (Delsuc et al. 2006), recent analyses have produced hundreds of genes with high levels of matrix completeness, especially when based on the latest generation of transcriptomic and genomic data (e.g. Andrade et al. 2014, Zapata et al. 2014, González et al. 2015, Laumer et al. 2015).

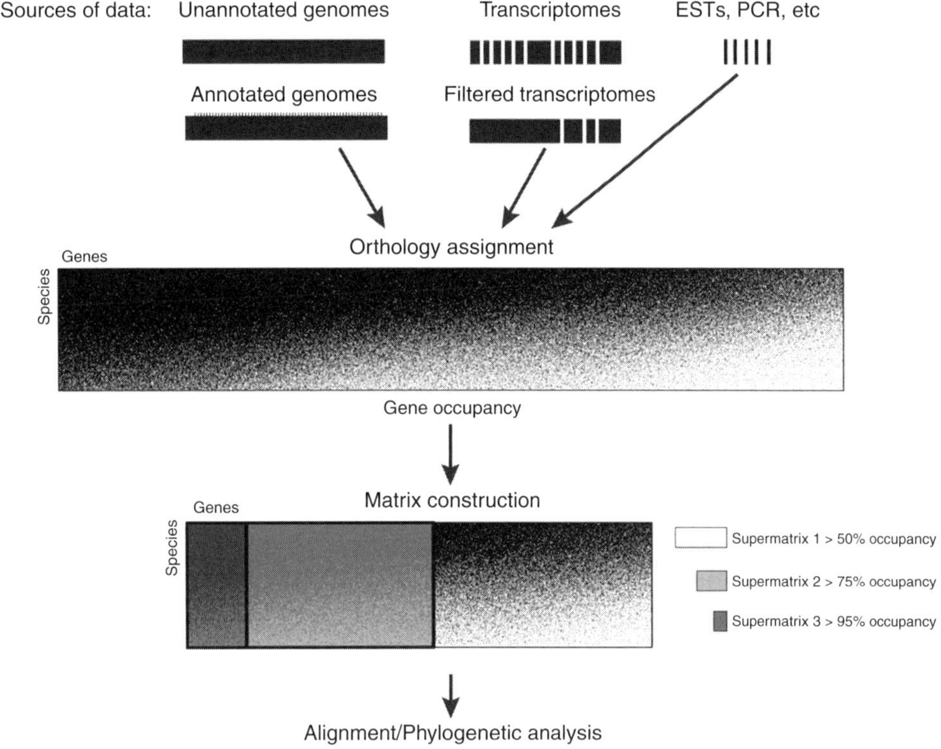

Fig 13.2 Schematic representation of the early steps of data analysis in a typical phylogenomic analysis up to the assembly of the data matrices for subsequent analyses. Input data may be genomes, transcriptomes, ESTs or PCR-amplified loci; all these data can be used for orthology assignment in a typical all-by-all approach. With the orthology matrix, using a criterion of gene occupancy, one may design one or more matrices.

13.3 Phylogenomics

While issues of homology in traditional systematics were often relegated to the nucleotide level during multiple sequence alignment (or fragment level in dynamic homology) (e.g. Wheeler 1995, 1996, 2005), the most critical step in phylogenomic analyses is orthology assignment (see Fig 13.2) – whether beginning from well-annotated genomes or from *de novo* transcriptome assembly. Another major difference is that phylogenomic analyses are often conducted with amino acid data instead of with the more common nucleotide data of traditional analyses.

During the orthology assignment step, all genomes or transcriptomes need to be evaluated to select the set of genes that will be compared downstream. Originally, sets of genes were usually identified and curated manually, although these approaches yielded a minute usable fraction of the available data (Philippe et al.

2004). More intelligent approaches have used all-by-all algorithms to automatically select sets of orthologous genes (Dunn et al. 2008, Smith and Dunn 2008, Hejnol et al. 2009). Several popular implementations include OMA (Altenhoff et al. 2011, 2013), first applied to phylogenomic analyses by Fernández et al. (2014b), HaMStR (Ebersberger et al. 2009), applied to phylogenomics by Kocot et al. (2011), or the all-by-all method implemented in Agalma (Dunn et al. 2013), originally developed by Dunn et al. (2008). Few studies have tested alternative methods of orthology assignment, however. Yet these have shown that phylogenetic results are robust to matrix construction (Zapata et al. 2014), and comparable results are obtained between studies using not only different orthology assignment algorithms, but even different species (e.g. compare Kocot et al. 2011, Smith et al. 2011). Solutions to the existence of multiple isoforms and paralogy (Spitzer et al. 2006) are discussed in several of these studies.

Once orthology is determined, researchers must select which orthogroups to use for downstream analyses. A common means to select orthogroups agnostically is by including all orthogroups present in four or more taxa for an initial matrix construction, and then only include those present above a certain threshold, e.g. 50% of taxa. Many recent studies typically investigate alternative matrices with varying levels of occupancy, such as 50%, 75%, 95% or even 100% (e.g. Andrade et al. 2014, Zapata et al. 2014, González et al. 2015, Lemer et al. 2015). These matrices are later subjected to multiple sequence alignment, concatenation and standard methods of phylogenetic analysis.

While in traditional Sanger-based phylogenetic analyses authors often evaluated incongruence among partitions using topological tests, recent developments in coalescent theory and a better understanding of non-tree-like evolution in some organisms have led to alternative methods to evaluate phylogenies. Some of these methods attempt to look at the 'species trees' as a whole, instead of as a sum of 'gene trees' (e.g. Liu et al. 2008, Liu et al. 2009, Heled and Drummond 2010, Liu et al. 2010, Boussau et al. 2013, Nakhleh 2013), and this has become so important as to seriously question concatenation approaches (Rokas et al. 2005, Xi et al. 2014, but see Dikow 2011, Tonini et al. 2015). Others have explored putative incongruence by using network methods (Fernández et al. 2014b, Laumer et al. 2015), or looked at possible conflict between partitions or sets of genes (Nosenko et al. 2013, Fernández et al. 2014a, Sharma et al. 2014). Many of these issues are restricted to the new forms of data and do not apply to older molecular phylogenetic studies.

While much discussion in the older systematic literature was spent on the virtues of favoured methods of phylogenetic analysis, few have evaluated the conditions and methods and their convergence on solutions. Phylogenomic analyses however have shown resolution and strong support for many relationships that were recalcitrant to resolution using older data (e.g. Kocot et al. 2011 and Smith et al. 2011 for molluscs; Weigert et al. 2014 for annelids). These results are likewise robust to data selection (method of orthology assignment) and taxon sampling.

At a smaller scale, recent work has shown that for densely sampled taxa, many methods of analysis converge on a stable solution, whether the data are analysed under dynamic or static homology criteria and whether parsimony or probabilistic methods are employed (Giribet and Edgecombe 2013, Giribet et al. 2014). This is supported by a number of theoretical analyses of method convergence (e.g. Tuffley and Steel 1997, Steel and Penny 2004; summarized in Wheeler 2012). However, the addition of data, especially in phylogenomic analysis can, at times, result in artificially supported nodes (Salichos and Rokas 2013), making it important to evaluate support with the addition of data (e.g. Fernández et al. 2014a, Sharma et al. 2014) or the existence of conflict between sets of data (Nosenko et al. 2013).

Accurate genome sequencing and assembly still relies in long-read technologies, which tend to be much more expensive than short-read Illumina sequencing. However, Illumina is now providing innovative long-read technology, sequencing up to 10 Kb fragments by combining highly accurate short reads, thus at lower cost. Long reads are generated by creating genomic DNA libraries of 10 Kb that are clonally amplified, sheared, and labelled with unique indices. The fragments are then sequenced using short sequence reads that are assembled separately into synthetic long reads, resulting in a full sequence for each long fragment. Such methods are currently under development and should greatly simplify the assembly process of whole (or nearly whole) genomes.

13.4 Dating the tree of life

While we have focused on estimating the tree of life for decades, dating the tree has become an important goal in recent times (Donoghue and Benton 2007). Generating such chronograms has recently reinvigorated the debate about total evidence, as a way to do accurate 'tip dating' or 'total evidence dating', as opposed to node dating – or the use of fossils to calibrate nodes (e.g. see Murienne et al. 2010, Pyron 2011, Wood et al. 2013, Garwood et al. 2014) (see an example of tip dating in Fig 13.3). Philosophical, methodological and theoretical developments for dealing with fossils are crucial (Parham et al. 2012, Clarke and Boyd 2015), especially when combining fossil data with large genetic data sets (e.g. Pyron 2011, Giribet 2015). Ideally, a complete and accurately dated tree of life, including extant and extinct diversity, could serve to answer many systematic and evolutionary questions. These could include meaningful precise comparisons of clades of comparable age, as opposed to using Linnaean ranks as proxies for such clades. For example, we would not need to count the number of families present in a particular unit of the geological time scale, as we could look at how many lineages originated during that time slice persisted, diversified, went extinct, etc.

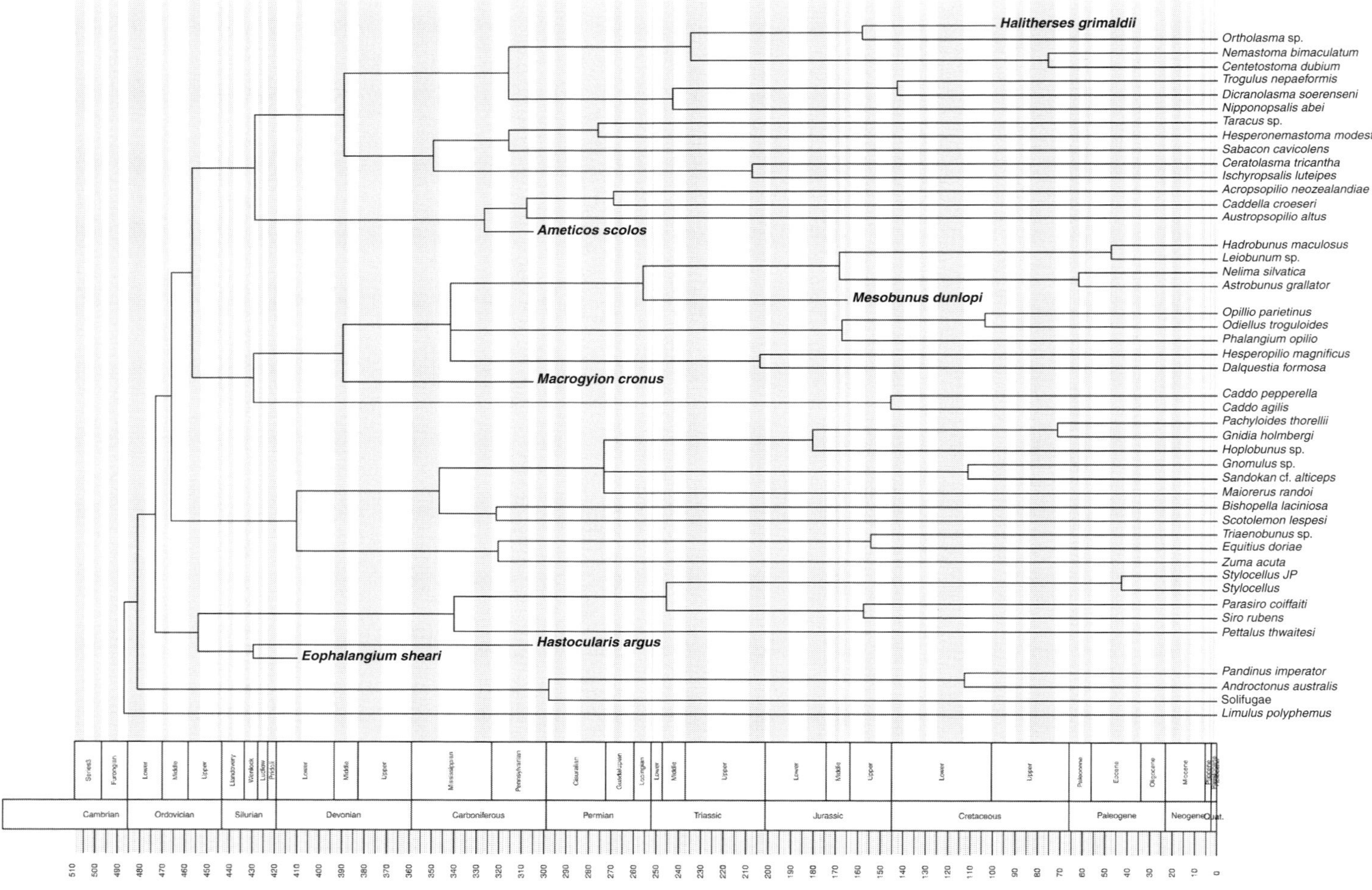

Fig 13.3 Chronogram of the arachnid order Opiliones obtained through 'tip dating' or using the fossils (in bold) as terminals in a combined analysis of morphology and molecular data. Tree modified from Sharma and Giribet (2014).

This achievable goal would have satisfied Hennig given his suggestion that taxonomic rank could be unambiguously determined by age of origin of taxa. The notion of taxonomic rank as something to worry about or not, could be resolved by modern node dating techniques. We would still have issues of sister (coeval) taxa requiring equal ranks (e.g. *Latimeria* and Tetrapoda), but we would have an empirical, falsifiable basis (at least for extant taxa and their ghost lineages, Norell 1992) for rank determination.

So as we sit here 100 years after Hennig's birth, what are the Hennigian elements that are with us today? Certainly the idea of monophyly and recency of common ancestry as fundamental natural comparable units flows through every aspect of modern analysis. The considerable technical advances that have made molecular data so easily available were unknown to Hennig, but all of these data types are still brought to bear on questions of whether groups are monophyletic or not. The Hennigian notion of synapomorphy as the only valid basis for establishing monophyly certainly exists in parsimony analyses of molecular data (e.g. Frost et al. 2006, Clouse and Wheeler 2014) and in a 'softer' form in other types of analysis (such as maximum average likelihood). What are the aspects of the most recent common ancestor of groups? How have these aspects transformed and are they linked to internal or external changes elsewhere? These were questions for Hennig and they are questions for us. It is a testament to the strength of his ideas that as we fathom more about life, the same concepts that Hennig applied to the anatomy he could observe are applied to molecular data he could never have imagined.

What does the future hold for systematics? Since we will, very soon, have all the genomic data available for most living taxa, is the end near? Clearly not. As mentioned above, only a very small fraction of Earth's life is alive today and we cannot understand even extant life without knowledge of what went before. An understanding of the totality of life requires anatomical analysis even in the presence of whole genomes (e.g. Pyron 2011, Giribet 2015). Time-consuming and painstaking work for sure, but necessary. Additionally, we are starting to see the beginnings of high-throughput morphological analysis (sometimes referred to as 'phenomics') in the form of CT-scanning and other forms of 3D, non-invasive imaging (e.g. Ziegler et al. 2010, Ziegler and Menze 2013). As these techniques mature (e.g. Ziegler et al. 2008, Ziegler and Menze 2013) and specific ontologies develop (Ramírez et al. 2007, Dahdul et al. 2010, Richter et al. 2010, Seltmann et al. 2012, Vogt et al. 2012, Deans et al. 2015), we will surely see new patterns emerge and new questions present themselves.

There remain types of questions systematics is only beginning to ask. Currently, we ask questions about what sort of changes have occurred in organisms, their relative timing, and correlation. However, as techniques such as RNA-interference have become available, we are able to examine the 'how' questions. How do molecular changes and gene regulation relate to morphological transformations? How does the developmental process map sequence data to anatomical pattern we can see in both extant and extinct creatures. Girded with this level of understanding, we can pursue even larger questions of how Earth history, in terms of environment and biogeography, has moulded life. Enormous changes in temperature, atmospheric gas composition, and periodic bolide impacts have occurred on Earth over the past 4 billion years. How have these forces affected life – its origin, extinction, and evolution?

Acknowledgements

We would like to thank Rosa Fernández who produced Fig 13.3 and Steven Thurston for expert aid with figures. We would also like to thank Gregory Edgecombe and David Williams for many helpful comments that improved the manuscript greatly.

References

Altenhoff, A.M., Gil, M., Gonnet, G.H. and Dessimoz, C. (2013). Inferring hierarchical orthologous groups from orthologous gene pairs. *PLoS One*, 8, e53786.

Altenhoff, A.M., Schneider, A., Gonnet, G.H. and Dessimoz, C. (2011). OMA 2011: orthology inference among 1000 complete genomes. *Nucleic Acids Research*, 39, D289–D294.

Andrade, S.C.S., Montenegro, H., Strand, M., et al. (2014). A transcriptomic approach to ribbon worm systematics (Nemertea): resolving the Pilidiophora problem. *Molecular Biology and Evolution*, 31, 3206–3215.

Andrade, S.C.S., Novo, M., Kawauchi, G.Y., Worsaae, K., Pleijel, F., Giribet, G. and Rouse, G.W. (2015). Articulating the "archiannelids": A phylogenomic approach to annelid relationships with emphasis on meiofaunal taxa. *Molecular Biology and Evolution*, 32, 2860–2875.

Arcila, D., Pyron, R.A., Tyler, J.C., Ortí, G. and Betancur-R, R. (2015). An evaluation of fossil tip-dating versus node-age calibrations in tetraodontiform fishes (Teleostei: Percomorphaceae). *Molecular Phylogenetics and Evolution*, 82, 131–145.

Boussau, B., Szollosi, G.J., Duret, L., et al. (2013). Genome-scale coestimation of species and gene trees. *Genome Research*, 23, 323–330.

Cannon, J.T., Kocot, K.M., Waits, D.S., et al. (2014). Phylogenomic resolution of the hemichordate and echinoderm clade. *Current Biology* 24, 2827–2832.

Clark, A.G., Eisen, M.B., Smith, D.R., et al. (2007). Evolution of genes and genomes on the *Drosophila* phylogeny. *Nature*, 450, 203–218.

Clarke, J.A. and Boyd, C.A. (2015). Methods for the quantitative comparison of molecular estimates of clade age and the fossil record. *Systematic Biology*, 64, 25–41.

Clouse, R.M. and Wheeler, W.C. (2014). Descriptions of two new, cryptic species of *Metasiro* (Arachnida: Opiliones:

Cyphophthalmi: Neogoveidae) from South Carolina, USA, including a discussion of mitochondrial mutation rates. *Zootaxa*, 3814, 177–201.

Cooper, A., Lalueza-Fox, C., Anderson, S., et al. (2001). Complete mitochondrial genome sequences of two extinct moas clarify ratite evolution. *Nature*, 409, 704–707.

Dahdul, W.M., Lundberg, J.G., Midford, P.E., et al. (2010). The teleost anatomy ontology: Anatomical representation for the genomics age. *Systematic Biology*, 59, 369–383.

Deans, A.R., Lewis, S.E., Huala, E., et al. (2015). Finding our way through phenotypes. *PLoS Biology*, 13, e1002033.

Delsuc, F., Brinkmann, H., Chourrout, D. and Philippe, H. (2006). Tunicates and not cephalochordates are the closest living relatives of vertebrates. *Nature*, 439, 965–968.

Dikow, R.B. (2011). Genome-level homology and phylogeny of *Shewanella* (Gammaproteobacteria: lteromonadales: Shewanellaceae). *BMC Genomics*, 12, 237.

Donoghue, P.C. and Benton, M.J. (2007). Rocks and clocks: calibrating the Tree of Life using fossils and molecules. *Trends in Ecology and Evolution*, 22, 424–431.

Dunn, C.W., Hejnol, A., Matus, D.Q., et al. (2008). Broad phylogenomic sampling improves resolution of the animal tree of life. *Nature*, 452, 745–749.

Dunn, C.W., Howison, M. and Zapata, F. (2013). Agalma: an automated phylogenomics workflow. *BMC Bioinformatics*, 14, 330.

Ebersberger, I., Strauss, S. and von Haeseler, A. (2009). HaMStR: Profile hidden markov model based search for orthologs in ESTs. *BMC Evolutionary Biology*, 9, 157.

Fernández, R., Hormiga, G. and Giribet, G. (2014a). Phylogenomic analysis of spiders reveals nonmonophyly of orb weavers, *Current Biology*, 24, 1772–1777.

Fernández, R., Laumer, C.E., Vahtera, V., et al. (2014b). Evaluating topological conflict in centipede phylogeny using transcriptomic data sets. *Molecular Biology and Evolution*, 31, 1500–1513.

Francis, W.R., Christianson, L.M., Kiko, R., et al. (2013). A comparison across non-model animals suggests an optimal sequencing depth for *de novo* transcriptome assembly. *BMC Genomics*, 14, 167.

Freedman, A.H., Gronau, I., Schweizer, R.M., et al. (2014). Genome sequencing highlights the dynamic early history of dogs. *PLoS Genetics*, 10, e1004016.

Frost, D.R., Grant, T., Faivovich, J., et al. (2006). The amphibian tree of life. *Bulletin of the American Museum of Natural History*, 297, 1–370.

Garwood, R.J., Sharma, P.P., Dunlop, J.A. and Giribet, G. (2014). A new stem group Palaeozoic harvestman revealed through integration of phylogenetics and development. *Current Biology*, 24, 1–7.

Giribet, G. (2015). Morphology should not be forgotten in the era of genomics – a phylogenetic perspective. *Zoologischer Anzeiger*, 256, 96–103.

Giribet, G. and Edgecombe, G.D. (2013). Stable phylogenetic patterns in scutigeromorph centipedes (Myriapoda: Chilopoda: Scutigeromorpha): dating the diversification of an ancient lineage of terrestrial arthropods. *Invertebrate Systematics*, 27, 485–501.

Giribet, G., McIntyre, E., Christian, E., et al. (2014). The first phylogenetic analysis of Palpigradi (Arachnida) – the most enigmatic arthropod order. *Invertebrate Systematics*, 28, 350–360.

González, V.L., Andrade, S.C.S., Bieler, R., et al. (2015). A phylogenetic backbone for Bivalvia: an RNA-seq approach. *Proceedings of the Royal Society, B Biological Science*, 282, 2014.2332.

Gravel, S., Zakharia, F., Moreno-Estrada, A., et al. (2013). Reconstructing Native American migrations from whole-genome and whole-exome data. *PLoS Genetics*, 9, e1004023.

Hejnol, A., Obst, M., Stamatakis, A., M., O., et al. (2009). Assessing the root of bilaterian animals with scalable phylogenomic methods. *Proceedings of the Royal Society, B Biological Science*, 276, 4261–4270.

Heled, J. and Drummond, A.J. (2010). Bayesian inference of species trees from multilocus data. *Molecular Biology and Evolution*, 27, 570–580.

Hennig, W. (1950). *Grundzüge einer Theorie der phylogenetischen Systematik.* Berlin: Deutscher Zentralverlag.

Hennig, W. (1966). *Phylogenetic Systematics.* Urbana: University of Illinois Press.

Kemp, C. (2015). The endangered dead. *Nature*, 518, 292–294.

Kocot, K.M., Cannon, J.T., Todt, C., et al. (2011). Phylogenomics reveals deep molluscan relationships. *Nature*, 447, 452–456.

Kocot, K.M., Citarella, M.R., Moroz, L.L. and Halanych, K.M. (2013). PhyloTreePruner: A phylogenetic tree-based approach for selection of orthologous sequences for phylogenomics. *Evolutionary Bioinformics*, 9, 429–435.

Laumer, C.E., Hejnol, A. and Giribet, G. (2015). Nuclear genomic signals of the "microturbellarian" roots of platyhelminth evolutionary innovation. *eLife* 4, e05503.

Lee, E.K., Cibrian-Jaramillo, A., Kolokotronis, S.-O., et al. (2011). A functional phylogenomic view of the seed plants. *PLoS Genetics*, 7, e1002411.

Lemer, S., Kawauchi, G.Y., Andrade, S.C.S., et al. (2015). Re-evaluating the phylogeny of *Sipuncula* through transcriptomics. *Molecular Phylogenetics and Evolution*, 83, 174–183.

Lemmon, A.R., Emme, S.A. and Lemmon, E.M. (2012). Anchored hybrid enrichment for massively high-throughput phylogenomics. *Systematic Biology*, 61, 727–744.

Liu, L., Pearl, D.K., Brumfield, R.T. and Edwards, S.V. (2008). Estimating species trees using multiple-allele DNA sequence data. *Evolution*, 62, 2080–2091.

Liu, L., Yu, L. and Edwards, S.V. (2010). A maximum pseudo-likelihood approach for estimating species trees under the coalescent model. *BMC Evolutionary Biology*, 10, 302.

Liu, L., Yu, L.L., Kubatko, L., Pearl, D.K. and Edwards, S.V. (2009). Coalescent methods for estimating phylogenetic trees. *Molecular Phylogenetics and Evolution*, 53, 320–328.

Mamanova, L., Coffey, A.J., Scott, C.E., et al. (2010). Target-enrichment strategies for next-generation sequencing. *Nature Methods*, 7, 111–118.

McCormack, J.E., Hird, S.M., Zellmer, A.J., Carstens, B.C. and Brumfield, R.T. (2013). Applications of next-generation sequencing to phylogeography and phylogenetics. *Molecular Phylogenetics and Evolution*, 66, 526–538.

Meyer, M., Kircher, M., Gansauge, M.T., et al. (2012). A high-coverage genome sequence from an archaic Denisovan individual. *Science*, 338, 222–226.

Misof, B., Liu, S., Meusemann, K., et al. (2014). Phylogenomics resolves the timing and pattern of insect evolution. *Science*, 346, 763–767.

Murienne, J., Edgecombe, G.D. and Giribet, G. (2010). Including secondary structure, fossils and molecular dating in the centipede tree of life. *Molecular Phylogenetics and Evolution*, 57, 301–313.

Nakhleh, L. (2013). Computational approaches to species phylogeny inference and gene tree reconciliation. *Trends in Ecology and Evolution*, 28, 719–728.

Norell, M.A. (1992). Taxic origin and temporal diversity: the effect of phylogeny. In *Extinction and phylogeny*, ed. M.J. Novacek and Q.D. Wheeler. New York: Columbia University Press pp. 89–118.

Nosenko, T., Schreiber, F., Adamska, M., et al. (2013). Deep metazoan phylogeny: when different genes tell different stories. *Molecular Phylogenetics and Evolution*, 67, 223–233.

Parham, J.F., Donoghue, P.C.J., Bell, C.J., et al. (2012). Best practices for justifying fossil calibrations. *Systematic Biology*, 61, 346–359.

Peloso, P.L.V., Frost, D.R., Richards, S.J., et al. (2016). The impact of anchored phylogenomics and taxon sampling on phylogenetic inference in narrow-mouthed frogs (Anura, Microhylidae). *Cladistics*, 32, 113–140.

Philippe, H., Snell, E.A., Bapteste, E., et al. (2004). Phylogenomics of eukaryotes: impact of missing data on large alignments. *Molecular Biology and Evolution*, 21, 1740–1752.

Pyron, R.A. (2011). Divergence time estimation using fossils as terminal taxa and the origins of Lissamphibia. *Systematic Biology*, 60, 466–481.

Ramírez, M.J., Coddington, J.A., Maddison, W.P., et al. (2007). Linking of digital images to phylogenetic data matrices using a morphological ontology. *Systematic Biology*, 56, 283–294.

Richter, S., Loesel, R., Purschke, G., et al. (2010). Invertebrate neurophylogeny: suggested terms and definitions for a neuroanatomical glossary. *Frontiers in Zoology*, 7, 29.

Riesgo, A., Andrade, S.C.S., Sharma, P.P., et al. (2012). Comparative description of ten transcriptomes of newly sequenced invertebrates and efficiency estimation of genomic sampling in non-model taxa. *Frontiers in Zoology*, 9, 33.

Riesgo, A., Farrar, N., Windsor, P.J., Giribet, G. and Leys, S.P. (2014). The analysis of eight transcriptomes from all Porifera classes reveals surprising genetic complexity in sponges. *Molecular Biology and Evolution*, 31, 1102–1120.

Rokas, A., Krüger, D. and Carroll, S.B. (2005). Animal evolution and the molecular signature of radiations compressed in time. *Science*, 310, 1933–1938.

Salichos, L. and Rokas, A. (2013). Inferring ancient divergences requires genes with strong phylogenetic signals. *Nature*, 497, 327–331.

Seltmann, K.C., Yoder, M.J., Mikó, I., et al. (2012). A hymenopterists' guide to the Hymenoptera Anatomy Ontology: utility, clarification, and future directions. *Journal of Hymenoptera Research*, 27, 67–88.

Sharma, P.P. and Giribet, G. (2014). A revised dated phylogeny of the arachnid order Opiliones. *Frontiers in Genetics*, 5, 255.

Sharma, P.P., Kaluziak, S., Pérez-Porro, A.R., et al. (2014). Phylogenomic interrogation of Arachnida reveals systemic conflicts in phylogenetic signal. *Molecular Biology and Evolution*, 31, 2963–2984.

Smith, B.T., Harvey, M.G., Faircloth, B.C., Glenn, T.C. and Brumfield, R.T. (2014). Target capture and massively parallel sequencing of ultraconserved elements for comparative studies at shallow evolutionary time scales. *Systematic Biology*, 63, 83–95.

Smith, S., Wilson, N.G., Goetz, F., et al. (2011). Resolving the evolutionary relationships of molluscs with phylogenomic tools. *Nature*, 480, 364–367.

Smith, S.A. and Dunn, C.W. (2008). Phyutility: a phyloinformatics tool for trees, alignments and molecular data. *Bioinformatics*, 24, 715–716.

Spitzer, M., Lorkowski, S., Cullen, P., Sczyrba, A. and Fuellen, G. (2006). IsoSVM–distinguishing isoforms and paralogs on the protein level. *BMC Bioinformatics*, 7, 110.

Steel, M. and Penny, D. (2004). Two further links between MP and ML under the poisson model. *Applied Mathematics Letters*, 17, 785–790.

Tonini, J., Moore, A., Stern, D., Shcheglovitova, M. and Orti, G. (2015). Concatenation and species tree methods exhibit statistically indistinguishable accuracy under a range of simulated conditions. *PLoS Currents Tree of Life*, 7.

Tuffley, C. and Steel, M. (1997). Links between maximum likelihood and maximum parsimony under a simple model of site substitution. *Bulletin of Mathematical Biology*, 59, 581–607.

Vogt, L., Grobe, P., Quast, B. and Bartolomaeus, T. (2012). Accommodating ontologies to biological reality – top- level categories of cumulative-constitutively organized material entities. *PLoS One*, 7, e30004.

Weigert, A., Helm, C., Meyer, M., et al. (2014). Illuminating the base of the annelid tree using transcriptomics. *Molecular Biology and Evolution*, 31, 1391–1401.

Wheeler, W.C. (1995). Sequence alignment, parameter sensitivity, and the phylogenetic analysis of molecular data. *Systematic Biology*, 44, 321–331.

Wheeler, W.C. (1996). Optimization alignment: the end of multiple sequence alignment in phylogenetics? *Cladistics*, 12, 1–9.

Wheeler, W.C. (2005). Alignment, dynamic homology, and optimization. In *Parsimony, phylogeny, and genomics*, ed. V.A. Albert. Oxford: Oxford University Press, pp. 71–80.

Wheeler, W.C. (2012). *Systematics: A Course of Lectures*. Chichester: Wiley-Blackwell.

Wood, H.M., Matzke, N.J., Gillespie, R.G. and Griswold, C.E. (2013). Treating fossils as terminal taxa in divergence time estimation reveals ancient vicariance patterns in the palpimanoid spiders. *Systematic Biology*, 62, 264–284.

Xi, Z.X., Liu, L., Rest, J.S. and Davis, C.C. (2014). Coalescent versus concatenation methods and the placement of amborella as sister to water lilies. *Systematic Biology*, 63, 919–932.

Zapata, F., Wilson, N.G., Howison, M., et al. (2014). Phylogenomic analyses of deep gastropod relationships reject Orthogastropoda. *Proceedings of the Royal Society, B Biological Science*, 281, 2014.1739.

Zhan, S., Zhang, W., Niitepõld, K., et al. (2014). The genetics of monarch butterfly migration and warning colouration. *Nature*, 514, 317–321.

Ziegler, A., Faber, C., Mueller, S. and Bartolomaeus, T. (2008). Systematic comparison and reconstruction of sea urchin (Echinoidea) internal anatomy: a novel approach using magnetic resonance imaging. *BMC Biology*, 6, 33.

Ziegler, A. and Menze, B.H. (2013). Accelerated acquisition, visualization, and analysis of zoo-anatomical data. In *Computation for Humanity*, eds J. Zander and P.J. Mosterman. Information Technology to Advance Society, Boca Raton: CRC Press, pp. 233–260.

Ziegler, A., Ogurreck, M., Steinke, T., et al. (2010). Opportunities and challenges for digital morphology. *Biology Direct*, 5, 45.

Hennig, Løvtrup, evolution and biology

ROBIN BRUCE

14.1 Introduction

> Biology is a science of antitheses. If we make a general survey of biological science
> we find that it suffers from cleavages of a kind and to a degree which is unknown in
> such a well unified science as, for example, chemistry. (Woodger 1929: 11)
>
> In no other science are the contrasts and struggle for survival among subdivisions
> of the field so strong as in biology. This is at least partly because the problems,
> and therefore the methods, are more varied in biology than in any other science.
> (Hennig 1966: 1)

Both quotations above, by Joseph Henry Woodger and Willi Hennig, point to a lack
of unity in biological thought, and instead point to the schisms and factions which
fracture past and current biology. Today, an evolutionary perspective is usually
taken to allow and enable a unity for this fractured diversity. But is it that simple? Is
evolution, in some theory, form or model, the unifying principle for biology that it is
usually claimed to be by many biologists?

Hennig created a theory of systematics based on the modern view of evolution
being just such a principle. Despite the triumphs and ascendancy of phylogenetic
systematics over the last decades, doubts remain on the primacy of evolution
over systematics, as several chapters in this collection demonstrate (e.g. Brower,
Chapter 5), and the debate still evades satisfactory resolution. One way of progress-
ing might be to expand the debate beyond the comparative biology of diversity and
include the nature of the biological individual. Edward Stuart Russell (1887–1954;
see Lauder 1982 for a brief biography) noted that in the long history of biological

The Future of Phylogenetic Systematics: The Legacy of Willi Hennig, eds. D. Williams,
M. Schmitt and Q. Wheeler. Published by Cambridge University Press.

thought only two ideas emerged to explain the development of individuals: preformation and epigenesis (Russell 1930) – and beyond development, what else is there in the world of the biology of individuals to explain? For does not development lead to growth, and growth to reproduction, and reproduction to renewal, and renewal to development, in sum, the continuity of living things?

Hennig, I believe, can be fairly recognized as a preformationist: that characters persist and allow for the proposed alignments of presumed descent. Søren Løvtrup (1922–2002) was an epigeneticist: for him, completely functional novelties do arise, see for example his reflections 'the creative aspect' in Løvtrup (1984: 166–169). Løvtrup, despite his efforts and sympathies in epigenetics and trenchant criticisms of the modern (synthetic) theory of evolution, also took a phylogenetic perspective. That two hugely able, productive and well-read biologists, but with profoundly different views on development, align themselves with phylogenetics, suggests either that phylogenetics can indeed comfortably enclose development, or that one or the other or both biologists have been led astray by the supposed efficacy of evolution, in this respect. That is to say, can history as phylogeny enclose development and thus then direct the present into the future, or can and does development transcend history? This chapter tries to find answers to these suggestions and asks a number of questions which evolutionary theory currently ignores but that seem fundamental for a more complete biology.

Løvtrup, for example, recognized four theories of evolution: the theory on the reality of evolution, the theory on the history of evolution, the epigenetic theory on the mechanism of evolution and the ecological theory on the mechanism of evolution (Løvtrup 1984). He tried to disentangle their confluences and divergences with some success which led to his more critical exposure of many evolutionary fallacies (Løvtrup 1987).

Gareth Nelson stopped listing evolutionary theories after identifying ten: Darwinism, neo-darwinism, neutralism, panbiogeography, macromutationalism, sociobiology, punctuationism, non-equilibrium thermodynamicalism, hierarchicalism and structuralism (Nelson 1989a). So might one be permitted to ask – how *exactly* does evolution *unify* biological thought?

Hennig used phylogeny as the organizing principle for his systematics, and by so doing has sparked a five-decade long review, revision and re-analysis of the relations between classification and evolution. As yet there does not appear to be any unity for a possible resolution for what has grown into yet another biological antithesis. Donn Rosen (1984) quoted Theodosius Dobzhansky's distinction between *evolution as process*, and *evolution as product*, before he (Rosen) stated that the product of evolution is phylogeny (Rosen 1984: 92). Out of these times somehow arose the false dichotomy of process and pattern which has much affected biological thought, and not in a progressive way. But Rosen (and Dobzhansky) recognized the dichotomy as process and product, both actual realities; how *product*, the empirical reality was transformed into *pattern*, an abstraction of that reality, is a curious thread in the writings of these times. For a perennially fresh and compelling review of the problems of trying to resolve classification with evolution and vice versa see Colin Patterson (2002) and comments by Peter Forey (2002). Patterson's text benefits from never having been intended for publication (it was for oral presentation only, but bad manners triumphed in the short term) and thus has a tentativeness which would surely have been edited out in review for publication.

14.2 Evolution as idea and cosmogony

In the history of European thought there have been three periods of constructive cosmological thinking; three periods, that is to say, when the idea of nature has come into the focus of thought, become the subject of intense and protracted reflection, and consequently acquired new characteristics which in their turn have given a new aspect to the detailed science of nature that has been based upon it. (Collingwood 1945: 1)

Modern cosmology could only have arisen from a widespread familiarity with historical studies, and in particular with historical studies of the kind which placed the conception of process, change, development in the centre of their picture and recognized it as the fundamental category of historical thought. (Collingwood 1945: 10)

Robin George Collingwood (1889–1943; see Inglis 2009) as both a historian and philosopher was well placed to try to unravel the idea of nature and its history through time in European thought. This work was published posthumously in 1945 as *The Idea of Nature*. Collingwood could discern three distinct periods of the idea of nature in western thought – Greek, post-Renaissance and modern. The microcosm-macrocosm of the Greeks with their emphasis on wholeness – Humphrey Davey Findley Kitto (1983), for example, a classics scholar, quoted Matthew Arnold's verse about Sophocles 'who saw life steadily, and saw it whole' (*To A Friend* 1849), which may serve as a suitable aphorism for our current needs. The second cosmology was that of materialism and mechanism of the post-Renaissance, the world of Galileo, Descartes and Newton, with nature now analogous to a machine. The third and current cosmology is the evolutionism of the modern world. The legacies of the earlier cosmologies linger on into the later attempts at system building, and the modern period suffers obviously from being active, current and therefore unsettled. Sadly, Collingwood's early death stopped any further reflections he might have provided us with respect to our modern concept of nature.

Our current view of nature is, then, an evolutionary one, a view I do not think many biologists would disagree with. Collingwood locates the beginnings of this idea in the mid-eighteenth century with certain historical studies, especially by Turgot and Voltaire which put process, change and development at the centre of their world; the next half-century saw the idea morph into 'progress' and thence into 'evolution' by the efforts of Erasmus Darwin and Lamarck; most biologists now like to fix the date as 1859. Collingwood here seems to be the more perceptive.

I have no problem with evolution as an idea. But unfortunately ideas have a long habit of taking on a life of their own and becoming realities should we allow them the latitude to so do. Therefore I must confess to being uncomfortable with evolution being portrayed as fact or reality, rather than the idea of nature at the centre of our modern cosmology and this is where I believe Collingwood wished it placed. Evolution, the word, as a mere abstract noun, now seems hugely over-burdened with hope and expectation, not to say process, progress, pattern, theory, method, fact, explanation and reality in modern biology, and a return to the simpler notion of *evolution as an idea of nature* might help clarify much biological thought.

14.3 On biological antitheses especially preformation and epigenesis

> Men can do nothing without the make-believe of a beginning. (George Eliot, the opening lines of *Daniel Deronda* 1876)

> Another factor which undoubtedly helped in the establishment of the preformationist view was the Biblical doctrine of creation [...] No new creation ever took place [...] Preformation was essentially a derivative of the philosophical doctrine of determinism; whether this original predetermination was due to a Divine act of creation or to 'blind chance' (as in the materialistic view) amounted to very much the same thing. (Russell 1930:27)

Woodger's *Biological Principles* (1929) was an attempt to refine, resolve or remove the antitheses which he saw being part of the obstructions in the way of a more unified biology. The antitheses which he recognized and reviewed were *vitalism and mechanism* (covering 44 text pages), *structure and function* (5), *organism and environment* (3), *preformation and epigenesis* (95), *teleology and causation* (23 excluding the appendix) and *mind and body* (19). Perhaps, as an embryologist, it could be expected that Woodger would devote so many pages to preformation and epigenesis. Perhaps an ecologist or physiologist would make much more of organism and environment, an anatomist or physiologist more of structure and function, a psychologist more of mind and body, and philosophers more of the remaining two. But perhaps the issues of preformation and epigenesis lie at the very core of biology – where the historical, preformed materials containing encoded past achievements are seized, fused, encapsulated and become the developing, living being of the present, which then molds them into a steady and whole functionality, in step and rhythm with temporal generation. Woodger returned to the problems of preformation and epigenesis again, perhaps in recognition of the problems therein which he had not resolved in *Biological Principles*. With particular relevance to the purpose of our present volume, Woodger's final words in his paper 'On Biological transformations' are worthy of quoting in full:

> That biology must aim at some such goal as this [analysis of elementary developmental processes] there seems to be little reason to doubt, but we are obviously only at the beginning of the long journey that must be accomplished before it is reached, and we can only dimly discern the route to be followed. But if it were reached we can see how completely taxonomy and morphology would be transformed, and how many of their puzzles would disappear. At present we are in a position analogous to chemistry before the advent of atomic theory. When biological theory is advanced along the lines suggested we shall be able to replace the classification of adults by a classification of zygote types, and phylogenetic speculation will be transferred from the plane of taxonomic transformations between adults to the plane of actual or possible genetic transformations between zygotes. (Woodger 1945: 120)

Rather than phylogenetic systematics, Woodger here seems to be pressing for the development of an ontogenetic systematics. One can, by a certain oblique logic, argue that phylogeny does indeed create ontogeny, by a sort of historical precedence of the continuation of dismembered characters – preformation in short. Garstang (1922), and Løvtrup (1984) quoting Garstang, present cogent summaries that ontogeny creates phylogeny – epigenesis in short.

Woodger has been largely ignored by most biologists and philosophers of biology over recent decades. But Hennig (1966) understood his importance (see Varma, Chapter 16), as did Løvtrup (1974). Nicholson and Gawne (2014) review and critique Woodger's huge contribution to biology. They pose a series of questions about current philosophy of biology and its persistent suppression and misrepresentation of Woodger's role in the discipline, which I believe cannot be ignored by any open-minded readers. Whether these questions will ever be answered is a separate issue.

Russell (1930) like Woodger (1929), had the misfortune of being published just as biology was embarking on its current genetic and molecular trajectory propelled by the fuel of post-renaissance material forces but for sincere contemporary review of both of these works see Haldane (1931).

Russell had already clearly recognized the critical importance of recapitulation theory to both phylogenetic reconstruction and the understanding of heredity:

> From the point of view of the pure morphologist the recapitulation theory is an instrument of research enabling him to reconstruct probable lines of descent; from the standpoint of the student of development and heredity the fact of recapitulation is a difficult problem whose solution would perhaps give the key to a true understanding of the real nature of heredity. (Russell 1916:312–313)

14.4 Hennig: a preformationist?

> Epigenesis and preformation represent two different attitudes to the problem of development, arise from two fundamentally different philosophies. The epigenetic view is dynamic, vitalistic, physiological; the preformationist is static, deterministic and morphological. The one stresses time or process, the other space and momentary state – the one emphasizes function, the other concentrates on form. (Russell 1930: 29)

Is it fair to pin the label of preformationist on Hennig? I believe it is, and I do not believe the term is in any way meant as pejorative. As Russell's quote above suggests, it refers to a mode of thought, an outlook on the problems of biological development. The idea of a continuity of characters through time, the idea of the semaphoront, the very idea of phylogeny itself, are all preformationist ideas. In short, I find it difficult not to see in Hennig's works an endlessness of preformations with little or no epigenesis. One particular circle of an evolutionary hell?

14.5 Løvtrup: the epigeneticist

Løvtrup's career and life's work sat across the antithesis of preformation and epigenesis, and the allied, euphonious twins of ontogeny and phylogeny. He was an early convert to the hope and promise of Hennig's phylogenetic systematics (see, for example, Løvtrup 1977). Løvtrup was an embryologist rather than a systematist, and an embryologist with a particular interest in epigenesis, at a time long before the topic had reached its current level of fashion.

Løvtrup's life and work covered descriptive and experimental embryology, explorations in epigenetics, attempts to create a phylogeny of Vertebrata, recognition of at least four theories of evolution, and a broad and trenchant critique of Darwinism – one of the most substantial scientific assaults on Darwinism of the late twentieth century (Løvtrup 1987). Despite these huge achievements I have been unable to find any comprehensive bibliography or biography of Løvtrup's life or his efforts towards furthering biological understanding. This is unfortunate, to say the least, if indeed not careless, if we are to have any hope of furthering biological thought, and the notion (now forgotten?) of building for the future. Mattsson and Løvtrup-Rein (1995) reflect on Løvtrup's contributions to amphibian embryology, but this was just one aspect of a multi-faceted life. Perhaps old biologists, like old soldiers, just fade away. It may be that we have now lost any cultural ability to build on past achievements and are content with the generation of endless novelty, but lack all sense of history – a world all of epigenesis but no preformations. Another circle of an evolutionary hell?

14.6 On the nature of organisms: the biological continuity

> The 'concept of organism' seems to be forcing itself upon our attention from three directions: First, from that of embryology and physiology [...] Secondly, there is the Gestalt-theorie beginning in psychology and extending to physics. Thirdly, there are those philosophers of evolution – Hobhouse, Alexander, Lloyd Morgan, Smuts and Whitehead – all of whom employ the concept of organism in various ways and degrees not exclusively biological. Perhaps this concept will enable us to understand evolution. Has anyone observed a machine that was capable of evolution *without a mechanic*? (Woodger 1930: 21)

> On the ground of philosophy, I think it is fair to say that the conception of vital process as distinct from mechanical or chemical change has come to stay, and has revolutionized our conception of nature. That many eminent biologists have not yet accepted it need cause no surprise. In the same way, the anti-Aristotelian physics which I have described as the new and fertile element in sixteenth-century cosmology was rejected by many distinguished scientists of that age; not only the futile pedants, but by men who were making important contributions to the advancement of knowledge. (Collingwood 1945: 136)

> Yet de Queiroz and Donoghue claim that yes, a population is an individual, even a population that is paraphyletic by the usual standard. They offer cohesion (interbreeding) as justification for the claim, again as if organisms + interbreeding = population = species = taxon (why do not organisms + interbreeding = more organisms?). (Nelson 1989a: 288)

To answer Nelson's question above – why not indeed. Despite the hopes, expectations and cautions expressed in the quotes of Woodger and Collingwood above, biological thought has still resisted moving forward to an outlook of organism. Does Nelson's equation above suggest a way forward for biology? Namely that *organisms + interbreeding = more organisms*, this may be only a self-evident truth, but it is the biological actuality. The suggestion amounts to growth, development *and* reproduction being imperatives located within the organism, reminiscent of Arber's leaf, that is a '*partial-shoot [that] has an inherent urge towards the development of whole-shoot characters*' (Arber 1950: 78, italic in original). The collectives of organisms that are posited by current evolutionary theory – the populations and the species – are given powers by that theory to be the active agents of evolution. In biological actuality, the active agents of life (and evolution?) are the organisms in themselves. Does current evolutionary theory's need to hold onto paraphyletic groups, as agents for change, arise from the failure of that theory to grant the organisms a right to independent lives? Paraphyly, viewed from a perspective of classification, is just a problem to be resolved by taxonomic effort. But paraphyly, viewed from a perspective of current evolutionary theory, remains a necessity to be retained – a necessity without which there can be no evolution. Out of the non-ness of paraphyly evolution emerges (see Nelson 1989b for cogent reflections on the non-ness of paraphyly).

Is the unity that biology seeks to be found in the concept of organism and the nexus between preformation and epigenesis rather than in the idea of evolution?

14.7 On leaving evolutionary materialism behind

I think biologists are required to address and attempt to solve three problems which are deeply rooted in most current presentations. The first problem is the necessity for a creative rank – the species – which allows organisms to change – but to my understanding it is organisms that enable species to change, not species that enable organisms to change. A second problem is a denial of the creative powers within the organisms and thus a denial of any concept of the individuality of organisms in themselves. These problems arise from accepting organisms as objects, materials and ignoring their true subjective nature as active agents, players and participants and not just passive particles moved by the forces of evolution. The third problem is that of the generative nature of life forms, and its source and relates back to the first problem.

14.8 The eternal species problem

> The rank is but the guinea's stamp. (Burns, 'A Man's a Man for a' that', 1795)
>
> At the outset I confess a disbelief in species, as the word is commonly understood to refer to the basic taxonomic unit of evolution. There seem to be no basic taxonomic unit and no particular taxonomic unit of evolution [...] I hesitate to suggest that if there are taxonomic units of evolution, the units are taxa generally, for the suggestion contradicts all that I have been taught by my teachers in school. (Nelson 1989b: 60–61)

The special reality of species is a particular aspect of our modern concept of evolution which just never seems to be capable of meaningful resolution. Nelson's criticisms, as cited above, seem to have largely been unable to penetrate a wider biology, and make that wider biology reflect on its own preconceptions. Many taxonomists have of course shed the notion of a special reality for species prompted in a large part by explicit cladistic efforts. But beyond taxonomy, the idea that species are real but that higher taxa are abstractions is firmly embedded in the modern view of evolution. The need, and I believe it to be a need, for a special creative rank in modern evolution remains – a rank that does something. But surely it is the individual organisms that do things; they grow, develop and reproduce. What do the collectives do? Do they by magic enable the changes that we want to discover and explain? Do they create the much searched for but never found transitions – as if functioning organisms were just the ideas of their would be creators and not independent, steady and whole in themselves. Have we allowed the collectives to fill the space in the modern synthesis where organisms and their continuity should lie?

Perhaps future cultural historians will be able to tease apart the idea of evolution from the idea of a special reality for species; the latter looks uncomfortably like a residue of the special creation of species from pre-Darwinian days. And it looks like evolution as explanation, as usually currently expressed, has largely failed to move on from that idea of fixity and special creation – in modern biology the species rank is the rank of special creation, and that special creation remains until another special creation is postulated. But is not change born out of the individuality of the organisms, not by way of the abstractions of the collectives?

14.9 The problem of individuality

> Each organism would have its own individual view-point upon the universe, and instead of the abstract, lifeless, quantitative universe of the physical sciences, which is itself merely a facet of reality seen from a generalized human standpoint and remaining within the bounds of the conceptual objective world of the human monad, we should have to think of a qualitative universe containing a multitude of active individualities, each mirroring in its own way that aspect of reality which is accessible to it, and has meaning for it. (Russell 1924: 31)

> The main thesis of this book is that all living organisms are subjects; that all but the simplest organisms (and possibly even these also) are organizations or nexus of subjects; that the characteristic activity of a subject is the act of perception; and that perception is the establishment by the subject of its causal relation with its external world. (Agar 1943: 7)

Objective materialism can have no operational framework for subjective organism.

Rather than still clutching a restricting and retreating science of materials, a residue of the post-Renaissance view of nature as recognized by Collingwood, here Russell and Agar are reaching towards an expanding science of organisms which includes within it the idea of their evolution.

Karl R. Popper is usually regarded, and regarded himself, as an evolutionary epistemologist, but these words of his point towards an outlook on the organism which could perhaps be shared by Alfred North Whitehead, Russell, Woodger and Agar:

> In fact, we, the organisms, are most active in our acquisition of knowledge – perhaps more active than in our acquisition of food. Information does not stream into us from the environment. Rather, it is we who explore the environment and suck information from it actively, like food. And humans are not only active but sometimes critical. (Popper 1984: 243)

14.10 The multiplicity problem

> On one of their walks in Phoenix Park, Dury mentioned Hegel. 'Hegel seems to me to be always wanting to say that things which look different are really the same', Wittgenstein told him. 'Whereas my interest is in showing things which look the same are really different.' [...] His [Wittgenstein's] concern was to stress life's irreducible variety. The pleasure he derived from walking in the Zoological Gardens had much to do with his admiration for the immense variety of flowers, shrubs and trees and the multitude of different species of birds, reptiles and animals. A theory which attempted to impose a single scheme upon all this diversity was, predictably, anathema to him. Darwin had to be wrong: his theory 'hasn't the necessary multiplicity'. (Monk [1990] 1991: 537)

Is there a multiplicity problem at the heart of our modern conception of evolution? To my reading there is – in fact to my reading there is largely no concept of the generative nature of life in the form of evolution which we have come to subscribe to. Does not this multiplicity arise out of the individuality of the organisms, their generative capacities, their spatial and temporal coincidences, and subsequently, but only subsequently, canalization by selection? But I can find no echo of this in the modern concept of evolution.

So, in conclusion, phylogenetics is but one view of life – a preformationist view – but life is also epigenetic, and truly and completely functional, novel organisms continually arise. Biological thought, which only recognizes one aspect, is obviously limited, and I have to add that the preformationist view is the one which usually claims now, falsely, to speak for all of biology. Wave-particle duality does not seem to have hindered physicists and the development of their discipline. Preformation-epigenesis duality should similarly not hinder biological thought provided biologists are not ideologically committed to one view to the total exclusion of the other. The doubts which remain over phylogenetics are valid I believe and I confess to the worry that these doubts might just in the end be ignored in the push towards a final phylogenetic dissolution of biology.

Woodger grappling with the preformation-epigenesis antithesis quoted Alfred North Whitehead's words from his *Principles of Natural Knowledge,* which I believe are a useful summation:

> There are two sides to nature, as it were, antagonistic the one to the other, and yet each essential. The one side is development in creative advance, the essential becomingness of nature. The other side is the permanence of things, the fact that nature can be recognized. Thus nature is always a newness relating objects which are neither new nor old. (Whitehead 1925: 98 in Woodger 1929:376)

To this Woodger added a somewhat cryptic footnote:

> But if the doctrine of evolution has any truth in it, and if evolution has been an epigenetic process then it seems that *some* objects will have been 'new'! (Woodger 1929:376)

I leave the last word to Collingwood:

> Similarly, I must say now, 'That is as far as science has reached'. All that has been said is a mere interim report on the history of the idea of nature down to the present time. If I knew what further progress would be made in the future, I should already have made that progress. Far from knowing what kind of progress it will be, I do not know it will be made at all. I have no guarantee that the spirit of natural science will survive the attack which now, from so many sides, is being made upon the life of human reason. (Collingwood 1945: 175)

Acknowledgements

David Williams kindly asked me to contribute to this volume and encouraged, edited and criticised my earlier efforts, and with good humour and resolve controlled my indolence. Tony Gill and Michael Heads read and commented on an earlier draft of this essay. I thank them all for their patience and criticism. All remaining misconceptions and errors are of course my responsibility.

References

Agar, W.E. (1943). *A Contribution to the Theory of the Living Organism*. Melbourne, Australia: Melbourne University Press. London: Hamilton, Kent and Co. Ltd.

Arber, A. (1950). *The Natural History of Plant Form*. Cambridge: Cambridge University Press.

Collingwood, R.G. (1945). *The Idea of Nature*. Oxford: Oxford University Press.

Forey, P. (2002). Systematics and creationism. *The Linnean*, 18, 13–14.

Garstang, W. (1922). The theory of recapitulation: a critical restatement of the biogenetic law. *Zoological Journal of the Linnean Society of London*, 35, 81–101.

Haldane, JS. (1931). *The Philosophical Basis of Biology*. Donnellan Lectures, University of Dublin, 1930. London: Hodder and Stoughton Ltd.

Hennig, W. (1966). *Phylogenetic Systematics*. Urbana, IL: University of Illinois Press.

Inglis, F. (2009). *History Man: the life of R. G. Collingwood*. Woodstock, NJ: Princeton University Press.

Kitto, H.D.F. (1983). That famous Greek 'wholeness'. In *Art, Science and Human Progress; the Richard Bradford Trust lectures given between 1975 and 1978 under the auspices of the Royal Institution*, ed. R.B. McConnell. London: John Murray, London. pp. 44–61

Lauder, G.W. (1982). Introduction. In *Form and Function*, E.S. Russell [1916; 1982 reprint], Chicago, IL: University of Chicago Press.

Løvtrup, S. (1974). *Epigenetics: A Treatise on Theoretical Biology*. London: Wiley.

Løvtrup, S. (1977). *The Phylogeny of Vertebrata*. London: Wiley.

Løvtrup, S. (1984). Ontogeny and phylogeny. In *Beyond Neo-Darwinism*, ed. Mae-Wan Ho and Peter T. Saunders. London: Academic Press, pp. 159–190.

Løvtrup, S. (1987). *Darwinism: The Refutation of a Myth*. London: Croom Helm.

Mattsson, M-O. and Løvtrup-Rein, H. (1995). Swedish contributions to the understanding of amphibian embryogenesis: a phenomenon of the past? *International Journal of Developmental Biology*, 39, 703–704.

Monk, R. (1990 [1991]). *Ludwig Wittgenstein, the duty of genius*. London: Vintage.

Nelson, G. (1989a). Cladistics and evolutionary models. *Cladistics*, 5, 275–289.

Nelson, G. (1989b). Species and taxa; systematics and evolution. In *Speciation and its Consequences*, ed. D. Otte and J.E. Endler, Sunderland, MA: Sinauer, pp. 60–81.

Nicholson, DJ and Gawne, R. (2014). Rethinking Woodger's legacy in the philosophy of biology. *Journal of the History of Biology*, 47, 243–292.

Patterson, C. (2002). Evolution and Creationism. *The Linnean*, 18, 15–33.

Popper, K.R. (1984). Evolutionary epistemology. In *Evolutionary Theory: Paths into the Future*, ed., J.W. Pollard. Chichester: John Wiley & Sons Ltd., pp. 239–255.

Rosen, D.E. (1984). Hierarchies and history. In *Evolutionary Theory: Paths into the Future*, ed. J.W. Pollard. Chichester: John Wiley & Sons Ltd., pp. 77–97.

Russell, E.S. (1916; 1982, reprint). *Form and Function*, Chicago, IL: University of Chicago.

Russell, E.S. (1924). *The Study of Living Things*. London: Methuen & Co.

Russell, E.S. (1930). *The Interpretation of Development and Heredity, A Study in Biological Method*. Oxford: Oxford University Press.

Woodger, J.H. (1929). *Biological Principles*. London: Kegan Paul, Trench, Trubner & Co.

Woodger, J.H. (1930). The "concept of organism" and the relation between embryology and genetics. *The Quarterly Review of Biology*, 5 (1), 1–22.

Woodger, J.H. (1945). On biological transformations. In *Essays on growth and form presented to D'Arcy Wentworth Thompson*, ed. W.E. Le Gros Clark and P.B. Medawar. Oxford: Oxford University Press, pp. 95–120.

15

Willi Hennig as philosopher

Olivier Rieppel

15.1 Introduction

There is something quixotic about Hennig the philosopher. His parents aimed for higher education, sending their son Willi to a boarding school, a grammar school (Gymnasium) in the Klotzsche district of Dresden, with mandatory classes in Latin and philosophy. It was not Latin and philosophy that fired up Hennig's enthusiasm, however, but extracurricular work he pursued in biological systematics instead. In fulfillment of a homework assignment, Hennig delivered at the age of 18 an essay on *The Position of Systematics in Zoology* (posthumously published in Hennig 1978), in which he demonstrated quite remarkable insights and skills in this discipline. In the same year, 1931, he published his first faunistic paper on the insects of his school district. In 1950, Hennig published his *Grundzüge einer Theorie der phylogenetischen Systematik* (*Foundations of a Theory of Phylogenetic Systematics*), making it clear in the title that he was seeking to develop not just a novel methodology in systematics, but also a theoretical justification of that new methodology. Ernst Mayr found Hennig's philosophical excursions "rather turgid," or worse, "virtually unintelligible" (Mayr 1982: 226), while the Hennig biographer Michael Schmitt explained:

> Hennig's original interests were certainly not merely methodological. He was a taxonomist with the ambition to provide a better scientific foundation for the order, into which the investigated taxa were to be brought [...] The ambitious metaphysical introductions to the *Grundzüge* and to *Phylogenetic Systematics* were probably included to further that goal, although they are totally superfluous for practical work, and only of limited usefulness as a theoretical foundation [...] It is possible

The Future of Phylogenetic Systematics: The Legacy of Willi Hennig, eds. D. Williams, M. Schmitt and Q. Wheeler. Published by Cambridge University Press.

> that the recognition of the novelty of his method drove Hennig to the belief that he
> had to provide a philosophical foundation for its exposition. (Schmitt 2001: 339)

Schmitt (2013: 163) noted that Hennig attended only one course in philosophy during his university education, whereas his correspondence with his wife, as also in the letters he exchanged with his friend and colleague Klaus Günther, reveal a deep interest in philosophy. His overall assessment of Hennig's philosophical underpinning of systematics is rather sobering, however: 'My impression is that Willi Hennig was interested in philosophy, but that he was not really a trained philosopher [...] He certainly did not purposefully join a certain school of philosophy, but rather eclectically took ideas from here and there' (Schmitt, 2013: 166). This starkly contrasts with Tremblay's assessment that:

> In his elaboration of this genealogical system, Hennig was exceptionally concerned
> with metaphysical issues [...] It was not uncommon for twentieth century German
> biologists to be concerned with philosophical issues. Some of this was a result of
> the interdisciplinary tendency in the academic culture of Germany at that time.
> (Tremblay 2013: 57)

Tremblay (2013: 56) commented extensively on 'Hennig's adoption and application of some of [Nicolai] Hartmann's ideas.'

A complicating factor in the assessment of the philosophical perspectives noticeable in the *Grundzüge* is the way the first draft of this book-length manuscript was generated. It was handwritten in a voluminous Italian notebook, bound in thick cardboard, Hennig's writing neatly filling page after page. The provenance of the notebook indicates that Hennig composed the text while serving in Northern Italy, first in the German army, later in British detainment: the manuscript "gives the impression of having been written in a continual steady flow" (Schlee 1978/79: 381). The *Grundzüge* was not the only manuscript Hennig completed while serving in the army, supported by his wife who sent him pertinent literature, and helped organize proofreading. How many philosophical tomes would his wife have been likely to send him to the military camp? It might be revealing in that context to investigate how Hennig cited philosophical works in the *Grundzüge*. When citing biological literature, Hennig routinely provided page numbers, thereby precisely indicating his sources. The same is not always true for philosophers he cited.

In the case of Aristotle, Hennig (1950: 17f) does not refer to any primary source or scholarly edition of Aristotle's biological writings, rendering it likely that his reception of Aristotle was entirely based on tertiary sources. Bernhard Bavink's work (Bavink 1941), in contrast, is cited by year and page number, revealing Hennig's interest in what the author had to say about a *mathesis universalis* of natural sciences (Hennig 1950: 152); indeed, Hennig also cited Bavink in his essay on phylogenetic systematics published in 1949 (Hennig 1949: 136). With 27 citations, most of which including page numbers, Ludwig von Bertalanffy is the philosopher-biologist most frequently referred to by Hennig, who in various respects adopted Bertalanffy's organicist conception of nature. Rudolf Carnap, one of Bertalanffy's mentors, is cited once only, when Hennig (1950: 153) referred to his doctoral thesis published in 1922 – with important consequences for an understanding of Hennig's phylogenetic system. Two works of Hugo Dingler emphasizing the historicity and consequent reality of the world of objects are cited once each (Hennig 1950: 23), the first with page numbers but missing in the list of references. Wilhelm Dilthey's justification of the hermeneutic circle as a valid, i.e. non-circular form or inference is mentioned in passing only by Hennig (1950: 26), without any indication of primary sources. Hans Driesch's holism is mentioned once in passing only, without indication of any primary source (Hennig 1950: 119). Arnold Gehlen is referred to twice, his defense of a coherence theory of truth highlighted by Hennig (1950: 190) with the indication of a page number. Max Hartmann earned six quotations, whereby the indications of page numbers refer to philosophical perspectives Hartmann offered in his textbook of biology (Hennig 1950: 2–5; 301). Nicolai Hartmann, in contrast, earned four quotations, but if page numbers are indicated, they always refer to quotations of Hartmann from secondary sources (Hennig 1950: 5, 115). It is highly likely that Hennig at some point had read Hartmann's compendium on *Systematic Philosophy* of 1942 (see his reference to Wein's chapter in this compendium below), but he seemed content to quote Hartmann either from secondary sources, or from memory: 'e.g. similar thoughts also in Nicolai Hartmann' (Hennig 1950: 299). Heinrich Rickert's expositions on nomothetic versus idiographic sciences are referred to twice by Hennig (1950: 2, 4), without any citation of the original source. The philosopher who indeed first introduced that distinction was Wilhelm Windelband, whom Hennig cited from a secondary source (Hennig 1950: 4). This leaves us with Theodor Ziehen, who, in spite of having earned only five citations – most with page numbers – in Hennig's *Grundzüge* (Hennig 1950: 4, 6, 23, 114, 115), must be considered a major influence on Hennig's epistemology again.

It seems unlikely that Hennig had all the original works for which he cited page numbers at hand when he laid down the first draft of the *Gründzüge* while serving in Northern Italy. After the war, the publication of his book was delayed due to a shortage of paper, a circumstance that motivated Hennig to publish his core ideas in the form of two essays that appeared in 1947 and 1949, respectively. This gave him ample time to add the relevant edits before the *Grundzüge* went to press. A clarification of this issue could be achieved by comparing the published *Grundzüge* with the original manuscript of its first draft, preserved and in possession of his family (Schlee 1978/79). Whatever the case may have been, Hennig did not bother to follow up on Nicolai Hartmann's original works, in contrast to – for example – Bertalanffy and Ziehen, whose work Hennig referenced with great care. Indeed, an argument can be made that of all philosophers he cited, Bertalanffy and Ziehen may have exerted the greatest influence on Hennig.

Hennig the systematist and philosopher grew up in the tradition of German biology as practiced in the first half of the twentieth century. This tradition is much more alive in his *Grundzüge* of 1950, than it is in his *Phylogenetic Systematics* of 1966. The latter book is a stripped down version of the *Grundzüge*, emphasizing methodological aspects of Hennig's phylogenetic systematics, underpinned with Ziehen's epistemology and Woodger's logic (see below). The *Grundzüge* is much richer in philosophical content, but for that same reason also harder to read. Hennig's *Grundzüge* aims at a broad synthesis of contemporary work in systematics and evolutionary theory, to be used as the foundation on which to build his phylogenetic systematics. However, during the first half of the twentieth century, German biology was torn between two major philosophical schools – holism/organicism on the one hand with Hans Driesch as one of its major exponents, logical positivism on the other hand, propagated by the Vienna Circle of which Rudolf Carnap was a leading representative. The tension between these two philosophical schools pulling systematics and evolutionary theorizing in opposite directions is perhaps best exemplified by Walter Zimmermann's polemic against the fellow botanist Wilhelm Troll. Hennig (1950: 14; 1966: 10) praised Zimmermann as 'one of the best modern theoreticians of systematic work' (see also Donoghue and Kadereit 1992), who in his campaign against the *Gestaltler* Troll, emphatically endorsed the neo-positivist philosophy of the Vienna Circle (Zimmermann 1937/38).

15.2 Hennig's major philosophical sources: Bertalanffy and Ziehen

As detailed above, Hennig made passing reference to a number of philosophers in different contexts, but two authors stand out having exerted major influence on Hennig. Von Bertalanffy instilled in Hennig a sense for an organicist conception of nature married to a process philosophy, and offered a discussion of biological hierarchies that Hennig could latch on to. Ziehen in turn espoused an epistemology based on a neo-positivist phenomenology that allowed Hennig to comply with a fundamental request earlier issued by Walter Zimmermann (1937/38): the strict separation of object and subject in empirical scientific research. Some background on Bertalanffy and Ziehen is thus helpful for an understanding of where Hennig came from when he laid down his *Grundzüge*.

Today Ludwig von Bertalanffy is mostly remembered as the father of systems biology (systems theory). In a 1941 paper cited by Hennig (1950), however, Bertalanffy (1941: 337) proclaimed himself, in a somewhat presumptuous way, the 'father of organicism'. The book that made Bertalanffy the most cited author in Hennig's *Grundzüge* (1950) was volume one of his *Theoretische Biologie*, published in 1932. Bertalanffy had earned his PhD in 1926 at the University of Vienna, with

Moritz Schlick, the founder of the Vienna Circle, as his thesis supervisor. The thesis analysed the work of the physicist and philosopher Gustav Theodor Fechner (1801–87). After his promotion, Bertalanffy worked at the II. Zoological Institute of the University of Vienna under the Dutch paleontologist and functional anatomist Jan Versluys. Bertalanffy also participated in the *Studiengruppe für wissenschaftliche Zusammenarbeit*, a section of the *Verein Ernst Mach* (the public arm of the Vienna Circle) that was chaired by Rudolf Carnap (Stadler 1997). It was during these years that Bertalanffy worked on his *Theoretische Biologie*, with the aim of clarifying whether a hypothetico-deductively structured theoretical biology could be developed, matching the challenging standards that characterize the most mature of all natural sciences, theoretical physics. On recommendation from Schlick, Versluys, and the neo-Kantian philosopher Robert Reininger, the book earned Bertalanffy the *venia legendi*, and with it the qualification to become eligible for appointment at the professorial level. It is fair to say that Bertalanffy's *Theoretische Biologie* was the most comprehensive synthesis of the theoretical foundations of biology of the time, which may explain Hennig's fascination with it. Its major messages were of a holistic/organicist nature: the proper ontology on which to base a modern comprehension of nature was one not of sets and their elements, but of complex wholes and their parts; one not of static material things, but of processes; one not of objects that instantiate properties, but of causal relations (Rieppel 2007). An organism is thus seen as a processual system, a hierarchically structured complex whole the parts of which are tied together through dynamic (processual) causal relations.

Theodor Ziehen[1] studied medicine, specializing in psychiatry. In 1903, he was appointed Professor of Psychiatry and Neuropathology at the University of Halle / Saale. In 1904 he was appointed head of the psychiatric section at the Charité Hospital in Berlin. In 1912, he asked for his dismissal so he could devote himself entirely to work in philosophy. In 1917, Ziehen was appointed Full Professor for Philosophy at the University of Halle, where his courses and lectures attracted a large audience. Ziehen again rejected a substance – predicate structure of the world, i.e. a world that is composed of objects that instantiate properties. Following veteran positivists like Ernst Mach and Richard Avenarius, Ziehen reduced matter to energy, and hence again embraced an ontology built on processual causal relations. Unlike Bertalanffy, but similar to Kant and Mach, Ziehen denied the epistemic accessibility of putative material objects (the controversial 'things in themselves' in Kantian terms) that take part in these processual relations. The result is a Machian phenomenologism that treats not the object per se, but the *perception* of an 'object' (whatever the latter is) as 'The Given'. All human cognition starts from what is immediately given in perception, which is a world that is represented in a multidimensional multiplicity of perceptions. Such chaos

---

[1] www.catalogus-professorum-halensis.de/ziehentheodor.html

cannot be the stuff of systematization, or scientific explanation, however. To systematize the multidimensional multiplicity of perceptions means to recognize the constant relations that prevail among them, for it is such regularity of relations that allows the inference of underlying lawfulness of natural processes, and thus renders those amenable to scientific explanations. For Ziehen, the substantial world that lies behind the perceptions is one of a multidimensional multiplicity of fields of energy of variable density, subject to complete causal determination. What is revealed in perceptions is the regularity of causal relations that reflect such causal determination. Yet perception reveals not only causal relations, but involves the human mind as well. There are, thus, two components to every perception, a physical (causal) and a psychological (mental) one. Relevant to scientific explanation of the extramental world are the physical (causal) relations. In order to conduct proper science, one has thus to separate the physical from the psychological component of perception. What is immediately given, i.e. the perception itself, Ziehen called a *Gignomen*. The physical (causal) relations relevant for scientific explanation obtain from the *reduction* of the *Gignomen* into its two components: its *reduction part* (R), revealing physical laws of nature, and its *parallel part* (N), revealing the psychological laws that govern the mind. For science to be successful, R and N must come together in the sense that the psychological laws, revealed in the N-part, run parallel to the physical laws revealed in the R-part. That parallelism is guaranteed because, according to Ziehen, the means of human cognition are an evolutionary adaptation to the physical structure of nature: the laws of logic represent an evolutionary adaptation to the structure of the world. The psychiatrist turned philosopher Theodor Ziehen may be largely forgotten today, but he was a prominent scholar whom contemporaries compared to Leibniz and Einstein (Gerhard and Blanz 2004: 1369). He was the much-celebrated hero of one of his most famous students, and one of the most important peers of Hennig, the zoologist Bernhard Rensch (1979), who together with Bertalanffy and Zimmermann ranks among the most frequently cited authorities in Hennig's *Grundzüge*. Indeed, the philosophical pronouncements in Hennig's *Phylogenetic Systematics* that the translators found hardest to translate, and which Mayr found downright incomprehensible, are quotations taken straight out of Ziehen that also appear in Hennig's *Grundzüge*.

The following sections will sketch some of the ways in which Hennig put this intricate philosophical machinery to work.

15.3 The *mathesis universalis* of systematics

Hennig considered systematics, i.e. the systematization of nature, as an essential part of all natural sciences, not just biology. The reason is that for Hennig, as for most German natural scientists of the first half of the twentieth century (and before), natural sciences are primarily tasked with the discovery of laws of nature. Furthermore, laws of nature should preferentially be captured in the timeless language of mathematics. The *mathesis universalis* is a philosophical ideal that goes back Leibniz, one which seeks the description of the universe in purely mathematical terms. For Kant (1786: viii), the arm of natural science reached only as far as it is capable to capture nature in mathematical relations. The young Carnap sought a *mathesis universalis* of the world, understanding logic – following his mentor Gottlob Frege – as a special branch of mathematics (Carus 2007: 129). For Hennig (1953: 6; note his adoption of Ziehen's terminology), 'the way in which a multidimensional multiplicity is to be treated classificatorily (i.e. "systematically"), is taught by set theory, which is a subdiscipline of mathematics, or of mathematical logic respectively.' This follows his conviction, which he expressed in his 1949 essay, that 'a mathematization in the sense of a *mathesis universalis* must be the goal of every true science' (Hennig 1949: 138). Hennig may well have encountered in Bavink (1933: 389) the idea that modern logic might deliver the tools to aspire to a '*mathesis universalis*, as had been envisaged by Leibniz.' Hennig's appeal to logic, and set theory in particular, was bound to create tensions with holistic/organicist concepts he was to incorporate in his phylogenetic systematics. Beyond that particular issue to be discussed below, it is remarkable how German biologists of the first half of the twentieth century – Hennig as well as his most important peers such as Bernhard Rensch and Adolf Remane – were unwilling or incapable to accept the radical historical contingency that characterizes not only Darwin's theory of evolution, but also the burgeoning population genetics as developed, for example, by Theodosius Dobzhansky (Dobzhansky 1937; see Gliboff 2008, Rieppel 2013). As late as 1965, Mayr felt the need to explain the radical historical contingency that underlies the modern synthesis of evolutionary theory 'especially for the German reader' (Mayr, cited in Kraus 1984: 280, Kraus and Hoßfeld 1998: 162) in a paper he published in German in *Die Naturwissenschaften* (Mayr 1965).

Hennig found the distinction of nomothetic versus idiographic sciences, initially drawn by Windelband and further refined by Rickert, discussed in Max Hartmann's (1933) textbook on biology, but rejected that contrast as one characterizing distinct types of sciences. Nomothetic and idiographic are two perspectives, he argued, that are common to all natural sciences. The search for laws of nature of increasing universality Hennig (1950: 3) recognized as the domain of nomothetic research, while the explanation of the present state of being of a particular object of nature as the lawful consequence of an antecedent state of being is part of idiographic research. But, argued Hennig (1950: 3) following von Bertalanffy: what is a causal explanation based on lawfulness at a lower level of generality becomes a mere descriptive statement relative to lawfulness of a higher level of generality. This insight further blurs the nomothetic-idiographic distinction, as it ties levels of generality of natural laws to levels of a hierarchically structured reality as was recognized by Nicolai

Hartmann. Hennig's phylogenetic system is, of course, tied to time and space, to history quite generally and hence would fall into the purview of idiographic science. While nomothetic sciences deal in universals, i.e. in generalities as expressed in laws of nature, idiographic sciences deal with particulars, i.e. individuals and their history. As argued by Tremblay (2013) in great detail, Hennig followed Nicolai Hartmann when he anchored individuality in time. The phylogenetic system as a whole, as well as its parts – i.e. species – thus become individuals, each embedded in its particular history. Conceived of as individuals, each species has its unique evolutionary origin, and travels along its unique evolutionary trajectory. But given such radical particularity of the phylogenetic system, there seems to be no room left for generalities to obtain that could reveal underlying lawfulness. Hennig solved this conundrum along two lines of argumentation.

Firstly, he concluded that no science is either purely nomothetic, or purely idiographic, but that every science, phylogenetic systematics included, represents a mix of nomothetic and idiographic components. Windelband himself, Hennig (1950: 4) insisted, had already emphasized that in biology, both the idiographic and the nomothetic perspective are implied as organisms are investigated from different vantage points, e.g. historical (evolutionary) and functional (physiological). Secondly, he followed Ziehen with his claim that historicity did not imply a radical particularity, i.e. 'unrepeatability' of all historical events (Hennig 1950: 23), thus allowing room for generalities even in historical sciences. Like Kant before him, Ziehen (1934: 58) rejected Leibniz's *principium identitatis indiscernibilium*, which states that two fully identical objects cannot exist, or conversely, that two objects located in different space regions must differ in additional properties as well. For Hennig this meant that the inference of an 'absolute historicity of all that exists' from the 'unrepeatability and all-pervading individuality of all that exists' had to be "critically assessed" (Hennig 1950: 23). 'To comprehend things, phenomena and events "systematically"', Hennig (1950: 4) emphasized, 'means to understand them not just as phenomena amenable to description, but as parts of a lawfully determined system of order of things, phenomena or events.' However, the generalities dealt with by the phylogenetic systematist could not be as universal as the generalities revealed by physics, but are more restricted in their occurrence: everything that is subject to Darwinian evolution is also subject to gravity, but not everything that is subject to gravity is also subject to Darwinian evolution. There was thus room for lawfulness in phylogenetic systematics. The only difference – one of degree, not of kind – was that phylogenetic laws would be of more restricted scope than physical laws (Hennig 1950: 306).

15.4 The enkaptic hierarchy

An enkaptic hierarchy is a nested hierarchy, dichotomously structured, that is subject to the part–whole relation. It is, in other words, a nested hierarchy of complex wholes or individuals of greater or lesser inclusiveness. A key feature of the enkaptic hierarchy is that at each level of inclusiveness, the complex whole cannot be reduced to the sum of its parts, but is characterized by emergent properties (Rieppel 2009). The concept of enkapsis had first been worked out by the Tübingen comparative anatomist and histologist Martin Heidenhain (e.g. Heidenhain 1907, 1923), but it was soon co-opted in functional anatomy (Benninghoff 1930), developmental biology (Dürken 1936), ecology (Friederichs 1927, Thienemann 1939, 1941), and phylogenetics (Beurlen 1937). The enkaptic hierarchy thus became a central concept of the holism/organicism characteristic of German biology of the first half of the twentieth century. Hennig followed Beurlen (1937) not only by calling his phylogenetic system an enkaptic hierarchy, but also by calling the complex wholes at each level of complexity, each characterized by emergent properties, phyletic *Gestalten*.

While Hennig cited Beurlen, Friederichs and Thienemann, he did not cite Heidenhain. He appears to have taken the concept of enkapsis not from its original author, but from Bertalanffy's discussion of Heidenhain's ideas in his *Theoretische Biologie* of 1932. For Heidenhain, each organism forms an enkaptic hierarchy, a hierarchically structured complex whole. Its ultimate fundamental part is the cell, and the fundamental property of the cell is the same as the fundamental property of all life: the potential to multiply by subdivision. Cell division forms rudiments of organs, the division of organ rudiments results in the formation of organ complexes, etc. The enkaptic hierarchy that is an organism is thus built up from the subdivision of its parts, the cell being the simplest and most fundamental one. The general term introduced by Heidenhein for those divisible parts that constitute the enkaptic hierarchy was 'histomeres'; the division of histomeres would result in the formation of histosystems. Histomeres and histosystems are strictly relational terms, as a histomere turns into a histosystem and vice versa as one moves up and down the enkaptic hierarchy, respectively. In his 1947 essay, Hennig explicitly drew a parallel between Heidenhain's histosystems and the phylogenetic system built up from subdividing species (Hennig 1947: 279). Just as the subdivision of histomeres results in an enkaptic hierarchy that is the organism, so does the subdivision of species result in an enkaptic hierarchy that is the phylogenetic system (Hennig 1949: 136). The fact that the enkaptic hierarchy is subject to the part – whole relation matches well with the idea that species as historical entities are individuals. It matches less well with Hennig's (1953: 6) appeal to set theory as a tool for systematics, since set theory is subject to the membership relation, although logicians have ascertained that either system – parts within wholes versus boxes within boxes – can be translated into one another.

Hennig's conception of the phylogenetic system as an enkaptic hierarchy has a number of important corollaries, one of which is the fact that all taxonomic units, from the species on up to more inclusive taxa, are complex wholes, i.e. individuals subject to the part-whole relation. Species, as much as monophyletic taxa, are individuals tied together through the relation of descent in the hierarchically structured phylogenetic system: 'In the phylogenetic system, the name of groups of animals at all levels of inclusiveness [...] are *proper names*' (Hennig 1953: 3). The enkaptic hierarchy is a dynamic, relational system: it is a system of species lineages splitting and splitting again. The corresponding phylogenetic system is, according to Hennig, strictly dichotomously structured (Rieppel 2011a). In his *Grundzüge* of 1950, Hennig went to great length to provide a biological basis to the dichotomously structured hierarchy rooted in theories of speciation, which in hindsight proved flawed, not the least for their orthogenetic implications. This may well be the reason why in his *Phylogenetic Systematics*, Hennig (1966: 210) reduced the 'cladistic principle of dichotomy' (Hennig 1974: 292) to a merely methodological one. There prevails, however, in the *Grundzüge* already a tendency for Hennig to endow the enkaptic hierarchy with properties of a formal, i.e. logical hierarchy, as he likewise used Venn diagrams, i.e. tools of set theory, to graphically represent the enkaptic hierarchy of species lineages splitting and splitting again (Hennig 1950: Fig 37). From a dichotomous splitting of a stem species into two daughter species results a dichotomously structured phylogenetic system. This he valued, at the same time, as the most precise and least ambiguous system as well, a point that Adolf Naef, another author frequently cited by Hennig (1950), had likewise argued (Rieppel 2012a). The reason is that each organism has to be classified either on one, or on the other side of the fork: 'there is no third alternative' (Naef 1919: 25). For Hennig (1950: 144f), the consequence was: 'We as well hold the view, to express it in Beurlen's (1937: 193–194) terms, that the common stem form of crossopterygians and stegocephalians must always have been either a fish, or a stegocephalian.' There is no truly intermediate form possible and hence also no true ancestor. Accordingly, Hennig's phylogenetic systematics does not sort species into sequences of ancestors and descendants, but into sister group relationships instead. In that respect, the 'principle of dichotomy' also acquired for Hennig a significant heuristic value. Even if a strictly dichotomously structured phylogenetic system could neither empirically, nor theoretically be 'justified in the strict sense', it motivates a particularly assiduous character analysis in all those cases where sister group relationships remained elusive (Hennig 1974: 293).

When he wrote the *Grundzüge*, Hennig had no problem with representing phylogenetic history, a system of species lineages splitting and splitting again, by means of a logically structured hierarchy as captured by Venn diagrams. After all, and according to Ziehen, (1934: 86), the capacity of logical thought is an evolutionary adaptation to the structure of the world. Ziehen's disciple Bernhard Rensch (1968: 232) put it like this: "Th. Ziehen (1934: §22) was the first to have ascertained that through phylogeny, thought processes evidently became adapted to the logical lawfulness of the extra-mental world." Hennig cited Max Hartmann, who in turn had adopted from Hermann von Helmholz the premise that order prevails in nature, since order in nature is the prerequisite for the comprehensibility of nature. But to Hartmann, science is not just concerned with the description of order in nature, but seeks to rationalize the world of phenomena. To rationalize the world of phenomena is to capture the world in categories of rational, i.e. logical thought. Hennig (1950: 5) explained: '"Systematics" – also in biology – in that sense means order, the rationalization of the world of phenomena, and thus represents the very essence of nomothetic science in general.'

15.5 The cladogram, a Carnapian structure description

Hennig (1950: 279) thought it impossible to graphically represent the phylogenetic system in its multidimensional complexity. He turned to Bavink again in support of his claim that the phyletic *Gestalten*, which in their multidimensional complexity populate the enkaptic hierarchy at its various levels of inclusiveness, defy precise measurement and computability. Bavink (1933: 388; 1941: 462) had called for a new calculus, a *mathematics of form* that still remained to be developed – and Hennig (1949: 136; 1950: 152) concurred. Species, Hennig (1949: 136) explained, differ in manifold ways, or dimensions, from one another: morphologically, physiologically, behaviorally, genetically, geographically, etc. These differences align – to speak with Ziehen – along different dimensions: a species thus forms a multidimensional multiplicity. Assessing the relations among species means to locate species in the multidimensional space of their properties (*multidimensionaler Eigenschaftsraum*: Hennig 1950: 152). Such a multidimensional property space is a theoretical conception of space that cannot be graphically represented, but one that Hennig (1950: 153) found fruitfully developed in Carnap's PhD thesis published in 1922. With reference to his concept of multidimensional property space Hennig (1950: 153) noted: 'Incidentally, on the concept of the formal space as invoked in the present context see Carnap (1922).'

The natural system in its multidimensional complexity is composed of 'natural objects' and 'events', 'whereby the latter categories should not be construed as pure opposites' (Hennig 1950: 5). This system exists, according to Hennig, independent of human intention to represent it graphically. What can be graphically represented is not, however, the multidimensional complexity of the property space within which species are located, but rather the *formal structure* of the phylogenetic system, and the nature of such a formal structure description Hennig found explained in Carnap (1922). Of fundamental importance in this context is the nature of the Carnapian structure description as a relational structure, i.e. a relational system. What are to be captured in such a structure description are not intrinsic, but relational properties that tie together the entities which populate the system. What Hennig wanted his structure description to be based on, therefore, was not similarity or dissimilarity in characters of any kind (morphological, physiological etc.), but phylogenetic relations as *indicated* by similarity or dissimilarity in properties. Characters are thus mere indicators (*Steckbriefmerkmale*: Hennig 1957: 52), pointing to a deeper reality that is anchored in the phylogenetic relations that tie the phylogenetic system together. Given its multidimensional nature, the phylogenetic system could in principle be represented along any one of its dimensions – morphological, physiological, etc. Given its relational structure, and the fact that Hennig wanted the system to be based exclusively on phylogenetic relations, he chose the dimension of time along which to graphically represent the system. In contrast to character evolution, which in principle is reversible as in wingless insects (e.g. fleas), time (at least experienced time, as Beurlen 1936, insisted) is not reversible, and for that reason the best, most unambiguous dimension to choose along which to represent the phylogenetic relations. The graphic representation of the formal structure of the phylogenetic system then depicts, in terms of a dichotomously structured hierarchy, the succession of species lineage splitting events along the time axis. Using another yardstick, or dimension, as a means to critique the phylogenetic system that is organized along the time axis means to commit the *metabase*, as Hennig's friend Klaus Günther (1956: 38) put it. The reconstruction of the phylogenetic system requires 'the strict unity of a point of view' (Hennig 1974: 280), which renders a critique of the system based on a different point of view invalid. Following the sequence of speciation events through time, crocodiles must be considered more closely related to birds than to other reptiles. To classify crocodiles with other reptiles in a group that excludes birds on morphological, physiological or ecological grounds uses a different measure for the reconstruction of phylogenetic relationships than the one chosen by Hennig, and hence gains no purchase in a critique of the sister group relationship of crocodiles and birds among extant amniotes that was argued for by Hennig (1974).

The dichotomously structured graphic that Hennig used as a representation of the formal structure of the phylogenetic system he called a *Strukturschema* (Hennig 1949: 136), or also *Strukturbild* (Hennig 1957: 58). Such a structural scheme, or structure description, Carnap characterized as follows:

> We call a 'general system of order' ('*allgemeines Ordnungsgefüge*') a system of relations not between certain identified objects [...] but between un-interpreted relational entities ('*unbestimmte Beziehungsglieder*'); the only thing known of this system is that conclusions can be drawn from relations of one certain kind to relations of another certain kind [...] this is therefore a system of meaningless relational entities for which the most diverse things can be substituted (such as numbers, colors, *degrees of relationship*s, circles, judgments, humans) in so far as there exist between them relations which satisfy certain *formal requirement*. (1922: 5f; emphasis added)

Carnap's characterization leaves the structure description undetermined, i.e. empirically empty. It provides a purely formal (logical) framework, which can be populated by all kinds of entities, both material (degrees of relationships) as well as abstract (judgments). The basic entity of Hennig's phylogenetic system is the biological species, capable of division through speciation: 'The generalized structural scheme that represents the phylogenetic relations that exist between species results from what we know about the origin of new species. New species originate through the splitting of species into daughter species' (Hennig 1949: 136). The process of species lineage splitting thus satisfies the formal requirement that determines the relational structure of the system and its graphic representation: a dichotomously structured hierarchy. Hennig recognized then, on the one hand, the *material* enkaptic hierarchy of species lineages splitting and splitting again that exists in nature independent of human intention to represent it graphically, and the *formal* representation of the phylogenetic system in terms of a *Strukturbild* that specifies relative degrees of relationships as determined by the sequence of species splitting events through time.

In Bertalanffy's *Theoretische Biologie*, the discussion of Heidenhain's enkaptic hierarchy (1932: 261–263) immediately precedes the discussion of Joseph H. Woodger's division hierarchy (1932: 263ff). Woodger, originally a developmental biologist from the University College London, had embarked on the project of disambiguating biology using the new logic – a calculus of relations – developed by Bertrand Russell and Alfred North Whitehead in their famous *Principia Mathematica* (Nicholson and Gawne 2014). Bertalanffy had pursued similar ideas under the tutelage of Carnap, grudgingly acknowledging Woodger's priority (Bertalanffy, 1932: 29, n.1; on the collegial friendship that developed between Bertalanffy and Woodger see Nicholson and Gawne, 2014). In 1952, Woodger presented, in the formal language of logic, a definition of a 'biological hierarchy' that would accommodate a hierarchy of squares within squares, or the Linnean hierarchy of taxonomic groups, just as much as the hierarchy that results from a zygote splitting and cells dividing again and again. His definition became the object of a thorough interpretation in set-theoretical terms by Gregg (1954). Hennig must have become aware of this work in the mid-1950s, when he immediately seized upon it (Hennig 1957). Hennig thus moved from the material enkaptic hierarchy subject to the part-whole relation, to the formal hierarchy of the cladogram satisfying a Carnapian structure description, to the set-theoretical hierarchy formulated by Woodger and Gregg. He might have thought this move to be a legitimate one, as all three hierarchies he considered are relational systems. Indeed, to merge Heidenhain's enkaptic hierarchy with Woodger's division hierarchy Hennig (1966: 80) found justified on the basis of a remark issued by Gregg (1954: 70): 'The author [i.e. Woodger] develops a simple language, with a structure entirely different from that of set theory, in which taxonomic group names may be construed as names of individuals.' The consequence for Hennig was:

> Woodger (1952) proceeds from the example of a square, which can be subdivided into smaller squares. If 'X' names each of these smaller squares, then 'ΣX' names the larger square of which they are parts [Woodger 1952: 11]. Consequently this

> is a hierarchy (division hierarchy), as in the system of phylogenetic systematics [...] the higher categories are not 'sets of organisms', but the subordinate categories are 'parts'. (Hennig 1966: 80)

Woodger and Gregg had finally pointed in the direction that would seem to promise a *mathesis universalis* of phylogenetic systematics:

> In my view, the project of Woodger and Gregg is of utmost importance, because it explains with methods, which preclude any ambiguity or contradiction, the properties of the hierarchical type of system by incorporating it in the science of set theory. Every systematist should occupy himself with these questions. (Hennig 1957: 55–56)

15.6 The semaphoront

The theoretical, uninterpreted entities with which Carnap populated his structure description he called character bearers (Carnap 1922: 23). To disambiguate phylogenetic systematics, Hennig wanted to replace natural language with a 'precision language' (Hennig 1949: 138), and rendered the character bearer as *Merkmalsträger-Semaphoront* (Hennig 1950: 9; 'character-bearing semaphoront', Hennig 1966: 6). In everyday practice, and certainly in Hennig's practice as an entomologist working with holometabolous insects, the semaphoront is taken as a stage in the life cycle of an organism. More precisely, Hennig wanted the semaphoront to function as the ultimate element of biological systematics, i.e. to correspond to an individual organism during a 'very small temporal duration' of its existence (Hennig 1950: 9), or to a 'theoretically infinitely small period of its life' (Hennig 1966: 6). Such a theoretical definition of a semaphoront reflects Hennig's commitment to a process ontology (Rieppel 2007), which caught Hennig, the systematist, in a conundrum, however. The *locus classicus* for process philosophy is Arthur N. Whitehead's *Process and Reality*, the essence of which Whitehead (1920: 14) summed up elsewhere: 'There is no holding nature still and looking at it.' But this is exactly what a systematist must do, and Hennig did so through the use of the semaphoront.

Ziehen denied the epistemic accessibility of the material world. The substantial world for him was one of fields of energy of different density. Whatever caused that which is given, i.e. the perceptions, he called 'somethings' (*Etwasse*). Organisms more specifically become 'animated natural things' ('*belebte Naturdinge*', Hennig 1950: 6), which Hennig, following Nicolai Hartmann, equated with 'systems of causal interactions' (Hennig 1950: 5, 23), a view he found likewise articulated by Bertalanffy (1941: 251): 'the organic form must be comprehended as a cross-section through a spatio-temporal flow of events' (approvingly cited by Hennig 1950: 5). The organism with its many properties thus becomes a processual system, a multidimensional multiplicity of causal interactions that determine the spatiotemporal flow of events that is the organism. The semaphoront then becomes a cross-section through that flow of events, the still image of a transitional phase in the organism's existence.

'The Given', the perception, the *Gignomen* sensu Ziehen is to be decomposed into its two components, the reduction part (*R*-part) that reveals the causal relations prevalent in nature, and the parallel part (*N*-part) that reveals parallel psychological laws governing the mind. Assume that the *Gignomen* is a perception of an 'animated natural thing', i.e. of an organism that is a system of causal interactions. Reducing such a perception to its *R*-part will disclose the causal relations prevalent at that cross-section through the spatiotemporal flow of events that is an organism. Assume this organism to be a holometabolous insect that is inspected at different stages of its life cycle. This will result in a series of *R-parts* that disclose the causal relations prevalent through that series of ontogenetic stages. If intermediate stages are not available, either due to lack of inspection or lack of preservation of such stages in museum collections, they can be inferred from the antecedent and consequent stages that are available in the series of *R*-parts: 'intermediate stages need not always to be observed, but can be inferred from causal laws' (Ziehen 1934: 60). In Hennig's rendition, this reads:

> We ascribe substantial sameness to two consecutive R's, R_1 and R_2 if the difference between R_1 and R_2 is causally understandable through a continuous sequence of intermediate stages' (Ziehen, I). Only on that basis can we recognize, for example, the acorn and the oak tree that develops from it as one and the same individual: 'thus two phases of the same thing may be completely different in their R-components [in this sense, peculiarities or characters], and yet on the basis of the continuous causal connection we speak of a single thing' (Ziehen, I). (1950: 114, Hennig 1966: 81)

On this account, the semaphoront corresponds to the *R*-part of a *Gignomen* of an animated natural thing, i.e. the *R*-part of a perception of an organism, where the perception slices through the spatiotemporal flow of causally interconnected events that is the organism. This is a thoroughly phenomenological account of the semaphoront, and of the organism reconstructed from the seriability of semaphoronts derived from it. The organism is logically reconstructed from what is given in perception, or rather, from the *R*-parts of what is given in perception. Such a phenomenological interpretation of Hennig's concept of semaphoront might seem to clash, however, with his contention that the semaphoront is an individual, located in three-dimensional space and time. Hennig spoke of the characteristics of the 'three-dimensional body' of the semaphoront (Hennig 1950: 9), but the *R*-part of a *Gignomen* cannot possibly be a three-dimensional body (but see Tremblay, 2013, for an argument that Hennig took the semaphoront to be a spatiotemporally located individual). How can Hennig claim a three-dimensional reality for the semaphoront? What the *R*-part of a *Gignomen* does is to reveal the causal relations captured by perception, in case of organisms the causal relations that determine the spatiotemporal flow of events that is the organism. Causal relations are existence implying: it is hard to understand how non-existents could enter into causal relations, could take part in events that tie cause to effect. On Ziehen's account, the organism is not, indeed cannot be directly apprehended, and yet, the perception of the organism is a perception of 'something', of an 'animated natural thing' that really exists in time and space.[2]

15.7 Conclusions

Hennig reproduced Ziehen's definition of *order* as the essence of systematics:

> Ziehen (1939) defines 'order' as 'the totality of progressively graduated vicinal similarities of more or less determined positional relationships of several or many, even an infinite number, maximally all, 'somethings' within a finite or infinite 'whole'. As 'position' he understands 'a more or less defined relation of a simple or complex something to other somethings that belong to a unified whole in regard to quality, intensity, locality, temporality, or number'. (Hennig 1950: 6; 1966: 3f; see Ziehen 1939: 3f)

It comes as no surprise that one of the translators of Hennig (1966), Rainer Zangerl, complained about the 'great many linguistic difficulties' he encountered (Zangerl in Hennig 1966: v). It is also no surprise that Mayr (1982: 226) judged such pronouncements pure gibberish; he 'grew up in a very philosophically-minded family', he conceded on one occasion, but got 'an overdose of it and have reached the point where I consider philosophy a rather sterile enterprise.'[3] What the preceding discussion shows, however, is that Hennig brought a quite sophisticated philosophical machinery to bear on the theoretical foundations of his phylogenetic systematics. Tremblay (2013) is certainly correct when he locates in Hennig's writings a deep concern for the ontological dimension of his phylogenetic system. One could well say that Hennig's methodological innovations in systematics represent his empirical project, whereas his philosophical excursions represent his concerns about the *nature* of the objects dealt with in phylogenetic systematics (metaphysical concerns), as well as questions of *epistemic access* to these objects (epistemological concerns).

Hennig was an outspoken supporter of the idea that species, as well as supraspecific taxa, are complex wholes, or individuals (Rieppel 2011b). This is certainly

---

[2] Whereas this is a possibility of interpretation, it is not necessarily Hennig's own understanding of a semaphoront. His writing on semaphoronts might also suggest that he did not fully grasp the implications of Ziehen's phenomenologism (M. Schmitt, pers. comm.).

[3] Harvard University Archives, Ernst Mayr Correspondence, HUG(FP) 14.7, Box 5, Folder 253.

a topic that, in contemporary philosophical literature, continues to be heavily debated. Other positions adopted by Hennig are equally uncontroversial, such as the way in which he anchored individuality, indeed reality in general, in time and causality. Rather more controversial is Ziehen's claim for a causally fully determined universe, in the light of which Ziehen (1939: 218f) even rejected Heisenberg's uncertainty principle, and with it the implications quantum mechanics had brought to discussions concerning the nature of laws of nature, and the validity of the claim to a hypothetico-deductive structure for a mature science such as theoretical physics. Hennig, like many of his peers – including Bernhard Rensch and Adolf Remane – principally accepted a causal determination of the world, but worked with scope restriction for laws of nature to make that demand meet the requirements of phylogenetic systematics. Hennig's acceptance of an evolutionary epistemology as such is hardly controversial either, although acceptance of Ziehen's claim that the logical structure of thought matches the historically conditioned structure of the universe as a consequence of evolutionary adaptation certainly is at odds with the modern synthesis of evolutionary theory, which emphasizes the historical contingency of at least the biological world.

The most intriguing aspect of Hennig's philosophical project certainly is his adoption of Ziehen's elaborate phenomenologism, where it is the *Gignomen*, i.e. the perception, not the object, that is immediately given to the cognizing mind. What must have attracted Hennig to this philosophical construct is Ziehen's claim that the *Gignomen* can be reduced to a *reduction part*, which reveals causal relations prevalent in the extramental world, and the *parallel part*, which captures the laws of the human mind at play in the process of cognition. At the same time, the causal laws captured by the *R*-part of the *Gignomen* guarantees the spatiotemporal location and hence reality of the objects underlying the perceptions. Hennig's motivation may well have been his quest for unbiased observation. He was clearly concerned about the 'participation of human activity in the process of perception' (Hennig 1950: 15, 1966: 11), and cited a chapter by Hermann Wein in the *Systematische Philosophie* of 1942 edited by Nicolai Hartmann, in which Wein talked about the 'new "actional" structure of the subject-object relation' (Hennig 1950: 15, 1966: 11). Hennig's adoption of Ziehen's epistemology becomes explicable in the context of his writing. Hennig understood his phylogenetic systematics as a superior alternative to idealistic morphology and the systematics based on it. Among German biologists of the first half of the twentieth century there existed two strands of idealistic morphology: one built on logic and epitomized by the zoologist Adolf Naef (Rieppel 2012a, Rieppel et al. 2013), the other committed to neo-Romanticism and epitomized by the botanist Wilhelm Troll (Rieppel 2012b). The Tübingen botanist Walter Zimmermann was not only considered a leading pioneer of phylogenetic systematics by Hennig, but also pursued a campaign against Troll, whom he accused of spreading 'myths and fairy-tales' (Zimmermann 1943: 29). Zimmermann (1930: v) vehemently opposed the 'irrational tendencies that are currently becoming apparent' in biology, 'mostly in opposition to phylogenetic theories'. It is in this context that Zimmermann invoked the philosophy of the Vienna Circle, citing in particular Moritz Schlick, Rudolf Carnap and Karl Popper (Zimmermann 1937/38; Popper was not a member, but rather a satellite of the Circle, Stadler 1997) in his plea for a strict separation of subject and object as the basis of sound scientific biology. Hennig might well have found in Ziehen an epistemology that promised to fulfill exactly that requirement.

Acknowledgments

I thank the editors of this book for having invited me to contribute a chapter. Gary Nelson kindly shared his index of authors for Hennig (1950, 1966) with me. Sophie Pécaud, Michael Schmitt, Frederic Tremblay and an anonymous reviewer offered comments that greatly improved an earlier draft of this chapter.

References

Bavink, B. (1933). *Ergebnisse und Probleme der Naturwissenschaften. Eine Einführung in die heutige Naturphilosophie. Fünfte, neu bearbeitete und erweiterte Auflage.* Leipzig: S. Hirzel.

Bavink, B. (1941). *Ergebnisse und Probleme der Naturwissenschaften. Eine Einführung in die heutige Naturphilosophie. 7. Auflage.* Leipzig: S. Hirzel.

Benninghoff, A. (1930). *Beiträge zur Anatomie funktioneller Systeme.* Frankfurt a.M.: Akademische Verlags-Gesellschaft.

Bertalanffy, L. v. (1932). *Theoretische Biologie, Band I.* Berlin: Gebr. Bornträger.

Bertalanffy, L. v. (1941). Die organismische Auffassung und ihre Auswirkungen. *Der Biologe*, 10, 247–258, 337–345.

Beurlen, K. (1936). Der Zeitbegriff in der modernen Wissenschaft und das Kausalitätsprinzip. *Kant-Studien*, 41, 16–37.

Beurlen, K. (1937). *Die stammmesgschichtlichen Grundlagen der Abstammungslehre.* Jena: G. Fischer.

Carnap, R. (1922). Der Raum. Ein Beitrag zur Wissenschaftslehre. In *Kant-Studien. Ergänzungshefte im Auftrag der Kant-Gesellschaft, Nr. 56.* ed. H. Vaihinger, M. Frischeisen-Köhler, Liebert, A. (Eds.), Berlin: Reuther & Reichard.

Carus, A.W. (2007). *Carnap and Twentieth-Century Thought.* Cambridge: Cambridge University Press.

Dobzhansky, T. (1937). *Genetics and the Origin of Species.* New York: Columbia University Press.

Donoghue, M.J. and Kadereit. J.W. (1992). Walter Zimmermann and the growth of phylogenetic theory. *Systematic Biology*, 41, 74–85.

Dürken, B. (1936). *Entwicklungsbiologie und Ganzheit. Ein Beitrag zur Neugestaltung des Weltbildes.* Leipzig: B.G. Teubner.

Friederichs, K. (1927). Grundsätzliches über die Lebenseinheiten höherer Ordnung und den ökologischen Einheitsfaktor. *Die Naturwissenschaften*, 15, 153–157; 182–186.

Gerhard, U.-J. and Blanz, B. 2004. Theodor Ziehen, M.D., Ph.D., 1862–1950. *The American Journal of Psychiatry*, 161, 1369.

Gliboff, S. (2008). *H.G. Bronn, Ernst Haeckel, and the Origins of German Darwinism. A Study in Translations and Transformation.* Cambridge, MA: The MIT Press.

Gregg, J.R. (1954). *The Language of Taxonomy. An Application of Symbolic Logic to the Study of Classificatory Systems.* New York: Columbia University Press.

Günther, K., (1956). Systemlehre und Stammesgeschichte. *Fortschritte der Zoologie*, N.F. 10, 33–278.

Hartmann, M. (1933). *Allgemeine Biologie. Eine Einführung in die Lehre vom Leben, zweite Auflage.* Jena: Gustav Fischer.

Hartmann, N. (Ed.) (1942). *Systematische Philosophie.* Stuttgart: Kohlhammer.

Heidenhain, M. (1907). *Plasma und Zelle. Erste Abteilung. Allgemeine Anatomie der lebendigen Masse.* Jena: Gustav Fischer.

Heidenhain, M. (1923). *Formen und Kräfte in der lebendigen Natur.* Berlin: Julius Springer.

Hennig, W. (1931). Einiges über die Insketen des Landesschulgebietes. *Mitteilungen aus der Landesschule Dresden*, 8, 1–6.

Hennig, W. (1947). Probleme der biologischen Systematik. *Forschungen und Fortschritte. Nachrichten der deutschen Wissenschaft und Technik*, 21/23, 276–279.

Hennig, W. (1949). Zur Klärung einiger Begriffe der phylogenetischen Systematik. *Forschungen und Fortschritte. Nachrichten der deutschen Wissenschaft und Technik*, 25, 136–138.

Hennig, W. (1950). *Grundzüge einer Theorie der phylogenetischen Systematik.* Berlin: Deutscher Zentralverlag.

Hennig, W. (1953). Kritische Bemerkungen zum phylogenetischen System der Insekten. *Beiträge zur Entomologie*, 3 (Beilageband), 1–85.

Hennig, W. (1957). Systematik und Phylogenese. In *Bericht über die Hundertjahrfeier der Deutschen Entomologischen Gesellschaft*, ed. H.-J. Hannemann. Berlin: Akademie-Verlag, pp. 50–71.

Hennig, W. (1966). *Phylogenetic Systematics.* Urbana, IL: University of Illinois Press.

Hennig, W. (1974). Kritische Bemerkungen zur Frage "Cladistic analysis or cladistic classification?" *Zeitschrift für Zoologische Systematik und Evolutionsforschung*, 12, 279–294.

Hennig, W. (1978). Die Stellung der Systematik in der Zoologie. *Entomologica Germanica*, 4, 193–199.

Kant, I. (1786). *Metaphysische Anfangsgründe der Wissenschaft.* Riga: J.F. Hartknoch.

Kraus, O. (1984). Die Veranstaltung "Phylogenetisches Symposion": Rückblick auf 25 Tagungen (1955–1982). *Verhandlungen des Naturwissenschaftlichen Vereins in Hamburg*, (NF) 27, 277–289.

Kraus, O. and Hoβfeld, U. 1998. 40 Jahre "Phylogenetisches Symposium" (1956–1997): eine Übersicht – Anfänge, Entwicklung, Dokumentation und Wirkung. *Jahrbuch für Geschichte und Theorie der Biologie*, 5, 157–186.

Mayr. E. (1965). Selektion und die gerichtete Evolution. *Die Naturwissenschaften*, 52, 173–180.

Mayr, E. (1982). *The Growth of Biological Thought. Diversity, Evolution, and Inheritance*. Cambridge, MA: The Belknap Press of Harvard University Press.

Naef, A. (1919). *Idealistische Morphologie und Phylogenetik (zur Methodik der systematischen Morphologie)*. Jena: Gustav Fischer.

Nicholson, D.J. and Gawne, R. (2014). Rethinking Woodger's legacy in the philosophy of biology. *Journal of the History of Biology*, 47, 243–292.

Rensch, B. (1968). *Biophilosophie auf erkenntnistheoretischer Grundlage (Panpsychistischer Identismus)*. Stuttgart: Gustav Fischer.

Rensch, B. (1979). *Lebensweg eines Biologen in einem turbulenten Jahrhundert*. Stuttgart: Gustav Fischer.

Rieppel, O. (2007). The metaphysics of Hennig's phylogenetic systematics: substance, events and laws of nature. *Systematics and Biodiversity*, 5, 345–360.

Rieppel, O. (2009). Hennig's enkaptic system. *Cladistics*, 25, 311–317.

Rieppel, O. (2011a). Willi Hennig's dichotomization of nature. *Cladistics*, 27, 103–112.

Rieppel, O. (2011b). Species are individuals: the German tradition. *Cladistics*, 27, 629–645.

Rieppel, O. (2012a). Adolf Naef (1883–1949), systematic morphology and phylogenetics. *Journal for Zoological Systematics and Evolutionary Research*, 50, 2–13.

Rieppel, O. (2012b). Wilhelm Troll (1897–1978): idealistic morphology, physics, and phylogenetics. *History and Philosophy of Life Sciences*, 33, 321–342.

Rieppel, O. (2013). Styles of scientific reasoning: Adolf Remane (1898–1976) and the German evolutionary synthesis. *Journal of Zoological Systematics and Evolutionary Research*, 51, 1–12.

Rieppel, O., Williams, D.M. and Ebach, M.C. (2013). Adolf Naef (1883–1949): On foundational concepts and principles of systematic morphology. *Journal of the History of Biology*, 46, 445–510.

Schlee, D. (1978/79). In Memoriam Willi Hennig 1913–1976. Eine biographische Skizze. *Entomologica Germanica*, 4, 377–391.

Schmitt, M. (2001). Willi Hennig (1913–1976). In *Darwin & Co. Eine Geschichte der Biologie in Portraits*, Vol. 2, ed. J. Jahn and M. Schmitt. Munich: C.H. Beck, pp. 157–175.

Schmitt, M. (2013). *From Taxonomy to Phylogenetics: Life and Work of Willi Hennig*. Leiden: Brill Academic Publishers.

Stadler, F. (1997). *Studien zum Wiener Kreis. Ursprung, Entwicklung und Wirkung des logischen Empirismus im Kontext*. Frankfurt a.M.: Suhrkamp.

Thienemann, A. (1939). Grundzüge einer allgemeinen Ökologie. *Archiv für Hydrobiologie*, 35, 267–285.

Thienemann, A. (1941). *Leben und Umwelt*. Leipzig: Johann Ambrosius Barth.

Tremblay, F. (2013). Nicolai Hartmann and the metaphysical foundation of phylogenetic systematics. *Biological Theory*, 7, 56–68.

Wein, H. (1942). Das Problem des Relativismus. In *Systematische Philosophie*, ed. N. Hartmann. Stuttgart: Kohlhammer, pp. 431–559.

Whitehead, A.N. (1920). *The Concept of Nature. Tarner Lectures delivered in Trinity College November 1919*. Cambridge: Cambridge University Press.

Woodger, J.H., (1952). From biology to mathematics. *British Journal for the Philosophy of Science*, 3, 1–21.

Ziehen, Th. (1934). *Erkenntnistheorie. Zweite Auflage. Erster Teil. Allgemeine Grundlegung der Erkenntnistheorie.*

Spezielle Erkenntnistheorie der Empfindungstatsachen einschliesslich Raumtheorie. Jena: Gustav Fischer.

Ziehen, Th. (1939). *Erkenntnistheorie. Zweite Auflage. Zweiter Teil. Zeittheorie. Wirklichkeitsproblem. Erkenntnistheorie der anorganischen Natur (erkenntnisheoretische Grundlagen der Physik). Kausalität.* Jena: Gustav Fischer.

Zimmermann W. (1930). *Phylogenie der Pflanzen. Ein Überblick über Tatsachen und Probleme.* Jena: Gustav Fischer.

Zimmermann, W. (1937/38). Strenge Objekt/Subjekt-Scheidung als Voraussetzung wissenschaftlicher Biologie. *Erkenntnis*, 7, 1–44.

Zimmermann W. (1943). Die Methoden der Phylogenetik. In *Die Evolution der Organismen. Ergebnisse und Probleme der Abstammungslehre*, ed. G. Heberer. Jena: Gustav Fischer, pp. 20–56.

16

Hennig and hierarchies

Charissa S. Varma

16.1 Introduction

The German entomologist Willi Hennig (1913–76) is remembered for his impressive contribution to taxonomy's methodological reform, a contribution that had a profound, lasting effect on taxonomic practice and earned him the title "father of cladistics." The methodological details of Hennig's reform work were backed by a strong empirical programme, and have been vigorously explored both historically and scientifically. However, when Hennig crafted his solution to the methodological crisis, he constructed a complex philosophical foundation to support his methodological innovations, including a logical account of hierarchies. Hennig was not alone in exploring the relationship between taxonomy and philosophy (their logic in particular). During the methodological reform debates, philosopher David Hull captured the complex relationship between biology and philosophy examined by many taxonomists in his 1964 statement in *Systematic Zoology*:

> The three factors in phylogenetic taxonomy are phylogeny, the taxonomic schema, and the relation between the two. Of the three factors in the phylogenetic program, only phylogeny is of an empirical nature. The structure of the taxonomic schema is entirely a matter of logic, and the relation which this schema is to have to phylogeny is primarily a concern of the purposes of taxonomy. (Hull 1964: 1)

In spite of framing issues in taxonomy to include investigations into the relationship between taxonomy and philosophy, close examinations of Hennig's philosophical project have been generally overlooked in the literature in favour

The Future of Phylogenetic Systematics: The Legacy of Willi Hennig, eds. D. Williams, M. Schmitt and Q. Wheeler. Published by Cambridge University Press.
© The Systematics Association 2016.

of his methodological project. Entomologist Lars Brundin, who introduced the English-speaking world to Hennig's work, was among the earliest to excise much of Hennig's philosophical programme from his methodological programme (Brundin 1966). Exploring Hennig's philosophical position and its relationship to his methodological ideas not only clears up some confusion in Hennig's work (specifically how to understand his account of hierarchies), but provides a more comprehensive picture of Hennig's complete project and the role it played in the history of taxonomy. Hennig approached his account of hierarchies from a few angles in his writing from the late 1940s onward, but one point of departure began by revisiting the concept of relations to build up an account of hierarchies.

Hennig's account of hierarchies is often told as a tale of confusion and inconsistency as a result of his alleged appeal to a set-theoretic treatment of such structures. Each version of this tale seemed to be variations on a common theme: in *Phylogenetic Systematics* (1966) Hennig either made a mistake in his discussion of hierarchies or he was inconsistent in his treatment of hierarchies. In *Science as a Process*, Hull suggested that the confusion could be the result of editorial decisions.[1] Later, however, Hull claimed:

> One common error in the early years of cladistics was to read too much into cladograms, an error committed by all sides of the dispute, but eventually the distinction became clear. Scott-Ram claims that the "roots of this misconception can be traced back to Hennig's own discussion of hierarchical systems, which is based on the Woodger- Gregg model" (p. 89). Scott-Ram is right that the Woodger–Gregg model of hierarchical structure is mistaken, but these errors did not influence Hennig because Hennig's German book appeared before Gregg published and it contains no reference to Woodger's analysis of hierarchical organization. Hennig added his discussion of Woodger and Gregg to his English version, possibly to make his system appear more quantitative and up-to-date. I doubt that he understood Woodger and Gregg's set-theoretic reconstructions. (Hull 1990: 422)

Here Hull changed his mind about Hennig and hierarchies, from confusions on hierarchies as a result of editorial decisions to confusions on Hennig's part. Hull maintained that there was no reference to the analysis of hierarchies by English cell

---

[1] Hull wrote: "However, these two men [Davis and Zangerl] did not just translate the manuscript. They also heavily edited it, eliminating what they took to be repetitive passages, simplifying Hennig's Teutonic sentences, and clarifying his ideas. In this midst of this undertaking, Davis died, and Zangerl had to carry on alone. As fate would have it, both men who translated Hennig's manuscript were trained in the very philosophical tradition that Hennig attacked in the book – idealistic morphology. Both Zangerl (1948) and Davis (1949) emphasized the necessary role that the hierarchies of morphological types of neoclassical morphology play in phylogeny construction, a position that Hennig himself adamantly opposed. Only a German scholar studying the relevant manuscripts can say how much the idealist presumptions of Davis and Zangerl influenced their translation. Hennig himself was unable to help much in the project because he was in the midst of fleeing from East to West Germany. An accurate translation of his manuscript was the least of his worries" (Hull 1988: 134).

biologist Joseph Henry Woodger (1894–1981) in Hennig's early work, and that he doubted that Hennig understood Woodger and the set-theoretic reconstructions found in the taxonomic writing of American cell biologist John R. Gregg (1917–2009).

Patricia Williams argued in her paper "Confusion in cladism" that Hennig conflated two types of "hierarchy" in *Phylogenetic Systematics,* which she called the Linnaean hierarchy and the "divisional hierarchy." She claimed:

> What seems to have happened is that Hennig carefully began with a particular concept of "hierarchy", the divisional "hierarchy", his interpretation of Gregg. Because Gregg was discussing the Linnaean hierarchy, Hennig thought his interpretation of Gregg to be a correct understanding of the Linnaean hierarchy. In this, he was wrong. The divisional "hierarchy" in biology is a linear, temporal phylogenetic tree; the Linnaean hierarchy is an atemporal, inclusive hierarchy which can be constructed from a phylogenetic tree, but is not the same thing as one. Hennig shifted back and forth between the two "hierarchies" and, in the process of doing so, developed a self-contradictory system of biological taxonomy. (Williams 1992: 151, see Williams et al. 1996)

As a result, she concluded, Hennig made the school of biological taxonomy known as "cladism" philosophically confused.

Olivier Rieppel drew attention to confusion in Hennig's work on hierarchies. Rieppel argued that by accepting Gregg's account in *Language of Taxonomy* (1954), of what was presumably Woodger's (1952) set-theoretical concept of hierarchies, Hennig landed himself in a conundrum. On Gregg's set-theoretic interpretation, sets imposed sharp boundaries that resulted in a classification system of nested sets. This had consequences on the logical understanding of a semaphoront. According to Rieppel, semaphoronts (semaphoront complexes) were not only the character bearers, but logically speaking they were members of sets, they instantiated their respective species (by instantiation of the characters that were the membership criteria for the respective sets). Herein lay the problem for Rieppel. On this reading of Hennig's species concept, Rieppel claimed that "based on (lawful) continuity through time rather than on certain (necessary and sufficient) properties, species become individuals," and consequently the hierarchy for species was a division hierarchy. Since species were parts of the genealogical nexus, they were not members of a set. So, for Rieppel, according to Hennig, semaphoronts (semaphoront complexes) were not the objects of the investigation of phylogenetic relationships: "In order to answer the question of whether the hierarchic system is rightfully used in biological systematics we must investigate whether semaphoronts can be substituted" for the argument spaces in Gregg's *Language of Taxonomy's* (1954) formal language (Hennig 1966: 18). As Rieppel pointed out, Hennig responded: "Obviously they cannot" (Rieppel 2006: 487). But a further problem remained.

Rieppel agreed with Hull's assessment of Woodger as an "idealist morphologist" based on Woodger's account of homology in "On biological transformations"

(Woodger 1945; Rieppel 2006: 487). Given Hennig's objections to idealist morphology throughout *Phylogenetic Systematics*, Rieppel puzzled over reconciling that with Hennig's overt praise of Woodger's treatment of homology in terms of "symbolic logic" and with his appeal to Woodger's claim that species and supraspecific taxa are real, and as such are individuals, individuated by their genealogical relations. Rieppel's solution was that Hennig ran together two arguments that should have been kept separate (Rieppel 2006).

Knox also worried about Hennig on hierarchies, in particular Hennig's reading of Gregg in *Phylogenetic Systematics*, given Hennig's position on genealogical groups. Knox wrote:

> It is difficult to determine whether Hennig misunderstood the work of Woodger and Gregg, or simply misrepresented it, because their definition of a hierarchy is antithetical to Hennig's stated intention of developing a truly historical approach to systematics. In Gregg's model, genera are sets of species, not genealogical groups. (Knox 1998: 9)

Knox also indicated that by 1952 Woodger had changed his position from a set-theoretic account (the Linnaean hierarchy) to a more "evolutionary phylogenetic scheme," and he was not certain that Hennig was aware of that change (Knox 1998: 9). So, sorting out if Hennig did make any of these errors requires further investigation into his philosophical programme. By examining Hennig's thoughts on hierarchies from the late 1940s, revisiting Gregg's work on hierarchies and ontology, and Woodger's work on hierarchies and ontology from the 1930s to the 1950s (specifically Woodger's change in position when it came to taxonomy), it becomes easier to see that Hennig held a clear and consistent position on hierarchies that drew from his discussion of relations, from the 1940s through the 1960s.

16.2 Relations and hierarchies

The role of relations in taxonomic methodology puzzled Hennig right from the start. As a student, Hennig raised a skeptical eye to the use of similarity relationships as the basis of taxonomic relationships; a practice he believed was standard in German idealistic morphology. Some of Hennig's published objections to using the concept of similarity as a reliable choice for estimating evolutionary relationships can be traced to this early work. In 1936 he recognized that using similarity to classify larvae and adults of insects in the same taxa often resulted in different classifications even though they shared the same phylogenetic history (Hennig 1936).[2] Later, leaning on an analysis of relations by Swiss zoologist Adolf Naef (1883–1949),

---

[2] Schmitt (2010) traced the history of this idea in Hennig and looking at Hennig's publications, I am
   inclined to agree. For examples of Hennig making claims about similarity at this point in his career,
   see Hennig (1936, 1943).

Hennig redefined "relationship." Hennig did not agree with all of Naef's ideas or his agenda, but Naef's attempt to differentiate between a notion of similarity and the relationships that fall under phylogeny, rather than discussing them together under the umbrella of "similarity," struck a chord. Discussing Naef and Hennig on this point, Breidbach stressed their differences:

> For Naef, idealistic morphology is the scientific foundation on which phylogenesis has to be based, and not the other way round. The historical approach towards systematics as formed by phylogenesis does not offer new material. It is just an interpretation of the results presented by idealistic morphology. Thus, idealistic morphology will be the basic science in comparative biology. It will be the only science that allows clarification of the systematic relations between various species. (Breidbach 2003: 187)

In contrast, for Hennig phylogenetic relations were primary, but this would be a claim he would have to defend. Even though they were primary, Hennig, just like Naef, knew that gaining direct epistemic access to phylogenetic relations would not be possible.[3] Even with epistemic limitations, Hennig sketched an approach to analysing relationships. In practice, relationships should be defined genealogically and genetically and this move enabled him to distinguish the ontological concept of "relationship" from a practical definition of "similarity."[4] The distinction between similarity relations and phylogenetic relations also provides some insight as to why Hennig might have focused part of his investigation in "Probleme der biologischen Systematik" (Hennig 1947), a short theoretical paper discussing the troubled state of taxonomy, on exploring a more philosophical idea of relations.

16.3 Relations and division hierarchies

In "Probleme der biologischen Systematik" (Hennig 1947), Hennig looked at relations in more detail: what they were and how they fit together. Hennig was concerned with what he called "morphological" or "typological" systematics because he believed that such systems arranged organisms according to similarity of form or "Gestalt." He noted that after the rise of evolutionary theory, taxonomists were not reinterpreting morphological similarity genetically. However, Hennig worried that after a while biologists were not reinterpreting systems so much as they were combining systems. As a result the different organizing principles were becoming more and more detached

---

[3] Rieppel made this point about epistemic access in his discussion of Naef and Hennig's use of Naef (Rieppel 2012).

[4] Schmitt notes Hennig's early appeal to Naef (Schmitt 2003: 371); Rieppel warns against reading too much into Naef's influence on Hennig, emphasizing the important ways they differ philosophically (Rieppel 2012).

from each other, so much so that systematics as a discipline seemed to be eliminated.[5] Systematics, Hennig maintained, needed to be brought back.

Hennig talked about hierarchies in taxonomy more generally, saying that although there were other types of systems available to taxonomists, for example the "periodic system" for butterflies and the network structure of family relations, the most suitable system for representing the phylogenetic system was a hierarchical system (Hennig 1947: 278; Hennig made a similar claim in his *Phylogenetisch Systematik*, Hennig 1950). Hennig went on to specify the kind of hierarchical system he had in mind when he was sketching out the structure of genetic relationships in reproductive communities when they split in two. Hennig wrote:

> All these attempts are based on mistaken ideas about the structure of the genetic relationships. Having recognized that all organisms break down into a number of reproductive communities (species) and that these reproductive communities break down through fission, and having determined that the resulting "phylogenetic" relations among the successor species are natural continuations of the existing genetic relations within the species – which continuations should be represented in the phylogenetic system and thus in the universal reference system of "taxonomy" – it is also certain that in taxonomy only the hierarchical system type can be used. The system describes a "division hierarchy" with the species as the dividing unit, much like the hierarchy of the Histosystem in the individual organism is based on the cell as the dividing unit. (Hennig 1947: 278, translated)

Hennig used the term "Teilungshierarchie" (or "division hierarchy" in English) to highlight a specific kind of hierarchy he believed would formally capture a particular kind of relation he had in mind, namely the formal nature of the phylogenetic relationship defined with respect to reproductive communities.

---

[5] Hennig wrote: "In dieser Frage stehen sich heute namentlich die Vertreter einer genetischen und einer morphologischen Systematik mit voneinander abweichenden Anworten gegenüber. Von vielen wird die Auffassung vertreten daß aus historischen und logischen Gründen dem morphologischen oder typologischen System in dem die Lebewesen nach ihrer Gestalt ähnlichkeit geordnet sind, vor anderen der Vorzug gebühre. Aus historischen Gründen, weil die älteren Systeme durch weg - nach diesem Prinzip aufgestellt und dann erst, nach dem Aufkommen der Deszendenztheorie, von einer genetischen Systematik genetisch "umgedeutet" worden seien. Aus logischen Gründen, weil das genetische System wenigstens für die höheren, auf entfernterer genetischer Verwandischaft ihrer Komponenten beruhenden Kategorien ausschließlich undeine auf eine Umdeutung der morphologischen Befunde angewisesen sei. Beide Begründungen beruhern auf irrigen Voraussetzungen Historische haben auch die ältesten Systeme bereits genetische Kriterien verwertet. Sie waren nicht Systemme der morphologischen Ähnlichkeitsbeziehungen, die später erst genetisch umgedeutet wurden, sondern vielmehr kombinierte Systeme, in denen die verschiedensten Prinzipien, nach denen Organismen geordnet werden können und müssen (s.oben) miteinander kombiniert wurden. Hier hat wissenschaftgeschichtlich eine "Ent-wicklung" im eigentlichen Sinne dieser Wortes stattgefunden, indem die verschiedenen Ordnungsprinzipien immer mehr voneinander gelöst bzw. teilweise auf [sic! aus] dem Aufgabengebiet der eigentlichen "Systematik" ausgeschieden wurden" (Hennig 1947: 276).

Although "Probleme der biologischen Systematik" (Hennig 1947) lacks a bibliography, examining the bibliographies of Hennig's other work published at around this time and the people he mentioned in the text, it is reasonable to assume he pulled the term "Teilungshierarchie" from Austrian biologist Karl Ludwig von Bertalanffy (1901–72). Bertalanffy, in turn, borrowed the term (with explicit references) from Woodger, who used the English term "division hierarchy." "Teilungshierarchie" is one of a number of technical terms with ontological import found in *Theoretische Biologie* (Bertalanffy 1932, 1942), a work Hennig cited frequently in *Grundzüge einer Theorie der phylogenetischen Systematik* (Hennig 1950). Other terms with a similar ontological agenda include German anatomist Martin Heidenhain's (1864–1949) terms "Histosystem" and "Enkapsis." All these terms can be found in the cell biology literature promoting a position called "organicism."

According to Bertalanffy, organicism is best described as a reaction to the mechanism/vitalism debate in biology, and their explanations centred on notions of processes (such as formation, segregation, differentiation, growth, polarity, symmetry, regeneration). However, what set their investigation apart was a particular strategy: an attempt to unravel the complexities of these processes using mathematics, logic, and concepts such as fields and force. Methodologically speaking, organicism was not against all forms of reductionism, even in terms of disciplines. In this sense, organicists distanced themselves from a more familiar logical positivism agenda. Even with the emphasis on mathematics and logic, there was a robust empirical side to their programme. Bertalanffy highlighted work done in experimental embryology that provided promising experimental evidence to support organicism.

Promoters of organicism registered disappointment with German biologists August Weismann's and Wilhem Roux's mechanistic attempt to describe living organisms through physico-chemical explanations, as well as German embryologist and (later philosopher) Hans Driesch's idea of entelechy or a vital force that governed the development and subsequent life of the organism. Their solution, the definition of organicism broadly speaking, began with the idea that a living biological system was more than the sum of its parts, and that somehow the "wholeness" of the living organism controlled and regulated key processes (see Bertalanffy 1928: 67–72 on the problems with developmental embryology, and pp. 73–77 for his discussion of Weismann, Roux, and Driesch). Many of the organicists developed the conceptual tools they needed, but not a formal language to accommodate such intuitions. Except Woodger.

16.4 Woodger

It is not surprising that Hennig found Woodger's position attractive. In the Preface of *The Axiomatic Method of Biology*, Woodger maintained:

> In every growing science there is always a comparatively stable, tidy, clear part, and a growing, untidy confused part. I conceive the business of theoretical science to be to extend the realm of the tidy and systematic by the application of the methods of the exact or formal sciences, i.e. pure mathematics and logistic. (Woodger 1937: vii)

Woodger proposed formal sciences, such as logic, be called upon to clean up the biological sciences. Woodger repeated this idea when he aimed to "provide an exact and perfectly controllable language by means of which biological knowledge may be ordered" (Woodger 1937: vii). Like Woodger, Hennig thought relations are best expressed using the language of set theory, a *mathesis universalis* (Hennig 1949: 138).[6] Using a universal language (like set theory) fit with Hennig's aim for a more unified taxonomic theory. In Hennig (1953), he explained how such a language could be a useful tool in methodological practice for insects. He wrote:

> [T]he departure point of all discussions must be the fact that organisms, in particular insects, form a multidimensional multiplicity...they differ from each other in many respects, which cannot be traced to each other. How such a multidimensional multiplicity is to be treated taxonomically is taught by set theory [Mengenlehre], a discipline of mathematics or of mathematical logic. (Hennig 1953: 9, translated)

But this does not explain why Hennig thought a mereological calculus would be a good fit.

Looking beyond sets was not unusual for biologists at this time. Early twentieth-century cell biologists, such as Woodger, saw biological organization as posing a troubling philosophical problem. Woodger understood biological organization as more than just a geometrical relation. A group of scattered coins on the ground did not make an organized whole, nor would simply arranging them in a pattern. Woodger explained:

> What we have to do is to make clear the difference between, say, a mass of frog's spawn and a frog blastula. Both are "analysable into cells" but there is clearly a difference between them which we express by saying that the latter is a single whole organism and the former is not. The example of the coins shows that it is

---

[6] Hennig wrote: "Man darf nun die Frage aufwerfen, ob es notwendig war, so viele neue Bezeichnungen einzuführen, nachdem die phylogentische Systematik bisher anscheinend ohne sie ausgekommen ist. Wenn es aber richtig ist, daß "Mathematisierung" im Sinne einer "Mathesis universalis" das Ziel jeder echten Wissenschaft sein muß, und wenn die Schaffung einer Präzisionssprache eine unbedingt notwendige Voraussetzung dafür ist, dann wird die phylogenetische Systematik auf die Dauer nicht ohne die vorgeschlagenen oder ähnliche, von der Vulgärsprache abgehobenen Bezeichnungen auskommen können (Hennig 1949 : 138).

Woodger wrote: "I began the study of the *Principia Mathematica* of Whitehead and Russell, which seemed to provide the necessary framework for a language which would enable us to calculate in biology. I have accordingly made use of that work in the present book in order to construct a biological calculus" (Woodger 1937: viii).

> not sufficient to say that in the case of the blastula there is some relation between the cells with respect to which they constitute a system and that in the case of the mass of frog's spawn there is no such relation. This is evidently important but there is something more required. And one further requisite seems to be that this relation should be an *internal* relation in the sense that a given term (e.g. a cell) is different when it is in this relation to the other terms from what it is when it is not in this relation. The whole will then change in its properties when it is deprived of a part, and a part will have different properties when removed from the whole from what it has in its place as a part. The whole will have "Gestalt" properties in the sense of Köhler. (Woodger 1930: 449)

Woodger insisted that the relations in a biological whole were different, thus the resulting hierarchy would also be different. Woodger aimed to provide a language to describe this type of situation, a mereological calculus. Where traditional set-theoretic calculi dealt in lists and elements, mereology looked at wholes and parts, and this difference usually implies a more fully fledged ontology because the components of a mereological calculus – fusions and overlaps – produce new entities, which is not the case for sets. Woodger's philosophical ideas and calculus found their way into the work of other cell biologists in addition to Bertalanffy, such as Theodore Torrey (1945–86), both of whom Hennig cites on this topic.[7]

In "Organisms in time", Torrey acknowledged biologists' reluctance to consider mathematics and physics in their analysis, specifically their reluctance to consider the biological world in four dimensions:

> It is most surprising how little attention has been paid to the significance of time by biologists. Here is a factor that has played an important part in the development of physical thought, and there is reason to believe its serious consideration would lead to consequences of equal importance in biology. Much of the tardiness on the part of biologists in thinking of organisms in terms of four dimensions seems to be based on a preconceived fear of the unknown or perhaps merely a stubborn literal adherence to Huxley's tenets of objective science. In any event the conception of the "four-dimensional continuum" which makes up a part of the every day thinking processes of the physicist or mathematician is a mental stumbling block over which the biologist trips, and having tripped, despairs. Actually it requires no special kind of intellectual feat to think in terms of four dimensions; the task is only one of logically building upward from a basis of familiar spatial concepts. (Torrey 1939: 275)

Not only did Torrey urge biologists to consider physics, but it became clear that when he said "logically building upward from a basis of familiar spatial concepts" he meant that statement literally, using tools from formal logic. Torrey showed how treating the organism as an organized whole and formally building time into

---

[7] Rieppel has noted that Bertalanffy was one of the most often cited authors in Hennig (1950); he also noted that Hennig cited Torrey (1939) in his discussions on space and time (Rieppel 2003: 169–71, 172).

the analysis helped solve some of the problem faced by biologists (Torrey 1939).[8] Hennig noted this appeal to the language and concepts of mathematics and logic in all editions:

> Trotz dieser heute nicht selten stark betonten Erkenntnis, daß die Unterscheidung zwischen Struktur und Prozeß nur konventionellen und anthropzentrischen Charakter habe (v. Bertalanffy I, p. 249) bleibt Torrey 1939 im Recht, wenn er es beklagt, daß der Biologe noch immer viel zu wenig mit dem Physiker und Mathematiker täglich vertrauten Begriffe des vierdimensionalen Kontinuums von Raum und Zeit arbeite. (Hennig 1950: 8; Hennig 1982: 13).

Although the insight, often strongly emphasized, that the distinction between structure and process has only conventional and anthropocentric character (von Bertalanffy), Torrey (1939) is correct when he complains that the biologist still works far too little with those concepts, familiar to the physicist and mathematician, of the four-dimensional continuum of space and time (Hennig 1966: 6).

Torrey's analysis of an organism was similar to that found in Hennig's work, from his concept of a semaphoront to his discussion of species. Torrey's organism was understood as an organized event manifested in time. Torrey discussed abstracting stages from the organism's multidimensional whole, coupled with understanding the organism (or in Hennig's case, the semaphoront), as a "division hierarchy" as described by Woodger:

> So far in our discussion, we have assumed a very comprehensive view of living things, looking upon any stage of an organism's ontogenetic or adult history as only a part abstracted from a hyperdimensional whole. Sight must not be lost of the fact, however, that such abstracted parts are of themselves also wholes, albeit wholes of a lower order. Any cross-section that one might choose, whether it be a 25 somite embryo or thirty-year-old adult, is also a unit, an organized, integrated going concern, it itself made up of parts subsidiary to the whole which in turn is a part of a greater whole. Thus an individual of any historical stage may be analysed progressively downward into systems, organs, tissues, cells, and cell components which, although they exhibit a certain degree of ontogenetic and adult independence, are

---

[8] Torrey wrote: "Proceeding to the question of how this 'time factor' is applied in conjunction with living organisms, we are immediately brought face to face with the almost insurmountable difficulty of attempting to describe or define what we mean by a living organism. Analyzed chemically and physically, organisms reveal no characteristics serving to distinguish them from non-living entities. They are something more than static, three-dimensional objects differentiated by some structural peculiarity or featuring a particular kind of chemical make-up. Such physical and chemical heteronomy is simply not found. We know living things not by how they are built, but by what they do. The materials comprising an organism are engaged in a multitude of activities, all directed towards the maintenance of both the individual and the tribe to which it belongs. How these separate activities are integrated into an organized whole is a problem in itself not to be considered here. Rather, the point to be made is that living things are truly organized events, and events manifest themselves only in time" (Torrey 1939: 277–278).

> entities of a lower order than the whole and bent essentially upon the service of the whole.
>
> In this connection attention should be drawn to the logistic system developed by Woodger (1930) in a study of the relation between embryology and genetics. Woodger conceives of the organism as a system of separate entities, ranging from the lowest cell component to the complete individual, arranged in hierarchal order. Every part is a whole, in a sense, but each is only a part of the whole of the systemic level above it and includes the whole of the level below it. To illustrate, the endoderm of a gastrula is a whole of one level, but only a part of a greater whole, the gastrula itself. Likewise the cells making up the endoderm are parts of the part of the whole, the nucleus of one such cell a part of the part of the part of the whole and so on. Woodger also conceives of a greater whole, the "division hierarchy" which includes the entire ontogenetic history of the organism. A given embryonic stage of development, therefore, becomes a "short temporal slice" of the division hierarchy. Though clothed in logistic language, Woodger's notions would seem to be essentially comparable to my own. (Torrey 1939: 279–280).

Following Torrey, Hennig understood the categories of the phylogenetic system to be individuals because the relations that bound them together were not similarity, but phylogenetic relations. Key to understanding the logical and ontological import of this claim is Hennig's idea that "they are all [...] segments of the temporal stream of successive breeding populations. As such they have a beginning and end in time" (Hennig 1966: 81). This notion of time demanded a different type of hierarchy (namely a mereological hierarchy). The idea of having a beginning and end in time is more important to a notion of biological individuality than functional or material coherence. From this, Hennig argued for the reality of species by claiming that if species are real and during a speciation event a parent species give rise to daughter species, the daughter species is no less real, provided it stood in a phylogenetic relation (not a similarity relation). He suggested a Woodger type of mereological approach:

> It would be of general epistemological interest and also of great importance for the theory of biological systematics to examine the extent to which this fact – that certain relationships of similarity existing between discrete natural objects are best expressed by ordering things bound together by such relationships of similarity into a hierarchical system of groups – suffices in and of itself to prove a historical development of these things from the way in which this development is hypothesized for the types of biological classification by the theory of evolution. That is, it would be of interest to explore to what extent the conclusion "hierarchy of similarity relations = division hierarchy" (Woodger, see Bertalanffy, 1932, p. 265 ff) is justified in such cases. (Hennig 1950: 20)

Hennig simplified this claim in 1966, saying instead:

> Wenn aber zwischen Naturkörpern Beziehungen bestehen, die offenbar nicht vom Menschen gesetzt sind, deren Struktur aber der eines hierarchischen

> Systems entspricht, so scheint sich als einzige annehmbare Erklärung für das Zustandekommen dieser Struktur die Annahme einer 'Teilungshierarchie' (woodger, siehe v. bertalanffy 1932, p. 265 ff.) anzubieten. (Hennig 1982: 28)

> But if there are relationships between natural bodies that obviously were not instituted by man but whose structure corresponds to that of a hierarchic system, then the only acceptable explanation for the occurrence of this structure seems to be the assumption of a "hierarchy of partition" (Woodger, see von Bertalanffy, 1932). (Hennig 1966: 21)

It is unfortunate that *Phylogenetic Systematics* has "Teilungshierarchie" translated as "hierarchy of partition" rather than "division hierarchy," but in all editions there is an explicit reference to Woodger through Bertalanffy. Key to this description was Hennig's thought that semaphoronts or species didn't form just any kind of hierarchy; they formed a Teilungshierarchie or division hierarchy. As Bertalanffy tried to articulate, the technical and formal logical discussion of hierarchies was not a strange formal footnote in the history of cell and developmental biology. So, given the context in which the term is found, coupled with Hennig's familiarity with Bertalanffy and Torrey's work, it is not unreasonable to assume Hennig maintained the logical and ontological understanding of "Teilungshierarchie."

However, when Woodger designed his mereological calculus, he very clearly *did not* have taxonomy in mind. Taxonomic hierarchies, for Woodger, were based on a set-member relation (Woodger 1932: 199–200). Briefly, and with disclaimers, in his *The Axiomatic Method in Biology* (1937), Woodger discussed taxonomy, applying the abstract definition of hierarchies. His abstract definition of hierarchies applied to sets and classes, not to individuals. So, what Hennig did in his work during the 1940s and *Phylogenetisch Systematik*, appeared not just radical, but remarkable. Rather than apply Woodger's formal analysis of a taxonomic hierarchy, Hennig extended Woodger's mereological notion of a division hierarchy to taxonomy. From a logical perspective, Hennig built his abstract semaphoront complexes and his species in much the same way Woodger built structures in cell biology.

Woodger (1937, 1952) defined an abstract hierarchy, such as the Linnaean hierarchy, where R is "relationship,": R is a hierarchy if and only if R is one-many and if the converse domain of R is identical with the set of all terms to which the first term of R stands in some power of R.[9]

This kind of hierarchy, the usual kind of set-theoretic hierarchy, was an organizational model with three noteworthy features. First, the highest level of organization consisted of a single entity. Second, any entity at a lower level of organization was related to only one entity at the next higher level; however, it could be related to more than one entity at the next lower level. Third, entities at all lower levels were

---

[9] Gregg defined hierarchical relations: HR = the set in which any relation z is a member if and only if (z∈ One-Many) and (ž "UN=žp0"{B′z}), Gregg (1954: 26).

related by extension to the single entity at the highest level of organization. A division hierarchy was different. For Woodger, division hierarchies were not fully nested and nor was it the case that the sum of all entities or the relations at one level of organization were equal to the sum of all entities or the relations at some other level. Simply put, division hierarchies did not necessarily display the property of summativity. This gives division hierarchy its distinctive feature – the whole is *more* than the sum of its parts, implying a more fully fledged ontology. Another significant feature of Woodger's mereological calculus, and why Hennig would have picked a division hierarchy over an abstract hierarchy, is Woodger's mereological calculus built in a notion of time which, among other things, allowed him to define parthood, organized entities, fission and fusion, which in turn would result in new entities. These were precisely the sort of features Hennig was looking for in a hierarchy for his new methodology. Given Hennig's appreciation of other aspects of technical philosophy, it is not unreasonable to assume he was aware of these important logical distinctions even if he did not burden us with the details of the formal calculus.

16.5 Division hierarchies pictured

In "Zur Klarung einiger Begriffe der phylogenetischen Systematik" (1949), Hennig provided what would be a pair of familiar diagrams in his work illustrating phylogenetic relations. With respect to his first two figures (Abb. 1 and 2 in Hennig 1949, reproduced here as Figs 16.1 and 16.2), Hennig wrote:

> In contrast to the morphological (and the same applies to the ecological), the phylogenetic relationships of the species should always be correctly and accurately represented in a dimension (t in Figure 1), so long as in practice the actual incompleteness of our knowledge does not also make the representation of these relationships imprecise. The phylogenetic relationships correspond in a different manner of representation to the hierarchical or enkaptic system type (Figure 2), in which the sequence of boundary lines following one another from the inside outwards corresponds to the one following, in which the bounded types have separated from the ancestral species common to them and only them. (Hennig 1949: 136)

Hennig wanted to show that there are two different, but equally correct ways to visualize phylogenetic relationships: one that shows a time dimension and one that shows a part-whole relation.

It is tempting to interpret Hennig's figure 2 as a Venn diagram, and this would be consistent with reading Hennig as providing a traditional set-theoretic treatment of species. Generally speaking, Venn diagrams were used to teach traditional set theory by illustrating logical relations that exist among a finite collection of sets. Such relations include: union, intersection, relative and absolute complement, symmetric difference. There are reasons to reject this reading of Hennig's figures. Hennig discussed an account of division hierarchies in "Probleme der

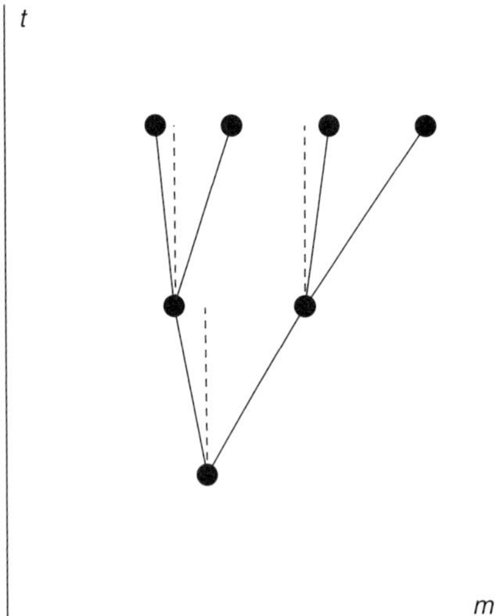

Fig 16.1 Phylogenetic relations as illustrated as a tree. After Hennig (1949: Fig 1).

Fig 16.2 Phylogenetic relations as illustrated as a hierarchical enkaptic system. After Hennig (1949: Fig 2).

biologischen Systematik" and he ended that article by arguing for the reality of species (Hennig 1947: 279).[10] He made a similar claim in *Grundzüge einer Theorie der*

---

[10] Hennig wrote: "Mit der Definition der Gruppenkategorien höherer Ordnung im phylogenetischen System hängt auch die Frage ihrer realen Existenz zusammen. Vielfach hat man sie als bloße Konstruktionen des ordnenden Menschengeistes angesehen. Die Erörterung dieser

phylogenetischen Systematik when he said, ontologically speaking, the two daughter species of a stem species form a group category of a *higher order* and it had to be viewed as identical with their stem species because it continued to exist through them (Hennig 1950: 115).[11] Brundin emphasised this idea:

> We are able to state that the recent biota, because of the nature of the evolutionary process, forms a hierarchy and that the strictly monophyletic groups of that hierarchy *(nota bene*, only supraspecific groups of that special kind) have reality and individuality. (Brundin 1972: 111).

These claims about the individuality and reality of species seem to suggest that Hennig did not understand his figure 2 as a Venn diagram, which would involve seeing sets rather than individuals. Further, this way of picturing relationships can be found in other works Hennig read, such as Handlirsch (1925) and Zimmermann (1931).

Hennig referenced fellow entomologist Handlirsch's discussion of the "Stammbaum" (as opposed to the phylogenetic system) as circles within circles (Hennig 1950: 62, 75).[12] Rieppel notes that Handlirsch was friends with another author with whom Hennig was familiar, Othenio Abel. Rieppel notes that Handlirsch belonged to Abel's evening discussion group, a group that discussed and later summarized these issues (see Abel 1909). In addition to Handlirsch (1925), Hennig credited Zimmermann (1931) and Bigelow for his definition of monophyly, specifically, in "Arbeitsweise der botanischen Phylogenetik" when Zimmermann defined the "phylogenetic relationship" in reference to Figure 172 in *Phylogenetic Systematics* (1966). Donoghue and Kadereit argue that in this figure, Zimmermann discussed two plants or organs being more closely related to a third, and the common ancestor of those two existed more recently that the ancestor of all three (Donoghue and Kadereit 1992: 79). Zimmermann claimed the relative age relationship of the ancestors was the only direct measure of a phylogenetic relationship, and went on to argue that no other representation was consistent with phylogenetic grouping:

> A statement about phylogenetic relationship which cannot be expressed in the basic scheme of Figure 172b does not exist [...] Whoever believes that he cannot illustrate relationships in this basic scheme [...] does not have a phylogenetic but

---

Fragen hat unter einer unglücklichen. Verquickung mit dem Universalienstreit um die Seinsweise der Allgemeinbegriffe gelitten, die noch dadurch gefödert worden ist, daß in der Bezeichnung "Gattung" für eine der höheren Gruppenkategorien der phylogenetischen Systems ein Terminus auftritt, der auch in der Logik ("Gattungsbegriffs") Verwendung findet" (Hennig 1947: 279).

[11] Hennig wrote: "zwei Arten, die aus einer gemeinsamen Stammart hervorgehen und damit eine Gruppenkategorie hoherer Ordnung bilden, [mütissen] als mit ihrer Stammart identisch angesehen werden, denn diese besteht ja in ihnen und zwar in ihnen gemeinsam weiter" (Hennig 1950: 115).

[12] I am grateful to Oliver Rieppel for pointing out this article to me.

an "idealistic" or purely systematic "relationship" in mind. (Donoghue and Kadereit 1992: 78)

What is significant about this diagram is that in both Zimmermann's and Hennig's case, in order to make sense of this illustration was to assume this was not a Venn diagram showing nested sets which depict timeless ideal types. Instead, both argued that the diagram illustrates relatedness. For Hennig, if there were three things described in a diagram, say x, y, and z, then the diagram would illustrate x as more closely related to y than z. This was a relation that could be measured. This was the type of relation Hennig was looking to illustrate. By abandoning the idea that the figure is a Venn diagram, and instead seeing it as a tree rotated and viewed from above with circles around the objects more closely related, the right interpretation is achieved. For Hennig, as for Zimmermann (or Heidenhain), this figure was supposed to illustrate a feature of an enkaptic system.

Although Hennig (1949) did not provide much in the way of background or detailed explanation for enkaptic systems, he did cite his earlier paper (Hennig 1947), and in that paper he referenced ideas found in Bertalanffy's *Theoretische Biologie* (1932, 1942) specifically where Heidenhain's system was discussed (for further details see Rieppel 2009, Hueck 1926). On Heidenhain's account, the components of form (what he called "Histosystem") ranged from parts that could be as small as the cell to as large as the whole body, and could be regularly divided and arranged in an increasing series. This was done in such a way that everything from the hypothetically smallest elements to the whole organism would be included in a superordinate system, which he called "Enkapsis" (von Bertlanffy 1928: 111).

The example Heidenhain used to illustrate this was the muscle. The fibrillae, columns, muscle fibres, muscle bundles, up to macroscopic muscle, combine to form "histosystems," and each consist of "encapsulated" histosystems of the order immediately below. Heidenhain stressed that ontogenesis included both a division and synthesis of histosystems, a synthesis because the descendants remained connected to form a system of higher order able to divide again. This ontological picture fit nicely with what Hennig argued with respect to creating new species, so it seems hardly a coincidence that he would unintentionally cite someone who made such statements (von Bertlanffy 1928: 111).[13] Although Heidenhain was not cited by Hennig, Bertalanffy wrote about Heidenhain just before he outlined Woodger's formal system for development in embryology, which included Woodger's concept of a division hierarchy in *Theoretische Biologie*, drawing a connection between enkaptic systems and division hierarchies.[14] So, although the stem species' symbolic

---

[13] Bertalanffy wrote: "While biology has hitherto sought to conceive the body as an aggregate of cells whose total function is to be regarded as a sum of cell-functions, and so leaves out of account the problem of how the unity of the body can result from a mere sum of cells, Heidenhain (1923, 1929) has rejected the tyranny of the cell theory."

[14] Hennig used the term "enkaptic" in Hennig (1949) and Hennig (1950).

boundary lines were drawn around their successor species, that it did not imply a summative property, as was the case with a set-theoretic hierarchy, where a set was defined by its members. Hennig mentioned enkaptic systems in his discussion of relations, distinguishing between enkaptic system that reflected phylogenetic relations and morphological relations that could be reduced to similarity relations.[15]

16.6 Gregg, Woodger, and sets

The years between *Grundzüge einer Theorie der phylogenetischen Systematik* and *Phylogenetic Systematics* afforded Hennig more information on formal accounts of hierarchies, including an investigation by Gregg in the *Language of Taxonomy* (1954), as well as a new investigation of the taxonomic hierarchical system by Woodger. *Phylogenetic Systematics* was not the first place Hennig mentioned Gregg and Woodger together on hierarchies and formal logic. Rieppel translated Hennig in 1957 as saying:

> [T]he choice of the type of system has to correspond to the structure description [Strukturbild] of certain relations, which exist between the entities that are to be components of the system [...] The hierarchical type of system has most recently been investigated by Woodger and Gregg [...] We therefore have to ask the question whether there exist relations between animal species that satisfy the requirements invoked by Woodger's definition of "hierarchy"; in addition, these relations that exist between animal species must exist objectively, i.e. independent of any human being that may or may not recognize them. (Rieppel 2011: 108, translated from Hennig 1957: 55, 57)

---

[15] Hennig wrote: "These considerations make it clear how the mutually conflicting views of different taxonomists ought to be assessed, among whom some believe that the relations of the species ought to be thought of in a net-like manner and should only be represented in the family tree diagram or hierarchical enkaptic system rudimentarily, while others deem the tree scheme and hierarchical system appropriate for correctly expressing the relationships of the species to one another. In addition, the reticulate relationships are frequently referred to in this controversy by the representatives of the first-mentioned opinion as "phylogenetic" relationships. However, what is actually being referred to here is obviously the "morphological relationships – phylogenetic to a certain degree – of interrelated species," which belong to Dimension m in the schema of Figure 1, and not the phylogenetic relationships in the strict sense, which only ever operate in the dimension t. These ambiguities are encouraged by the inconsistent application of the concept of "relatedness," which on the one hand is used in broad agreement with the concept of chemical affinity for the description of shape similarity as well as similarity in the chemical structure of the body ("protein family" in the Serumdiagnostik), and on the other hand is used in the purely genealogical sense to refer to the lineage relationships. To avoid any misunderstanding and confusion, the designation "family" should be used exclusively for genealogical relationships operating in dimension t, and the term "similarity" (shape similarity, similarity in the way of life) should be used to identify the relationships that can be reduced to the dimension m in the sense discussed" (Hennig 1949:138).

In addition, Rieppel recorded Hennig's praise of their efforts: "In my view the work of Woodger and Gregg is enormously important [...] every systematist should look into it" (Rieppel 2011: 108). Hennig mentioned them together, favourably, in *Phylogenetic Systematics*, when he wrote:

> We consider the investigations of Woodger and Gregg extraordinarily important because they clarify, with methods that exclude all confusion and contradiction, the peculiarities of the hierarchical system, and so create exact prerequisites for investigating questions of whether and why it deserves the favour it enjoys in biological systematics. (Hennig 1966: 17).

It is tempting to assume that Hennig was endorsing Gregg's position in this passage, but it becomes clear further on in *Phylogenetic Systematics* that he has something else in mind when it comes to hierarchies, and it was not Gregg's system.

At first glance, it is easy to see why Gregg's system might have been attractive. Although Gregg did not draw from the same German literature that Hennig did, before he published *The Language of Taxonomy*, he discussed the role of formal symbolic logic, specifically on the question of whether species were objectively real (Gregg 1950) – a similar concern about ontological issues. Gregg situated his argument in the 1949 debate on the ontological status of species concept in the evolutionary studies flagship journal *Evolution*, between paleontologist Benjamin Burma and zoologist Ernst Mayr (Burma 1949, Mayr 1949). Gregg seized the opportunity to use formal logic to clean up the conceptual mess of philosophy of language, metaphysics, and biology generated by Burma and Mayr. Paleontologist George Gaylord Simpson agreed with Gregg's assessment of the debate:

> By a ponderous application of symbolic logic, Gregg [in reference Gregg 1954] sought to show that the issue raised by Burma and Mayr is not a genuine taxonomic problem or, at least, that if it does relate to a taxonomic problem it does so in the wrong words. It is, of course, important that words be used as accurately as possible and that they do not obscure properly taxonomic questions. Nevertheless, Burma and Mayr (as well as subsequent discussants) were considering a genuine taxonomic problem, in words perhaps not logically impeccable but, taken in context, adequately performing their main semantic function, that of communicating understandably among colleagues. (Simpson 1951: 285)

Although Burma and Mayr's discourse may not have been "logically impeccable," Simpson believed they tackled a genuine taxonomic problem, and perhaps more importantly, they had effectively communicated this problem among their own colleagues. Simpson also believed that Gregg did not have the last word on the matter.

Gregg managed to get his foot in the door likely because, as Simpson aptly noted, Burma and Mayr presented hopelessly bad philosophical arguments for their positions. Although not met warmly by the taxonomic community, the virtue of Gregg's position lay in capitalizing on existing bad ontological arguments by countering with his own, stronger philosophical arguments. Gregg's logical system might have been deeply flawed when viewed against the taxonomic evidence, but his *ontological* arguments for species as sets proved more persuasive than Mayr or Burma's ontological arguments. When Gregg entered the debate he thought he could clear up some of the misunderstandings by revealing them as a misuse of taxonomic language. As history shows, it was not until Hull and Ghiselin were able to formulate a sound rebuttal with their philosophical argument for species-as-individuals thesis that one could begin to think about slamming the philosophical door on Gregg.

One problem with taxonomic language, Gregg argued, is it gave rise to what he called the "pseudo-problem of existence" (Gregg 1950: 424).[16] He began by asking whether taxonomic groups should be thought of as "abstract" entities or whether they could be discovered in nature, like organisms. If species could be thought of as classes of organisms, then the question of whether or not taxonomic groups were abstract entities would be part of a more general question of whether classes were abstract or spatiotemporal entities. Gregg argued classes could not be spatiotemporal entities, and here is why.

Classes of organisms, such as species, cannot be spatiotemporal entities in nature because it violated the notion of the class-member relation and led to nonsensical statements – objects cannot be members of other objects. It just does not make sense. Objects are *parts* of a whole, and *members* of classes. Gregg recognized that some taxonomists objected to this claim, asserting that this was nothing more than word games. To that he wrote:

> Objections that it is not nonsense to say this seem based upon the notion (advanced independently by two different taxonomists reading an earlier version of this paper) that species are composed of organisms just as organisms are composed of cells: according to this argument a species is just as much a concrete, spatiotemporal thing as is an individual organism, though it is of a less integrated, more spatiotemporally scattered sort. This argument sounds fairly plausible, until one reflects that it contains an equivocation upon two common meanings of "is composed of." It is true that an organism is composed of cells; it is also true, but in a different sense, that species and other taxonomic groups are composed of organisms. The relation of a cell to the organism in which it is located is the relation of

---

[16] This was a puzzle Gregg introduced to highlight what he took to be a pseudo-problem around claims such as as: "Species exist" or "Genera exist." He claimed statements of the form "F's exist" where "F" stood for any taxonomic category was represented as "There is an A such that A is an F." Such statements he claimed were true whenever "F" was substituted by any taxonomic category name. He wrote "In such statements, ostensible reference to existence as a characteristic of taxonomic groups does not occur. Certainly, no problems worthy of many decades of heated debate are posed by assertions such as this." This was why he saw this as a pseudo-problem (Gregg 1954: 424).

> part to whole. But the organism–species relation is that of member to class; and these are entirely different sorts of relations. (Gregg 1950: 424)[17]

Was Gregg just putting too fine a point on a semantic quibble? Is it not true that species are composed of organisms just as organisms are composed of cells?

Gregg brought in formal logic to answer this objection. Using transitivity, Gregg argued for the difference between the two relevant compositionality relations. Gregg said given a cell, an organ, and an organism, the following could be claimed: a cell is a part of an organ, and an organ is a part of an organism. It follows that the cell is part of the organism. The part-whole relation is a transitive relation, like "less than." The class-member relation, however, is not a transitive relation (Gregg 1950: 425). Gregg explained: "It should be noted here that given '$x = y$' and '$y = z$', it is always correct to infer '$x = z$'; but given '$x \in A$' and '$A \in F$' it is *not* correct to infer '$x \in F$'. For example, given 'Jones $\in$ *Homo sapiens*' and '*Homo sapiens* $\in$ Zoological Species', it is not correct to infer 'Jones $\in$ Zoological Species'." This was because for Gregg there was a relevant ontological difference among the following objects: Zoological Species, *Homo sapiens*, and Jones. *Homo sapiens* was a *class* of organisms that consequently may contain the organism Jones *as a member*, while Zoological Species was a class of *classes* of organisms of which *Homo sapiens* was only one, and consequently did not contain as a member the organism Jones, who was not a class of organisms. According to Gregg, one-way membership differed from identity and from part-whole in that it was a non-transitive relation (Gregg 1950: 429).

Although his contemporaries did not call him out on it, this particular objection rested on shaky logical ground. On this logical point, he stood in opposition to most logicians at the time, as well as Woodger, who generally took the relevant difference between the part-whole relation and the set-member relation (specifically on the notion of a proper part in set theory) to be not transitivity, but reflexivity. Nevertheless, for Gregg, species were sets, not individuals. With this idea in hand, Gregg began a much more comprehensive logical treatment of species and classification systems. In *The Language of Taxonomy*, Gregg wrote: "The purposes of this book are two. First, it has been written to suggest that the symbolic methods of modern formal logic are useful and appropriate tools for the prosecution of methodological research in the foundations of taxonomy" (Gregg 1954: viii). In particular, Gregg chose set theory because he believed it provided "a systematization of some highly general ideas centering on the notion of set, class, aggregate, manifold, or collection that seem made to order for use in the methodological treatment of taxonomic classificatory systems" (Gregg 1954: viii). As it turned out, many taxonomists, Hennig included, did not share Gregg's assumption that this particular calculus was tailored to taxonomic systems. Second, Gregg hoped to illustrate "the

---

[17] Ernst Mayr claimed to be one of the two taxonomists Gregg cited. He was unsure if the other was A.J. Cain or T. Dobzhansky; see Mayr (1987: 152).

methodological utility of set-theoretical methods by using them to reconstruct the neo-Linnaean concept of taxonomic classificatory system" (Gregg 1954: ix). Gregg was motivated by what was had become a common concern among taxonomists, namely the lack of suitable methodological terminology. Rather than develop new terms, Gregg embarked on what he took to be a necessary conceptual housekeeping, arguing that "the available descriptions of such systems are nearly all deficient in clarity and rigor" (Gregg 1954: ix). The last three sections of the book were offered as a first step to improve this situation. At first glance, this goal appeared to be in line with what many taxonomists were aiming to do. For many taxonomists, however, Gregg's execution left much to be desired.

The first two chapters of Gregg's short book outlined the basics of set theory. In chapter 3 Gregg introduced the formal notion of a hierarchy, which he borrowed from Woodger. Woodger discussed two types of hierarchies in his work: abstract hierarchies and division hierarchies, the latter applied to individuals. Gregg defined a hierarchy as:

> A hierarchical relation or HR: HR = the set in which any relation z is a member if ($z \in$ One-many) ($\check{z}''\mathrm{UN} = \check{z}_{p0}''\{B\ 'z\}$). (Gregg 1954: 26)

In other words, a hierarchy was any one-many relation z whose converse domain was identical with the set of all *first* constituents of $\check{z}_{p0}$ pairs whose *second* constituent is the beginner of z.[18]

Gregg then provided the following definition of a taxonomic classificatory system:

A relation z is to be regarded as a taxonomic classificatory system just in case these four conditions are satisfied:

1. The field of z is included in G;
2. There exists a member of the field of z in which every member of the field of z is included;
3. Given any members x and y of the field of z, x bears z to y if y is included in but not identical with x and there exists no third member of the field of z in which y is included and which is included in x; and
4. Given any members x and y of the field of z, either x bears z to y or else y bears z to x or else the overlap of x and y is identical with the empty set.

---

[18] Assume the following notation:

EM = the empty set
$\langle x, y \rangle$ = notion for ordered pairs
$\{x, y\}$ = notion denoting a set
$\in$ = membership
& = and
$\cap$ = union
$\subset$ = inclusion

According to Gregg, if the postulate was adopted that the empty set was not a member of G, then $(<x,w> \in IC(u))$ and $(y \neq w) \to ((y \cap w) = \mathbf{EM}))$ was a logical consequence of the above four conditions together with the definitions of "$IC(u)$" and "HR," where each were defined respectively as:

> "$IC(u)$" designates a relation that appeared in such contexts as "$(<x,y> \in IC(u))$," which translated to "the set x immediately contains the set y, with respect to the set u," or more briefly, "the set x immediately u-contains the set y." The use of the sign is fixed by the following rule: $(<x,w> \in IC(u)) \leftrightarrow (x \in u)$ and $(y \in u)$ and $(y \subset x)$ and $(y \neq x)$ and $\sim\exists z\ (z \in u)$ and $(z \neq x)$ and $(y \subset z)$ and $(z \subset x)$
>
> HR = the set in which any relation z is a member if $(z \in$ One-Many$)$ and $(\check{z}\,{''}\,UN = \check{z}_{p0}\,{''}\{B'z\})$
>
> z_{p0} = the relation in which any pair $<x,y>$ is a member if $(<x,y> \in z) \vee (z \mid z)) \vee (<x,y> \in z \mid z \mid z)) \vee (<x,y> \in z \mid z \mid z \mid z)) \vee \dots)$
>
> For example, if z is the relation of parent to child, then $(<x,y> \in z_{p0})$ if and only if x is a parent of y, or else x is a grandparent of y, or else x is a great-grandparent of y, or else x is a great-great-grandparent of y, or … (Gregg 1954: 23)

With this definition of a taxonomic classification system, Gregg presented his analysis of the Linnaean hierarchy.

The resulting hierarchies were fully *nested* and display the property of summativity. Gregg was not alone in thinking of taxonomic groups as sets, numerical taxonomists supported this idea. Although Simpson had some problems with the execution, he did not shun the idea entirely, as seen in *Principles of Animal Taxonomy*, where he seemed to promote a "species taxa as sets" position (Simpson 1961, see as well Sokal and Sneath 1963). Simpson wrote:

> *Nevertheless, an individual never is and cannot be classified.* Classification involves only groups; no entity possible in classification is an individual. An individual may be referred to or placed in a given group. That is often called "classifying," but that is a misnomer. That process is *identification*, which is not the same as classification. (Simpson 1961: 18)[19]

Simpson thought that groups were important in classification hence a set-theoretic calculus similar to this one would not be out of the question. Hennig, however, had something different in mind.

> $\to$ = conditional
>
> $''$ = image relation. The image of a set x by a relation z, briefly $z\,{''}\,x$ is the set of all first constituents of pairs in z whose second constituents are members of x: $(z\,{''}\,x$ = the set in which any y is a member iff there is a w such that $(<y,w> \in z)$ and $(w \in x))$. For example, if z is the relation between parasite and host, and x is the set of all mammals, then $z''x$, the z-image of x, is the set of all parasites of mammals.
>
> $'$ = image relation of $\{x\}$ when $\{x\}$ has exactly one member.

[19] This was not an uncommon position to hold. Jardine, for example, included that passage in his attempt to provide a set-theoretic account of a taxonomic hierarchy that would improve upon Gregg's, see Jardine (1969: 38).

In 1952, Woodger took a second look at taxonomic hierarchies because he was concerned with some of his claims in *The Axiomatic Method in Biology* (Woodger 1937) – the departure point for Gregg's *Language of Taxonomy* (Gregg 1954). Woodger noted where he provided his "purely abstract definition of hierarchy" that "[a]t one time I found myself applying the term 'hierarchy' to very diverse objects and without being able to state at all in words why I did so" and he went on to describe division hierarchies exemplified by zygotes dividing, and hierarchies composed of nesting sets of mutually exclusive subclasses exemplified the Linnean classification of animals and plants. He then provided the following definition:

> R is a hierarchy if and only if R is one-many and if the converse domain of R is identical with the set of all terms to which the first term of R stands in some power of R. (Woodger 1952: 11)

In the corresponding footnote, Woodger made a point of stating:

> It does not suffice to define "hierarchy" as denoting the set of all one-many relations which have one and only one beginner. In the first of the above illustrations of the notion of hierarchy the generating relation is the relation in which a square stands to each of its four quarter squares [the division hierarchy]; in the second it is the relation D explained above (p. 9) [Woodger's example of cell division], and in the third it is the converse of the relation of immediate inclusion of classes [Linnean hierarchy]. (Woodger 1952: 11)

Gregg only looked at the last type of hierarchy, and consequently his definition was more restrictive. Woodger stipulated "if the converse domain of R is identical with the set of all terms to which the first term of R stands in some power of R" in contrast to Gregg "whose converse domain was identical with the set of all first constituents of $\check{z}_{p0}$ pairs whose second constituent is the beginner of z."

These observations about taxonomic hierarchies in 1952 led Woodger to distinguish between the Linnaean system and what he understood as an "evolutionary" system. The evolutionary system, Woodger suggested, required a different compositionality relation than the Linnaean system, since species under the evolutionary interpretation were not "atemporal." Woodger distinguished between "evolutionary species" which he believed were concrete entities that had a beginning and end in time, and Linnean species, which were abstract and timeless (Woodger 1952: 19). Woodger went on to propose a solution in terms of species names where species names were treated like general names. Near the end of his 1952 paper, Woodger entertained the possibility that species could be something like logical individuals. Although he did not explore this concern in any more detail, he did have the logical calculus to provide such an analysis if he wanted to – the division hierarchy he outlined in *The Axiomatic Method in Biology*. Looking at the German manuscript shows that Hennig was well aware that Woodger made this argument.

16.7 Phylogenetic systematics

The revised section on hierarchies in Part I of *Phylogenetic Systematics* began with Gregg's definition of an abstract hierarchy and Gregg's graphic representation of it (Hennig 1966: Fig 2).[20]

Hennig explained his Figure 1 (redrawn here as Fig 16.3) by saying that the elements represented by X_0 through X_9 are paired by relations that extended in only one direction and such relationships exist, for example, "between mother and child, father and son, employer and employee" (Hennig 1966: 17). Of course, claimed Hennig, relationships exist that could be organized non-hierarchically, such as those between brothers, but those would not be included in this figure. Hennig then asked: what in biology could be substituted for X_0 through X_9?

Although Hennig offered Gregg's set-theoretical hierarchy, he was aware that such hierarchies are not designed to deal with objects like semaphoronts or species that exist in time and have relations (such as ontogenetic and genealogical relations). Accordingly, Hennig did not believe this kind of hierarchy could account for organisms in their ontogenetic development from zygotes to adults, and said that "the structure of these ontogenetic relations does not correspond to the conditions of a hierarchic system" (Hennig 1966: 18). Instead, Hennig promoted a radically different ontological picture than Gregg. Where Gregg argued for sets or classes, Hennig argued for individuals, a position he made clearer in Part II. This is why Hennig suggested the right kind of hierarchy was Woodger's division hierarchy, a position he argued for in *Phylogenetisch Systematik*. In fairness, Gregg did not intend this type of hierarchy to organize the kind of objects Hennig had in mind.

At least part of the blame for difficulties in understanding Hennig's position on hierarchies can be blamed on inconsistent translation. In Part II, where Hennig explored in greater detail his logical and ontological ideas regarding hierarchies and claimed phylogenetic relationships were arranged in terms of a division hierarchy, the term "Teilungshierarchie" was inconsistently translated. For example, when Hennig opened the floor to a discussion of hierarchies, "Teilungshierarchie" was translated as "partitioning hierarchy" instead of "division hierarchy":

> In den vorstehenden Kapiteln ist der Nachweis geführt worden, daß die Gesamtheit aller Arten, die in der Gegenwart und Vergangenheit unterschieden werden können geordnet nach den phylogenetischen Beziehungen, die zwischen ihnen bestehen der Definition einer Hierarchie (*Teilungshierarchie*) im Sinne von woodger und GREGG entspricht. Die adäquate Darstellungsform ist dementsprechend das hierarchische System. Die Fragen, denen wir nun im einzelnen nachgehen müssen, ergeben sich aus der Struktur eines solchen hierarchischen Systems. (Hennig 1982: 75; [my italics])

[20] In the text Hennig ascribed the definition to Woodger, but that was because Gregg ascribed the definition to Woodger. It was not Woodger's definition, but Gregg's.

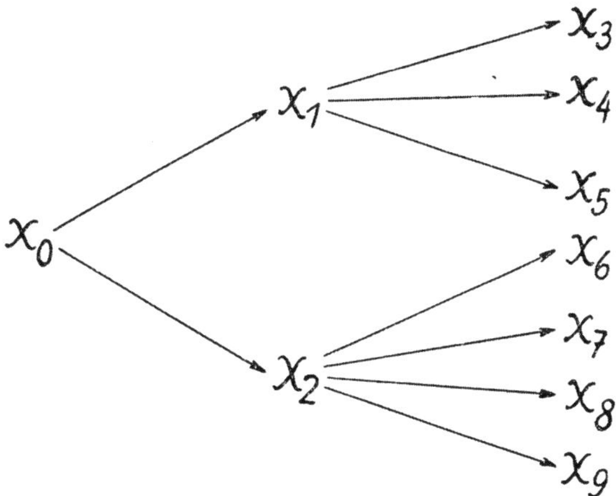

Abb. 1. Struktur der zwischen den Elementen einer
,,Hierarchie" bestehenden Beziehungen.
Nach GREGG

Fig 16.3 Hennig's graphic representation of Gregg's abstract hierarchy. After Hennig (1957: 56, Fig 1).

In the preceding chapters it was shown that all species that have ever lived, when arranged according to their phylogenetic relationships, correspond to the definition of a hierarchy (*partitioning hierarchy*) in the sense of Woodger and Gregg. Accordingly the adequate form of presentation is the hierarchic system. The questions that we must now pursue in detail result from the structure of such hierarchic system. (Hennig 1966: 70 [my italics])

This mistranslation could make it difficult to appreciate Hennig's point. Woodger (in his early work) and Gregg do mention division hierarchies, but division hierarchies were *not* part of their vision of a taxonomic system. However, for Hennig, division hierarchies were equipped to deal with the biological relations he had in mind. If Hennig's deviation rule is followed, a stem species ceases to exist when it split into two daughter species, just as a cell ceases to exist when it divided. When the two descendant daughter species formed a new entity at a higher level of complexity (just as the daughter cells formed part of an organ) this new entity "encapsulated" the two daughter species, making it a monophyletic taxon at its lowest level of complexity. To drive this point home, Hennig recycled the figure he used in 1949 in which he discussed "enkaptic systems."

In both the German and the English translation of the *Phylogenetic Systematics*, the term "Teilungshierarchie" (or "partitioning hierarchy" as it was translated) was used in conjunction with this diagram, because Hennig wanted to make clear that this type of hierarchy was importantly different from the usual set-member

Hennig, W. (1949). Zur Klärung einiger Begriffe der phylogenetischen Systematik. *Forschungen und Fortschritte* 25, 137–139.

Hennig, W. (1950). *Grundzüge einer Theorie der phylogenetischen Systematik.* Berlin: Deutscher Zentralverlag

Hennig, W. (1953). Kritische Bemerkungen zum phylogenetischen System der Insekten. *Beiträge zur Entomologie, 3, Sonderheft Festschrift Sachtleben*: 1–85.

Hennig, W. (1957). Systematik und Phylogenese. In *Bericht über die Hundertjahrfeier der Deutschen Entomologischen Gesellschaft Berlin*, ed. H. -J. Hannemann. Berlin: Akademie-Verlag, pp. 50–71.

Hennig, W. (1966). *Phylogenetic Systematics.* Urbana, IL: University of Illinois Press.

Hennig, W. (1982). *Phylogenetische Systematik.* Berlin: Paul Parey.

Hueck, W. (1926). Die Synthesiologie von Martin Heidenhain als Versuch einer allgemeinen Theorie der Organisation. *Naturwissenschaften*, 14, 149–158.

Hull, D.L. (1964). Consistency and monophyly. *Systematic Zoology*, 13, 1–11.

Hull, D.L. (1965). The effect of essentialism on taxonomy: two thousand years of stasis. *British Journal for the Philosophy of Science* 15: 314–326, 16, 1–18.

Hull, D.L. (1967). The metaphysics of evolution. *British Journal for the Philosophy of Science*, 3, 309–37.

Hull, D.L. (1976). Are species really individuals? *Systematic Zoology*, 25, 174–191.

Hull, D.L. (1978). A matter of individuality. *Philosophy of Science*, 45, 335–360.

Hull, D.L. (1988). *Science as a Process: An Evolutionary Account of the Social and Conceptual Development of Science.* Chicago, IL: University of Chicago Press.

Hull, D.L. (1990). The descriptive attitude: transformed cladistics, taxonomy and evolution by N.R. Scott-Ram. *Systematic Zoology*, 39, 420–423.

Jardine, N. (1969). A logical basis for biological classification. *Systematic Zoology*, 18, 37–52.

Knox, E.B. (1998). The use of hierarchies as organizational models in systematics. *Biological Journal of the Linnean Society*, 63, 1–49.

Mayr, E. (1949). The Species Concept: A discussion. The species concept: Semantic verses semantics. *Evolution*, 3, 371–372.

Mayr, E. (1987). The ontological status of species: Scientific progress and philosophical terminology. *Biology and Philosophy*, 2, 145–166.

Michener, C.D. (1957). Some bases for higher categories in classification. *Systematic Zoology*, 6, 160–173.

Rieppel, O. (2003). Semaphoronts, cladograms and the roots of total evidence. *Biological Journal of the Linnean Society*, 80, 167–186.

Rieppel, O. (2006). On concept formation in systematics. *Cladistics*, 22, 474–492.

Rieppel, O. (2009). Hennig's enkaptic system. *Cladistics*, 25, 311–317.

Rieppel, O. (2011). Willi Hennig's dichotomization of nature. *Cladistics*, 27, 103–112.

Rieppel, O. (2012). Adolf Naef (1883–1949), systematic morphology and phylogenetics. *Journal of Zoological Systematics and Evolutionary Research*, 50, 2–13.

Schmitt, M. (2003). Willi Hennig and the rise of cladistics. In *New Panorama of Animal Evolution*, ed. A. Legakis, S. Stenthourakis, R. Polymeni and M. Thessalou-Legaki. Sofia: Pensoft Publishers, pp. 369—379.

Schmitt, M. (2010). Willi Hennig, the cautious revolutioniser. *Palaeodiversity* 3, Supplement, 3–9.

Simpson, G.G. (1951). The species concept. *Evolution*, 5, 285–298.

Simpson, G.G. (1961). *Principles of Animal Taxonomy* New York: Columbia University Press.

Sokal, R.R., and Sneath, P.H.A. (1963). *Principles of Numerical Taxonomy*. London: W.H. Freeman and Co.

Torrey, T.W. (1939). Organisms in time. *The Quarterly Review of Biology*, 14, 275–288.

Williams, D.M., Scotland, R.W., Humphries, C.J. and Siebert, D.J. (1996). Confusion in philosophy: a comment on Williams (1992). *Synthese*, 108, 127–136.

Williams, P.A. (1992). Confusion in Cladism. *Synthese*, 91, 135–152.

Woodger, J.H. (1929). *Biological Principles*. London: K. Paul, Trench, Trubner.

Woodger, J.H. (1930). The 'concept of organism' and the relation between embryology and genetics. Part I. *Quarterly Review of Biology*, 5, 1–22.

Woodger, J.H. (1932). The 'concept of organism' and the relation between embryology and genetics. Part III. *Quarterly Review of Biology*, 6, 178–207.

Woodger, J.H. (1937). *The Axiomatic Method in Biology*. Cambridge: Cambridge University Press.

Woodger, J.H. (1945). On biological transformations. In *Essays on Growth and Form, presented to D' Arcy Wentworth Thompson*, ed. W.E. Le Gros Clark and P.B. Medawar, Oxford: Oxford University Press, pp. 95–120.

Woodger, J.H. (1952). From biology to mathematics. *The British Journal for the Philosophy of Science*, 3, 1–21.

Zimmermann, W. (1931). Arbeitsweise der botanischen Phylogenetik und anderer Gruppierungswissenschaften. In *Handbuch der biologischen Arbeitsmethoden, Abteilung IX: Methoden zur Erforschung der Leistungen des tierischen Organismus, Teil 3: Methoden der Vererbungsforschung, Heft 6 (Lieferung 356)*, ed. E. Abderhalden. Berlin: Urban and Schwarzenberg, 941–1053.

17.2 Classification and phylogeny in biological systematics: Zimmermann and Hennig

In 1931, the renowned German botanist Walter Max Zimmermann (1892–1980) wrote a long article on botanical phylogenetics (Zimmermann 1931). He clearly stated the reason why it was necessary for humans to classify diverse objects (Zimmermann 1931: 942). He pointed out that we cannot but make groups ("Wir müssen gruppieren") only because natural objects and their parts do not exist as clearly demarcated groups ("wohlabgegrenzte Gruppen") but as individuals, unique phenomena. Because it is impossible for humans to grasp those objects separately, one by one, his conclusion was that there was a need to order and systematize them for economy of memory. It is worth noting in his view that biological classification and taxonomy were viewed as one of "grouping sciences" ("Gruppierungswissenschaften").

Zimmermann's second question was how to make groups ("Wie wollen wir gruppieren?"). In this regard, he took the position that any classificatory system should be strictly consistent with a phylogenetic tree (see Fig 17.1). He advocated that a strictly phylogenetic system of a group of organisms could be constructed if closely related subgroups (monophyletic taxa) were ordered hierarchically on the basis of the phylogenetic tree estimated for the group.

After Zimmermann's pioneering discussion there was an endless dispute concerning the principles and methods of building classificatory systems. In particular, during the 20 years, throughout the 1960s and late 1970s, a severe controversy was fought among schools of systematists (Hull 1988, Minaka 1997, Yoon 2009). The focal point of the controversy, at the time, was the criteria for constructing classificatory systems of organisms. The cognitive root of scientific taxonomy is deeply related to the intuitive, common-sense domains of folkbiological taxonomy (Atran 1990, 1998). A good example of human cognitive universals is "psychological essentialism" (Kornblith 1993: 70) which is the implicit assumption that all objects (e.g. species or higher taxa) are natural kinds with invisible essences or underlying natures that make them as they are. Another example is "Umwelt" (Yoon 2009: 15) which is the shared, inherited perception of the phenomenal world by human beings.

Scott Atran emphasized that this cognitive root formed the conceptual framework for modern systematics:

> Although the relative autonomy of common-sense notions of living kinds is underscored by the fact that folk treat them essentially while biology does not, historically the difference wasn't always so clear cut. Until Darwin – and arguably still among some taxonomists today – scientific classifications shared a presumption with lay classifications that the constituent taxa are natural kinds with underlying natures. (Atran 1990: 80)

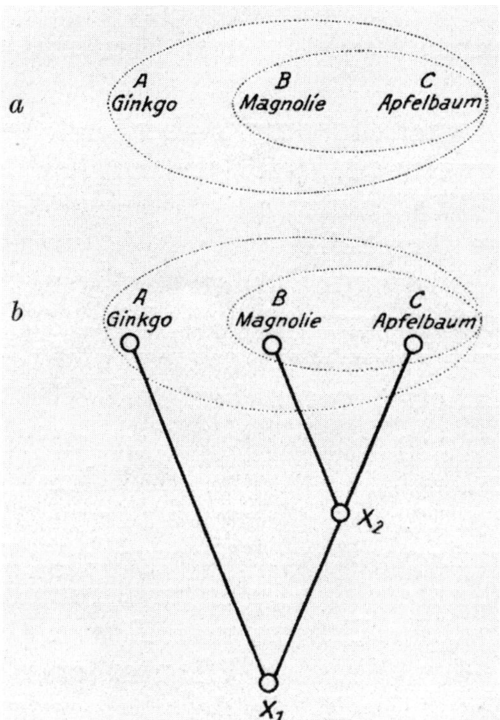

Fig 17.1 The relationship between classification (above) and phylogeny (below). Reproduced from Zimmermann (1931: 990, figure 172).

Traditional taxonomists use both logic and intuition. For example, the famous paleontologist and evolutionary systematist George Gaylord Simpson (1902–84) stated in his classical textbook, *Principles of Animal Taxonomy*:

> Like many other sciences, taxonomy is really a combination of a science, most strictly speaking, and of an art. Its scientific side is concerned with reaching approximations, hopefully believed to be successively closer as the science progresses, toward understanding of relationships present in nature. One of the dictionary definitions of "art" is "human contrivance or ingenuity," and taxonomy becomes largely artistic, in that sense, when applied to construction of classifications. (Simpson 1961: 110)

Many systematists accepted Simpson's view that biological classification cannot be constructed solely by pure logic. However, in opposition to Simpson and other evolutionary systematists, Willi Hennig (1913–76), a German entomologist, insisted on the supremacy of the strictly phylogenetic system over the other kinds of classificatory systems (Hennig 1950, 1957, 1966). From the end of the 1940s to the 1950s, Hennig clarified the logical relationship between phylogeny and classification that

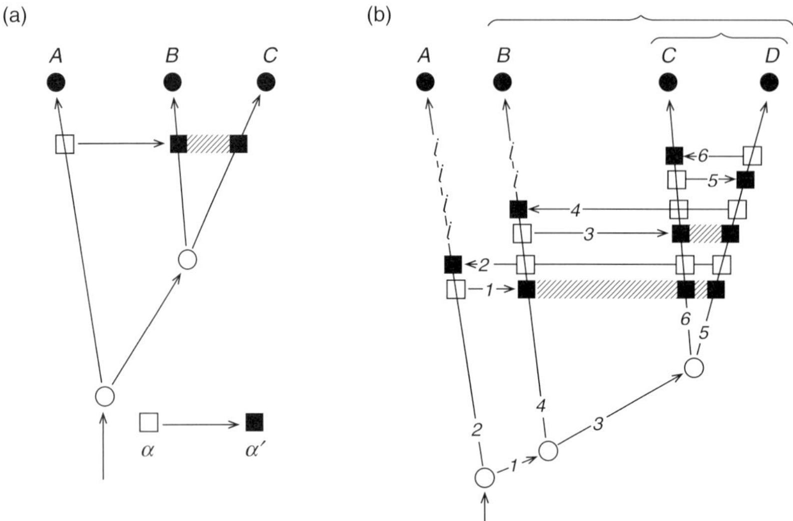

Fig 17.2 The argumentation scheme of Hennig's phylogenetic systematics. (a) For a character with a primitive state (a) and derived state (a'), organisms sharing the derived state (a') form a monophyletic group. (b) A phylogenetic system can be constructed for a group of organisms by repeating this procedure for all characters. Reproduced from Hennig (1957: 66, figs 8 and 9).

reinforced and generalized the basic ideas Zimmermann had developed earlier (Fig 17.2).

After Hennig's work was translated into English and published as *Phylogenetic Systematics* (Hennig 1966), his theory was discussed among the English-speaking countries with the result that the "Cladistic Revolution" spread throughout the systematics community during the following 20 years (Hull 1988, Schmitt 2013).

For Hennig the German word "System" ("system" in English) implies something more fundamental than the German word "Klassifikation" ("classification" in English). Hennig aimed at constructing the one and the only "system" of organisms in order to comprehend biodiversity on the Earth. The "system" is not merely "classification" for some practical use. What, then, is "system" for Hennig? He used a metaphor to illustrate the point:

> Let me begin with an example. If an archaeologist discovers potsherds in a tomb, he might begin by ordering, or classifying, them in some way: according to their material (clay or metal), their color, their decorations, etc. Subsequently, he might attempt to reconstruct the original vessels (vases, urns, etc.), of which the potsherds are fragments. This reconstruction is another kind of ordering. One might call it a system, but one need not call it a classification. For another example, I refer to the rivers of Europe. These may be classified according to their navigability, water management, the conditions they offer for the settling of organisms, etc. But one

> might seek to determine the drainage (Danube, Rhine, Elbe, etc.) to which each belongs, in order to construct a different kind of system of rivers. Similarly, the construction of a cladogram in accordance with the principles of phylogenetic systematics results in a system rather different in principle from various kinds of possible classifications. Although my original perception of this distinction was somewhat unclear, I have nevertheless avoided speaking of phylogenetic "classification," preferring instead phylogenetic "system" – but I have sometimes used "classification" under the influence of English usage. (Hennig 1975: 245–246 from the English translation of Hennig 1974: 281)

In the paragraph cited above, Hennig gave as examples of "systems" not "classification" the reconstruction of archeological objects and the determination of the drainage of rivers. There exist many practical classifications each with a specific purpose, but there may also exist a unique and unitary "system" with a general purpose. This viewpoint was applied to biodiversity:

> Similarly, the construction of a cladogram in accordance with the principles of phylogenetic systematics results in a system rather different in principle from various kinds of possible classifications. (Hennig 1975: 246 for English translation of Hennig 1974: 281–282)

Hennig claimed that the phylogenetic tree composed of all organisms on Earth, extant and extinct, is nothing more than "The Tree of Life" which is the unique phylogenetic system. This unique phylogenetic system is, according to Hennig's view, a general reference system ("allgemeines Bezugssystem", Hennig 1950), which should be superior to all other special-purpose classificatory systems. For the standpoint of cladistics, systematic pattern means this general reference system represented as a phylogenetic diagram called "cladogram." A cladogram is defined as a branching diagram showing only sister group relationships among organisms without any connotations of evolutionary rates, progressiveness, phenetic similarity, etc. Every branching point in a cladogram corresponds to a monophyletic group based on one or more shared derived character states (synapomorphies).

The controversy over the principles of constructing classificatory systems began in the 1960s and continued throughout the 1970s. During the 1980s there was a discussion on how systematic patterns can be defined and analysed independent of the evolutionary framework. Among other things, whether a cladogram depicting a pattern ("being") among organisms can be separated from any assumptions of evolutionary processes ("becoming") was much discussed in this period.

Some cladists who claimed a strict consistency of phylogenetic relationships and classificatory system insisted on a radical proposal of decoupling cladogram and cladistics from the idea of evolution itself. This "transformation" (Platnick 1979) formed a new school of cladistics called "transformed cladistics" or "pattern cladistics" (Nelson and Platnick 1981). Transformed (pattern) cladistics was in opposition to "phylogenetic cladistics", which asserts the evolutionary assumptions are required in both cladistic theory and practice.

17.3 Generalized pattern cladistics as a science of trees and networks: Nelson's legacy

The theory of pattern cladistics (Nelson 1979) was established during 1970s as a general system of tree diagrams including cladogram and phylogenetic tree. David Hull (1979) pointed out that wider application of cladistics to any evolving objects necessarily required a temporal dimension.

In general, cladistic analysis can be used to discover the cladistic relations between any entities that change by means of modification through descent (Platnick 1979). As general as this notion of cladistic analysis is, it still retains a temporal dimension. Transformation series must be established for characters rather than just an abstract atemporal series like the cardinal numbers or the periodic table, but a series of actual transformation in time (Hull 1979: 418). And Hull insisted, the transformation of cladistics split into the following two "schools":

> From the beginning, Gareth Nelson (1973) seems to have been developing two notions of cladistic analysis simultaneously, one limited to historically developed patterns (cladism with a small "c"), the other a more general notion applicable to all patterns (Cladism with a large "C"). His method of component analysis is a general calculus for discerning and representing patterns of all sorts. (Hull 1979: 418)

Pattern cladistics aims to discover and represent patterns of any kinds of objects, regardless of their origins (Williams and Ebach 2008). In this sense pattern cladistics can be regarded as being a part of discrete mathematics that studies the partial-order relationships whose structures are visualized as trees and networks (Minaka 1993, Davey and Priestely 2002, Semple and Steel 2003, Papavero and Llorente 2008, Dress et al. 2012).

This generalized version of pattern cladistics, or Nelson's cladistic component analysis as Hull called "Cladism with a large 'C'", was not limited to biology. Discerning patterns of any kinds of objects is the inferential basis for the discussion of causal processes that brought about those patterns.

Due to this generality, pattern cladistics can be widely applied not only to biological taxa but to other non-biological objects (Platnick and Cameron 1977). In fact, cladistics and related parsimony-based methods have been widely and independently used for phylogeny reconstruction not only for organisms but for languages and manuscripts, archeological materials, historical styles, and other cultural constructs (Hodson et al. 1971, Hoenigswald and Wiener 1987, Atkinson and Gray 2003, O'Brien and Lyman 2003, Lipo et al. 2005, Moretti 2005, Schmidt-Burkhardt 2005, Forster and Renfrew 2006, Nakao and Minaka 2012).

17.4 Phylogeny estimation in manuscript stemmatics and historical linguistics: from Lachmann to Maas

As discussed in the previous section, the severe controversy in biological systematics in 1960s and 1970s was concerned with how to construct classification systems and how to estimate phylogenetic relationships among organisms. Pattern cladistics, which arose as a new school of systematics during this period, had sufficient generality to be applied to objects other than organisms. At this point it is worth reviewing the history of historical linguistics and manuscript stemmatics (for more details see Timpanaro 1971, 2005 for stemmatics, Alter 1999 for linguistics).

Methods for estimating genealogical relationships among languages in historical linguistics and among manuscripts in comparative philology have been associated in their historical development. Henry M. Hoenigswald (1915–2003) succinctly pointed out the existence of a parallel connection between linguistics and stemmatics:

> A number of languages descended separately from one ancestor (a number of manuscripts copied from one model) may be expected to share innovations (common errors) of a non-trivial sort only by accident. If accident can be excluded and if innovations (errors) can be independently recognized as such, subancestors (hyperarchetypes) and the ancestor (archetype) may be reconstructed and the resulting tree may be historically interpreted in terms of separation, migration, etc. (in terms of monastic history, of the history of writing and printing, etc.). (Hoenigswald 1973: 25)

Systematizing solely based on shared innovations in linguistics and shared errors in stemmatics is in principle equivalent to the cladistic method that discovers monophyletic taxa using shared derived character states (synapomorphies) (Hennig 1950, 1966, Baum and Smith 2013).

Procedures for manuscript stemmatics were originally established through revising and editing the Christian Bible and the classics from Greek and Roman times (Timpanaro 2005). For example, the Biblical scholar Johann Albrecht Bengel (1687–1752) wrote in 1734 about the "genealogical relationship" between manuscripts: "Manuscripts are closely related to one another if they have the same ancient arrangements of text on the page, subscriptions, and other subsidiary features" (cited in Timpanaro 2005: 65).

In addition Bengel proposed that it would be possible to trace the genealogy of all manuscripts to their root, summarized as "tabula genealogica" (Timpanaro 2005: 65). Bengel had discussed the estimation of relationships among manuscripts about a century before the appearance of evolutionary thinking in biology.

Carl Johan Schlyter (1795–1888) in Sweden drew for the first time this "tabula genealogica" of manuscripts in the form of a phylogenetic tree (Holm 1972, Ginzburg 2004). In 1827, Schlyter estimated a manuscript stemma of ancient Swedish legal

(a)   (b)

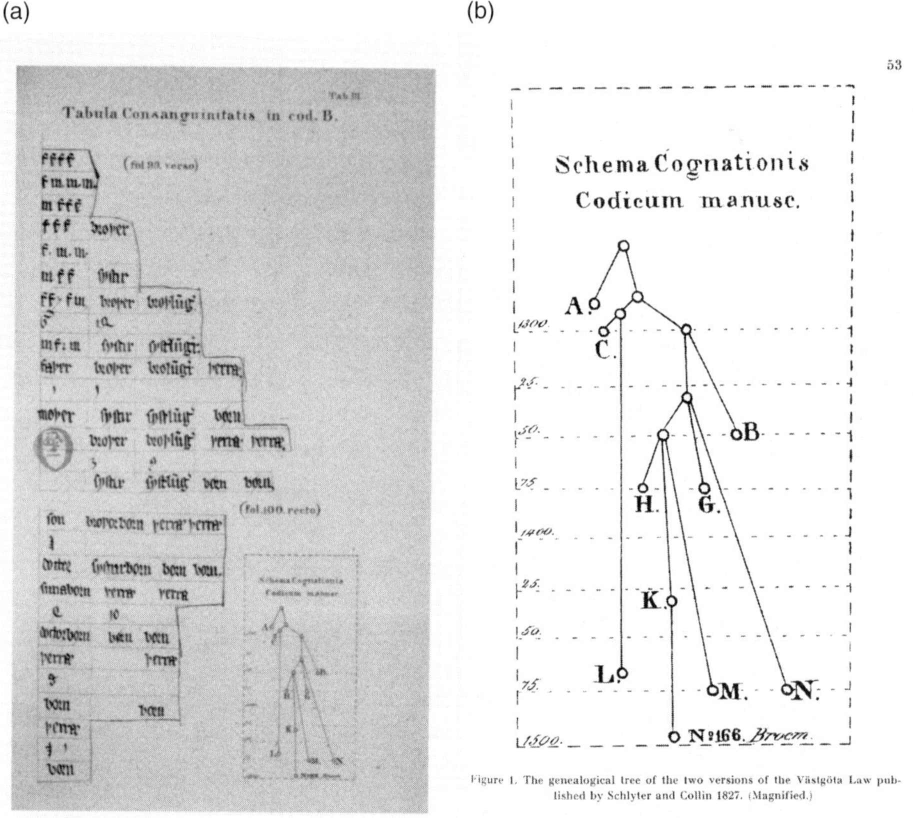

Figure 1. The genealogical tree of the two versions of the Västgöta Law published by Schlyter and Collin 1827. (Magnified.)

Fig 17.3 Tabula consanguinitatis (a) and genealogical tree (b: enlarged) of ancient Swedish legal manuscripts "*Västgöta*" drawn by Carl Johan Schlyter (1827). Reproduced from Holm (1972) and Ginzburg (2004).

texts "*Västgöta*" to show their genealogical relationships as a branching "tree" (Fig 17.3). His tree was the origin of the style of drawing manuscript genealogy upside down by placing the root (archetype) at the top of the tree with the descendants hanging downward.

Schlyter drew his manuscript genealogy ("schema cognationis codicum manusc [riptorum]", Fig 17.3, right) based on the table of cognates ("tabula consanguinitatis", Fig 17.3, left). Absolute age scale is added to this genealogy. His tabula consanguinitatis can be associated with older forms called "arbores consanguinitatis" or "arbores affinitatis" which had been widely used since the Middle Ages in Europe. Many iconographical studies have accumulated in the past with respect to this "tree (arbor)" graphical representation (Schadt 1982, Barsanti 1992, Klapisch-Zuber 2000, 2003, Minaka and Sugiyama 2012).

In 1831, the Latin scholar Carl Gottlob Zumpt (1792–1849) coined the word "stemma codicum" for manuscript genealogy. In this era when the idea of biological evolution was not yet popular, comparative philologists had already established a phylogenetic theory of reconstructing manuscript relationships as a research program. Karl Lachmann (1793–1851), a classical scholar of the nineteenth century, compiled major methods and techniques of manuscript stemmatics. Since then the genealogical method of stemmatics has been called "Lachmann's method" (Timpanaro 1971, 2005).

Here I will discuss how textual mutations (that is, transcription errors) in descendent manuscripts are used to estimate the stemma in Lachmann's method. Paul Maas (1880–1964), a German comparative philologist, published in 1927 a short but highly influential textbook on manuscript stemmatics, *Textkritik* (Maas 1927 [1950]). His book briefly summarized the basic concepts and the practical method for reconstructing manuscript stemma.

Maas sorted out the books into three categories of informative errors, other than those uninformative errors that can occur at any time only by chance (Maas 1937):

Significant errors (*errores significativi*): informative errors that cannot occur by chance.

Separative errors (*errores separativi*): informative errors that are specific to a particular manuscript.

Conjunctive errors (*errores coniunctivi*): informative errors that are shared by multiple manuscripts.

The original ancestral text (prototype) contains, in principle, no errors, thus any errors that occur in descendent manuscripts can be considered derived character states. However, among various kinds of errors, only conjunctive errors have evidential value for estimating the manuscript stemma. The reason is that shared derived similarities can discriminate between competing phylogenetic trees while the other similarities cannot.

The conceptual scheme (Fig 17.4) proposed by Maas indicates a definite correspondence with Hennig's phylogenetic theory discussed in the previous section because separation errors are uniquely derived character states, equivalent to "autapomorphies," and conjunctive errors are shared derived character states, equivalent to "synapomorphies," in cladistic terminology. In this way the methods for phylogeny reconstruction in manuscript stemmatics, historical linguistics, and biological systematics were independently established and converged on the same methodology based on the principle of parsimony (Platnick and Cameron 1977, Cameron 1987, O'Hara 1996, Atkinson and Gray 2003).

On the other hand Maas's formulation of manuscript stemmatics is related to graph theory or "discrete mathematics." In fact Maas was interested in graph-theoretical

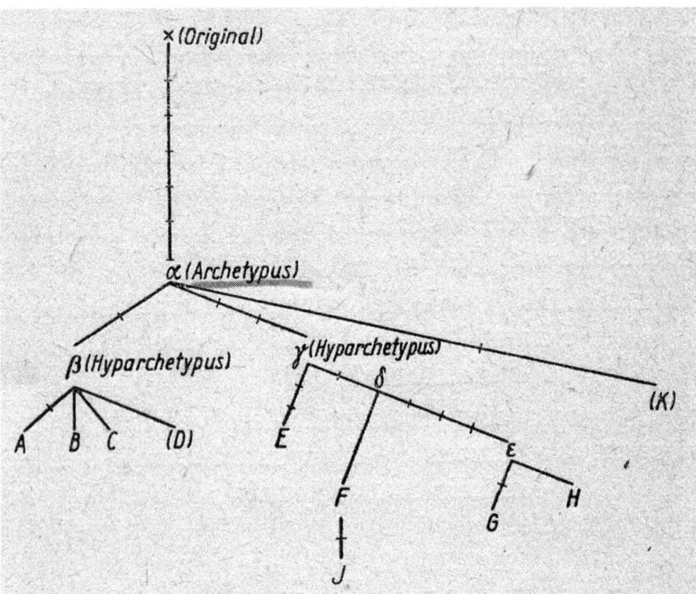

Fig 17.4 Argumentation scheme of manuscript stemmatics formulated by Paul Maas. Reproduced from Maas (1927 [1950]: 7).

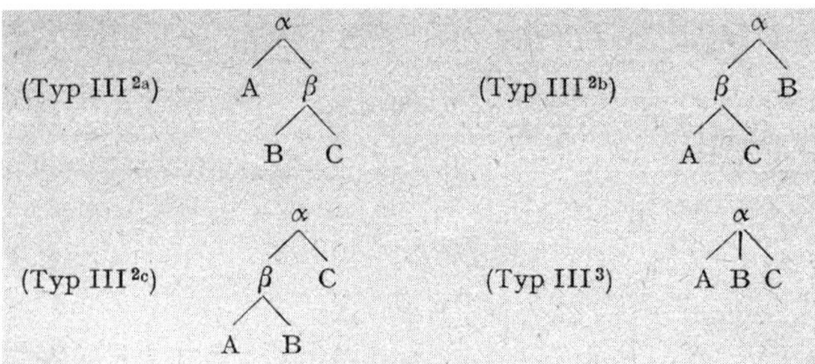

Fig 17.5 Enumeration and categorization of manuscript stemma. Reproduced from Maas (1937).

properties of manuscript stemma and tried to enumerate all possible trees for a set of manuscripts (Fig 17.5).

Maas's intellectual legacy was inherited by Henry M. Hoenigswald, a comparative linguist, who published in 1973 a book entitled as *Studies in Formal Historical Linguistics* (Hoenigswald 1973). In this theoretical work Hoenigswald developed a general graph theory of linguistic and stemmatic trees in detail. These mathematical treatments of tree structures in phylogeny reconstruction found successors in the pattern cladistics of the 1970s and molecular phylogenetics of the 1980s.

17.5 Axiomatic methods in systematics: *Principia Mathematica* and its followers

Principia Mathematica (1910–13), a monumental work published by Alfred North Whitehead (1861–1947) and Bertrand Russell (1872–1970), was a turning point through which a movement of axiomatizing all sciences based on pure logic was born. This movement for "axiomatization" also affected systematics communities in biology, stemmatics, and linguistics. For example, Joseph Henry Woodger (1894–1981) proposed "axiomatic biology" (Woodger 1937) in the late 1930s and promoted the strict axiomatization of several concepts and terms used in taxonomy, genetics, embryology, etc. In 1954, John Gregg (1916–2009) applied Woodger's logical system to the Linnaean classificatory system (Gregg 1954).

However, about 10 years prior to this axiomatization movement in biology, another thread had been independently applied to manuscript stemmatics. In 1927, a slim book entitled *The Calculus of Variants* had been published by a philologist Walter Wilson Greg (1875–1959). The word "calculus" in the title was a synonym for "axiomatic system" in accordance with *Principia Mathematica* by Whitehead and Russell. Greg worked on the axiomatization of some key concepts required to describe manuscript relationships. Woodger and Gregg tried to axiomatize basic concepts and terms in biological classification, whereas Greg's goal was not classification but the phylogeny of manuscripts. Greg was interested in the axiomatization of phylogenetic relationship based on the "ancestor relation R^* (the ancestral relation)" defined in Volume 1 of *Principia Mathematica* (Part 2, section E, * 90: 576).

Consider a simple example. Assume that there are six manuscripts a – f. A "common ancestor" for manuscripts c – f is represented as $A'cdef$ according to the notation in *Principia Mathematica*. While there may be one or more "common ancestors" for a group of manuscripts in transmission, the "exclusive common ancestor" is determined uniquely among other common ancestors. The exclusive common ancestor, denoted as $(x)A'cdef$, is such that a unique direct common ancestor of c, d, e, and f, yet it does not lead to any descendent other than c, d, e, and f. Greg's concept of manuscripts sharing an exclusive common ancestor is essentially the same as Willi Hennig's monophyletic taxa in his phylogenetic systematics since the 1950s.

By using this exclusive common ancestor, Greg represented any hierarchical structure of phylogenetic relationships among manuscripts as a mathematical formula. For example, when there are four monophyletic groups, $(x)A'cdef$, $(x)A'def$, $(x)A'ef$, and $(x)A'ab$, we can combine these as follows:

$$(x)A'cdef + (x)A'def + (x)A'ef + (x)A'ab = (x)A'(ab)(c(d(ef))).$$

As a result of this operation, the hierarchical structure of a monophyletic group of manuscripts that share the exclusive common ancestor $(x)A'$ can be denoted as $(ab)(c(d(ef)))$. Greg showed that the topological structure of phylogenetic relationships

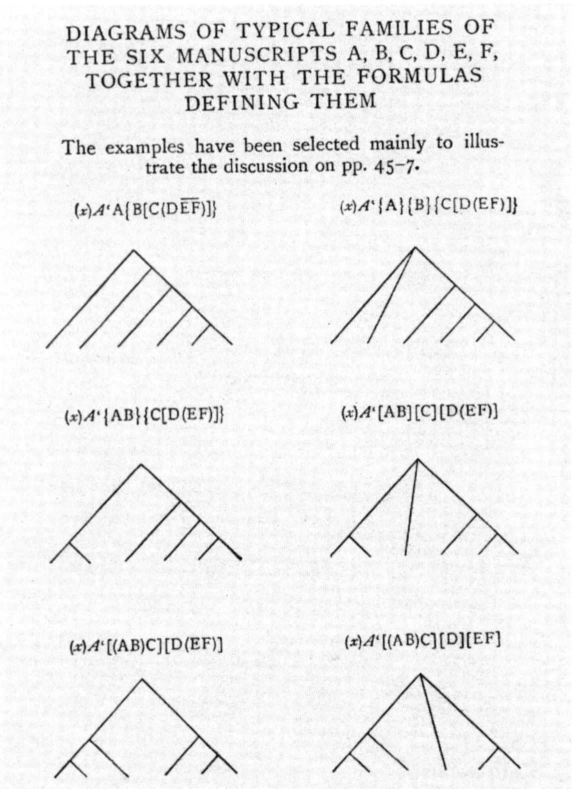

Fig 17.6 Axiomatic notation of manuscript relationships by Greg. Reproduced from Greg (1927: 60).

could be represented accurately according to the notation of *Principia Mathematica* (Fig 17.6).

Greg's formal method for representing hierarchical structures in phylogeny was later followed by the cladogram–tree system in Nelson's "cladistic component analysis" (Nelson 1979, Nelson and Platnick 1981) and the "Newick format" of coding phylogenetic trees now widely used in molecular phylogenetics (Swofford et al. 1996, Felsenstein 2004).

17.6 Abductive inference in systematics in biology, stemmatics, and linguistics: the wider domain of phylogenetic systematics

The meaning of "ancestor" differs in manuscripts, languages, or organisms because the processes of evolution (transmission) from ancestor to descendent are quite different among these three kinds of objects. A variety of "mutations" such as misspelled words or erroneously deleted sentences that could occur in the course of handwriting by scribers will be transmitted from an ancestral manuscript to descendent ones. Similarly another kind of "mutations" such as phonological transitions in words or morphological changes in sentences will also be inherited from a protolanguage to daughter languages. All of these shared "mutations" in stemmatics and linguistics are historical markers of lost routes of phylogenetic history.

The concept of shared derived similarity (that is "synapomorphy") (Hennig 1966) can be equally applied to biology, stemmatics, and linguistics when our aim is to reconstruct phylogenetic history based on inherited similarities. From the point of synapomorphy, there is no essential difference in the logic and methodology of estimating phylogenetic history. An example is stemmatic study of *The Canterbury Tales* by Geoffrey Chaucer (1343?–1400). Phylogenetic analysis of descendent manuscripts using maximum parsimony method estimated the best tree and network calculated by computer software of molecular phylogenetics (Robinson and O'Hara 1992, Barbrook et al. 1998).

Tucker (2004) proposed a new category of "historiographic sciences," which is composed of those sciences that share the purpose of reconstructing phylogenetic relationships among objects and the search for common causes for historical processes. Historically speaking, theory, methodology, and conceptual system for phylogeny estimation in stemmatics, linguistics, and biology showed considerable convergence (Atkinson and Gray 2003). The fact that these disciplines had independently established the same procedures for tree building confirms that they focus on an identical problem of the historiographic sciences separately. The identical, common problem is "abduction" of past historical events (Fitzhugh 2006). Historiographic abduction is to estimate the best systematic pattern based on the observed data at present. The goal of historiographic sciences is reconstructing genealogical traces based on observed data regardless of whether the objects under study are manuscripts, languages, or organisms, etc. From the point of view of abduction in historiographic sciences, manuscript stemmatics and historical linguistics have been tackling the common problem of how to estimate relationships with high accuracy by limited amount of information about the past.

Needless to say, the trends of thought and background knowledge changed from era to era. For example, Greg (1927) attempted to construct an axiomatic logical system of textual criticism based on logical positivism of *Principia Mathematica* (Whitehead and Russell 1910–1913). However, Greg's logical system was not widely accepted by most contemporary philologists (Rosenblum 1998: Preface). For another example, Woodger (1937) and Gregg (1954) built an axiomatic system for biological taxonomy but could find very few sympathizers. Strict logical systems in historiographic sciences might not have been sufficiently appreciated.

Sebastiano Timpanaro (1923–2000) pointed out the parallel relationship of phylogenetic reconstruction methods in historical linguistics and comparative philology:

> There is an undeniable affinity between the method with which the Classical philologist classifies manuscripts genealogically and reconstructs the reading of the archetype, and the method with which the linguist classifies languages and as far as possible reconstructs a lost mother language, for example, Indo-European. In both cases inherited elements must be distinguished from innovations, and the unitary anterior phase from which these have branched out must be hypothesized on the basis of various innovations. The fact that innovations are shared by certain manuscripts of the same text, or be certain languages of the same family, demonstrates that these are connected by a particularly close kinship, that they belong to a subgroup: a textual corruption too is an innovation compared to the previously transmitted text, just like a linguistic innovation. On the other hand, shared "conservations" have no classificatory value: what was already found in the original text or language can be preserved even in descendants that are quite different from one another. (Timpanaro 2005: 119[1])

That is, shared innovations are evidence of monophyletic groups of manuscripts, languages, or organisms but shared conservations have no evidence of phylogenetic relationships. This means that the distinction between shared innovations and conservations had been made in philology and linguistics before Hennig (1966) coined the pair of new words synapomorphy and symplesiomorphy. Moreover, it suggests that a universal historiographic method will be feasible beyond the differences among manuscripts, languages, and organisms.

The wider domain of phylogenetic systematics is still expanding not only in biology but also in other historiographic sciences including stemmatics, linguistics, archaeology, etc. A historical landscape of contemporary systematics during past 80 years from 1930s to 2010s is pictorially summarized in Figs 17.7 and 17.8. As a modern descendent of the old concept of *arbor scientiae* by Raimundus Lulls, phylogenetic trees have provided us with a useful infographic tool for visualizing object diversity (Lam 1936, Ragan 2009, Pietsch 2012, Archibald 2014, Lima 2014, Minaka 2015). Phylogenetic systematics will continue to evolve and will witness future developments in various fields of scientific research as well as in other literary, social, and cultural studies.

Acknowledgments

I cordially thank David Williams for inviting me to contribute a chapter to this volume and for painstakingly improving the English of my manuscript. I would like to acknowledge valuable comments and suggestions of an anonymous reviewer. This

[1] Timpanaro (2005) is an English translation of *La genesi del metodo del Lachmann* (Timpanaro 1963)

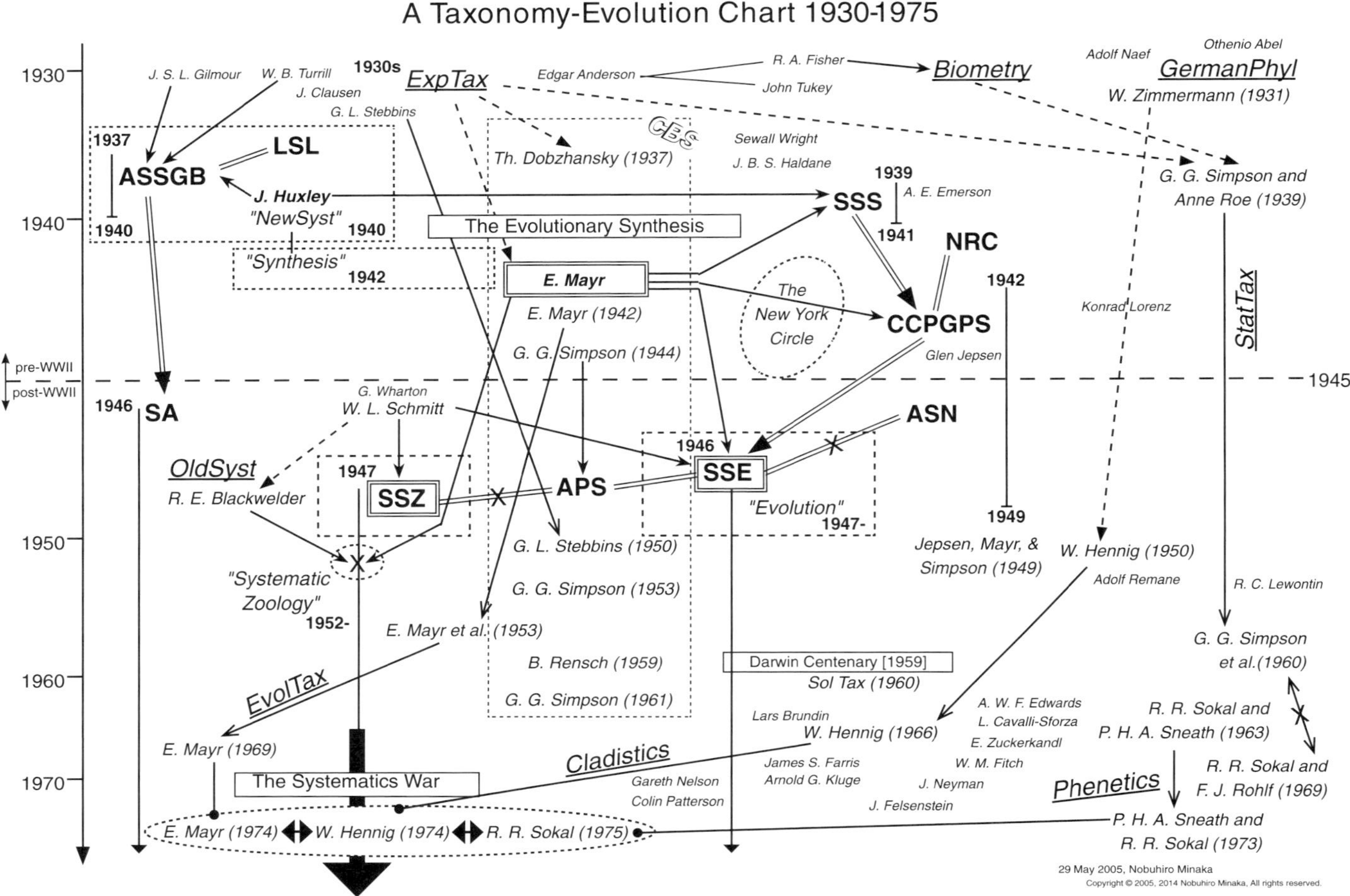

Fig 17.7 Pictorial chart of contemporary history of systematics 1930–1975. Adapted from Minaka (2005).

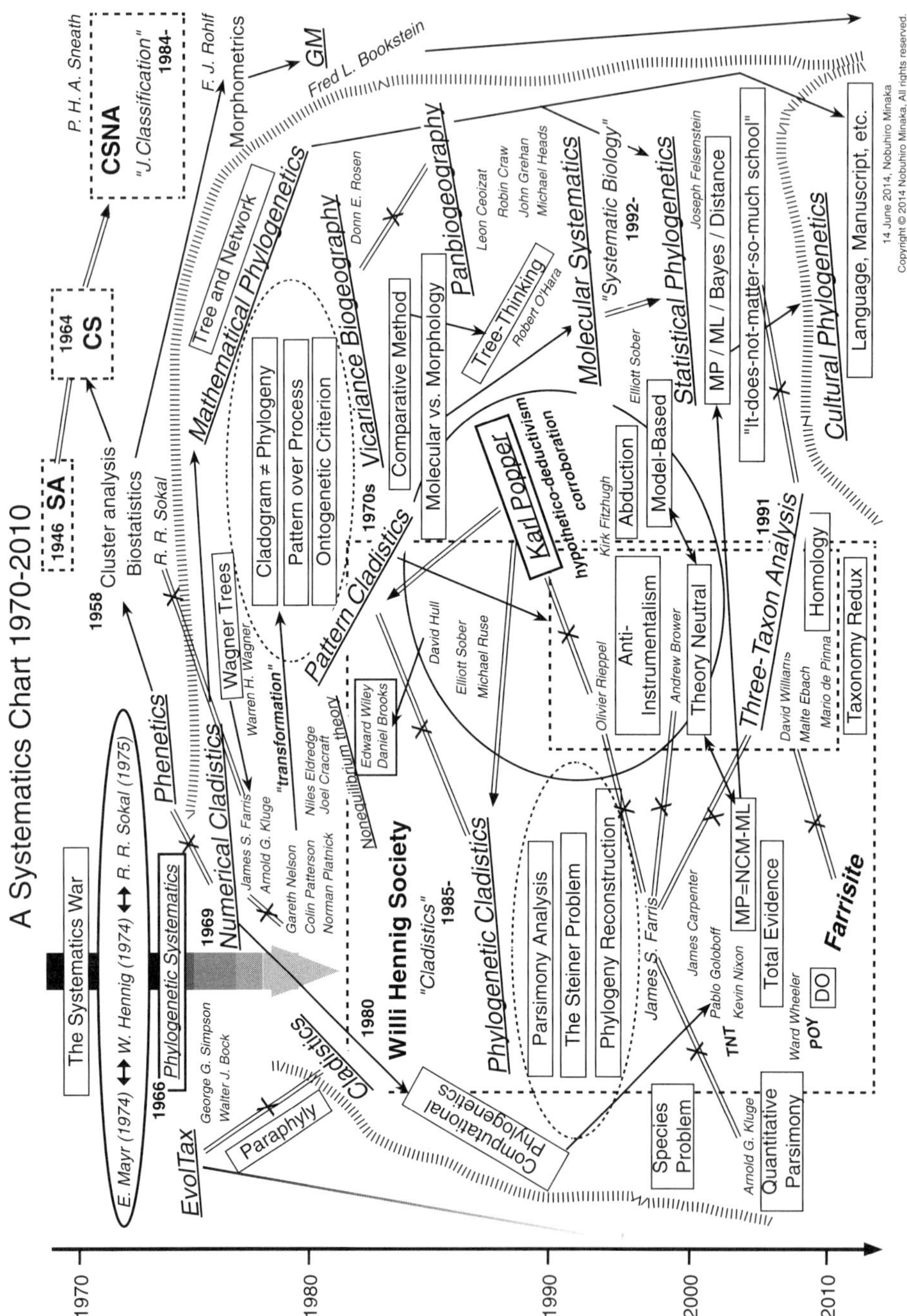

Fig 17.8 Pictorial chart of contemporary history of systematics 1970–2010.

study was supported in part by Topic-Setting Program to Advance Cutting-Edge Humanities and Social Sciences Research Area Cultivation (Synthesizing historical sciences: New perspectives to cultural evolutionary studies), Japan Society for the Promotion of Science (JSPS).

References

Archibald, J.D. (2014). *Aristotle's Ladder, Darwin's Tree: The Evolution of Visual Metaphors for Biological Order.* New York: Columbia University Press.

Atkinson, Q.D. and R.D. Gray (2003). Curious parallels and curious connections: phylogenetic thinking in biology and historical linguistics. *Systematic Biology*, 54, 513–526.

Atran, S. (1990). *Cognitive Foundations of Natural History: Towards an Anthropology of Science.* Cambridge: Cambridge University Press.

Atran, S. (1998). Folk biology and the anthropology of science: cognitive universals and cultural particulars. *The Behavioral and Brain Sciences*, 21, 547-609.

Barbrook, A.C., C.J. Howe, N. Blake and P. Robinson (1998). The phylogeny of *The Canterbury Tales. Nature*, 394, 839.

Barsanti, G. (1992). *La scala, la mappa, l'albero: immagini e classificazioni della natura fra Sei e Ottocento.* Florence: Sansoni Editore.

Baum, D. and S. Smith (2013). *Tree Thinking: An Introduction to Phylogenetic Biology.* Greenwood Village, TX: Roberts and Company.

Cameron, H. Don (1987). The upside-down cladogram: problems in manuscript affiliation. In *Biological Metaphor and Cladistic Classification: An Interdisciplinary Perspective*, ed. by H.M. Hoenigswald and L.F. Wiener, Philadelphia, PA: University of Pennsylvania Press, pp. 227–242.

Davey, B.A. and H.A. Priestely (2002). *Introduction to Lattices and Order, Second Edition.* Cambridge: Cambridge University Press.

Dress, A., K.T. Huber, J. Koolen, V. Moulton, and A. Spillner (2012). *Basic Phylogenetic Combinatorics.* Cambridge: Cambridge University Press.

de Queiroz, K. and M.J. Donoghue (1990). Phylogenetic systematics or Nelson's version of cladistics? *Cladistics*, 6, 61–75.

Felsenstein, J. (2004). *Inferring Phylogenies.* Sunderland, MA: Sinauer Associates.

Fitzhugh, K. (2006). The abduction of phylogenetic hypotheses. *Zootaxa*, 1145, 1–110.

Forster, P. and C. Renfrew (eds.) (2006). *Phylogenetic Methods and the Prehistory of Languages.* Cambridge: The McDonald Institute for Archaeological Research.

Ginzburg, C. (2004). Family resemblances and family trees: two cognitive metaphors. *Critical Inquiry*, 30, 537–556.

Gould, S.J. (1986). Evolution and the triumph of homology, or why history matters. *American Scientist*, 74, 60–69.

Greg, W.W. (1927). *The Calculus of Variants: An Essay on Textual Criticism.* Oxford: Clarendon Press.

Gregg, J.R. (1954). *The Language of Taxonomy: An Application of Symbolic Logic to the Study of Classificatory*

Systems. New York: Columbia University Press.

Hennig, W. (1950). *Grundzüge einer Theorie der phylogenetischen Systematik*. Berlin: Deutscher Zentralverlag.

Hennig, W. (1957). Systematik und Phylogenese. In *Bericht über die Hundertjahrfeierder Deutschen Entomologischen Gesellschaft Berlin, 30. September bis 5. Oktober 1956*, ed. H. -J. Hannemann, 50–71. Berlin: Akademie-Verlag.

Hennig, W. (1966). *Phylogenetic Systematics*. Urbana, IL: University of Illinois Press.

Hennig, W. (1974). Kritische Bemerkungen zur Frage "Cladistic analysis or cladistic classification?" *Zeitschrift für Zoologische Systematik und Evolutionsforschung*, 12, 279–294.

Hennig, W. (1975). "Cladistic analysis or cladistic classification?": a reply to Ernst Mayr. *Systematic Zoology*, 24, 244–256.

Hennig, W. (1982). *Phylogenetische Systematik*. Berlin: Verlag Paul Parey.

Hoenigswald, H.M. (1973). *Studies in Formal Historical Linguistics*. Dordrecht: D. Reidel.

Hoenigswald, H.M. and L.F. Wiener (eds.) (1987). *Biological Metaphor and Cladistic Classification: An Interdisciplinary Perspective*. Philadelphia, PA: University of Pennsylvania Press.

Hodson, F.R., D.G. Kendall, and P. Tăutu (eds.) (1971). *Mathematics in the Archaeological and Historical Sciences*. Edinburgh: Edinburgh University Press.

Holm, G. (1972). Carl Johan Schlyter and textual scholarship. *Saga och Sed (Kungliga Gustav Adolf Akademiens Aarsbok)*, 1972, 48–80.

Hull, D.L. (1979). The limits of cladism. *Systematic Zoology*, 28, 414–438.

Hull, D.L. (1988). *Science as a Process: An Evolutionary Account of the Social and Conceptual Development of Science*. Chicago, IL: The University of Chicago Press.

Klapisch-Zuber, C. (2000). *L'ombre des ancêtres: essai sur l'imaginaire médiéval de laparenté*. Paris: Librairie Arthème Fayard.

Klapisch-Zuber, C. (2003). *L'arbre des familles*. Paris: Éditions de la Martière.

Kornblith, H. (1993). *Inductive Inference and Its Natural Ground : An Essay in Naturalistic Epistemology*. Cambridge, MA: The MIT Press.

Lam, H.J. (1936). Phylogenetic symbols, past and present (being an apology for genealogical trees). *Acta Biotheoretica, Series A*, 2, 153–194

Laudan, R. (1992). What's So Special about the Past? In *History and Evolution*, ed. by M.H. Nitecki and D.V. Nitecki, 55–67, Albany, NY: State University of New York Press.

Lima, M. (2014). *The Book of Trees: Visualizing Branches of Knowledge*. New York: Princeton Architectural Press.

Lipo, C.P., M.J. O'Brien, M. Collard, and S.J. Shennan (eds.) (2005). *Mapping our Ancestors: Phylogenetic Approaches in Anthropology and Prehistory*. New Brunswick: Transaction Publishers.

Maas, P. (1927). *Textkritik*, 2nd edition [1950]. Leipzig: B.G. Teubner.

Maas, P. (1937). Leitfehler und stemmatische Typen. *Byzantinische Zeitschrift*, 37, 289–294.

Minaka, N. (1993). Cladistics and biogeographic methods: cladograms, components, and parsimony. *Bulletin of the Biogeographical Society of Japan* 48(2), 1–27. [In Japanese with English abstract]

Minaka, N. (1997). *Systematics, Phylogenetics, and the Tree of*

Life: A Cladistic Perspective. Tokyo: University of Tokyo Press. [In Japanese]

Minaka, N. (2005). Ernst Mayr and Willi Hennig: Revisiting the Systematics Controversy in the 1970s. *Taxa: Proceedings of the Japanese Society of Systematic Zoology*, 19, 95–101. [In Japanese]

Minaka, N. (2006). *Tree-Thinking: An Introduction to Phylogenetic World View.* Tokyo: Kodansha. [In Japanese]

Minaka, N. (2015). Translator's note: The great tree of knowledge – Past, present, and future. In M. Lima (2015). *The Book of Trees.* Translated by N. Minaka. Tokyo: BNN, Inc., pp. 202–207. [In Japanese]

Minaka, N. and K. Sugiyama (2012). *Phylogeny Mandala: Chain, Tree, and Network.* Tokyo: NTT Publishing. [In Japanese]

Moretti, F. (2005). *Graphs, Maps, Trees: Abstract Models for a Literary History.* London: Verso.

Nakao, H. and N. Minaka (eds.) (2012). *An Introduction to Cultural Phylogenetics: An Unified Framework for Studying Patterns in Cultural Evolution.* Tokyo: Keiso Shobo. [In Japanese]

Nelson, G. (1973). Classification as an expression of phylogenetic relationship. *Systematic Zoology*, 22, 344–359.

Nelson, G. (1979). Cladistic analysis and synthesis: Principles and definitions, with a historical note on Adanson's Familles des plantes (1763–1764). *Systematic Zoology*, 28, 1–21.

Nelson, G. and N. Platnick (1981). *Systematics and Biogeography: Cladistics and Vicariance.* New York: Columbia University Press.

O'Brien, M.J. and R. Lee Lyman (2003). *Cladistics and Archaeology.* Salt Lake City, UT: The University of Utah Press.

O'Hara, R.J. (1996). Trees of history in systematics and philology. *Memorie della Societa Italiana di Scienze Naturali e del Museo Civico di Storia Naturale di Milano*, 27, 81–88.

Papavero, N. and J. Llorente (eds.) (2008). *Principia Taxonomica: Una Introducción a los Fundamentos Lógicos, Filosóficos y Metodológicos de las Escuelas de Taxonomía Biológica, Volumen I – IX.* Ciudad Universitaria, Mexico: Universidad Nacional Autónoma de México.

Pietsch, T.W. (2012). *Trees of Life: A Visual History of Evolution.* Baltimore, MD: The Johns Hopkins University Press.

Platnick, N.I. (1979). Philosophy and the transformation of cladistics. *Systematic Zoology*, 28, 537–546.

Platnick, N.I. and H. D.Cameron (1977). Cladistic methods in textual, linguistic, and phylogenetic analysis. *Systematic Zoology*, 26, 380–385.

Ragan, M.A. (2009). Trees and networks before and after Darwin. *Biology Direct*, 4, 43.

Robinson, P.M.W. and R.J. O'Hara (1992). Report on the textual criticism challenge 1991. *Bryn Mawr Classical Review*, 3, 331–337.

Rosenblum, J. (ed.) (1998). *Sir Walter Wilson Greg: A Collection of His Writings.* New York: The Scarecrow Press.

Schadt, H.(1982). *Die Darstellungen der Arbores Consanguinitatis und der Arbores Affinitates: Bildschemata in juristischen Handschriften.* Tübingen: Verlag Ernst Wasmuth.

Schmidt-Burkhardt, A. (2005). *Stammbäume der Kunst: Zur Genealogie der Avantgarde.* Berlin: Akademie-Verlag.

Schmitt, M. (2013). *From Taxonomy to Phylogenetics: Life and Work of Willi Hennig.* Leiden: Brill

Semple, C. and M. Steel (2003). *Phylogenetics*. Oxford: Oxford University Press.

Simpson, G.G.(1961). *Principles of Animal Taxonomy*. New York: Columbia University Press.

Swofford, D.L., G.J. Olsen, P.J. Waddell and D.M. Hillis (1996). Phylogenetic inference. In *Molecular Systematics, Second Edition*, ed. by D.M. Hillis, C. Moritz and B.K. Mable, 407–514. Sunderland, MA: Sinauer Associates.

Timpanaro, S. (1963). *La genesi del metodo del Lachmann*. Florence: Le Monnier.

Timpanaro, S. (1971). *Die Entstehung der Lachmannschen Methode*. Translated by D. Irmer. Hamburg: Hermut Buske.

Timpanaro, S. (2005). *The Genesis of Lachmann's Method*. Translated by G.W. Most. Chicago, IL: The University of Chicago Press.

Tucker, A. (2004). *Our Knowledge of the Past: A Philosophy of Historiography*. Cambridge: Cambridge University Press.

Whewell, W. (1847). *The Philosophy of the Inductive Sciences, Founded upon Their History. Second Edition. 2 Volumes.* London: John W. Parker. [Reprinted in 1967: F. Cass, New York]

Whitehead, A.N. and B. Russell (1910–13). *Principia Mathematica (Three Volumes)*. Cambridge: Cambridge University Press.

Williams, D.M. and M.C. Ebach (2008). *Foundations of Systematics and Biogeography*. New York: Springer-Verlag.

Woodger, J.H. (1937). *The Axiomatic Method in Biology*. Cambridge: Cambridge University Press.

Yoon, C.K. (2009). *Naming Nature: The Clash between Instinct and Science*. New York: W.W. Norton.

Zimmermann, W. (1931). Arbeitsweise der botanischen Phylogenetik und anderer Gruppierungswissenschaften. In *Handbuch der biologischen Arbeitsmethoden, Abteilung IX: Methoden zur Erforschung der Leistungen des tierischen Organismus, Teil 3: Methoden der Vererbungsforschung, Heft 6 (Lieferung 356)*, ed. E. Abderhalden, 941–1053. Berlin: Urban and Schwarzenberg.

18

The relational view of phylogenetic hypotheses and what it tells us on the phylogeny/classification relation problem

STÉPHANE PRIN

18.1 Introduction

Classifications are usual in biology. They are found not only in biological systematics[1] (taxonomic classifications) but also in biochemistry (classifications of amino acids, metabolites, etc.), cytology (classifications of cells into muscular, nervous, stem, blood cells, etc.), ecology (trophic classifications, i.e. classifications of organisms as producers, herbivores, carnivores, scavengers, etc.), and so on. Among them, taxonomic classifications (i.e. classifications of organisms into taxa[2]) have a particular status.[3] On the one hand, they seem to classify organisms for what they are, by virtue of their physical constitution and behavior. On the other hand, while their origin goes back to antiquity, either from Aristotle (Tassy 2005: 64) or from

---

[1] By "biological systematics," I mean both biological taxonomy (the description, classification, and identification of organisms in terms of structural and behavioral properties) and phylogenetics.

[2] By "taxon," I primarily mean any generalization of organisms (empirically) based on at least one structural (i.e. morphological, anatomical, or molecular) or behavioral property.

[3] Actually, taxonomic classifications are at least of two kinds (Jardine 1969): (1) the partitioned classification of organisms into "basic taxa" (*sensu* Jardine); and (2) the hierarchical classifications of the latter into inclusive taxa. Hereafter, it will mainly be question of taxonomic classifications in this second sense.

The Future of Phylogenetic Systematics: The Legacy of Willi Hennig, eds. D. Williams, M. Schmitt and Q. Wheeler. Published by Cambridge University Press.

his successor Theophrastus (Lecointre and Le Guyader 2001: 11), it is only since Darwin (1859: Chapter 13) that we have a causal theory allowing for their interpretation. According to the latter, taxonomic classifications are thus also phylogenetic classifications. That is, classifications where the membership of a given organism into this or that class (i.e. the extension of a given taxon) has something to do with its kinship with the other members of this class. In other words, these classifications express, at least partially, phylogeny.

Darwin's interpretation naturally led to the question of how taxonomic classifications and phylogeny are connected. Following an intense debate within biological systematics during the 1960s, 1970s, and 1980s (Hull 1970, 1986, 2001, Nelson 1971, 1974, Mayr 1974, 1981, 1986, Hennig 1975, Sokal 1975, Ashlock 1979, Farris 1979, Ridley 1989), it is generally admitted that three "schools" of systematics/taxonomy can be distinguished (Hull 1970, Mayr 1974, 1981, 1986, Matile et al. 2004):

1. *Phenetic/numerical systematics/taxonomy* (or more simply "phenetics": Sokal and Sneath 1963, Sneath and Sokal 1973). According to this school, taxonomic classifications have to be both (a) founded on the quantification of the overall resemblance between taxonomic units (TUs),[4] and (b) be completely independent from any knowledge of phylogeny. Thus, the inclusive taxa resulting from phenetic analyses primarily appear as clusters of resemblances. On this basis, phylogeny is then either completely ignored (because it is considered too conjectural) or inferred from taxonomic classifications (which implies that degree of overall resemblance can be a good approximation of degree of kinship).

2. *Phylogenetic systematics/taxonomy* (also "Hennigian cladistics": Hennig 1965, 1966). According to this school, taxonomic classifications have to be founded only on knowledge of phylogeny, this last being seen as a (fully ramified) tree-like genealogy of species. In other words, it is question of translating a series of cladogenetic speciation events into a hierarchical classification of inclusive clades (monophyletic taxa *sensu* Hennig),[5] or, if you prefer, to only represent the cladistic relationship (degree of kinship relationship), between species (and more generally between TUs).

---

[4] By "taxonomic unit" ("TU"), I mean any taxon defined prior to a given phylogenetic/cluster analysis, and taken as an object of this analysis. The counterparts of TUs are hypothetical taxa (HTs), i.e. taxa highlighted and/or deduced from such phylogenetic/cluster analyses. In phylogenetic analyses of degree of kinship (i.e. cladistic phylogenetic analyses and, more generally, tree-oriented phylogenetic analyses of any kind), TUs and HTs are respectively terminal and inclusive taxa.

[5] For Hennig (1965: 98), a clade is a group of species such that these species are related compared to any species that does not belong to it. In more modern terms, this means that any set of TUs having a common and exclusive origin corresponds to the extension of an inclusive clade and conversely (i.e. any inclusive clade always consists of all and only the descendant TUs of a given ancestor).

3. *Evolutionary/eclectic systematics/taxonomy* (Mayr 1974, 1981, Ashlock 1979). According to this school, taxonomic classifications must be founded on phylogeny as well but not in the same way as in phylogenetic systematics. Indeed, for proponents of this approach, taxonomic classifications must take into account not only the cladistic relationship but also the patristic relationship, i.e. non-bifurcating evolution or, if you prefer, anagenesis (degree of divergence). Irrespective of what the term refers exactly (either an anagenetic speciation event[6] or simply the accumulation of change from a given ancestor), it is thus a question here of grouping into grades[7] rather than clades.

In addition to these three classical "schools," there is also another version of cladistics named "pattern cladistics" by Beatty (1982) but also known as "transformed cladistics" (Platnick 1979; Nelson and Platnick 1981; Williams and Ebach 2008). As I understand it, the question here is to adopt something between the phenetic and phyletic (i.e. phylogenetic and evolutionary) "schools" of systematics/taxonomy. Indeed, for pattern cladists, prior taxonomic knowledge consists of cladograms (the basic product of cladistic analysis), the latter being seen as "synapomorphy schemes." Then, and only then, we can infer both phylogenetic hypotheses (i.e. phylogenetic trees) and taxonomic classifications from these cladograms.

As we can see, all four "schools" have stated the problem of the phylogeny/classification link in terms of priority and dependence. At least two problems follow:

1. For the four viewpoints sketched above, classification and phylogeny are always a priori considered as distinct, regardless any priority relationship between them. In other words, it would be impossible for any phylogenetic hypothesis to also be a taxonomic classification, i.e. impossible for it to also be of classificatory kind. For the phyletic "schools", this implies that all the phylogenetic information conveyed by taxonomic classifications is already conveyed by phylogenetic hypotheses. As for the non-phyletic "schools" (phenetics and pattern cladistics), this implies that either taxonomic classifications are, at best, a (necessarily incomplete) prerequisite for phylogenetic hypotheses or, at worst, they express no phylogenetic information at all. But, if true, then taxonomic classifications

---

[6] By "anagenetic speciation," I simply mean a process where a species gives rise to a single new species.

[7] By "grade," I mean a taxon such that its members have a common origin and, at least theoretically, the same level of evolution/adaptation (i.e. the same ecological niche). Thus, and in contrast to clades, the origin of the members of a given grade is not necessarily exclusive (i.e. the extension of a grade does not necessarily contain all the descendant of a given ancestor), although it remains unique.

cannot have any true theoretical function regarding phylogeny and evolution. Hence the following question: in what way can taxonomic classifications be *theoretically* interesting for phylogenetics (and more generally evolutionary biology) if they are not phylogenetic hypotheses? On this point, I agree with Felsenstein (2004: 145–146): if taxonomic classifications are just a matter of names and ranks, then the "it doesn't matter very much" school of taxonomy is welcomed.

2. The formal aspect of the problem had largely been ignored. On the one hand, not only has the use of graph theory in phylogenetic methodology been neglected, but, in addition, the discussions never involve any general theory on the formal structure of phylogenetic hypotheses. On the other hand, the concept of classification is, in large part, irrelevant. Indeed, even if taxonomic classifications have correctly been defined as sets/structures of classes (Buck and Hull 1966: 97), they were nevertheless essentially discussed either as (ordered) indices (Ashlock 1979: 448, Mayr 1981: 511, Mayr 1986: 148) or (ranked) lists of names (Hull 1970: 22, Hull 1986: 168). In other words, even if a correct (although insufficient[8]) definition was provided, it was more often (if not always) a question of a definition confusing classifications (as structures) with a particular way for representing some of them (i.e., the hierarchical ones).[9]

In summary, all of these "schools" have curiously ignored the following simple question: *intrinsically speaking*, is it possible for any given phylogenetic hypothesis to also be a classification? It is a shame because the latter exactly determines the theoretical importance of classifications in phylogenetics and more generally in evolutionary biology. The object of this paper is therefore to address this issue. More precisely, I will argue that phylogenetic hypotheses are also of the classificatory kind, at least for the three concepts of kinship considered below. My argument will roughly be the following:

- Formally speaking, any phylogenetic hypothesis can be view as a simple relational structure involving this or that concept of kinship.

---

[8] Indeed, this definition does not specify what kinds of sets/structures of classes are involved. Furthermore, there is also the question of what a class is. However, given that Buck and Hull (1966) characterize the organism-taxon relation in terms of set-theoretic membership relation (i.e. taxa are classes having organisms as members), it follows that any class is also a set.

[9] It is even possible that such definitions confuse taxonomy (the creation/discovery of classes and their arrangement into a general system) with nomenclature (the name attribution of these classes, something necessarily *a posteriori* to the previous activities). Indeed, the (ordered) indices and (ranked) lists of names are nomenclatural-based representations of (hierarchical) classifications.

- From a mathematical point of view, some kinds of classificatory structures are equivalent, i.e. connected by a particular bijection, to some kinds of simple relational structures.
- At least under some concepts of kinship, phylogenetic hypotheses can therefore be formalized not just as simple relational structures (which is what they fundamentally are), but also as classificatory structures.

I will proceed in five parts. First, I will introduce the mathematical concepts considered necessary. Second, I will argue the relational view of phylogenetic hypotheses. Third, I will (a) formalize phylogenetic hypotheses as relational structures, and (b) see in what they are also classificatory structures. Fourth, I will address some consequences and related topics. Finally, I will specifically discuss a particular problem regarding the degree of kinship phylogenetic hypotheses.

18.2 Mathematical concepts

On the symbolism

Though this is not a pure logical and/or mathematical paper, I will use several symbols from logic and set theory:

- "$\neg$", "$\wedge$", "$\vee$", "$\rightarrow$", and "$\leftrightarrow$" for, respectively, "not" (negation), "and" (conjunction), "and/or" (inclusive disjunction), "if...then/only if" (conditional) and "if and only if" (bi-conditional).
- "$\forall$" and "$\exists$" for, respectively, "for any/for all" (universal quantifier) and "there is at least one" (existential quantifier, only used in the Appendices).
- "$=/\neq$", "$\in/\notin$", "$\subseteq/\supseteq$", "$\subset/\supset$", and "$\leq/<$" for, respectively, "being equal to/being not equal to" (equality/identity relation), "belong to/do not belong to" (membership relation), "being included into/include" (non-strict set-theoretic inclusion), "being properly included into/properly include" (strict set-theoretic inclusion), and "being lower than or equal to/being lower than" (only used in Appendices).
- "$\cap$", "$\cup$", "$\setminus$", "$\varnothing$", and "$|A|$" for, respectively, the set-theoretic intersection, union, and difference operators (the latter is only used in the Appendices), the empty set, and the cardinal (number) of any set A.
- "$\bigcup$" for the great union operator (only used in the Appendices).

Relation and relational structure

- k-tuple: any "collection" of k objects in a given order. For k objects $x_1, x_2, ..., x_k$ and k objects $y_1, y_2, ..., y_k$, the k-tuple $\langle x_1, x_2, ..., x_k \rangle$ is therefore equal to the k-tuple $\langle y_1, y_2, ..., y_k \rangle$ if and only if $x_1 = y_1$, $x_2 = y_2$, ..., and $x_k = y_k$. An *ordered pair* is a 2-tuple (and conversely).

- k-ary relation: any set R such that for any x, if x is in R, then x is a k-tuple. *Binary*, *ternary*, and *quaternary* relations are then (and respectively) nothing but 2-ary, 3-ary, and 4-ary relations.
- k-ary Cartesian product of A: given a set A, the set A^k of all k-tuples $\langle x_1, x_2, ..., x_k \rangle$ such that $x_1, x_2, ...,$ and x_k belong to A.
- k-ary relation on A: given a non-empty set A, any subset of A^k, i.e. any set R such that $R \subseteq A^k$.
- Reflexive/anti-reflexive binary relation on A: respectively, any binary relation R on A such that $\langle x, x \rangle \in R / \langle x, x \rangle \notin R$ for any x of A.
- Symmetric/asymmetric/anti-symmetric binary relation on A: respectively, a binary relation R on A is symmetric/asymmetric if and only if for all x and y of A, if $\langle x, y \rangle \in R$, then $\langle y, x \rangle \in R / \langle y, x \rangle \notin R$. In contrast, R is anti-symmetric if and only if for all x and y of A, if $\langle x, y \rangle \in R$ and $x \neq y$, then $\langle y, x \rangle \notin R$. Note that asymmetry implies anti-symmetry.
- Transitive binary relation on A: any binary relation R on A such that for all x, y, and z of A, if $\langle x, y \rangle \in R$ and $\langle y, z \rangle \in R$, then $\langle x, z \rangle \in R$. Note that transitivity and anti-reflexivity imply asymmetry.
- Simple k-ary relational structure: any ordered pair $\langle A, R \rangle$ such that A is a non-empty set and R is a k-ary relation on A.
- Relational structure of equivalence: any simple binary relational structure $\langle A, R \rangle$ such that R is a *relation of equivalence* on A, i.e. a reflexive, symmetric, and transitive binary relation on A.
- (Relational structure of) strict order: any simple binary relational structure $\langle A, R \rangle$ such that R is a *relation of strict order* on A, i.e. an anti-reflexive and transitive binary relation on A. Thus, any strict order is asymmetric and so, anti-symmetric too. Note also that to any strict order $\langle A, R_< \rangle$ corresponds a non-strict order $\langle A, R_\leq \rangle$ (a reflexive, anti-symmetric, and transitive simple binary relational structure) – and conversely – such that for any $\{x, y\} \subseteq A$, $\langle x, y \rangle \in R_\leq$ if and only if $\langle x, y \rangle \in R_<$ or $x = y$ (and reciprocally, $\langle x, y \rangle \in R_<$ if and only if $\langle x, y \rangle \in R_\leq$ and $x \neq y$). Among orders (i.e. anti-symmetric and transitive simple binary relational structures), non-strict orders are thus the reflexive counterparts of strict orders.[10]

Function and bijection

- Binary Cartesian product of A and B: given two sets A and B, the set $A \times B$ of all ordered pairs $\langle x, y \rangle$ such that $x \in A$ and $y \in B$.
- Binary relation from A to B: given two non-empty sets A and B, any subset of the binary Cartesian product of A and B, i.e. any set R such that $R \subseteq A \times B$.

---

[10] What is understood as a "(relational structure of) non-strict/strict order" corresponds to what the majority of English-speaking mathematicians calls a "(non-strict)/strict partially ordered set."

- Function (of a single variable): a binary relation f from A to B is also a function if and only if for any $x \in A$ there is exactly a single $y \in B$ such that $\langle x, y \rangle \in f$. If appropriate, we can write "$f(x) = y$" instead of "$\langle x, y \rangle \in f$".
- Injection from A to B: any function f from A to B such that for any $\{x, y\} \subseteq A$ and any $z \in B$, $f(x) = f(y) = z$ only if $x = y$. If appropriate, it follows that $|A| \leq |B|$.
- Bijection from A to B: any function f from A to B such that for any $y \in B$ there is exactly a single $x \in A$ such that $f(x) = y$. If appropriate, it follows that $|A| = |B|$.

Classification and classificatory structure

- Classification: given a finite non-empty set A, any set C of non-empty subsets of A – hereafter called the *clusters* of C – such that any object of A belongs to at least one cluster of C.
- Classificatory structure: any ordered pair $\langle A, C \rangle$ such that A is a finite non-empty set and C is a classification of A.
- Partitioned classificatory structure: any classificatory structure $\langle A, P \rangle$ such that P is a *partition* of A, i.e. a classification such that the intersection between any two distinct clusters X and Y of P is always empty.
- Hierarchical classificatory structure: any classificatory structure $\langle A, H \rangle$ such that H is a *hierarchical classification* of A, i.e. a classification such that: (1) $A \in H$; (2) $\{x\} \in H$ for any x of A; and (3) the intersection between any two clusters X and Y of H is always either empty or one of the two (i.e. no overlapping).

(Informal) Graph theory

- Graph: any structure G compounded of discrete entities (at least one), called "(the) vertices (of G)", and of (a finite number of) links between some of them, called "(the) edges (of G)."
- Undirected/directed graph: any graph with undirected/directed edges only.
- Simple graph: any graph with no loop and no parallel edges.
- [Undirected/directed] tree: any acyclic and connected [undirected/directed] simple graph. Undirected trees involve two kinds of vertices: leaves (vertices connected with a single other vertex) and nodes (vertices connected with at least two other vertices).
- Fully ramified undirected tree (often called "unrooted tree"): any undirected tree such that each of its node is connected with at least three other vertices.
- Fully ramified undirected n-tree: finite fully ramified undirected tree with n leaves.
- Rooted tree: directed tree with a single reading direction. Rooted trees involve three kinds of vertices: leaves (terminal vertices), nodes (intermediate vertices), and roots (initial vertices).

- Hierarchical tree (often called "rooted tree"): fully ramified rooted tree with a single root.
- Hierarchical *n*-tree: finite hierarchical tree with *n* leaves.

18.3 Evolution, phylogeny, and phylogenetic hypotheses

The explanation of the diversity of life

One of the main tasks of contemporary biology is to account for (i.e. describe and explain) resemblances and differences between organisms (i.e. the diversity of life). For its explanative part, the basic principle is to consider this diversity as the result of a filiation process (by reproduction of organisms) combining both heredity and modification mechanisms. Indeed, the notions of filiation, heredity, and modification respectively imply those of kinship, resemblance, and difference. In other words, such a process implies a set of kinship relationships between organisms, and this last allows us to understand why these and those organisms show these or those resemblances and differences.

The intra-LITU/inter-LITU distinction

As I understand it, the concept of "least inclusive taxonomic unit" (LITU; Pleijel and Rouse 2000: 629) implies a non-trivial epistemological distinction between the intra-LITU level and the inter-LITU level. Indeed, any LITU is a taxon for which we cannot define sub-taxa within it. Consequently, While it is only a question of reversible variations (i.e. of allelic frequency changes) at the intra-LITU level, it is also (and mainly) a question of irreversible modifications (i.e. of apparitions and fixations of new properties) at the inter-LITU level. In other words, the explanation of the diversity of life by a filiation process needs an evolutionary theory at the inter-LITU level, contrary to the intra-LITU level.

The principle of phylogenetics

The primary task of phylogenetics is to account the diversity of life at the inter-LITU level. Phylogenetics is necessary because if evolutionary theory – seen as a theory about a certain process of filiation (by reproduction), heredity, and irreversible modifications – helps to explain, in general, the existence of this diversity, it cannot conversely explain why this or that LITU satisfies this or that structural property.

Given the general explanatory principle of life's diversity seen above, phylogenetics consists of elaborating hypotheses of kinship relationships. However, the intra-LITU/inter-LITU distinction also has its consequences on the involved kinship relationships. Indeed, given that no evolutionary theory is needed at the intra-LITU level, it follows that the kinship relationships can be, at this level, reduced to a set of disconnected organism genealogies, i.e. each LITU has its proper genealogy of organisms. Conversely, such a reduction is impossible at the inter-LITU level because (a) the kinship relationships between organisms have to be extended to LITUs themselves, and (b) the kinship relationships between organisms of a same LITU are trivial at the inter-LITU level. Hence, the inter-LITU level needs a concept of kinship relationships having as basic objects not the organisms but the LITUs, and therefore is clearly distinct from the concept of organism genealogy. Such a concept is then nothing more than that of *phylogeny*.

Phylogeny and phylogenetic hypotheses

Let A be a finite non-empty set of LITUs having a same origin, i.e. such that for all LITUs X and Y of A and for all organisms x of X and y of Y, x and y have a same origin (i.e. there is at least one organism which is the ancestor of x and y). In this case, the phylogeny of A is the set of the kinship relationships between the LITUs of A.

This definition leads to the following remarks:

- Given that these LITUs exhibit resemblances and differences (*via* their respective organisms), their common origin combined with such relationships imply that the latter are necessarily the product of an evolutionary process, i.e. a process that involves reproduction, heredity, and irreversible modifications.
- It is possible to extend the above definition of a phylogeny to any kind of TUs. However, such a thing needs to consider only TUs having mutually exclusive extensions. In other words, if we denote "e[x]" the extension of any taxon x, then, for a given finite non-empty set A of TUs having a same origin and such that $e[x] \cap e[y] = \varnothing$ for all x and y of A, the phylogeny of A is the set of the kinship relationships between the TUs of A.
- In the context of current biology (and ever since Darwin), it is understood and accepted that all living beings have a common origin. Thus there is a single phylogeny for all LITUs; there is but one phylogeny of life.

Despite the importance of phylogeny for understanding the inter-LITU diversity, its knowledge can only be indirect. Indeed, any kinship relationship is abstract by nature and therefore not observable. Thus, even if knowledge of phylogeny has an empirical base (the organisms and their parts), this is also a matter of abstraction (the structural and behavioral properties involved in characters), inference (from these characters), and representation (of both these characters and the result of these inferences).

Usually, kinship relationships resulting from phylogenetic inferences are synthetically expressed in *phylogenetic hypotheses*, i.e. more or less formal representations of whole or part of the phylogeny of life. Phylogeny being a set of relationships, it follows that *phylogenetic hypotheses are fundamentally relational hypotheses.* In other words, any phylogenetic hypothesis basically involves a relation in the mathematical sense (i.e. a set of k-tuples). But as a set, any relation can also be viewed as the extension of a given relational property (within a Cartesian product). Hence, this formal approach immediately takes account of the fact that any phylogenetic hypothesis involves a concept of kinship.

In summary, two things remain to be examined: (1) what concepts of kinship are mobilized in phylogenetics; and (2) how these concepts precisely constrain the formal structure of phylogenetic hypotheses. After that, and only after that, will it be possible to see how a phylogenetic hypothesis can also be formalized as a classification.

18.4 Phylogenetics hypotheses: from relations to classifications

Preliminaries

On the concepts of kinship in phylogenetics

By "kinship," I generally mean a relationship between biological entities resulting from a historical process of filiation (i.e. a process where organisms produce new organisms by reproduction). A *concept of kinship* then corresponds to relationships of kinship of the same kind, i.e. having the same general properties. In particular, all instances of a given concept of kinship must have the same relational properties.

In phylogenetics, it seems that there are only a few of them. For example, Nelson (1970: 377, 1973: 344–345) considers that there are only two: the ancestor–descendant relationship and the degree of kinship relationship. In addition, he has, like Hennig (1965: 97), defined the second item in terms of the first.[11] As we will see in section 18.6, this dependency is then a possible source of difficulties when these concepts are restricted to species, i.e. when they involve ancestral species.

In one certain sense, there is no need to continue if "concepts of kinship" means only those that are non-trivial. However, one of the related topics addressed in the section 18.5 is the introduction of a third concept of kinship. Interestingly, one that was suggested by Hennig (1965: 97–98) in a very implicit, and even elliptic, way.

---

[11] "Degree of kinship relationship" is called "relationship of common ancestry" by Nelson (1970, 1973) and "phylogenetic relationship" by Hennig (1965).

Indeed, Hennig considered the degree of kinship relationship as a "relative" concept and as such does not consist of statements that at least one species is related to at least one other. But "relative" should not be confused with "relational." Indeed, "a is related to b" is also a relational judgment. The difference with degree of kinship relationship judgments (e.g. "a is related to b compared to c") comes from the fact that no third term is specified (even in an implicit way).

In summary, (at least) three concepts of kinship are involved in phylogenetics:

- *The simple kinship relationship* (SKR). It is a binary relational concept used to say that a biological entity (i.e. an organism, a species, a LITU, a TU...) x is related to a second y (because they have a common origin).
- *The ancestor–descendant relationship* (A–D relationship, ADR). It is a binary relational concept used to say that a biological entity x gave rise to a second y (alone or not, directly or not, regardless).
- *The degree of kinship relationship* (DKR). It is a ternary relational concept used to say that a biological entity x is related with a second y compared to a third z (because x and y have a common and exclusive origin compared to z).

On two fundamental mathematical equivalences

For our purposes, two very particular documented mathematical equivalences, each of them being characterized by a single (pair of reverse) bijection(s), will be very useful: (a) that between equivalence relations and partitions; and (b) that between what I call degree of equivalence relations and hierarchical classifications. Among them, the first is well-know and generally introduced in basic books on set theory (e.g. Suppes 1960, Stoll 1963). And the second, it has been established first by Colonius and Schulze (1981)[12] for finite sets and then extended by McMorris and Powers (2003) for infinite sets.[13]

Phylogenetic hypotheses as relational structures

The formalization of phylogenetic hypotheses as relational structures is relatively simple, even obvious, once the relational view of phylogenetics is admitted. Indeed, it can be summarized into the following five-step reasoning:

1. Formulate each k-ary concept of kinship r by a k-adic predicate "ρ". For the three concepts of kinship seen above, this gives:

---

[12] Colonius and Schulze (1981) talk about hierarchical n-trees, not about hierarchical classifications, but these two kinds of structures are also mathematically equivalent, which makes it less of an issue.

[13] It is interesting to note that before these mathematical papers, which come from areas out of biology (respectively psychology and social sciences), this equivalence had already been explored, albeit with a non-formal approach, in biological systematics (Nelson 1972, 1973).

SKR	"Being related to"	"σ"
ADR	"Being an ancestor of"	"α"
DKR	"Being related with compared to"	"δ"

2. Associate each k-adic predicate "ρ" to k individual variables x_1, x_2, ..., and x_k, in order to obtain a k-adic propositional function "ρ(x_1, x_2, ..., x_k)":

SKR	"x is related to y"	"σ(x, y)"
ADR	"x is an ancestor of y"	"α(x, y)"
DKR	"x is related with y compared to z"	"δ(x, y, z)"

3. Establish the relational properties of r, and formally express them with the corresponding propositional function (see below).

4. Apply r in a finite non-empty referential set[14] A of TUs, i.e. search the k-tuples $\langle x_1, x_2, ..., x_k \rangle$ of A^k for which "ρ(x_1, x_2, ..., x_k)" gives a true proposition, and this in order to define a finite k-ary relation R_ρ on A, i.e.:

SKR	$R_\sigma = \{\langle x, y \rangle \in A^2 : \sigma(x, y)\}$
ADR	$R_\alpha = \{\langle x, y \rangle \in A^2 : \alpha(x, y)\}$
DKR	$R_\delta = \{\langle x, y, z \rangle \in A^3 : \delta(x, y, z)\}$

5. Finally, given that A is not always deductible from R_ρ, it follows that **any phylogenetic hypothesis is**, from a formal and structural viewpoint, **a finite simple k-relational structure** $R_\rho = \langle A, R_\rho \rangle$, i.e.:

SKR	$R_\sigma = \langle A, R_\sigma \rangle$
ADR	$R_\alpha = \langle A, R_\alpha \rangle$
DKR	$R_\delta = \langle A, R_\delta \rangle$

Conclusion. Mathematically speaking, any phylogenetic hypothesis is an ordered pair $\langle A, R \rangle$ such that (a) A is a finite non-empty set of TUs, and (b) R is both a k-ary relation on A (i.e. $R \subseteq A^k$) and the extension (into A^k) of a given k-ary concept of kinship.[15]

---

[14] In biological systematics, it is only question of finite sets, even if some of them can be very large.

[15] Strictly speaking, a complete phylogenetic hypothesis not only consists into a relational structure, but also to the empirical arguments justifying such a structure. Note that this corresponds to Hennig's (1965: 98 and 105–106) distinction between what he calls a "phylogeny tree/dendrogram" and an "argumentation plan."

Relational properties of the Concepts of Kinship

Remarkably, the previous formalization implies that the properties of r entirely constrain those of R_ρ. Thus, there is a strict correlation between r and the type of relational structure instantiated by R_ρ. Below are the details.

Simple kinship relationship

The SKR satisfies the relational properties of reflexivity, symmetry, and transitivity. In other words (and respectively):

- Any TU x of A is related with itself, i.e.:

$$(\forall x \in A)\, \sigma(x, x)$$

- For all TUs x and y of A, if x is related to y, then y is related to x, i.e.:

$$(\forall x, y \in A)\, [\sigma(x, y) \rightarrow \sigma(y, x)]$$

- For all TUs x, y, and z of A, if x is related to y and y is related to z, then x is related to z, i.e.:

$$(\forall x, y, z \in A)\, [[\sigma(x, y) \wedge \sigma(y, z)] \rightarrow \sigma(x, z)]$$

Given that $\langle x, y \rangle \in R_\sigma$ if and only if $\sigma(x, y)$ for all x and y of A (comprehensive definition of R_σ from "$\sigma(x, y)$"), it thus (and respectively) follows for all x, y, and z of A that:

$$\langle x, x \rangle \in R_\sigma$$
$$\langle x, y \rangle \in R_\sigma \rightarrow \langle y, x \rangle \in R_\sigma$$
$$[\langle x, y \rangle \in R_\sigma \wedge \langle y, z \rangle \in R_\sigma] \rightarrow \langle x, z \rangle \in R_\sigma$$

Consequently, R_σ is a (finite and binary) relation of equivalence on A, and so R_σ **is a (finite) relational structure of equivalence**.

Ancestor-descendant relationship

The ADR relationship satisfies the relational properties of anti-reflexivity and transitivity (which implies asymmetry and so, anti-symmetry too). In other words (and respectively):

- No TU x of A is its own ancestor, i.e.:

$$(\forall x \in A)\, \neg\, \alpha(x, x)$$

- For all TUs x, y, and z of A, if x is an ancestor of y and y is an ancestor of z, then x is an ancestor of z, i.e.:

$$(\forall x, y, z \in A)\, [[\alpha(x, y) \wedge \alpha(y, z)] \rightarrow \alpha(x, z)]$$

Given that $\langle x, y \rangle \in R_\alpha$ if and only if $\alpha(x, y)$ for all x and y of A (comprehensive definition of R_α from "$\alpha(x, y)$"), it thus (and respectively) follows for all x, y, and z of A that:

$$\langle x, x \rangle \notin R_\alpha$$

$$[\langle x, y \rangle \in R_\alpha \wedge \langle y, z \rangle \in R_\alpha] \rightarrow \langle x, z \rangle \in R_\alpha$$

Consequently, R_α is a (finite and binary) relation of strict order on A, and so R_α **is a (finite relational structure of) strict order**.[16]

Degree of kinship relationship

Contrary to the two previous concepts of kinship, establishing the properties of the DKR is not so easy. Of course, some of its properties are relatively obvious. For example, it is not hard to see that this relationship is 1–2 symmetric, 1–2 restrictively reflexive, and 1–3 and 2–3 anti-reflexive. In other words (and respectively):

- For all x, y, and z of A, if x is related with y compared to z, then y is related with x compared to z, i.e.:

$$(\forall x, y, z \in A)\, [\delta(x, y, z) \rightarrow \delta(y, x, z)]$$

- For all x and y, if x is distinct from y, then x is related with itself compared to y, i.e.:

$$(\forall x, y \in A)\, [x \neq y \rightarrow \delta(x, x, y)]^{17}$$

- For all x, y, and z of A, if x is related with y compared to z, then x and y are both distinct from z, i.e.:

$$(\forall x, y, z \in A)\, [\delta(x, y, z) \rightarrow (x \neq z \wedge y \neq z)]^{18}$$

Likewise the biological literature can also help us. Indeed, the latter provides at least two other properties: 1–2 transitivity (Nelson and Ladiges 1992: 490), and 1–3 and 2–3 asymmetry (Kluge 2002: 589, Rieppel 2005: 15).[19] Respectively, this gives:

- For all w, x, y, and z of A, if x is related with y compared to w, and y is related with z compared to w, then x is related with z compared to w, i.e.:

$$(\forall w, x, y, z \in A)\, [[\delta(x, y, w) \wedge \delta(y, z, w)] \rightarrow \delta(x, z, w)]$$

- For all x, y, and z of A, if x is related with y compared to z, then z is not related with y compared to x and x is not related with z compared to y, i.e.:

$$(\forall x, y, z \in A)\, [\delta(x, y, z) \rightarrow \neg[\delta(z, y, x) \vee \delta(x, z, y)]]$$

---

[16] In contrast, Estabrook (1997: 3) considers the ADR as reflexive (i.e. any TU is its own ancestor), which implies, once combined with anti-symmetry and transitivity, a (relational structure of) non-strict order.

[17] 1–2 restricted reflexivity differs from 1–2 reflexivity in the sense that the latter works for all x and y of A.

[18] This is equivalent to say that $\neg\delta(x, y, x)$ and $\neg\delta(x, y, y)$ for all x and y of A.

[19] These authors do not use this terminology.

Unfortunately, this second approach has the same weakness as the first. Indeed, in both cases, even if they were combined, there is no guarantee that these properties are mutually independent, each necessary, and together sufficient.

An axiomatic approach is in fact necessary. It is assumed that this relationship can be formalized as a ternary relation (on a set) of a very particular kind that has axiomatically been studied both by Colonius and Schulze (1981) and McMorris and Powers (2003). What supports this idea are the following elements: (1) the informal definition given by Colonius and Schulze (1981: 170) also consists of saying (though in psychological terms, not phylogenetic) that, given three objects, two are closer to each other than to the third; (2) Colonius and Shulze (1981) and McMorris and Powers (2003), respectively, established their link with hierarchical n-trees and hierarchical classifications; but these mathematical structures respectively correspond to the usual graph-theoretic and classificatory formalizations of DKR phylogenetic hypotheses; and (3) all of the properties of the DKR seen above can be deduced (as axiomatic or derived properties) from the axiomatic definitions proposed by these authors.

Now I will label as a "(finite) *relation of degree of equivalence* (on A)" any (finite) ternary relation (on A) of this very particular kind. In corollary, I will also label as a "(finite) *degree of equivalence relational structure*" its corresponding simple ternary relational structure (i.e. any simple relational structure $\langle A, T \rangle$ for which RT is a degree of equivalence relation on A). Following the work of the previous authors on these relations, I now propose a third axiomatic definition. Let $R = \langle A, T \rangle$, be an ordered pair. In this case, R is a (finite) degree of equivalence relational structure if and only if (1) A is a (finite) non-empty set, and (2) T is a (finite) degree of equivalence relation on A, i.e. a (finite) ternary relation on A (i.e. $T \subseteq A^3$) such as for all w, x, y, and z of A:

- $\langle x, y, z \rangle \in T \rightarrow \langle y, x, z \rangle \in T$ 1–2 symmetry
- $\langle x, y, z \rangle \in T \rightarrow \langle x, z, y \rangle \notin T$ 2–3 asymmetry
- $x \neq y \rightarrow \langle x, x, y \rangle \in T$ 1–2 restricted reflexivity
- $[\langle x, y, w \rangle \in T \wedge \langle y, z, w \rangle \in T] \rightarrow \langle x, z, w \rangle \in T$ 1–2 transitivity
- $\langle w, x, z \rangle \in T \rightarrow [\langle w, x, y \rangle \in T \vee \langle w, y, z \rangle \in T]$ 2–3 negative transitivity

Beyond this axiomatic approach, the most important point regarding these relations is that Colonius and Schulze (1981) and McMorris and Powers (2003) both showed a mathematical equivalence (i.e. a particular bijection) between them and hierarchical classifications. In other words, just as any equivalence relation determines and is determined by a single partition (and conversely), **any degree of equivalence relation determines and is determined by a single hierarchical**

classification (and conversely). I propose a third proof of it for finite sets in the second appendix below.

Consequences in terms of classification

Simple and degree of kinship relationships

It is now very easy to see in what SKR and DKR phylogenetic hypotheses are also classificatory structures. Indeed:

- On the one hand, each SKR phylogenetic hypothesis s on A can be formalized as an equivalence relational structure $\langle A, R \rangle$ (where R is an equivalence relation on A). But (the) equivalence relations (on A) are mathematically equivalent to (the) partitions (of A) (i.e. connected with them by a particular pair of reverse bijections). Consequently, s can also be formalized as a partitioned classificatory structure, i.e. a classificatory structure $\langle A, P \rangle$ such that P is a partition of A (namely, the partition determining and determined by R).
- On the other hand, each DKR phylogenetic hypothesis d on A can be formalized as a degree of equivalence relational structure $\langle A, T \rangle$ (where T is a degree of equivalence relation on A). But (the) degree of equivalence relations (on A) are mathematically equivalent to (the) hierarchical classifications (of A) (i.e. connected with them by a particular pair of reverse bijections). Consequently, d can also be formalized as a hierarchical classificatory structure, i.e. a classificatory structure $\langle A, H \rangle$ such that H is a hierarchical classification of A (namely, the hierarchical classification determining and determined by T).

Ancestor-descendant Relationship

I did not find any documented bijection theorem connecting (strict) order relations with classifications. However, it is in fact possible to establish an analogous one between (strict) order relations and what I call "ordered classifications."[20] The key concept here is that of the *upper set*. Let R be a relation of strict order on a finite non-empty set A. Then, let $a \in A$. In this case, the R-relative upper set of a is the set $[a^{\uparrow}]_R$ of the elements x of A such that x is either strictly upper than a in R or identical to a. In symbols, this gives:

$$[a^{\uparrow}]_R = \{x \in A: \langle a, x \rangle \in R \vee a = x\}$$

This set being defined, we can then define the R-relative *descendant set* of A as the set of all upper sets induced by R, i.e. the set A^*R of all sets of kind $[x_{\uparrow}]$ such that $x \in A$. Formally:

---

[20] I do not exclude that such equivalence has already been proposed and that it has been missed because of terminological issues.

$$A^*R = \{[x^\uparrow]_R : x \in A\}$$

i.e.

$$A^*R = \{X \subseteq A : (\exists x \in A)\, X = [x^\uparrow]_R\}$$

On this basis, not only we can show that A^*R is necessarily a classification but also a classification of a very particular kind. Such classifications, which will be called *ordered classifications* for the next, indeed satisfy three very particular properties. Let $\langle A, O \rangle$ be a classificatory structure. In this case, $\langle A, O \rangle$ is also ordered if and only if O is an ordered classification of A, i.e. if and only if: (a) there is at least one $x \in A$ such that $\{x\} \in O$; (b) for any $X \in O$, (i) $|X| \geq 2$ only if there is at least one $Y \in O$ such that $Y \subset X$, and (ii) there is always one and only one $x \in X$ that belongs to none of the clusters of O properly included into X; and (c) for all X and Y of O, if X and Y overlap, then for any $x \in X \cap Y$ there is at least one $Z \in O$ such that $x \in Z$ and $Z \subseteq X \cap Y$.

Conversely, it is also possible to show that any ordered classification O of A determines a single relation of strict order R_O on A such that for all x and y of A, x is strictly lower than y in R_O if and only if there is at least one Y in O such that (a) y is in Y but not x, and (b) for any X of O, x is in X only if X properly includes Y. This second main determination being defined, it now remains to show three things. First, that the relation of strict order R_{A^*R} determined by the ordered classification A^*R of A is identical to the relation of strict order R on A that determines A^*R (as the R-relative descendant set of A). Then, that the ordered classification A^*R_O of A determined by the relation of strict order R_O on A is identical to the ordered classification O of A that determines R_O. Finally, that for any relation of strict order R on A and any ordered classification O of A, $R = R_O$ if and only if $A^*R = O$. I propose a proof for these five statements in the first appendix of this paper.

18.5 Concerning some consequences and related topics

On the methodology of phylogenetics

The fact that equivalence, strict order, and degree of equivalence relations are respectively mathematically equivalent with partitions, ordered classifications, and hierarchical classifications is interesting (and actually decisive) for the methodological perspective. Indeed, it allows us to establish relationships of kinship by grouping into clusters and consequently, from structural or behavioral properties. But precisely, such properties correspond to generalizations of parts of organisms (i.e. of "parts" of TUs at a more abstract level), which implies that beforehand these parts are compared to other parts. Consequently, these concepts of kinship allow a comparative procedure, i.e. a procedure where the studied organisms/TUs are compared at the level of their parts.[21] However, this does not mean that these concepts are methodologically equivalent. Indeed:

- For the SKR, the procedure will consist of comparing at least two distinct TUs and then to search if at least two (possibly all) of them have parts that can be considered the "same thing" (because they satisfy at least one common property).
- For the DKR, it will imply comparing of at least three distinct TUs and then, to search if at least two (but not all) of them have parts that can be considered as the "same thing" compared to all possible parts of the other TUs (because they satisfy at least one common property that the other ones do not satisfy).
- As for the ADR, this will consist in comparing at least two distinct TUs and then, to search if among n TUs having parts considered the "same thing" (because they satisfy at least one common property), (a) $n - 1$ of them also have parts that can be considered the "same thing" compared to all parts of the nth remaining TU (because they satisfy at least one property that none part of the nth TU satisfies), (b) there is no part in the nth TU which satisfies an exclusive property, i.e. any property satisfied by at least one part of the nth TU is also satisfied by at least one part in each of the $n - 1$ TUs, and (c) the origin of the nth TU is older than those of the n - 1 TUs.[22]

Simple kinship relationship and non-polarized characters

Remarkably, the fact that the SKR leads to (actually unordered) partitioned phylogenetic hypotheses implies an undesirable consequence on what we usually call "characters" in phylogenetic analyses. Indeed:

- On the one hand, and as we have already seen, it is currently accepted that life is monophyletic, i.e. all (past and present) living beings are related. For any referential set A of TUs and for all x and y of A, "$\sigma(x, y)$" is therefore always true. Given (1) the definition of R_σ from "$\sigma(x, y)$", and (2) the fact that R_σ determines (and is determined by) a single partition P_σ of A, this implies (a) that $R_\sigma = A^2$, and (b) a single possible phylogenetic partition of A, i.e. $P_\sigma = \{A\}$. Thus, the SKR concept is trivial: for any set A of TUs, the only thing that it is able to say is that A is a set of related TUs.

---

[21] Hence the importance of comparative biology in phylogenetics.

[22] Note that among these three methodologies, the third is the only one that cannot be reduced to comparative biology (because it also incorporates temporal consideration on TUs' age).

- On the other hand, transformation series constitute the most traditional evolutionary interpretation of characters in phylogenetics. According to the latter (Hennig 1965, Farris et al. 1970: 172), any character K is both a set of states and an evolutionary unit such that: (1) it always evolves from a single initial ancestral state; (2) some of its states gave birth, by a transformation process, to one or several other states; (3) each of its states can be attributed to (i.e. can describe) one or several TUs; and (4) each ancestral state is always separated from its descendant states. Given a set A of TUs, it follows from (2), (3), and (4) that if any TU of A can be described by at least one state of K, then K defines an ordered partition on A. From this, K then becomes an unordered partition when the sense, i.e. the polarity, of the transformations is not assumed.

Hence, we are left with the following question: if all SKR phylogenetic hypotheses are trivial (which implies unordered partitions into a single cluster) and if non-polarized characters generally correspond to unordered partitions of any kind (which implies most of the time unordered partitions in several clusters), how are such characters phylogenetic? In other words, how can we infer phylogenetic hypotheses from them?

Fully ramified undirected n-*trees and quaternary relations*

Fully ramified undirected n-trees are other mathematical structures commonly encountered in phylogenetics. Indeed, they often constitute the primary (and sometimes the only) result of tree-oriented phylogenetic analyses.[23] Yet these graph-theoretic structures are problematic as well. First of all, their historical interpretation is, at best, difficult. Indeed, while evolution, and therefore phylogeny, implies direction, these graphs have no particular way of reading, given the symmetry of their branches.

However, the formalization of phylogenetic hypotheses as relational structures highlights a more fundamental problem. Indeed, if any phylogenetic hypothesis involves a concept of kinship, then any fully ramified undirected n-tree, once interpreted as a phylogenetic hypothesis, has to necessarily express a relationship of kinship. Remarkably, such a thing is mathematically possible since these trees are in bijection with particular simple quaternary relational structures, i.e. particular quaternary relations on a set (Colonius and Schulze 1981, Bandelt and Dress

---

[23] Interestingly, such a fact is strongly related with the formalization of characters as unordered partitions. Indeed, tree-oriented phylogenetic analyses lead to fully ramified undirected n-trees when the used characters are not polarized, i.e. when they are, formally speaking, unordered partitions.

1986, McMorris and Powers 2003).[24] Unfortunately, the phylogenetic sense of these relations is problematical. Indeed, either they are relations of separation, and their phylogenetic meaning remains trouble, or they are relations of grouping, and then, they become redundant with degree of equivalence relations. Given that degrees of equivalence relations are not in bijection with fully ramified undirected n-trees[25], it follows that this second interpretation is false. As a consequence, only the first interpretation is possible. Thus, the phylogenetic meaning of these trees remains obscure.

18.6 On the degree of kinship relationship

There is a curious paradox in current phylogenetics. Indeed, though the traditional view of evolution implies true ancestral species (and more generally, true non-inclusive ancestral taxa of any kind), only "hypothetical ancestors" are proposed in most of contemporary phylogenetic hypotheses. For the specific case of the tree-like phylogenetic hypotheses, this even that means the latter are actually inter-TU DKR phylogenetic hypotheses, and not inter-TU ADR phylogenetic hypotheses. Should we then conclude that these hypotheses are not really phylogenetic or, at best, incomplete? In fact, it depends of what is understood by "evolution." To better understand it, consider two aspects of this problem.

The first concerns dendrograms, the usual dendritic representations of degree of kinship phylogenetic hypotheses. As seen already, degree of kinship phylogenetic hypotheses can also be formalized as hierarchical classificatory structures (because they are fundamentally degree of equivalence relational structures). But any hierarchical classification induces an inclusive order, given the properties of the set-theoretic inclusion relationship (namely anti-symmetry and transitivity). Consequently, any DKR phylogenetic hypothesis also corresponds to an inclusive order of a very particular kind (such orders can be called "fully ramified hierarchical orders" because of their defining properties: a single origin, no reticulation, and a complete divergent ramification). Remarkably, the "Hasse diagram" (i.e. the arrow diagram[26]) of the (reflexive and transitive) reduction of any order of this kind

---

[24]   This type of quaternary relation was first introduced, although implicitly, by Dobson (1974).

[25]   For example, whether/if/while any set compounded of three distinct elements implies only one possible fully ramified undirected n-tree (a star), it conversely implies four degrees of equivalence relations (one of them being the trivial one).

[26]   By "arrow diagram", I mean the following thing: given a finite binary relational structure $\langle A, R \rangle$, its arrow diagram corresponds to a graphic representation where the elements of A and R are respectively represented by dots and arrowed lines and such that for all x and y of A, an arrowed line

always corresponds to a dendrogram (because such a reduction actually corresponds to a hierarchical n-tree for which the n leaves correspond to the n TUs under the study).[27] Hence the problem. Indeed, though the nodes of these dendrograms fundamentally correspond to (the extension of) the inclusive taxa (i.e. the inclusive clades) highlighted by the degree of kinship phylogenetic analyses, we traditionally interpret them as ancestral species (and more generally as non-inclusive ancestral taxa of any kind). In other words, we interpret these dendrograms as if they were genealogies of species (which is very seductive, given that fully ramified tree-like species genealogies are isomorphic with DKR phylogenetic hypotheses when the latter are viewed as inclusive orders). But such ancestral species/non-inclusive taxa are purely supernumerary in respect to the (degree of kinship) analysis. Indeed, not only such additional taxa are not defined before it (contrary to the non-ancestral species or, if you prefer, the TUs) but, in addition, they are not highlighted/argued by it (contrary to the inclusive clades). Remarkably, this is directly correlated with the fact that there is no specimen for representing/instantiating them (hence their "hypothetical" status). In short, these ancestral species/non-inclusive taxa are generated after the analysis only for incorporating these dendrograms into the traditional view of evolution, i.e. for interpreting them as species genealogies.

The second aspect of this problem is correlated to the previous one and concerns the connection between the DKR and ADR. As I defined it, the DKR is indeed strongly connected with the ADR, i.e. saying that x and y are related compared to z is equivalent to say that x and y have a common ancestor that z has not. However, this definition says nothing about the nature of this ancestor. Conversely, if we consider the more precise definition of Hennig (1965: 97) in terms of ancestral species, then these exclusive common ancestors cannot be something other than species. Hence the problem (again). Indeed, the inter-TU DKR becomes less informational than (and finally redundant with) the inter-TU ADR when by "x and y are related compared to z", we actually mean that "w is an ancestor of x

connects x to y if and only if $\langle x, y \rangle \in R$. On this basis, a *Hasse Diagram* is then just a simplified arrow diagram of a finite order $\langle A, R \rangle$ where only the immediate links between distinct elements of A are depicted (if $\langle x, y \rangle \in R$ and $\langle y, z \rangle \in R$, then the link from x to z is not represented, although it also implies $\langle x, z \rangle \in R$) by non-arrowed lines (which implies an implicit sense of reading, generally from the bottom to the top). In other words, the Hasse diagram of $\langle A, R \rangle$ corresponds to the arrow diagram of the (reflexive and) transitive reduction of $\langle A, R \rangle$, i.e. of the greatest anti-reflexive and anti-transitive binary relation on A included in R, which is also known as the "covering relation" of R (Davey and Priestley 2002: 11–12, Roman 2008: 4 5).

[27] Note that the interpretation of cladograms (and more generally dendrograms) as Hasse diagrams has already been documented by Minaka (1993).

and y but not of z" (where w, x, y, and z are TUs). In other words, if we can deduce "$\delta(x, y, z)$ / $\delta(y, x, z)$" from "$\alpha(w, x)$", "$\alpha(w, y)$", and "$\neg\alpha(w, z)$", the converse is false. Under the traditional view of evolution, inter-TU ADR phylogenetic hypotheses are therefore always more informative than inter-TU DKR phylogenetic hypotheses (Dayrat 2005: 348).

In summary, the DKR is either (at best) an incomplete (when we restrict it to non-ancestral TUs), or (at worst) an inadequate (when we do not make such a restriction) concept of kinship when we consider evolution as a process of species transformation (since this last implies that phylogeny is a genealogy of species). On this point we could then follow Hull (1979), Williams (1992), and Dayrat (2005) in their criticism of cladistic phylogenetics (and more generally of most of the contemporary tree-oriented phylogenetics). Indeed, cladistic phylogenetics (i.e. most of the contemporary tree-oriented phylogenetic analyses) primarily consists to establish the DKR among TUs and not their ADR. Nevertheless, this criticism is valid only within the traditional view of evolution (and more generally, within any view of evolution where it is question of ancestral TUs). And do not question this view is somewhat equivalent to perform a *non sequitur*. Indeed, what this problem actually implies is the following alternative:

- Either evolution is really a process of species transformation. In this case, phylogeny is nothing other than a genealogy of species. Given that this implies true ancestral species, it follows that the search for the DKR without identifying ancestors among TUs (as in cladistic phylogenetics for example) is (at best) instrumentalist.
- Or the search for DKR without identifying ancestors among TUs is both complete and realistic (i.e. in conformity with the result of the evolutionary process). In this case, no ancestral TU is possible. But evolution implies true ancestral taxa. Consequently, the true ancestral taxa are nothing but the inclusive clades highlighted by the analysis, and so evolution is not a process of species transformation.

In other words, there are actually at least two views/interpretations (or models in the "pattern" sense) of evolution and so of phylogenetics. On the one hand, we can consider evolution as a process of transformation and succession of transitory species. In this case,

1a. phylogeny is nothing but a genealogy of species (these last being the only possible real taxa), and
1b. phylogenetics consists of establishing the ADR between species, which implies identifying (true) ancestral species.

On the other hand, we can consider evolution as a process of differentiation of new clades into older and persistent clades. Viewed this way,

2a. phylogeny is nothing but a genealogy of clades (these last being the only possible real taxa), and

2b phylogenetics consists of establishing the DKR between TUs (seen as non-inclusive clades), which then implies a (fully ramified tree-like) ADR between (non-inclusive and inclusive) clades.[28]

18.7 Conclusion

The phylogeny/classification relation problem has initially been stated as a problem of priority and dependency. In other words, it treated as distinct phylogenetic hypotheses and taxonomic classifications. Hence the question to know in what way can taxonomic classifications be *theoretically* interesting and/or relevant for phylogenetics and evolutionary biology. In contrast, I propose that the phylogeny/classification link is a problem of mathematical equivalency. On this basis, this approach shows that at least under certain concepts of kinship, phylogenetic hypotheses are not only relational structures but also classificatory structures. Leaving aside the question of what taxonomic classifications *actually* are (just a matter of taxa – in particular of clades – or also a matter of ranks, names, and – why not – of taxonomic descriptions?), it is therefore well intrinsically question of classification in phylogenetics (i.e. at least under these concepts of kinship, phylogenetic methods are methods of classifications as well as methods of linking). However, the most obvious non-trivial classificatory phylogenetic hypotheses, the DKR phylogenetic hypotheses, imply a particular problem concerning ancestral taxa, since they never contain ancestral TUs. By accepting the idea that any true phylogenetic hypothesis has to contain identified ancestral taxa (i.e. taxa for which we have specimens; either taxa defined before the analysis or taxa constructed from the latter.), it follows that we have to either reject (the realistic interpretation of) most of the contemporary tree-oriented phylogenetics (and especially those of cladistic phylogenetics), or change our view on phylogeny (and therefore the evolutionary process too).

---

[28] In spite of its heterodoxy, this second model is not really a novelty. Indeed, it has already been argued by Løvtrup (1988: 23–27), Nelson (1989), and more recently, Zaragüeta Bagils (2011: 50–54). Besides, it allows us to eliminate paraphyly (*sensu* Hennig 1965, 1966) from ancestral taxa, contrary to the first model.

References

Ashlock, P.D. (1979). An evolutionary systematist's view of classification. *Systematic Zoology*, 28, 441–450.

Bandelt, H-J. and Dress, A. (1986). Reconstructing the shape of a tree from observed dissimilarity data. *Advances in Applied Mathematics*, 7, 309–343.

Beatty, J. (1982). Classes and cladists. *Systematic Zoology*, 31, 25–34.

Buck, R.C. and Hull, D.L. (1966). The logical structure of the Linnaean hierarchy. *Systematic Zoology*, 15, 97–111.

Colonius, H. and Schulze, H.H. (1981). Tree structures for proximity data. *British Journal of Mathematical and Statistical Psychology*, 34, 167–180.

Darwin, C. (1859). *On the Origin of Species*. London: John Murray.

Davey, B.A. and Priestley, H.A. (2002). *Introduction to Lattices and Order*, 2nd edition. Cambridge: Cambridge University Press.

Dayrat, B. (2005). Ancestor–descendant relationship and the reconstruction of the Tree of Life. *Paleobiology*, 31, 347–353.

Dobson, A.J. (1974). Unrooted trees for numerical taxonomy. *Journal of Applied Probability*, 11, 32–42.

Estabrook, G.F. (1997). Ancestor–descendant relations and incompatible data. In *Mathematical Hierarchies and Biology*, ed. B. Mirkin, F.R. McMorris, F.S. Robert and A. Rzhetsky. DIMACS series in discrete mathematics and theoretical computer sciences, 37, 1–28. Providence, RI: American Mathematical Society.

Farris, J.S. (1979). The information content of the phylogenetic system. *Systematic Zoology*, 28, 483–519.

Farris, J.S., Kluge, A.G. and Eckardt, M.J. (1970). A numerical approach to phylogenetic systematics. *Systematic Zoology*, 19, 172—189.

Felsenstein, J. 2004. *Inferring Phylogenies*. Sunderland, MA: Sinauer Associates.

Hennig, W. (1965). Phylogenetic systematics. *Annual Review of Entomology*, 10, 97–116.

Hennig, W. (1966). *Phylogenetic Systematics*. Urbana, IL: University of Illinois Press.

Hennig, W. (1975). "Cladistic analysis or cladistic classification?" A reply to Ernst Mayr. *Systematic Zoology*, 24, 244–256.

Hull, D.L. (1970). Contemporaries systematic philosophies. *Annual Review of Ecology and Systematics*, 1, 19–54.

Hull, D.L. (1979). The limits of cladism. *Systematic Zoology*, 28, 416–440.

Hull, D.L. (1986). Les fondements philosophiques de la classification biologique. In *L'Ordre et la Diversité du Vivant*, ed. P. Tassy. Paris: Fayard, pp. 161–203.

Hull, D.L. (2001). The role of theories in biological systematics. *Studies in History and Philosophy of Sciences Part C: Biological and Biomedical Sciences*, 32, 221–238.

Jardine, N. (1969). A logical basis for biological classification. *Systematic Zoology*, 18, 37–52.

Kluge, A.G. (2002). Distinguishing "or" from "and" and the case for historical identification. *Cladistics*, 18, 585–593.

Lecointre, G. and Le Guyader, H. (2001). *Classification phylogénétique du vivant*, 1st edition. Paris: Belin.

Løvtrup, S. (1988). *La systématique et l'évolution de Lamarck aux théoriciens modernes*. Biosystema, 3. Paris: Société Française de Systématique.

Matile, L., Tassy, P. and Goujet, D. (2004). *Introduction à la systématique zoologique: concepts, principes, méthodes*, 2nd edition. Biosystema, 1. Paris: Société Française de Systématique.

Mayr, E. (1974). Cladistic analysis or cladistic classification? *Zeitschrift fuer Zoologische Systematik und Evolutionsforschung*, 12, 94–128.

Mayr, E. (1981). Biological classification: toward a synthesis of opposite methodologies. *Science*, 214, 510–516.

Mayr, E. (1986). La systématique évolutionniste et les quatre étapes du processus de classification. In *L'Ordre et la Diversité du Vivant*, ed. P. Tassy. Paris: Fayard, pp. 145–160.

McMorris, F.R. and Powers, R.C. (2003). The Arrovian Program from weak orders to hierarchical and tree-like relations. In *Bioconsensus*, ed. M.F. Janowitz, F.J. Lapointe, F.R. McMorris, B. Mirkin and F.S. Robert. DIMACS series in discrete mathematics and theoretical computer sciences, 61, 37–45. Providence, RI: American Mathematical Society.

Minaka, N. (1993). Cladistics and biogeographic methods: Cladograms, components, and parsimony. *Bulletin of the Biogeographical Society of Japan*, 48, 1–27.

Nelson, G. (1970). Outline of a theory of comparative biology. *Systematic Zoology*, 19, 373–384.

Nelson, G. (1971). "Cladism" as a philosophy of classification. *Systematic Zoology*, 20, 373–376.

Nelson, G. (1972). Phylogenetic relationship and classification. *Systematic Zoology*, 21, 227–231.

Nelson, G. (1973). Classification as an expression of phylogenetic relationships. *Systematic Zoology*, 22, 344–359.

Nelson, G. (1974). Darwin-Hennig classification: a reply to Ernst Mayr. *Systematic Zoology*, 23, 452–458.

Nelson, G. (1989). Species and taxa: systematics and evolution. In *Speciation and its consequences*, ed. D. Otte and J.A. Endler. Sunderland, MA: Sinauer Associates, pp. 60–81.

Nelson, G. and Ladiges, P.Y. (1992). Information content and fractional weight of three-item statements. *Systematic Biology*, 41, 490–494.

Nelson, G. and Platnick, N.Y. (1981). *Systematics and biogeography: cladistics and vicariance*. New York: Columbia University Press.

Platnick, N.I. (1979). Philosophy and the transformation of cladistics. *Systematic Zoology*, 28, 537–546.

Pleijel, F. and Rouse, G.W. (2000). Least-inclusive taxonomic unit: a new taxonomic concept for biology. *Proceeding of the Royal Society of London, series B: Biological Sciences*, 267, 627–636.

Ridley, M. (1989). Principles of classification. In *Philosophy of Biology*, ed. M. Ruse, New York: Macmillan, pp. 167–179.

Rieppel, O. (2005). Des racines de positivisme logique dans la philosophie empiriocritique de Hennig. In *Philosophie de la systématique*, ed. P. Deleporte and G. Lecointre, Biosystema, 24, Paris: Société Française de Systématique, pp. 9–22.

Roman, S. (2008). *Lattices and Ordered Sets*. New York: Springer.

Sneath, P.H.A. and Sokal, R.R. (1973). *Numerical Taxonomy. The Principles and Practice of Numerical Classification*. San Francisco, CA: W.H. Freeman and Co.

Sokal, R.R. (1975). Mayr on cladism-and his critics. *Systematic Zoology*, 24, 257–262.

Sokal, R.R. and Sneath, P.H.A. (1963). *Principles of numerical taxonomy*. San Francisco, CA: W.H. Freeman and Co.

Stoll, R.R. (1963). *Set theory and logic* (2nd ed.). San Francisco, CA: Freeman and Co.

Suppes, P. (1960). *Axiomatic set theory.* New York: Van Nostrand.

Tassy, P. (2005). Fait et théorie: Quelles connaissances de base pour la cladistique structurale? In *Philosophie de la Systématique*, ed. P. Deleporte and G. Lecointre, Biosystema, 24, Paris: Société Française de Systématique, pp. 63–74.

Williams, D.M. and Ebach, M.C. (2008). *Foundations of Systematics and Biogeography*. New York: Springer.

Williams, P.A. (1992). Confusion in cladism. *Synthese*, 91, 135–152.

Zaragüeta Bagils, R. (2011). Ancêtre. In *L'arbre du vivant existe-il?* ed. V. Malécot, N. Léger and P. Tassy, Biosystema, 28, Paris: Société Française de Systématique, pp. 41–62.

Appendix 18.1 (Strict) order relation and ordered classification

Strict order relations

Let R = $\langle A, R \rangle$ be an ordered pair. In this case, R is also a finite relational structure of strict order if and only if (1) A is a finite non-empty set, and (2) R is a strict order relation on A, i.e. a binary relation on A such that for all x, y, and z of A:

- $\langle x, x \rangle \notin R$ anti-reflexivity

- $[\langle x, y \rangle \in R \wedge \langle y, z \rangle \in R] \rightarrow \langle x, z \rangle \in R$ transitivity

Now, let S(A) be the set of all possible strict order relations on A. Given the axiomatic properties of strict order relations, it follows that any member of S(A) is asymmetric and so, anti-symmetric too. In corollary, any strict order relation $R_<$ on A determines and is determined by a single non-strict order relation $R_\leq$ on A (and conversely) as follows: for any $\{x, y\} \subseteq A$, (a) $\langle x, y \rangle \in R_\leq$ if and only if either $\langle x, y \rangle \in R_<$ or $x = y$, and (b) $\langle x, y \rangle \in R_<$ if and only if $\langle x, y \rangle \in R_\leq$ and $x \neq y$. Thus, $\langle A, R_\leq \rangle$ is not only anti-symmetric and transitive, but also reflexive. By respectively denoting "$x \leq y$" and "$x < y$" instead of "$\langle x, y \rangle \in R_\leq$" and $\langle x, y \rangle \in R_<$, we therefore have (here too, for all x and y of A):

$$x \leq y \leftrightarrow (x < y \vee x = y)$$

and

$$x < y \leftrightarrow (x \leq y \wedge x \neq y)$$

Ordered classifications

Let $C = \langle A, O \rangle$ be an ordered pair. In this case, C is a finite ordered classificatory structure if and only if (1) A is a finite non-empty set, and (2) O is a ordered classification of A, i.e. a set of non-empty subsets of A – hereafter called the "clusters" of O – such that: (a) any element of A belongs to at least one cluster of O; (b) there is at least one element $x \in A$ such that $\{x\} \in O$; (c) for any cluster $X \in O$, (i) $|X| \geq 2$ only if there is at least one $Y \in O$ such that $Y \subset X$, and (ii) there is always one and only one $x \in X$ that belongs to none of the clusters of O properly included into; and (d) for all X and Y of O, X and Y overlap only if for any $x \in X \cap Y$, there is at least one $Z \in O$ such that $x \in Z$ and $Z \subseteq X \cap Y$. In other words, a finite ordered classificatory structure is an ordered pair $\langle A, O \rangle$ for which (a) A is a finite non-empty set, and (b) O is a subset of $\wp(A)$ such that:

$$\varnothing \notin O$$
$$\cup\{X : X \in 0\} = A$$
$$(\exists x \in A)\{x\} \in O$$
$$(\forall X \in 0)((|X| \geq 2 \to (\exists Y \in 0)Y \subset X) \wedge |\cup\{Z \in 0 : Z \subset X\}| = |X|-1)$$
$$(\forall X, Y \in 0)(X \cap Y \notin \{\varnothing, X, Y\} \to \cup\{Z \in O : Z \subseteq X \cap Y\} = X \cap Y)$$

For the next, I will denote by "$O(A)$" the set of all ordered classifications of A.

Upper and descendant sets

Definitions

Let $R \in S(A)$ and $a \in A$. In this case, the R-relative upper set of a is the set $[a_\uparrow]_R$ of the elements x of A such that either a is R-lower than x or $a = x$. Formally:

$$[a\uparrow]_R =_{df} \{x \in A : \langle a, x \rangle \in R \vee a = x\}$$

i.e.

$$[a\uparrow]_R =_{df} \{x \in A : a < x \vee a = x\}$$

i.e.

$$[a\uparrow]_R =_{df} \{x \in A : a \leq x\}$$

In corollary, the R-relative descendant set of A is the set of all upper sets induced by R, i.e. the set A^*R of the upper sets $[x_\uparrow]_R$ such that $x \in A$. Formally:

$$A^*R =_{df} \{[x_\uparrow]_R : x \in A\}$$

i.e.

$$A^*R =_{df} \{X \subseteq A : (\exists x \in A) X = [x_\uparrow]_R\}$$

Properties of $[x_\uparrow]_R$

1. **Existence**. Follows from the comprehensive definition of $[x_\uparrow]_R$. Indeed, if $|A| \geq 1$, then there is at least one element $a \in A$ and so, $[a_\uparrow]_R$ exists.

2. **Uniqueness**. Results from the comprehensive definition of $[x_\uparrow]_R$ and the set-theoretic principle of extensionality.

3. **$[x_\uparrow]_R \subseteq A$**. Follows from the comprehensive definition of $[x_\uparrow]_R$.

4. **$x \in [x_\uparrow]_R$**. Results from the reflexivity of "$\leq$". Thus, $[x_\uparrow]_R \neq \varnothing$.

5. **If $y \in [x_\uparrow]_R$ and $z \in [y_\uparrow]_R$, then $z \in [x_\uparrow]_R$**. Follows from the transitivity of "$\leq$".

6. **$y \in [x_\uparrow]_R$ and $x \in [y_\uparrow]_R$ if and only if $x = y$**. Indeed, if $y \in [x_\uparrow]_R$ and $x \in [y_\uparrow]_R$, then $x = y$ (anti-symmetry of "$\leq$"). Conversely, if $x = y$, then $y \in [x_\uparrow]_R$ and $x \in [y_\uparrow]_R$ (property 4).

7. **$y \in [x_\uparrow]_R$ if and only if $[x_\uparrow]_R \supseteq [y_\uparrow]_R$**. Indeed, if $y \in [x_\uparrow]_R$, then for any $z \in A$, $z \in [y_\uparrow]_R$ only if $z \in [x_\uparrow]_R$ (property 5) and so, $[x_\uparrow]_R \supseteq [y_\uparrow]_R$. Conversely, if $[x_\uparrow]_R \supseteq [y_\uparrow]_R$, then the fact that $y \in [y_\uparrow]_R$ (property 4) implies that $y \in [x_\uparrow]_R$.

8. **$y \in [x_\uparrow]_R$ and $x \in [y_\uparrow]_R$ if and only if $[x_\uparrow]_R = [y_\uparrow]_R$**. Follows from (7).

9. **$[x_\uparrow]_R = [y_\uparrow]_R$ if and only if $x = y$**. Follows from (6) and (8).

10. **$[x_\uparrow]_R = \{x\}$ if and only if $(\forall y \in A)\ \langle x, y \rangle \notin R$**. Indeed, if $[x_\uparrow]_R = \{x\}$, then for any $y \in A$, $y \in [x_\uparrow]_R$ if and only if $y = x$ and so, $\langle x, y \rangle \notin R$. Conversely, if $\langle x, y \rangle \notin R$ for any $y \in A$, then $y \in [x_\uparrow]_R$ if and only if $y = x$ and so, $[x_\uparrow]_R = \{x\}$.

11. **$(\exists x \in A)\ [x_\uparrow]_R = \{x\}$**. Suppose that this does not hold. If true, then for any $x \in A$, there is always at least one $y \in A$ such that $\langle x, y \rangle \in R$. But this is impossible. Indeed, assume that $|A| = n$ with $\langle x_i, x_{i+1} \rangle \in R$ for any $i \in N$ such that $1 \leq i < n$. In this case, $\langle x_i, x_n \rangle \in R$ for any i such that $1 \leq i < n$ (transitivity). But this implies $\langle x_n, x_i \rangle \notin R$ for any i such that $1 \leq i \leq n$ (asymmetry). Consequently, $[x_{n\uparrow}]_R = \{x_n\}$ and so the property holds.

12. **If $[x_\uparrow]_R \cap [y_\uparrow]_R \notin \{\varnothing, [x_\uparrow]_R, [y_\uparrow]_R\}$, then $(\forall z \in A)\ z \in [x_\uparrow]_R \cap [y_\uparrow]_R$ only if $[z_\uparrow]_R \subseteq [x_\uparrow]_R \cap [y_\uparrow]_R$**. Indeed, if $[x_\uparrow]_R \cap [y_\uparrow]_R \notin \{\varnothing, [x_\uparrow]_R, [y_\uparrow]_R\}$, then for any $z \in A$, $z \in [x_\uparrow]_R \cap [y_\uparrow]_R$ only if (i) $z \in [x_\uparrow]_R$ and $z \in [y_\uparrow]_R$, and (ii) $z \neq x$ and $z \neq y$. But if true, then $[x_\uparrow]_R \supset [z_\uparrow]_R$ and $[y_\uparrow]_R \supset [z_\uparrow]_R$ and so, $[z_\uparrow]_R \subseteq [x_\uparrow]_R \cap [y_\uparrow]_R$.

Properties of $A*R$

13. **Existence**. Follows from the comprehensive definition of $A*R$.

14. **Uniqueness**. Results from the comprehensive definition of A/R and the set-theoretic extensionality principle.

15. **$A/R \subseteq \wp(A)$**. Follows from (3) and the comprehensive definition of $A*R$.

16. **$A/R \neq \varnothing$**. Indeed, if $|A| \geq 1$, then there is at least one element x in A, which implies the existence of $[x_\uparrow]_R$.

17. **$\varnothing \notin A/R$**. Follows from (4).

18. **$(\exists X \in A*R)\ |X| = 1$, i.e. $(\exists x \in A)\ \{x\} \in A*R$**. Follows from (11).

Strict order relation => ordered classification

Let $R \in S(A)$. In this case, the descendant set $A*R$ induced by R necessarily satisfies the following five axiomatic properties:

$$A^*R \subseteq \wp(A) \setminus \{\varnothing\}$$
$$\cup\{X : X \in A^*R\} = A$$
$$(\exists x \in A)\{x\} \in A^*R$$
$$(\forall X \in A^*R)\,((|X| \geq 2 \to (\exists Y \in A^*R)\,Y \subset X) \wedge |\cup\{Z \in A^*R : Z \subset X\}| = |X| - 1)$$
$$(\forall X, Y \in A^*R)\,(X \cap Y \notin \{\varnothing, X, Y\} \to \cup\{Z \in A^*R : Z \subseteq X \cap Y\} = X \cap Y)$$

Among them, the first exactly corresponds to the conjunction of (15) with (17). Then, the second follows from (4), the third exactly corresponds to (18), and the fifth follows from (12). Now, let $X \in A^*R$ and suppose that $|X| \geq 2$. In this case, there are necessarily at least two distinct elements x and y in A such that $X = [x_\uparrow]_R$ and $y \in [x_\uparrow]_R$. Given that this implies $[y_\uparrow]_R \subseteq [x_\uparrow]_R$ (property 7) and $x \notin [y_\uparrow]_R$ (property 6), it then follows $[y_\uparrow]_R \subset [x_\uparrow]_R$. But if true, then (a) there is necessarily at least one $Y \in A^*R$ such that $Y \subset X$ (namely, $[y_\uparrow]_R$), and (b) $x \notin Z$ for any $Z \in A^*R$ such that $Z \subset X$. Consequently, the fourth property also holds.

Conclusion: For any $R \in S(A)$, the descendant set A^*R induced by R is necessarily an ordered classification of A, i.e. a member of $O(A)$. There is therefore a single function f from $S(A)$ to $O(A)$ such that $f(R) = A^*R$ for any $R \in S(A)$.

Ordered classification => strict order relation

Let $O \in O(A)$. Then, let R_O be a binary relation on A (i.e. $R_O \subseteq A^2$) such that:

$$(\forall x, y \in A)\,[\langle x, y \rangle \in R_O \leftrightarrow (\exists Y \in O)\,[(x \notin Y \wedge y \in Y) \wedge (\forall X \in O)\,(x \in X \to X \supset Y)]]$$

i.e.

$$R_O = \{\langle x, y \rangle \in A^2 : (\exists Y \in O)\,[(x \notin Y \wedge y \in Y) \wedge (\forall X \in O)\,(x \in X \to X \supset Y)]\}$$

In this case, R_O is necessarily anti-reflexive and transitive (on A). Indeed, and respectively:

- For any $x \in A$, if $\langle x, x \rangle \in R_O$, then there is at least one cluster Y in O such that (1) $x \notin Y$ and $x \in Y$. But this is impossible. Hence, $\langle x, x \rangle \notin R_O$.
- For any $\{x, y, z\} \subseteq A$, if $\langle x, y \rangle \in R_O$ and $\langle y, z \rangle \in R_O$, then: (1) there is at least one cluster Y in O such that (i) $x \notin Y$ and $y \in Y$, and (ii) for any cluster X of O, $x \in X$ only if $X \supset Y$; and (2) there is at least one cluster Z in O such that (iii) $y \notin Z$ and $z \in Z$, and (iv) for any cluster W of O, $y \in W$ only if $W \supset Z$. But given that this implies that $x \in X$ only if $y \in X$ for any $X \in O$, it follows that there is at least one cluster Z in O such that (a) $x \notin Z$ and $z \in Z$, and (b) for any $X \in O$, $x \in X$ only if $X \supset Z$. As a result, $\langle x, z \rangle \in R_O$.

Conclusion: for any $O \in O(A)$, the binary relation $R_O \subseteq A^2$ defined from O is necessarily a strict order relation on A, i.e. a member of $S(A)$. There is therefore a single function g from $O(A)$ to $S(A)$ such that $g(O) = R_O$ for any $O \in O(A)$.

Strict order relation <=> ordered classification

$R_{A*R} = R$

Let $R \in S(A)$. In this case, the strict order relation R_{A*R} implied by the ordered classification $A*R$ is identical to R. Indeed:

- For all x and y of A, if $\langle x, y \rangle \in R$, then $x \notin [y_\uparrow]_R$ and $y \in [x_\uparrow]_R$ and so, $[x_\uparrow]_R \supset [y_\uparrow]_R$. But if true, then for any for any $z \in A$, $x \in [z_\uparrow]_R$ only if $y \in [z_\uparrow]_R$ and $z \notin [y_\uparrow]_R$ (transitivity of R), which implies $[z_\uparrow]_R \supset [y_\uparrow]_R$. There is therefore at least one $Y \in A*R$ such that (i) $x \notin Y$ and $y \in Y$, and (ii) for any $X \in A*R$, $x \in X$ only if $X \supset Y$. Hence, $\langle x, y \rangle \in R_{A*R}$ and so, $R \subseteq R_{A*R}$.
- For all x and y of A, if $\langle x, y \rangle \notin R$, then either (i) $\langle y, x \rangle \in R$, (ii) $x = y$, or (iii) $x \neq y$ and $\langle y, x \rangle \notin R$. But if (i) holds, then $\langle y, x \rangle \in R_{A*R}$ and so, $\langle x, y \rangle \notin R_{A*R}$ (asymmetry of R_{A*R}). Likewise, if (ii) holds, then $\langle x, y \rangle \notin R_{A*R}$ (anti-reflexivity of R_{A*R}). Now, suppose that (iii) holds. In this case, $x \notin [y_\uparrow]_R$ and $y \notin [x_\uparrow]_R$. But if true, then even if there is well a cluster Y in $A*R$ such that $x \notin Y$ and $y \in Y$ (namely, $[y_\uparrow]_R$), there is also at least one cluster X in $A*R$ such that $x \in X$ and $y \notin X$ (namely, $[x_\uparrow]_R$ and all its supersets). As a result, $Y \not\subset X$ and so, $\langle x, y \rangle \notin R_{A*R}$. This third case leading to the same conclusion as the previous ones, we can therefore conclude that $R_{A*R} \subseteq R$.

Conclusion: the function f from $S(A)$ to $O(A)$ is injective. But if true, then $|S(A)| \leq |O(A)|$.

$A*R_O = O$

Let $O \in O(A)$. In this case, the ordered classification $A*R_O$ implied by the strict order relation R_O is identical to O. Indeed:

- For any $X \subseteq A$, if $X \in O$, then there is exactly one $x \in A$ such that $x \in X$ and for any $Z \in O$, $x \in Z$ only if $Z \supseteq X$. Let $a \in X$ this element. For any $x \in A$, this means that if $x \in X$ and $x \neq a$, then there is necessarily at least one $Y \in O$ such that $x \in Y$, $a \notin Y$, and for any $Z \in O$, $Z \supset Y$ if $a \in Z$. But if true, then $\langle a, x \rangle \in R_O$. Hence, $x \in [a_\uparrow]_{RO}$ and so, $X \subseteq [a_\uparrow]_{RO}$. Conversely, if $x \notin X$, then $\langle a, x \rangle \notin R_O$ and $x \neq a$ and so, $x \notin [a_\uparrow]_{RO}$. Given that this implies $[a_\uparrow]_{RO} \subseteq X$, it follows $X = [a_\uparrow]_{RO}$. Therefore, $X \in A*R_O$ and so, $O \subseteq A*R_O$.
- For any $X \subseteq A$, if $X \in A*R_O$, then there is at least one $a \in A$ such that $X = [a_\uparrow]_{RO}$. For any $x \in A$, this means that $x \in [a_\uparrow]_{RO}$ only if either $x = a$ or $\langle a, x \rangle \in R_O$. But if $\langle a, x \rangle \in R_O$, then there are at least two clusters Y and Z in O such that $a \in Y$, $a \notin Z$, $x \in Z$, and $Z \subset Y$. Hence, $x \in Y$ and so, $[a_\uparrow]_{RO} \subseteq Y$. Reciprocally, if $x \in Y$, then either $x = a$ or $\langle a, x \rangle \in R_O$ and so, $x \in [a_\uparrow]_{RO}$. Given that this implies $Y \subseteq [a_\uparrow]_{RO}$, it then follows $[a_\uparrow]_{RO} = Y$. Thus, $X \in O$ and so, $A*R_O \subseteq O$.

Conclusion: the function g from $O(A)$ to $S(A)$ is injective as well. But if true, then $|O(A)| \leq |S(A)|$.

$R = R_O$ if and only if $O = A*R$

Since $|S(A)| \leq |O(A)|$ and $|O(A)| \leq |S(A)|$, it follows $|S(A)| = |O(A)|$. Hence f and g are both bijective. Now, let $R \in S(A)$ and $O \in O(A)$. First, suppose that $R = R_O$. In this case, $A*R = A*R_O$. But $A*R_O = O$ for any $O \in O(A)$. Therefore, $A*R = O$. Then, suppose that $O = A*R$. In this case, $R_O = R_{A*R}$. But $R_{A*R} = R$ for any $R \in S(A)$. Consequently, $R_O = R$. Thus not only f and g are bijective, but also g is the reverse bijection of f (and conversely).

Conclusion: as any equivalence relation (on A) determines and is determined by a single partition (of A) and conversely, any strict order relation (on A) determines and is determined by a single ordered classification (of A) and conversely. In other words, as equivalence relations and partitions are two different ways for saying that given two objects either they are the "same thing" or they are not (for a given parameter, context, etc.), strict order relations and ordered classifications are just two different ways for saying that among two objects, either one of them is respectively "lower/upper," "younger/older," "before/after," "lesser/greater," "smaller/bigger," etc., than the other one or they are not related in this way (here too, for a given parameter, context, etc.).

Appendix 18.2 Degree of equivalence relation and hierarchical classification

Degree of equivalence relation

Let $R = \langle A, T \rangle$ be an ordered pair. In this case, R is also a finite degree of equivalence relational structure if and only if (1) A is a finite non-empty set, and (2) T is a degree of equivalence relation on A, i.e. a ternary relation on A that satisfies the following five axiomatic properties for all w, x, y, and z of A:

- $\langle x, y, z \rangle \in T \rightarrow \langle y, x, z \rangle \in T$	1–2 symmetry
- $\langle x, y, z \rangle \in T \rightarrow \langle x, z, y \rangle \notin T$	2–3 asymmetry
- $x \neq y \rightarrow \langle x, x, y \rangle \in T$	1–2 restricted reflexivity
- $[\langle x, y, w \rangle \in T \wedge \langle y, z, w \rangle \in T] \rightarrow \langle x, z, w \rangle \in T$	1–2 transitivity
- $\langle w, x, z \rangle \in T \rightarrow [\langle w, x, y \rangle \in T \vee \langle w, y, z \rangle \in T]$	2–3 negative transitivity

Remarkably, such axiomatic properties then imply seven derived properties (here too, for all w, x, y, and z of A):

- $\langle x, y, z \rangle \in T \rightarrow \langle z, y, x \rangle \notin T$	1–3 asymmetry
- $\langle x, y, x \rangle \notin T \wedge \langle x, y, y \rangle \notin T$	(1+2)–3 anti-reflexivity

- $\langle x, x, y \rangle \in T \leftrightarrow x \neq y$	1–2 identity
- $\langle x, x, x \rangle \notin T$	anti-reflexivity
- $[\langle w, x, y \rangle \in T \wedge \langle w, y, z \rangle \in T] \rightarrow \langle w, x, z \rangle \in T$	2–3 transitivity
- $[\langle w, x, y \rangle \in T \wedge \langle y, z, w \rangle \in T] \rightarrow [\langle w, x, z \rangle \in T \wedge \langle y, z, x \rangle \in T]$	pair-separativity
- $[\langle w, x, y \rangle \in T \wedge \langle x, y, z \rangle \in T] \rightarrow [\langle w, x, z \rangle \in T \wedge \langle w, y, z \rangle \in T]$	crossed transitivity

Now, let D(A) be the set of all possible degree of equivalence relations on A. It is then possible to distinguish two kinds of 3-tuples for any $T \in$ D(A): (i) those that are trivial or non-informative (because they are necessary), i.e. 3-tuples $\langle x, x, y \rangle$ of A^3 such that $x \neq y$; and (ii) those that are non-trivial or informative (because they are contingent), i.e. 3-tuples $\langle x, y, z \rangle$ of A^3 such that $x \neq y$, $x \neq z$, and $y \neq z$. Thus, any finite non-empty set A always implies a single trivial degree of equivalence relation, i.e. a degree of equivalence relation T_{A0} on A such that

$$T_{A0} = \{\langle x, y, z \rangle \in A^3 : x = y \wedge x \neq z\}$$

i.e.

$$T_{A0} = \{\langle x, x, y \rangle \in A^3 : x \neq y\}$$

and

$$(\forall T \in \text{D}(A)) \, T_{A0} \subseteq T$$

Hierarchical classification

Let $C = \langle A, H \rangle$ be an ordered pair. In this case, C is also a finite hierarchical classificatory structure if and only if (1) A is a finite non-empty set, and (2) H is a hierarchical classification of A, i.e. a set of non-empty subsets of A – hereafter called the "clusters" of H – such that: (i) H contains both the exhaustive cluster of A (i.e. A itself) and all singletons $\{x\}$ such as x belongs to A; and (ii) there is no overlapping between the clusters of H, i.e. the intersection between any two clusters of H is either empty or one of the two. In other words, a finite hierarchical classificatory structure is an ordered pair $\langle A, H \rangle$ where (a) A is a finite non-empty set, and (b) H is a subset of $\wp(A)$ such that:

$$\varnothing \notin H$$
$$H \supseteq \{A\} \cup \{\{x\} : x \in A\}$$
$$(\forall X, Y \in H) \, X \cap Y \in \{\varnothing, X, Y\}$$

Now, let H(A) be the set of all possible hierarchical classifications of A. It then follows the possibility to distinguish two kinds of clusters in any $H \in$ H(A): (i) those that are trivial or non-informative (because they are necessary), i.e. clusters X of

H such that $X = A$ and/or $|X| = 1$; and (ii) those that are non-trivial or informative (because they are contingent), i.e. clusters X of H such that $2 \leq |X| < |A|$. Thus, any finite non-empty set A always implies a single trivial hierarchical classification, i.e. a hierarchical classification H_{A0} such that

$$H_{A0} = \{X \subseteq A: X = A \vee |X| = 1\}$$

i.e.

$$H_{A0} = \{A\} \cup \{\{x\}: x \in A\}$$

and

$$(\forall H \in \mathrm{H}(A))\, H_{A0} \subseteq H$$

Degree of equivalence class and propinquity set

Definitions

Let $T \in \mathrm{D}(A)$ and $\{a, b\} \subseteq A$. In this case, the T-relative degree of equivalence class of a and b is the set $[a\,\&\,b]_T$ of the elements x of A such that a is not T-equivalent to b compared to x. Formally:

$$[a\,\&\,b]_T =_{\mathrm{df}} \{x \in A: \langle a, b, x \rangle \notin T\}$$

In corollary, the T-relative propinquity set of A is the set of all degree of equivalence classes induced by T, i.e. the set $A//R$ of the degree of equivalence classes $[x\,\&\,y]_T$ such that $\{x, y\} \subseteq A$. Formally:

$$A//T =_{\mathrm{df}} \{[x\,\&\,y]_T: \{x, y\} \subseteq A\}$$

i.e.

$$A//T =_{\mathrm{df}} \{X \subseteq A: (\exists x, y \in A)\, X = [x\,\&\,y]_T\}$$

Properties of $[x\,\&\,y]_T$

1. **Existence.** Follows from the comprehensive definition of $[x\,\&\,y]_T$. Indeed, if $|A| \geq 1$, then there is at least one element $a \in A$ and so, $[a\,\&\,a]_T$ exist.
2. **Uniqueness.** Results from the conjunction of the comprehensive definition of $[x\,\&\,y]_T$ with the set-theoretic principle of extensionality.
3. $[x\,\&\,y]_T \subseteq A$. Follow from the comprehensive definition of $[x\,\&\,y]_T$.
4. $[x\,\&\,y]_T = [y\,\&\,x]_T$. Follows from the 1–2 symmetry of T.
5. **If** $z \notin [x\,\&\,y]_T$, **then** $y \in [x\,\&\,z]_T$ **and** $x \in [z\,\&\,y]_T$. Results from the (1+2)–3 asymmetry of T.
6. $\{x, y\} \subseteq [x\,\&\,y]_T$. Follows from the 1–3 and 2–3 anti-reflexivity of T. Thus, $[x\,\&\,y]_T \neq \varnothing$.

7. **If $w \notin [x \& y]_T$ and $w \notin [y \& z]_T$, then $w \notin [x \& z]_T$.** Results from the 1–2 transitivity of T.

8. **If $y \in [w \& x]_T$ and $z \in [w \& y]_T$, then $z \in [w \& x]_T$.** Follows from the 2–3 negative transitivity of T.

9. **If $y \notin [w \& x]_T$ and $z \notin [w \& y]_T$, then $z \notin [w \& x]_T$.** This is a consequence of the 2–3 transitivity of T.

10. **If $y \notin [w \& x]_T$ and $w \notin [y \& z]_T$, then $z \notin [w \& x]_T$ and $x \notin [y \& z]_T$.** Results from the pair-separativity of T.

11. **If $y \notin [w \& x]_T$ and $z \notin [x \& y]_T$, then $z \notin [w \& x]_T$ and $z \notin [w \& y]_T$.** Follows from the crossed transitivity of T.

12. **$[w \& x]_T \cap [y \& z]_T \in \{\varnothing, [w \& x]_T, [y \& z]_T\}$.** Suppose that this property does not hold. In this case, this means that there are at least three elements a, b, and c in A such that: (1) $\{a, b\} \subseteq [w \& x]_T$ and $c \notin [w \& x]_T$; and (2) $\{a, c\} \subseteq [y \& z]_T$ and $b \notin [y \& z]_T$. But, this is impossible. Indeed:

   - On the one hand, we have $\langle w, x, a \rangle \notin T$, $\langle w, x, b \rangle \notin T$, and $\langle w, x, c \rangle \in T$ (definition of $[w \& x]_T$). In this case, we also have (a) $\langle w, x, a \rangle \in T$ and/or $\langle w, a, c \rangle \in T$, and (b) $\langle w, x, b \rangle \in T$ and/or $\langle w, b, c \rangle \in T$ (2–3 negative transitivity from $\langle w, x, c \rangle \in T$). It thus follows $\langle w, a, c \rangle \in T$ and $\langle w, b, c \rangle \in T$. But $\langle w, a, c \rangle \in T$ leads to $\langle a, w, c \rangle \in T$ (1–2 symmetry). Consequently, $\langle a, b, c \rangle \in T$ finally holds (1–2 transitivity).

   - On the other hand, we have $\langle y, z, a \rangle \notin T$, $\langle y, z, c \rangle \notin T$, and $\langle y, z, b \rangle \in T$ (definition of $[y \& z]_T$). In this case, we also have (a) $\langle y, z, a \rangle \in T$ and/or $\langle y, a, b \rangle \in T$, and (b) $\langle y, z, c \rangle \in T$ and/or $\langle y, c, b \rangle \in T$ (2–3 negative transitivity from $\langle y, z, b \rangle \in T$). It thus follows $\langle y, a, b \rangle \in T$ and $\langle y, c, b \rangle \in T$. But $\langle y, a, b \rangle \in T$ leads to $\langle a, y, b \rangle \in T$ (1–2 symmetry). Consequently, $\langle a, c, b \rangle \in T$ holds too (1–2 transitivity) and so $\langle a, b, c \rangle \notin T$ (2–3 asymmetry).

13. **$[x \& x]_T = \{x\}$.** Indeed, on the one hand, $[x \& x]_T \subseteq \{x\}$, since $x \neq y$ only if $y \notin [x \& x]_T$ (1–2 restricted reflexivity). On the other, $\{x\} \subseteq [x \& x]_T$ (property 6).

14. **If $z \notin [x \& y]_T$, then $[x \& y]_T \subset [x \& z]_T$.** Indeed, on the one hand, if $z \notin [x \& y]_T$, then $[x \& y]_T \neq [x \& z]_T$ (property 6). On the other hand, if $z \notin [x \& y]_T$, then $y \in [x \& z]_T$ (property 5). But this implies for any w of A that if $w \in [x \& y]_T$, then $w \in [x \& z]_T$ (property 8). Hence, $[x \& y]_T \subseteq [x \& z]_T$ and so $[x \& y]_T \subset [x \& z]_T$.

15. **If $z \notin [x \& y]_T$, then $[x \& z]_T = [y \& z]_T$.** Indeed, on the one hand, if $z \notin [x \& y]_T$, then $x \in [z \& y]_T$ (property 5). But, for any w of A, if $w \in [x \& z]_T$, then $w \in [z \& x]_T$ (property 4). Consequently, $w \in [z \& y]_T$ (property 8) and so $w \in [y \& z]_T$ (property 4). Hence $[x \& z]_T \subseteq [y \& z]_T$. On the other hand, if $z \notin [x \& y]_T$, then $z \notin [y \& x]_T$ (property 4) and so $y \in [z \& x]_T$ (property 5). But, for any w of A, if $w \in [y \& z]_T$, then $w \in [z \& y]_T$ (property 4). As a result, $w \in [z \& x]_T$ (property 8) and so $w \in [x \& z]_T$ (property 4). Therefore, $[y \& z]_T \subseteq [x \& z]_T$ and so $[x \& z]_T = [y \& z]_T$.

16. **($\forall x \in A$)($\exists y \in A$) $[x \And y]_T = A$.** We already know that ($\forall x, y \in A$) $[x \And y]_T \subseteq A$ (property 3). Hence, we just have to show that ($\forall x \in A$)($\exists y \in A$) $[x \And y]_T \supseteq A$. First, suppose that T is trivial. Two cases are then possible: either $|A| = 1$ or $|A| \geq 2$. If $|A| = 1$, then there is a single x such that $A = \{x\}$. But $[x \And x]_T = \{x\}$ (property 13). Hence, $[x \And x]_T = A$. Conversely, if $|A| \geq 2$, then for any $x \in A$, there is at least one $y \in A$ such that $x \neq y$. But, since T is trivial, this implies $\langle x, y, z \rangle \notin T$ for any z of A. Hence, $z \in [x \And y]_T$ and so $[x \And y]_T \supseteq A$. Then, suppose that T is non-trivial. In the simplest case, this means that there are at least three distinct elements a, b, and c in A such that (a) $c \notin [a \And b]_T$, i.e. $\langle a, b, c \rangle \in T$, and (b) for any degree of equivalence class X induced by T, $X \neq A$ and $|X| \geq 2$ only if $X = [a \And b]_T$. In other words, there is a single set B such that: (1) $B \subset A$; (2) $|B| \geq 2$; and (3) for any $\{x, y, z\} \subseteq A$, $\{x, y\} \subseteq B$ and $z \notin B$ if and only if $\langle x, y, z \rangle \in T$. In this case, $[x \And y]_T \supseteq A$ for any distinct elements x and y of A such that $x \notin B$ and/or $y \notin B$. Indeed, if at least one of these elements belongs to $A \setminus B$, then for any $z \in A$, $\langle x, y, z \rangle \notin T$ and so $z \in [x \And y]_T$. This result can now be generalized. Let H be the set of all degree of equivalence classes X induced by T such that $X \subset A$ and $|X| \geq 2$, and assume $|H| \geq 1$. We already know that no overlapping is possible between any two X and Y of H (property 12). Thus, either (1) there is a single X in H such that $X \supseteq Y$ for any $Y \in H$, or (2) there is one set $I \subseteq H$ such that (i) $X \cap Y = \varnothing$ for all distinct X and Y of I, and (ii) there is no Z in H such that $Z \supseteq \cup J$ for any $J \subseteq I$. If (1) holds, then $[x \And y]_T \supseteq A$ for all distinct x and y of A if at least one of them belongs to $A \setminus X$. Conversely, if (2) holds, then $[x \And y]_T \supseteq A$ for all distinct x and y of A if $\{x, y\}$ is not included in X for any X of I, i.e. if either (a) $\{x, y\}$ is not included in $\cup I$, or (b) $x \in V$, $y \in W$, $V \subseteq X$, and $W \subseteq Y$ for any $\{V, W\} \subseteq H$ and any $\{X, Y\} \subseteq I$ such that $X \neq Y$.

Properties of *A//T*

17. **Existence.** Follows from the comprehensive definition of $A//T$.
18. **Uniqueness.** Follows from the conjunction of the comprehensive definition of $A//T$ with the set-theoretic principle of extensionality.
19. **$A//T \subseteq \wp(A)$.** Follow from (3) and the comprehensive definition of $A//T$.
20. **$A//T \neq \varnothing$.** Indeed, if $|A| \geq 1$, then there is at least one element x in A, which implies the existence of $[x \And x]_T$.
21. **$\varnothing \notin A//T$.** Results from (6).
22. **$HA_0 \subseteq A//T$.** Indeed, on the one hand, $A \in A//T$ (property 16). On the other hand, $[x \And x]_T = \{x\}$ (property 13). Since $[x \And x]_T$ is defined for any x of A, this means that $\{x\} \in A//T$ for any $x \in A$. Thus, $A//T \supseteq \{A\} \cup \{\{x\}: x \in A\}$.

Degree of equivalence relation => hierarchical classification

Let $T \in \mathrm{D}(A)$. In this case, the propinquity set $A//T$ induced by T necessarily satisfies the following three axiomatic properties:

$$A//T \subseteq \wp(A) \setminus \{\emptyset\}$$
$$A//T \supseteq \{A\} \cup \{\{x\}: x \in A\}$$
$$(\forall X, Y \in A//T)\, X \cap Y \in \{\emptyset, X, Y\}$$

Indeed:

- The first simply corresponds to the conjunction of (19) with (21).
- The second exactly corresponds to (22).
- The third exactly corresponds to (12).

Conclusion: for any $T \in \mathrm{D}(A)$, the propinquity set $A//T$ induced by T is a hierarchical classification of A, i.e. a member of $\mathrm{H}(A)$. There is therefore a single function f from $\mathrm{D}(A)$ to $\mathrm{H}(A)$ such that $f(T) = A//T$ for any $T \in \mathrm{D}(A)$.

Hierarchical classification => degree of equivalence relation

Let $H \in \mathrm{H}(A)$. Then, let T_H be a ternary relation on A (i.e. $T_H \subseteq A^3$) such that:

$$(\forall x, y, z \in A)\, [\langle x, y, z \rangle \in T_H \leftrightarrow (\exists X \in H)\, (\{x, y\} \subseteq X \wedge z \notin X)]$$

i.e.:

$$T_H = \{\langle x, y, z \rangle \in A^3 : (\exists X \in H)\, (\{x, y\} \subseteq X \wedge z \notin X)\}$$

In this case, T_H is necessarily 1–2 symmetric, 2–3 asymmetric, 1–2 restrictively reflexive, 1–2 transitive, and 2–3 negatively transitive (on A). Indeed, and respectively:

- For any $\{x, y, z\} \subseteq A$, if $\langle x, y, z \rangle \in T_H$, then there is at least one cluster X in H such that $\{x, y\} \subseteq X$ and $z \notin X$. But $\{x, y\} = \{y, x\}$. Hence, $\langle y, x, z \rangle \in T_H$.
- For any $\{x, y, z\} \subseteq A$, if $\langle x, y, z \rangle \in T_H$, then there is at least one cluster X in H such that $\{x, y\} \subseteq X$ and $z \notin X$. But if true, then no cluster Y of H can contain x and z but not y. Consequently, $\langle x, z, y \rangle \notin T_H$.
- For any $\{x, y, z\} \subseteq A$, if $x \neq y$, then $x \in \{x\}$ and $y \notin \{x\}$. Given that $\{x\} \in H$ if $x \in A$, it follows at least one cluster X in H such that $x \in X$ and $y \notin X$. But $\{x, z\} = \{x, x\} = \{x\}$ when $z = x$. Therefore, $\langle x, x, y \rangle \in T_H$.
- For any $\{w, x, y, z\} \subseteq A$, if $\langle x, y, w \rangle \in T_H$ and $\langle y, z, w \rangle \in T_H$, then there are at least two clusters X and Y in H such that (a) $\{x, y\} \subseteq X$ and $\{y, z\} \subseteq Y$, and (b) w belongs to none of them. But if true, then $X \cap Y \neq \emptyset$ and so, $X \subseteq Y$ and/or $Y \subseteq X$. Thus, $\{x, z\} \subseteq Y$ and/or $\{x, z\} \subseteq X$, and so $\langle x, z, w \rangle \in T_H$.
- For any $\{w, x, y, z\} \subseteq A$, if $\langle w, x, z \rangle \in T_H$, then there is at least one cluster X in H such that $\{w, x\} \subseteq X$ and $z \notin X$. In this case, either $y \in X$, which implies $\langle w, y, z \rangle \in T_H$, or $y \notin X$, which implies $\langle w, x, y \rangle \in T_H$. But there is a specific sub-case of the second above where $\langle w, y, z \rangle \in T_H$ also holds, namely that where there is also at least one cluster $Y \supset X$ in H containing w, x, y, but not z. Hence, $\langle w, x, y \rangle \in T_H$ and/or $\langle w, y, z \rangle \in T_H$.

Conclusion: for any $H \in$ H(A), the ternary relation $T_H \subseteq A^3$ defined from H is a degree of equivalence relation on A, i.e. a member of D(A). There is therefore a single function g from H(A) to D(A) such that $g(H) = T_H$ for any $H \in$ H(A).

Degree of equivalence relation <=> hierarchical classification

$T_{A//T} = T$

Let $T \in$ D(A). In this case, the degree of equivalence relation $T_{A//T}$ implied by the hierarchical classification $A//T$ is identical to T. Indeed:

- For all x, y, and z of A, if $\langle x, y, z \rangle \in T$, then $z \notin [x \,\&\, y]_T$. But if true, then there is at least one cluster X in $A//T$ such that $\{x, y\} \subseteq X$ and $z \notin X$. Consequently, $\langle x, y, z \rangle \in T_{A//T}$ and so, $T \subseteq T_{A//T}$.
- For all x, y, and z of A, if $\langle x, y, z \rangle \in T_{A//T}$, then there are (1) at least one cluster X in $A//T$ such that $\{x, y\} \subseteq X$ and $z \notin X$, and (2) two elements a and b in A such that $X = [a \,\&\, b]_T$. Therefore, $\langle a, b, x \rangle \notin T$, $\langle a, b, y \rangle \notin T$, and $\langle a, b, z \rangle \in T$. Then, $\langle a, b, z \rangle \in T$ implies both (a) $\langle a, b, x \rangle \in T$ and/or $\langle a, x, z \rangle \in T$, and (b) $\langle a, b, y \rangle \in T$ and/or $\langle a, y, z \rangle \in T$ (2–3 negative transitivity). Hence, $\langle a, x, z \rangle \in T$ and $\langle a, y, z \rangle \in T$. But $\langle a, x, z \rangle \in T$ implies $\langle x, a, z \rangle \in T$ (1–2 symmetry). Consequently, $\langle x, y, z \rangle \in T$ (1–2 transitivity) and so, $T_{A//T} \subseteq T$.

Conclusion: the function f from D(A) to H(A) is injective. But if true, then $|D(A)| \leq |H(A)|$.

$A//T_H = H$

Let $H \in$ H(A). In this case, the hierarchical classification $A//T_H$ implied by the degree of equivalence relation T_H is identical to H. Indeed:

- For any $X \subseteq A$, if $X \in H$, then there are at least two elements a and b in A such that (a) $\{a, b\} \subseteq X$ and (b) for any $Y \in H$, $Y \supseteq X$ if $\{a, b\} \subseteq Y$.[29] For any $x \in A$, this means that if $x \in X$, then $\langle a, b, x \rangle \notin T_H$. Therefore, $x \in [a \,\&\, b]_{TH}$ and so, $X \subseteq [a \,\&\, b]_{TH}$.

[29] This property meaning that X is the smallest cluster of H containing a and b, it follows that for any H of H(A) and any cluster X of H, there are necessarily at least two (distinct or identical) elements x and y in A for which X is the smallest cluster that contains them. For the singletons of H, this property is trivial (the singleton $\{x\}$ of H is necessarily the smallest cluster of H that contains x). Now, let X be a non-singleton cluster of H (i.e. either the exhaustive cluster or an informative cluster of H). Two cases are then possible: either (i) there is no informative sub-cluster properly included into X or (ii) there is at least one informative cluster of H properly included into X. If (i) holds, then the smallest cluster of H containing any two distinct elements x and y of X will be X itself (and so, if $X = A$, then $H = H_{A0}$). If (ii) holds, then either (iii) X is partitioned by its most inclusive informative proper sub-clusters, i.e. any element x of X necessarily belongs to at least one informative proper sub-cluster of X or (iv) there is at least one element x in X that belongs to none of the informative proper sub-cluster of X. If (iii) holds, then for any most inclusive informative proper sub-cluster Y of X, any element x of X, and any element y of Y, the smallest cluster of H containing x and y will be X if x does not belong to Y. Finally, if (iv) holds, then the smallest cluster of H containing any two

Conversely, if $x \notin X$, then $\langle a, b, x \rangle \in T_H$ and so, $x \notin [a \,\&\, b]_{TH}$. Given that this implies $[a \,\&\, b]_{TH} \subseteq X$, it follows $X = [a \,\&\, b]_{TH}$. Hence, $X \in A//T_H$ and so, $H \subseteq A//T_H$.[30]

- For any $X \subseteq A$, if $X \in A//T_H$, then there are at least two elements a and b in A such that $X = [a \,\&\, b]_{TH}$. For any $x \in A$, this means that if $x \notin [a \,\&\, b]_{TH}$, then $\langle a, b, x \rangle \in T_H$. Hence, there is a least one cluster Y in H such that $\{a, b\} \subseteq Y$ and $x \notin Y$, which implies $Y \subseteq [a \,\&\, b]_{TH}$. Reciprocally, if $\{a, b\} \subseteq Y$ and $x \notin Y$, then $\langle a, b, x \rangle \in T_H$ and so, $x \notin [a \,\&\, b]_{TH}$. Given that this implies $[a \,\&\, b]_{TH} \subseteq Y$, it thus follows $[a \,\&\, b]_{TH} = Y$. Therefore, $X \in H$ and so, $A//T_H \subseteq H$.

Conclusion: The function g from $H(A)$ to $D(A)$ is injective as well. But if true, then $|H(A)| \leq |D(A)|$.

$T = T_H$ if and only if $H = A//T$

Since $|D(A)| \leq |H(A)|$ and $|H(A)| \leq |D(A)|$, it follows $|D(A)| = |H(A)|$. Hence f and g are both bijective. Now, let $T \in D(A)$ and $H \in H(A)$. First, suppose that $T = T_H$. In this case, $A//T = A//T_H$. But $A//T_H = H$ for any $H \in H(A)$. Therefore, $A//T = H$. Then, suppose that $H = A//T$. In this case, $T_H = T_{A//T}$. But $T_{A//T} = T$ for any $T \in D(A)$. Consequently, $T_H = T$. Thus not only f and g are bijective, but also g is the reverse bijection of f (and conversely).

Conclusion: As for the [equivalence relations / partitions] and [(strict) order relations / ordered classifications] mathematical equivalences, any degree of equivalence relation (on A) well determines and is well determined by a single hierarchical classification (of A) and conversely. In other words, degree of equivalence relations and hierarchical classifications are just two different ways for saying that given three objects, either two of them are the "same thing" compared to the third or there is no possibility to separate one of them from the two other ones (here too, for a given parameter, context, etc.).

distinct elements x and y of X will be X itself if at least one of them belongs to none of the informative proper sub-clusters of X.

[30] Unlike the case where we assume $x \in X$, the fact that X must be the smallest cluster containing a and b is unnecessary when we assume $x \notin X$.

19

This struggle for survival: systematic biology and institutional leadership

QUENTIN WHEELER

If in this struggle for survival biological systematics has recently lost ground to other and, as is often heard, younger and more modern disciplines, this is not so much because of the limited practical or theoretical importance of systematics as because *systematics has not correctly understood how to present its importance in the general field of biology*, and to establish a unified system of instruction in its problems, tasks, and methods. (Hennig 1966: 1; italics mine)

Depending on your understanding of what systematics is, and is not, far more ground has been lost to younger fields since 1966 than before (see quote, above). In particular, the "more modern" DNA-based approaches to both "species" recognition and "phylogeny" reconstruction have largely supplanted curiosity-driven systematic biology with what is, by comparison, a pedantic identification and branching-diagram business. Where theory-rich species concepts and a diversity of character sources dominated before, today's diminished systematics consists disproportionately of isolated species descriptions and "phylogenetic" analyses divorced from monographs, revisions, classifications, and much visible interest in understanding individual complex characters for their own sake. I use quotation marks around the terms above because the meanings of species and phylogeny in their formerly rigorous evolutionary contexts are violated by DNA barcodes, metagenomic environment surveys, and so-called phylogenetic analyses based on measures of genetic similarity rather than theories of homology (Williams et al.

The Future of Phylogenetic Systematics: The Legacy of Willi Hennig, eds. D. Williams, M. Schmitt and Q. Wheeler. Published by Cambridge University Press.

2010). These are all symptoms of a greater problem: forgetting the fundamental importance, independence, and purpose of systematic biology (Wheeler 2009). Whether this is due to ignorance – after all, generations of professional biologists have received university degrees with little or no introduction to the theories, epistemology, or mission of systematics – or greed and hubris, is for historians to sort out. Today, the urgent need is to restore taxonomy[1] to college curricula, taxonomists to university faculties, and clarity of purpose to a field that has lost its way.

19.1 Systematic biology and scientific rigor

> In no other science are the contrasts and struggles for survival among subdivisions of the field so strong as in biology. This is at least partly because the problems, and therefore the methods, are more varied in biology than in any other science. (Hennig 1966: 1)

The greatest contrast exists between what Nelson and Platnick (1981) described as general and systematic biology. The former is experimental, seeking to understand how life functions through the study of "law-like" rules shared by many or most living things. By this term I mean that while generalities about functional systems exist in biology, they are rarely without exception. The latter is observational and comparative, seeking to discover and describe particulars and understand what makes each unique and how it came to be so. Like molecules of hydrogen, any two DNA cytosine bases are identical anywhere in time or space. They have no individual identity, no historical information content. In contrast, no two characters, species, or clades are the same and, due to complexity, from protein folding patterns to rhinoceros horns, each contains unique information about their historical origins.

General biology mostly focuses on processes, how things work in real time, while systematic biology is concerned with patterns that are the result of history. As a result, general biologists appropriately focus on living organisms, whereas systematists may derive much, often more, knowledge from dead specimens in collections and the fossil record. General biology is limited to a few study sites in the case of ecology, a few generations in the case of population genetics, or one or a few model organisms in the laboratory. By comparison, systematic biology is constrained only

---

[1] In this chapter, I use the terms taxonomy, systematic biology, systematics, and phylogenetic systematics synonymously (see also Wheeler 2008). Systematics in the narrow sense is sometimes used to refer to phylogenetic analyses, which are more precisely one part of systematic biology *sensu lato*, done primarily to inform formal classifications and secondarily as an historical context for other branches of biology. Molecular systematics is not synonymous but refers instead only to this one task of taxonomy carried out with a single data source. Methods of classification and nomenclature are other parts of the greater field. It is when all the pieces of taxonomy are combined into a whole, from species theories to cladistic analyses, phylogenetic classifications, and Linnaean names that the grandeur and power of the discipline are revealed.

by monophyly, the study of a single genus or family often spanning continents, eco-systems, and millions of years. The epistemologies of general and systematic biology are wholly different, contributing to misunderstandings (Gaffney 1979). Instead of educating general biologists about the epistemology of taxonomy, we have instead opted to search for ways to make taxonomy look more like general biology. Whether that was "population thinking" in the 1940s (e.g. Mayr 1942, Ross 1974), giving preference to evidence perceived to be "high tech," or massaging DNA data in ways inconsistent with Hennig's phylogenetic theory, such disciplinary mimicry is doomed to fail as a way to advance systematic biology.

The demarcation principle dividing science from non-science is simply testability (Popper 1959). It follows that the more easily a generalization about the world may be tested – shown to be false – the more stringently scientific it is. No claim is, therefore, more scientifically rigorous than an all-or-nothing statement. There is no number of white swans that may be observed sufficient to prove the claim that "all swans are white," yet the sighting of a single black swan suffices to falsify the assertion. Nearly all hypotheses in systematic biology are just this kind of elegant, all-or-nothing claim, including those of homology, synapomorphy, species status, and monophyly. Few experiments rise to this level of theoretical elegance, showing the folly in assertions that taxonomy is any lesser a science than the experimental fields.

No confusion in theoretical biology has been more damaging to systematics than the failure to distinguish between studies of *species*, as (constantly) diagnosable historical entities, and *speciation*, as a set of dynamic genetic processes. In the 1940s, the emerging science of population genetics was intentionally confounded with systematic biology (Mayr 1942) in a misguided effort to make taxonomy appear more modern and possibly divert a little funding and prestige away from genetics. Blurring the distinction between systematics and experimental sub-disciplines, however, worked decidedly better than any profit motive that may have been present (Wheeler 2008).

Systematists study fully established species and characters, and the phylogenetic relationships among them (Nelson and Platnick 1981, Wheeler and Platnick 2000). In contrast, population geneticists study populations as they are in the process of differentiating. Danish philosopher Kierkegaard famously observed that while history must be lived forward, it may only be understood looking backwards. The same is true of phylogeny. The tempo and modes of processes potentially leading to speciation may be studied in detail only by focusing on the developing genetic gaps between populations as they happen. But phylogeny may only be understood by studying species *after* speciation and character transformation are complete and the potential of introgression removed. We may speculate that 99% divergent populations are insipient species, and would probably be right, but until that process is complete the chance remains that the populations could coalesce. Insisting on complete divergence marked by equally complete character transformation (Platnick

1979, Wheeler and Platnick 2000) sets a very high bar for species, but this is the price of making species rigorously testable hypotheses.

19.2 Zombie science

There are those who claim that there is no taxonomic crisis, that systematics is alive and well. I wish they were right. The state of taxonomy is more like that of the undead, still walking around, but devoid of its soul. Real systematics is an audacious, fundamental, curiosity-driven science that dares to ask which and how many species exist on, under, and above the surface of an entire planet. And, further, to ask which evolutionary novelties make each species and clade unique and the sequence of transformations in a multi-billion-year history that explains their existence and relations. Real systematics has the soul shared by astronomy and cosmology, possessing the boldness to explore, know, and classify the millions of species that live or have lived since the origin of life some 3.8 billion years ago.

One version of the all-is-well argument is that molecular systematics is simply the modern version of systematics. If we accept this redefinition of what systematics is then the problem becomes nonexistent. This did not work for population thinking from the 1940s to 1970. It will not work for the "phylogeny thinking" that currently violates Hennig's (1966) distinction between phylogeny and tokogeny. And it will not work for the mistaken view that DNA is some kind of super-data that reveals, rather than simply provides data about, phylogeny (Nelson 2004). In short, any approach that ignores individual homologues and the theoretical advances in the work of Cuvier, Owen, Darwin, and Hennig is and should be suspect. Lessons, such as those that came from a focus on homology, and a distinction between cladogram and phylogram, have been forgotten and replaced by an infatuation with and religious-like adherence to modern technology. Taxonomy done for its own sake, out of pure curiosity, has been supplanted by rote procedures carried out not to explore the "Cosmos" of life on our planet, but merely in the service of other branches of biology. The needs to identify species, have names to refer to them, and some idea of their relationships are very real across the life sciences, but these are applications of taxonomy. Fundamental taxonomy is driven by comprehensive studies of monophyletic clades and its ultimate expression is in the form of phylogenetic classifications in the context of taxonomic monographs.

Another variation on the all-is-well argument is that the large number of authors associated with publications in which new species are described is indicative that there are more practicing taxonomists than ever (Costello et al. 2013, 2014). This is a silly misreading of facts (Wheeler 2013, Bebber et al. 2014). Anyone willing to follow a few rules can name a new species. But in doing so, she does not become a taxonomist. There are many species named by amateurs, some to high standards, others not. Many others are named by biologists who are not taxonomists, but who, in the absence of professional taxonomy, name species out of necessity. And there is a trend toward multiple-authored papers more fully acknowledging the contributions made by collectors, technicians, photographers, and others who are clearly co-authors but not taxon experts. While there are many and varied things that contribute to taxonomy, the taxonomic paradigm includes comprehensively comparative studies of all species in a monophyletic clade with special attention to individual original attributes and all their subsequent modifications (Platnick 1979).

Who is responsible for taxonomy's decline? There is no shortage of blame to go around. Taxonomists have been their own worst enemies failing to explain or advocate for the importance of their own science. Living on a species-rich planet undergoing a mass extinction event, it was inconceivable that taxonomy would not be supported. But no one foresaw the toxic combination of modernity, technology, and immediate gratification that would value quick data over deep knowledge. A frightening number of experimental biologists have taken a narrow view of science that excludes the comparative and historical and that is reinforced by peer pressure and trends in grant funding. Above all, I believe institutions deserve to be singled out for condemnation. Not only did they stop hiring taxonomists, teaching taxonomy, and expanding and curating natural history collections, they alone have the power to instantly reverse the trend.

There are explanations for why leaders of natural history museums, botanical gardens, and research universities have expunged taxonomy from their ranks, but no excuses. It is easier to attract funds to an institution by following popular trends than by leading an inventory of life on earth. It is easier to go along with fashions than to do the heavy lifting of explaining and defending why another direction is more important and urgent. Taking this to extremes, it is possible to maximize externally funded research by replacing taxonomists and natural historians with faculty representing "more modern" fields. This, of course, is further evidence of a lack of leadership. Universities should be as concerned about academic breadth and balance as they are funding competitiveness. And they should be as devoted to responding to the biodiversity crisis as they are appearing modern. Few deans, directors, or presidents today are natural historians or systematic biologists who fully appreciate the role of collections and taxonomy in biology as a whole. They are often hired and evaluated on the basis of fund raising ability rather than a clear vision of the disciplines under their control.

19.3 A challenge to institutional leaders

We are at a critical crossroads. As the biodiversity crisis worsens (Barnosky et al. 2011, Kolbert 2014), the time remaining to explore species shortens and opportunity costs mount: opportunities to conserve species and habitats; to discover models for sustainable designs, materials, and processes; and, to preserve irreplaceable evidence required to explore the origin and history of life of which we are a part. We have two choices. We can stay the current path, neglecting systematic biology and its collections, being complicit in imposing permanent ignorance about the phylogeny of life, or, we can reinvest in taxonomy in time to document life on earth, create options for adapting to a rapidly changing world, and secure evidence of evolutionary history.

For the first time in history, Hennig's phylogenetic systematics created the real possibility of a rigorous and testable reconstruction of the origin and history of species and their evolutionary novelties. It should have, and did for a while, reinvigorate systematic biology. But this progress was cut short (Nelson 2004). The greatest contributions of systematic biology and Hennig's ideas have yet to be realized, but they depend on access to the species that are being lost while we dither with technology instead of asking fundamental questions about species and characters. Rather than being supported to reveal the astonishing story of evolutionary history and the origin of the biosphere, systematic biology is being reduced to a mere service industry providing identifications for field biologists and branching diagrams purporting to show relationships among the tiny fraction of plants and animals already known.

It need not be this way. I organized a workshop asking whether it would be possible to complete a first-pass inventory of ten million species in 50 years or less. The conclusion was a resounding yes (Wheeler et al. 2012). This, however, requires vision and leadership at the institutional level as well as educated and employed taxon experts.

Seven challenges to department heads, deans, and presidents in our universities are given below that, if met, could return taxonomy to a positive path and resume the advances that logically followed Hennig's revolutionary ideas (Nelson 2004). The first five apply equally to leaders of natural history museums and botanical gardens.

1. *Hire taxonomists.* Bright young scientists of every generation are attracted to the intellectual challenges and rewards of becoming a taxon expert but the paucity of jobs is a strong deterrent. Universities and museums can spark a revival in taxonomy instantly by simply hiring faculty. Such experts are superb additions to any life science faculty, providing a complementary view of biological diversity to students and an organismic perspective in teaching. Hiring taxonomists will only address the urgent biodiversity problem if they are expected and allowed to do taxonomy full time. Many taxon experts must spend a percentage of their time doing some other kind of work in order to secure grants, diluting their impact in creating an inventory of life.

2. *Grow, develop, and maintain natural history collections.* While natural history museums have a special responsibility for building worldwide collections,

universities have an important part to play, too. Collectively, the world's thousands of collections add up to the most comprehensive physical representation of the species diversity of the biosphere available. Such university collections are an important part of life science training and the material basis for serious taxonomic research. Even when strongly biased toward the local flora and fauna or one taxon of special interest to faculty curators, such collections are a node in an international network that facilitates systematic biology research and creates a permanent record of biological diversity in the early Anthropocene.

3. *Be a leader and value knowledge streams over revenue streams.* The easiest path for a university leader is to simply respond to whatever kinds of science are currently popular and well-funded, but that is not leadership. Everyone shares the responsibility of identifying the knowledge most needed and becoming part of the solution by assuring appropriate research is happening. Given the pace and extent of species extinction, there is no objective assessment of needs that does not elevate systematics to a high priority. This cannot happen when success is gauged by money procured rather than knowledge produced. It is unquestionably more difficult to do what should be done rather than what can most easily be done, but why else should institutions employ leaders?

4. *Deny boundaries.* As scientists it is not enough to follow our curiosity about the natural world. Beyond completing an inventory of the flora and fauna and reconstructing its history in character-focused detail, we should share the wonders of our discoveries with the general public and awaken the innate curiosity about living things in children. Taxonomy at its core is a rigorous science, but that does not deny its role in enabling an emotional connection between humans and the natural world. There is unparalleled beauty in the nature that we take for granted, in part because we do not take the time to get to know it up close and personal on a species to species basis. Taxonomy is the key that unlocks these wonders by differentiating and making them accessible to anyone with enough humility and curiosity.

5. *Talk the talk.* The preceding half dozen measures ask institutions to walk the walk, enabling and doing taxonomy. It is equally important to talk the talk, to speak the language of biodiversity. The vocabulary of that language consists of Linnaean names for species and higher taxa while its syntax is found in phylogenetic classifications. In order that society make sacrifices and adjustments necessary for species conservation, it must see the inherent worth in other species and respect them as kinds of living things that are more than resources for our use. Without names, other species are no more than statistics. Honoring them with names is a first step toward knowing and valuing them (Kimmerer 2014). While such recognition is possible with common names, the ignorance of taxonomy's needs is nonetheless symptomatic of our society's neglect of biodiversity itself.

6. *Restore taxonomy to the curriculum.* Universities that once had several advanced taxonomy courses in botany and zoology most likely today offer no more than

one or two introductory courses. The advanced taxonomy classes I took at The Ohio State University are no longer offered, much to the detriment of the preparation of the next generation of entomologists. It is possible to complete a course of study in biology today and never take a single class in the natural history or identification of any group of plants or animals. A full curriculum should include a substantial module on taxonomy in introductory biology classes, one or more introductory taxonomy classes, advanced classes that teach students to identify species of some major taxon to some useful level, and advanced classes on theories and techniques of systematics. These are required for the preparation of professional taxonomists but are equally useful for any biologist who does field work or wants to truly understand the model organisms they study in their broader evolutionary and environmental context.

7. *Restore natural history to the curriculum.* Taxonomy is at the heart of natural history, but other aspects of natural history have suffered neglect equally and are no less important to exploring and understanding biodiversity. Teaching students to make scientific observations of plants and animals in their natural habitats is no longer expected in the training of biology majors, but ought to be. Autecology is trivialized because it may not involve precise measurements of any kind, yet the careful study of what each species does in its ecosystem, how it behaves and interacts with other species, builds a deeper foundation for ecosystem science. In addition to being an important skill for professionals, careful observation can enable serious citizen science and vastly increase the pace at which valuable natural history knowledge is acquired.

19.4 Conclusion

It is crunch time. Unless we restore positions, prestige, and support for taxonomists to do taxonomy we will have missed our last opportunity to achieve deep knowledge of the history of life on what is possibly the only biodiversity-rich planet we ever set foot on. Our hopes for conservation are unnecessarily constrained by our ignorance of what species exist to begin with. Our options to learn from adaptations favored by natural selection to discover more sustainable ways for humans to adapt to rapidly changing climate and environments are diminished with each species that goes extinct unknown and undescribed. And the depth with which we ultimately understand the living Cosmos is reduced if we fail to develop natural history collections as a permanent record of the number and kinds of living things in the twenty-first century.

Darwin gave to us the tantalizing knowledge that there exists a phylogenetic history of life, and the frustration that it remained out of the reach of rigorous science. A century later, the theoretical advances made by Hennig removed that obstacle opening the whole of evolutionary history to detailed exploration. But just as taxonomic knowledge is needed as never before, we have abandoned systematic biology and Hennig's advances in favor of fashionable technology and lucrative "contracts" to provide services to other biologists. This can and must be reversed immediately. Systematic biology must be restored in both museums and universities, and the tradition of Linnaeus, Darwin, and Hennig reengaged.

From the emergence of Modern humans, and probably before, we have pondered the meaning of our lives. What does it mean to be human? As Platnick (1979) explained, each evolutionary novelty is a modification of some pre-existing feature in an ancestral species. It is therefore the case that we cannot begin to fathom what it is to be human unless we know, too, what it is to be a primate, mammal, vertebrate, and metazoan. Hennig gave to humanity the opportunity to know ourselves as never before by making rigorous phylogenetic classifications within our reach. It is tragic that just as Hennig enabled science to pursue our innate and age-old curiosity about biodiversity and our place in it, we chose to derail his program, ignore natural history collection needs, and sideline taxonomy.

Armed with Hennig's ideas, systematic biology can fulfill its potential to be the cosmology of the life sciences. Just as 90% of the Universe consists of dark matter that we have not yet seen and do not understand, so too is the biosphere comprised of species most of which we have not yet recognized. The mission of systematic biology is no less audacious than that of cosmology, yet we embrace the wonder of the latter while denying basic support to the former. The irony is that the Universe will be available to cosmologists thousands of years from now, while the only living "cosmos" within our reach will be hugely diminished in just a few centuries. Systematic biology has always been necessary, from simply distinguishing food from foe in our local environment to addressing our burning curiosity to explain the biodiversity we see. But its work is now made urgent by the approaching mass extinction event. Unless we revitalize taxonomy and support it to explore individual clades, we risk permanent diminishment of both science and our humanity.

References

Barnosky, A.D., Matzke, N., Tomiya, S., et al. (2011). Has the earth's sixth mass extinction already arrived? *Nature*, 471, 51–57.

Bebber, D.P., Wood, J.R.I., Barker, C. and Scotland, R.W. (2014). Author inflation masks global capacity for species discovery in flowering plants. *New Phytologist*, 201, 700–706.

Costello, M.J., May, R.M. and Stork, N.E. (2013). Can we name earth's species before they go extinct? *Science*, 339, 413–416.

Costello, M.J., Houlding, B. and Joppa, L.N. (2014). Further evidence of more taxonomists discovering new species, and that most species have been named: response to Bebber et al. (2014). *New Phytologist*, 202, 739–740.

Gaffney, E. (1979). An introduction to the logic of phylogeny reconstruction. In *Phylogenetic Analysis and Paleontology*, ed. J. Cracraft and N. Eldredge, New York: Columbia University Press, pp. 79–111

Hennig, W. (1966). *Phylogenetic Systematics*. Urbana, IL: University of Illinois Press.

Kimmerer, R. (2014). Returning the gift. *Minding Nature*, 7, 18–24.

Kolbert, E. (2014). *The Sixth Extinction: An Unnatural History*. New York: Henry Holt & Co.

Mayr, E. (1942). *Systematics and the Origin of Species*. New York: Columbia University Press.

Nelson, G. (2004). Cladistics: Its arrested development. In *Milestones in Systematics*, ed. D.M. Williams and P.L. Forey, London: Taylor and Francis, London, pp. 127–147.

Nelson, G. and N. Platnick. (1981). *Systematics and Biogeography: Cladistics and Vicariance*. New York: Columbia University Press.

Platnick, N. (1979). Philosophy and the transformation of cladistics. *Systematic Zoology*, 28, 537–546.

Popper, K. (1959). *The Logic of Scientific Discovery*. New York: Hutchinson & Co.

Ross, H.H. (1974). *Biological Systematics*. Reading, MA: Addison-Wesley.

Wheeler, Q.D. (2008). Introductory: Toward the New Taxonomy. In *The New Taxonomy*, ed. Q.D. Wheeler, London: Taylor and Francis, pp. 1–17.

Wheeler, Q.D. (2009). Revolutionary thoughts on taxonomy: declarations of independence and interdependence. *Zoologia*, 26, 1–4.

Wheeler, Q. (2013). Are reports of the death of taxonomy an exaggeration? *New Phytologist*, 201, 370–371.

Wheeler, Q.D. and Platnick, N. (2000). The phylogenetic species concept. In *Species Concepts and Phylogenetic Theory: A Debate*, ed. Q. Wheeler and R. Meier, New York: Columbia University Press, pp. 55–69.

Wheeler, Q.D., Knapp, S., Stevenson, D.W., et al. (2012). Mapping the biosphere: exploring species to understand the origin, organization and sustainability of biodiversity. *Systematics and Biodiversity*, 10, 1–20.

Williams, D.M., Ebach, M.C. and Wheeler, Q.D. (2010). Beyond belief: the steady resurrection of phenetics. In *Beyond Cladistics: The Branching of a Paradigm*, ed. D.M. Williams and S. Knapp, Berkeley: University of California Press, pp. 169–197.

Index

Systematics Association Special Volumes

17. The Shore Environment: Methods and Ecosystems (2 volumes) (1980)*
    Edited by J. H. Price, D. E. C. Irvine and W. F. Farnham

18. The Ammonoidea(1981)*
    Edited by M. R. House and J. R. Senior

19. Biosystematics of Social Insects (1981)*
    Edited by P. E. House and J.-L. Clement

20. Genome Evolution (1982)*
    Edited by G. A. Dover and R. B. Flavell

21. Problems of Phylogenetic Reconstruction (1982)*
    Edited by K. A. Joysey and A. E. Friday

22. Concepts in Nematode Systematics (1983)*
    Edited by A. R. Stone, H. M. Platt and L. F. Khalil

23. Evolution, Time and Space: The Emergence of the Biosphere (1983)*
    Edited by R. W. Sims, J. H. Price and P. E. S. Whalley

24. Protein Polymorphism: Adaptive and Taxonomic Significance (1983)*
    Edited by G. S. Oxford and D. Rollinson

25. Current Concepts in Plant Taxonomy (1983)*
    Edited by V. H. Heywood and D. M. Moore

26. Databases in Systematics (1984)*
    Edited by R. Allkin and F. A. Bisby

27. Systematics of the Green Algae (1984)*
    Edited by D. E. G. Irvine and D. M. John

28. The Origins and Relationships of Lower Invertebrates (1985)‡
    Edited by S. Conway Morris, J. D. George, R. Gibson and H. M. Platt

29. Infraspecific Classification of Wild and Cultivated Plants (1986)‡
    Edited by B. T. Styles

30. Biomineralization in Lower Plants and Animals (1986)‡
    Edited by B. S. C. Leadbeater and R. Riding

31. Systematic and Taxonomic Approaches in Palaeobotany (1986)‡
    Edited by R. A. Spicer and B. A. Thomas

32. Coevolution and Systematics (1986)‡
    Edited by A. R. Stone and D. L. Hawksworth

33. Key Works to the Fauna and Flora of the British Isles and Northwestern Europe, 5th
    edition (1988)‡
    Edited by R. W. Sims, P. Freeman and D. L. Hawksworth

34. Extinction and Survival in the Fossil Record (1988)‡
    Edited by G. P. Larwood

35. The Phylogeny and Classification of the Tetrapods (2 volumes) (1988)‡
    Edited by M. J. Benton

36. Prospects in Systematics (1988)‡
    Edited by J. L. Hawksworth

37. Biosystematics of Haematophagous Insects (1988)‡
    Edited by M. W. Service

38. The Chromophyte Algae: Problems and Perspective (1989)‡
    Edited by J. C. Green, B. S.C. Leadbeater and W. L. Diver

---